Stereoselective Synthesis in Organic Chemistry

Enantioselective Synthesis β-Amino Organic Chemistry

Atta-ur-Rahman Zahir Shah

Stereoselective Synthesis in Organic Chemistry

With 85 Figures

Springer-Verlag
New York Berlin Heidelberg London Paris
Tokyo Hong Kong Barcelona Budapest

Atta-ur-Rahman
H.E.J. Research Institute
  of Chemistry
University of Karachi
Karachi-75270, Pakistan

Zahir Shah
H.E.J. Research Institute
  of Chemistry
University of Karachi
Karachi-75270, Pakistan

Library of Congress Cataloging-in-Publication Data
Rahman, Atta-ur-, 1942–
    Stereoselective synthesis in organic chemistry / Atta-ur-Rahman,
Zahir Shah.
        p.   cm.
    Includes bibliographical references and index.
    ISBN 0-387-94029-4 (New York). — ISBN 3-540-94029-4 (Berlin)
    1. Stereochemistry.   2. Organic compounds — Synthesis.   I. Shah,
Zahir.   II. Title.
QD481.R23   1993
547′.2 – dc20   93-284

Printed on acid-free paper.

Production managed by Hal Henglein; manufacturing supervised by Vincent R. Scelta.
Camera-ready copy prepared by the authors.
Printed and bound by Edwards Brothers, Inc., Ann Arbor, MI.
Printed in the United States of America.

9 8 7 6 5 4 3 2

ISBN 0-387-94029-4 Springer-Verlag New York Berlin Heidelberg
ISBN 3-540-94029-4 Springer-Verlag Berlin Heidelberg New York

Foreword

This monumental tome by Prof. Atta-ur-Rahman and Dr. Zahir Shah presents a broad overall perspective of stereoselectivity in the synthesis of organic molecules. Thus it treats a problem that is of fundamental importance and will be even more important in the future as the drug industry is required to supply 100% optically pure compounds.

After an exposition of general principles, the following subjects are treated: Catalytic Reductions, Heterogeneous Catalytic Hydrogenations, Stereoselective Non-Catalytic Reductions, Stereoselective Carbon–Carbon Bond Forming Reactions, Asymmetric Oxidations, Asymmetric Carbon–Heteroatom Bond Formations, Enzyme Catalyzed Reactions, Stereoselective Free Radical Reactions, and finally Miscellaneous Stereoselective Reactions. For each subject, a wealth of examples are given. The highly selective reactions are mentioned along with reactions that are not. This is helpful as it will teach the practical chemist what to avoid.

Much progress has been made in the last two decades in the design of new, very stereoselective reactions which can be applied in industry. For example, and in alphabetical order, we can mention (among other peers): H.C. Brown (hydroboration), D.C. Evans (carbon–carbon bond formation), R. Noyori (BINAP reagents for hydrogenation), and K.B. Sharpless (epoxidation and dihydroxylation of double bonds). Thus the field has completely changed since the 1950s, when optically pure compounds were always obtained by difficult resolutions of racemates and not by stereoselective reactions.

The book will provide a very useful introduction to stereoselective processes in organic synthesis. The large number of examples with the appropriate references will lead the practitioner, at any scientific level, to a rapid evaluation of his problem. From the point of view of all types of organic chemists, this book will be their first choice for information about stereoselective reactions.

D. H. R. Barton

D.H.R. Barton
Texas A&M University
October 7, 1992

Preface

The marvellous chemistry evident in living processes is largely enzyme-controlled stereoselective synthesis of highly complex organic molecules, designed to fulfill specific tasks in living organisms. Man continues to learn from nature, and in many fields he tries to mimic it. Stereoselective synthesis is one such area where organic chemists have tried to develop new reaction methodologies to transform substrates in stereoselective or enantioselective fashions.

There has been an explosive growth in the area of stereoselective synthesis in the last few decades, and the field has taxed the genius of some of the most distinguished chemists of today. These developments have created the need for a textbook that could present the salient features of the major developments in the field of stereoselective synthesis. The vast and rapidly growing literature in this important area prevents one from writing a "comprehensive" treatise on all the stereoselective synthetic methods developed to date, but it is hoped that the readers will find that the more important developments in the field up to mid-1992 are adequately covered.

We are grateful for the help extended to us by a number of persons in the writing of this book. Our thanks go to Mr. M. Rais Hussain and Mr. M. Asif for diligently typing the manuscript. We are also indebted to Mr. Abdul Hafeez and Mr. Ahmadullah Khan for structure drawing and to Mr. S. Ejaz Ahmed Soofi and Miss Farzana Akhtar for their assistance in preparing the subject index. Last, but not least, we are thankful to Mr. Mahmood Alam for secretarial assistance.

The first author dedicates this book to a magic lake in Dera Ismail Khan, and to a shining star which has given him eternal hope and enriched his life in a million ways.

Atta-ur-Rahman
Zahir Shah
Karachi
October, 1992

Contents

1 Stereochemical Principles

1.1 Introduction

Compounds having the same molecular formula may differ from one another in the nature or sequence in which the individual atoms are bound. Such compounds are known as *isomers* and they may differ significantly in their chemical and physical properties, depending on the structures. For instance ethylene oxide and acetaldehyde both have the formula C_2H_4O but they differ in their constitution. When substances have the same constitution but differ from one another in the manner in which the individual atoms (or groups) are arranged in space, then they are termed *stereoisomers*. When two stereoisomers are so related to each other that one is the nonsuperimposable mirror image of the other, then the two are said to be *enantiomeric* and each enantiomer is *chiral*. They differ from one another in having an equal and opposite optical rotation. Stereoisomers which are not enantiomers are called *diastereomers*. Diastereomers may therefore be defined as substances which have the same constitution, which are not mirror images and which differ from one another in having a different configuration at one or more asymmetric centers in the molecule [1-3]. Substances may also exist as *conformers*; the conformational isomerism results from the existence of discrete isomers due to barriers in the rotation about single bonds.

1.2 Chirality

Chirality refers to the property of nonsuperimposability of an object on its mirror image. Chiral compounds are optically active, the actual value of the optical rotation depending on the structure as well as on the experimental conditions, particularly on the temperature, solvent and wavelength of the incident light. The wavelength normally employed is 589 nm, the emission wavelength of sodium arc lamps (sodium D line) and optical rotations are therefore designated as $[\alpha]_D$ if measured at this wavelength.

Chirality can arise by the presence of an sp^3 hybridised carbon center which has four different atoms or groups bound to it. Such a carbon atom is then described as an *asymmetric center*. If two of the four substituents on this carbon atom are identical, then it would no longer be asymmetric. For instance compounds (1) and (2) are mirror images, and they are not superimposable upon one another. If however one replaces the chlorine by a hydrogen atom, the resulting substance (3) is no longer chiral since it has a plane of symmetry passing through the atoms Br - C(1) - C(2). It is not essential to have a tetracoordinate atom in order for a compound to be optically active and chirality can be encountered in sulfoxides (e.g. 4), phosphorus compounds (e.g. 5) etc.

(1)   (2)   (3)

(4)   (5)

The configuration at an asymmetric center may be described either in terms of the Fischer convention (employing the terms D and L), or the Cahn-Ingold-Prelog convention (employing the terms R and S). In the Fischer convention, (+)-glyceraldehyde (D-glyceraldehyde 6) is chosen as the reference standard and the configuration at any asymmetric center is described as D or L by relating it to this standard. This convention finds wide use in describing carbohydrates and amino acids. It is notable that the terms D- or L- used have no bearing on the sign of the optical rotation measured.

D-glyceraldehyde (6)   L-glyceraldehyde (7)

The Fischer convention is useful only as long as the substances can be readily correlated in their structures with glyceraldehyde but it becomes difficult to apply when the structures differ significantly from glyceraldehyde in the substituents attached to the asymmetric center. It has therefore been widely replaced by the Cahn-Ingold-Prelog convention which relies on determining the sequence of the groups attached to the asymmetric center in a decreasing priority order. According to the *sequence rule* [4], the atoms linked to the asymmetric center are initially ordered according to decreasing order of atomic numbers, the lowest priority being given to the atom with the lowest atomic number. If two identical atoms are attached to an asymmetric center then priority is assigned after considering the substituents on the attached atoms. Thus if H, CH_3,

CH$_2$OH and OH are attached to an asymmetric carbon, then the decreasing order of priority will be OH, CH$_2$OH, CH$_3$ and H.

It is convenient to look at the asymmetric center by holding the atom with the smallest priority away from the viewer and then determining the configuration as R (Latin: *rectus*, right) or S (Latin: *sinister*, left) by seeing whether the remaining three attached atoms or groups if considered in a decreasing order of priority appear clockwise or anti-clockwise. When a multiply bonded atom is present then it is counted as a substituent for each bond. Thus CH=O would be counted as (O,O and H), while C = N would be counted as (N,N and N).

For instance, in order to determine the configuration of the asymmetric center in D-glyceraldehyde, (6) we would view the structure from the side opposite to the hydrogen atom (which is the lowest priority atom present in the molecule) so that only the three groups shown in (6) are considered and the hydrogen atom is ignored. The atom with the highest priority is the oxygen atom of the OH group and it is therefore given the highest priority "a". The CHO group has the next priority "b" since the carbon atom of the CHO group has the attached atoms O,O and H (the doubly bonded oxygen in the aldehyde group being counted twice). The CH$_2$OH has the next lowest priority since it has the attached atoms H,H and O. Now going from "a" to "b" and then to "c" we find that we have to trace a clockwise path, affording us an R-configuration for the asymmetric carbon in D-glyceraldehyde. It is noteworthy that D-configuration in the Fischer projection may turn out to be either "R" or "S" in the Cahn-Ingold-Prelog convention since different principles are applied in the two conventions so that there is no direct correlation between them.

CHO

H ► C ◄ OH

CH$_2$OH

(6)

HO (a)   CHO
         (b)

CH$_2$OH
(c)

(8) (R)

In optically active sulfoxides in which the chiral sulfur is tricoordinate, or in chiral phosphorus compounds, three atoms are bound to the sulfur or phosphorus atom forming a cone. If the tip of this cone contains the sulfur or phosphorus and if the three attached atoms are directed towards the viewer then by seeing if a clockwise or anti-clockwise rotation is involved in going from the highest priority *via* the middle priority to the lowest priority group one can similarly assign the "R" or "S" configurations respectively.

CHO

H ▶ C ◀ CH₃

COOCH₃

(9)

⟶

(c) H₃C   (b) CHO
         ∨
      COOCH₃
      (a)

↻

(10) (S)

↓

↓

CHO

H ▶ C ◀ CH₃

CH₂OH

(11)

⟶

H₃C   CHO
    ∨
  CH₂OH

↻

(12) (R)

It may be noted that the assignment of configuration at an asymmetric center can be changed from "R" to "S" or vice versa by a chemical operation (e.g. reduction of an ester group) without actually affecting the stereochemical disposition of the groups at the asymmetric center. For instance consider the compound (9). According to the Cahn-Ingold-Prelog convention, the highest priority "a" is assigned to the ester group since the carbonyl carbon has the attached atoms (O,O,C). The next priority "b" is assigned to the aldehyde carbon (attached atoms: (O,O,H) while the lowest priority "c" is given to the CH₃ carbon (attached atoms: H,H,H) as shown in structure (10). Application of the sequence rules shows an anti-clockwise rotation leading to "S" configuration of the asymmetric center. Protection of the aldehyde function, reduction of the ester group and deprotection to regenerate the aldehyde affords alcohol (11) which possesses the "R" configuration shown in structure (12) as opposed to the "S" configuration of the starting material. This illustrates the point made above that one should not presume that an inversion of configuration has occurred if an "R" configuration is changed to an "S" configuration through *chemical conversion* since this might correspond to a mere transformation of attached groups with the accompanying redesignation of the configuration according to the Cahn-Ingold-Prelog convention, rather than an actual inversion at the asymmetric center.

It is possible for molecules to be chiral without having an asymmetric center, due to the presence in them either of a chiral plane or of a chiral axis. Molecules dissymmetric due to the presence of a chiral axis may be exemplified by optically active biphenyls [e.g. S-(+)-1,1'-binaphthyl (13)] or allenes [e.g. R-(-)-1,3-dimethylallene (14)] [6] while those dissymmetric due to a chiral plane may be exemplified by a *trans*-cycloalkene e.g. R-(-)-*trans*-cyclooctene (15) [7]. These may be assigned R S configuration by the Cahn-Ingold-Prelog convention according to special rules [4,8].

(13)   (14)   (15)

As mentioned in section 1.1 diastereomers are stereoisomers which are not related to each other as an object with its mirror image. Consider, for instance, all the possible stereoisomers of 2,3,4-trihydroxybutanoic acid (16). Since the compound has two asymmetric centers, four different stereoisomers are possible: 2R,3R (16) with its mirror image 2S, 3S (17), and 2S, 3R (18) with its mirror image 2R,3S (19). The 2R, 3R isomer (16) is diastereomeric with respect to the 2R, 3S isomer (19) as well as with the 2S, 3R isomer (18). Each enantiomer has a specific optical rotation and it can have only one mirror image which will have an equal and opposite optical rotation with its asymmetric centers in the opposite configurations. The mirror image of the 2S, 3R isomer (18) for instance is the 2R, 3S compound (19).

(16) (2R, 3R)   (17) (2S, 3S)

Diastereomers   Diastereomers   Diastereomers   Diastereomers

(18) (2S, 3R)   (19) (2R, 3S)

Diastereomers, unlike enantiomers, differ in their physical and chemical properties. They can be separated chromatographically from one another and usually have different melting points, boiling points, optical rotations, solubilities, refractive indices, dipole moments etc. Enantiomers in a racemic mixture may be separated by resolution into the individual mirror images. This may be carried out by preferential crystallization, by chromatography on a chiral column, or more commonly by converting them into a diastereomeric mixture through treatment with an optically active reagent [9,10]. Since the resulting diastereomers will have different physical and chemical properties, they can be separated by standard methods and the individual enantiomers then regenerated. Thus if a racemic mixture of an optically active acid A consists of two enantiomers A_1 and A_2 then treatment of this mixture with an optically base B will afford a mixture of two *diastereomers* A_1B and A_2B. Separation of the two diastereomers and regeneration of the free acid will afford the pure enantiomers, A_1 and A_2.

Alternatively enantiomers can be separated by the process of kinetic resolution. This exploits the differences in transition state energies when two different enantiomers react with an optically active reagent resulting in the selective reaction of one of two mirror images. This is because the two transition states (i.e. enantiomer 1 + chiral reagent, and enantiomer 2 + chiral reagent) have a diastereomeric relationship to each other so that the rates at which the individual enantiomers react with the chiral reagent differ, thereby allowing their separation.

As stated earlier, enantiomers in a racemic mixture may be separated by chromatography on a chiral column. This is on account of the different magnitudes of non-covalent bonding between the individual enantiomers with the chiral column material so that the two enantiomers pass through the column at different rates, thereby allowing their separation.

In order to determine the optical purity of a compound, its optical rotation may be measured. However the optical rotation can only be used to determine the optical purity if the optical rotation of the pure enantiomer has been previously reported accurately. In the case of new compounds measurement of the value of the optical rotation will not allow the determination of optical purity of a substance. One way to do this is to prepare a derivative of the partially optically pure molecule with a chiral reagent, whereby an additional asymmetric center is introduced and the unequal mixture of enantiomers is converted into the two correspondingly different diastereomers. These diastereomers (unlike the enantiomers from which they were prepared) will have different physical properties including differences in NMR chemical shifts. If an NMR spectrum of this mixture of diastereomers is recorded, the spectrum obtained will be a superimposition of the NMR spectra of the two individual diastereomers and a doubling of those peaks will be observed in which the chemical shift differences are significant. By measuring the relative peak intensities of such protons one can determine the percentage of each diastereomer (and hence of the enantiomers from which they were derived). Mosher's reagent **(20)** [11] is one of many reagents now available for preparing such derivatives.

(structure 20: benzene ring—C*(OCH₃)(CF₃)—CO₂H)

$$\text{C}_6\text{H}_5 - \overset{\overset{\displaystyle OCH_3}{|}}{\underset{\underset{\displaystyle CF_3}{|}}{C^*}} - CO_2H$$

(20)

Optically active materials can be prepared by employing chiral catalysts so that the complex formed between the substrate and the chiral catalyst will be asymmetric. The reacting molecule approaching this complex may therefore preferentially attack it from one side leading to "asymmetric synthesis" . A structure having "n" asymmetric centers can have a maximum of 2^n enantiomers and half as many *racemates*, though in compounds which have a plane or axis of symmetry, this number can be reduced. Thus if three asymmetric centers are present, there will be $2^3 = 8$ enantiomers and four racemates, each racemate comprising a mixture of two enantiomers which have a mirror image relationship to each other. If all 8 enantiomers exist, then chromatography of such a mixture on an achiral column could lead to the separation of the four racemates, but not of the enantiomers since chiral chromatographic materials are required to separate enantiomers. If two identically substituted chiral centers are present in a molecule then instead of four enantiomers, there will be only two enantiomers, while the presence of a mirror plane will give rise to an optically inactive *meso* form which is superimposible on its mirror image, as illustrated in the D- , L- and *meso*-tartaric acids, (21), (22) and (23) respectively.

CO₂H	CO₂H	CO₂H
(Fischer projection)	(Fischer projection)	(Fischer projection)

D-tartaric acid (21)   L-tartaric acid (22)   *meso*-tartaric acid (23)

As stated above diastereomers differ from each other in several respects, one important difference being the internuclear distance between one or more selected pairs of atoms or groups. For instance the two diastereomers of 2,3-dichlorobutene, (24) and (25) differ from each other in the internuclear distance between the methyl groups (or of the chlorine atoms). The prefix "Z" (for "zusammen" (German: meaning together) indicates that the two higher priority substituents lie close to one another, while the prefix "E" (for "entgegen" (German: meaning opposite) indicates that they lie further apart. To determine the group priorities the sequence rule is again applied. If the two substituents having the higher atomic numbers of the atoms attached to the olefinic bond lie on the same side, then the prefix "Z" is used whereas if they lie on opposite side, then the prefix "E" is applied. If the atoms directly attached to the olefin are of the same atomic number, then priorities are assigned based on the atom attached to these atoms. If only three substituents are present (as in oximes) then the fourth substituent is assumed to be a "ghost" atom with atomic number zero.

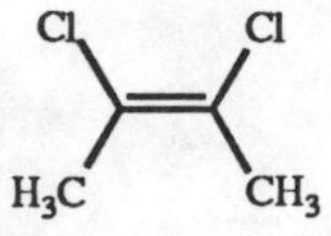

(Z)-2,3-dichlorobutene (24 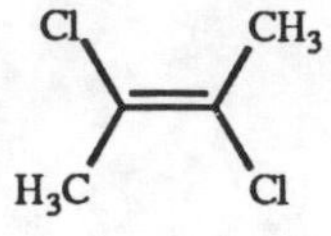)   (E)-2,3-dichlorobutene (25)

   The prefixes Z- and E- should not be used to describe the arrangements of groups in rings, for which the notations *cis* or *trans* are more appropriate. The terms *syn* and *anti* are employed to describe addition and elimination reactions occurring from the same or opposite sides respectively.

   There has been a considerable degree of confusion [12] in the literature in describing the relative configuration in acyclic molecules. Historically, when two groups occurred on the same side in a Fischer projection then they were termed *erythro* but if they were on opposite sides, they were termed *threo*. In fact the Fischer projection gives a false impression of the actual stereochemical disposition of the groups since when the chain is represented in the actual zig-zag form, the substituents which were *erythro* (on the same side) in the Fischer projection are now seen to be actually on opposite sides, while the *threo* groups appear on the same side. It was therefore suggested that the conventional system of nomenclature should be reversed, and this suggestion was widely adopted which led to the existence of both opposing systems of nomenclature in the literature, causing much confusion. A number of revised systems of assigning relative configuration in acyclic molecules have been proposed [13-16] (Scheme 1).

   Another convention often used is that of *syn* or *anti*, depending on whether the two groups on adjacent carbon atoms are both pointing in the same direction or whether they point in opposite directions as shown in structures (28)-(31).

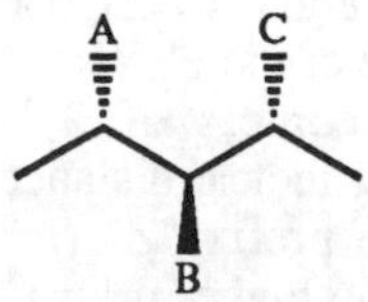

anti,anti   syn,anti   syn,anti   syn, syn

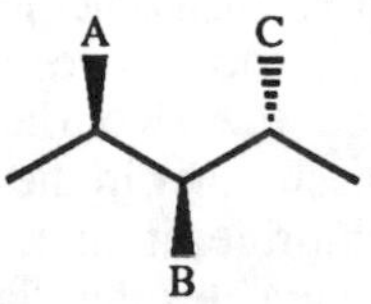

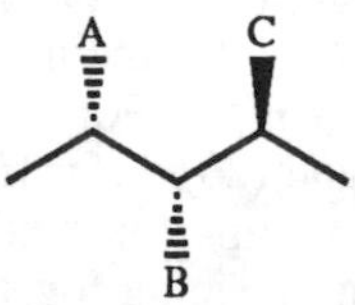

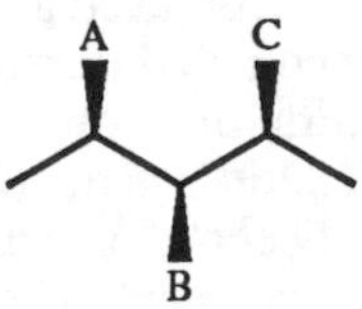

(28)   (29)   (30)   (31)

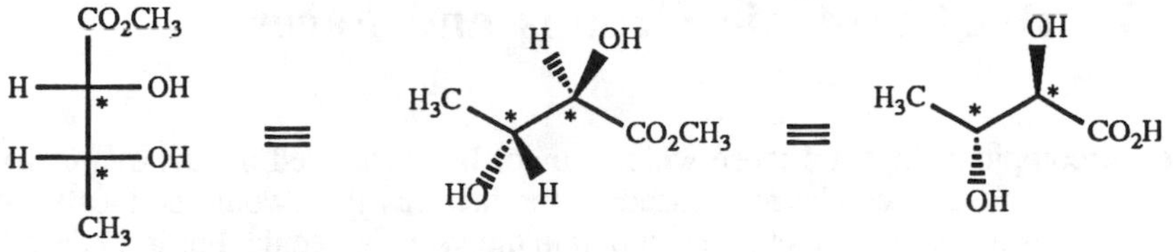

(26)	(2R, 3R)	anti
Classical	Heathcock (aldol)	
nomenclature	nomenclature *threo*	
erythro		

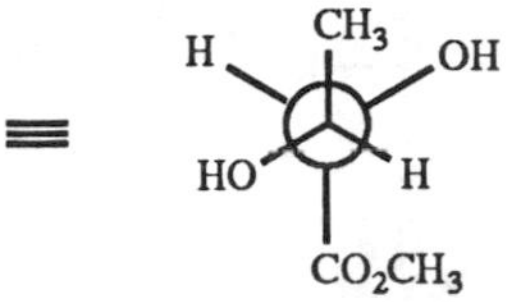

Newman
convention

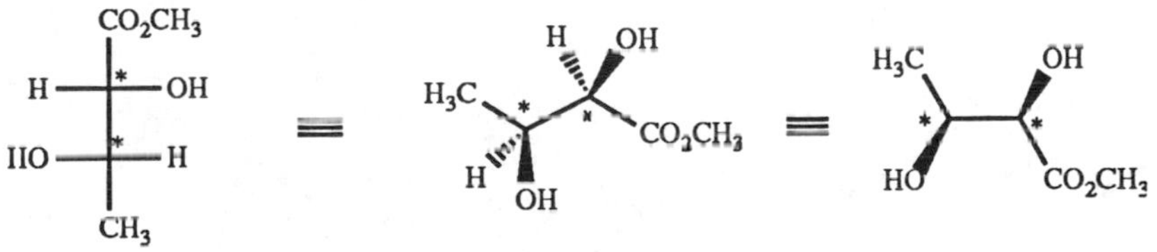

(27)	(2R, 3S)	syn
Classical	Heathcock (aldol)	
nomenclature	nomenclature	
threo	*erythro*	

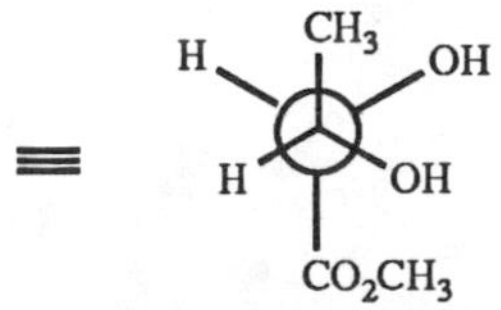

Newman
convention

Scheme 1.1. Comparison of various systems of naming chiral compounds.

1.3   Diastereotopic Groups and Faces

Diastereotopic groups are those which cannot be exchanged by any symmetry operation, and they can be recognised by the fact that they would be located at different distances from a reference group in the same molecule. For instance the two methylene groups of the acrylic acid derivative (32) are diastereotopic since they lie at different distances from the carboxylic group. Similarly the two hydrogen atoms in (33) or the two Br atoms in (34) are diastereotopic.

(32)   (33)   (34)

If a plane divides a molecule such that it is not a plane of symmetry and the molecule does not possess a coplanar axis of symmetry, then the two faces of such a molecular plane are diastereotopic, e.g. the faces of an asymmetric molecule such as (35).

(35)

1.4   Enantiotopic Groups and Faces

Those groups which can be exchanged by rotation across a plane or center of symmetry are said to be enantiotopic and the presence of such a symmetry element in the molecule results in the molecule being achiral. Thus the two aldehyde groups in (36) and the chlorine atoms in (37) are enantiotopic.

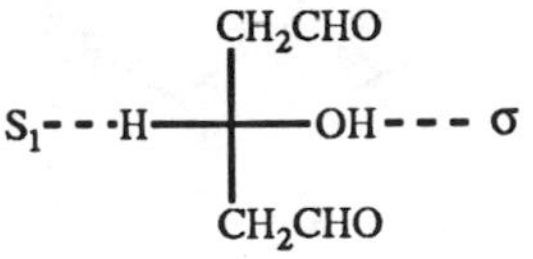

CH₂CHO

S_1- - ·H——OH - - - σ   (σ = S_1)

CH₂CHO

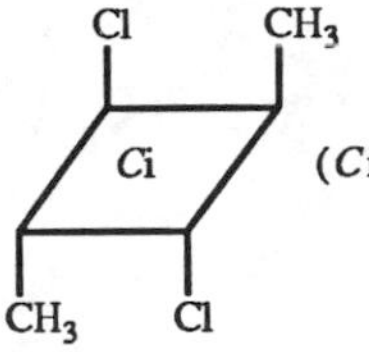

$(Ci = S_2)$

(36)   (37)

The two faces of a molecule such as acetyl chloride (38) are also enantiotopic. Such faces are divided by a molecular plane of symmetery but not by a coplanar axis of symmetry.

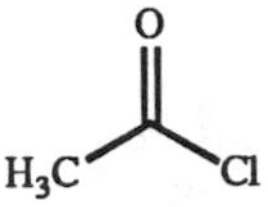

(38)

1.5   Homotopic Groups and Faces

Homotopic groups are those which can be exchanged by rotation at an axis of symmetry. For instance the six hydrogens of benzene (39) or the two hydrogens of dibromomethane (40) are homotopic, having C_6 and C_2 axes of symmetry respectively.

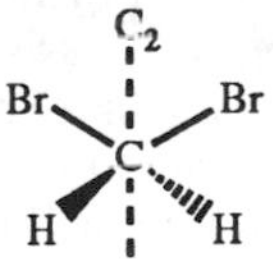

(39)   (40)

Consider 1,3-dibromopropane (41). If one were to substitute one of the hydrogen atoms attached to C-2 by a different group R, the product generated would be identical to the one obtained if the *other* hydrogen attached to C-2 was to be substituted by the some group R. The two hydrogen atoms at C-2 are topologically (as well as chemically) equivalent and they are said to be homotopic.

(41)

The two faces of a molecular plane which contains a coplanar axis of symmetry are said to be homotopic, e.g. the two faces of the acetone molecule.

1.6   Homochiral Relationships

If we take the example of 1,3-dibromopropane again and consider the relationship of the two hydrogens attached to C-1 then substitution of the hydrogen atoms with group R can give rise to two different products since these hydrogens are topologically non-equivalent. The two hydrogen atoms at C-1 of 1,3-dibromopropane are therefore *heterotopic* and since substitution of either results in the formation of enantiomers they are also termed *enantiotopic*. It is apparent that the two products of substitution have an enantiomeric relationship to one another, and centers in which replacement of a ligand by another ligand,

(41)

(e.g. H by R in the above example) gives rise to a new chiral center are said to be *prochiral*. The heterotopic groups attached to such a prochiral center may be designated as pro-R and pro-S. This is done by arbitrarily assigning one of these heterotopic ligands a higher priority than the other and then applying the sequence rule. If application of the sequence rule results in assignment of R to the prochiral center, then the heterotopic ligand to which the higher priority was assigned is designated pro-R but if the prochiral center is assigned S then the ligand is designated pro-S, and the ligands are labelled by a subscript R or S. Enantiotopic atoms such as H_R and H_S in 1,3-dibromopropane interact differently with chiral reagents such as enzymes. If the molecule already has another asymmetric center then replacement of the prochiral ligands will lead not to enantiomers but to diastereoisomers and the prochiral ligands will in that case be diastereotopic e.g. the diastereotopic protons of phenylalanine (42).

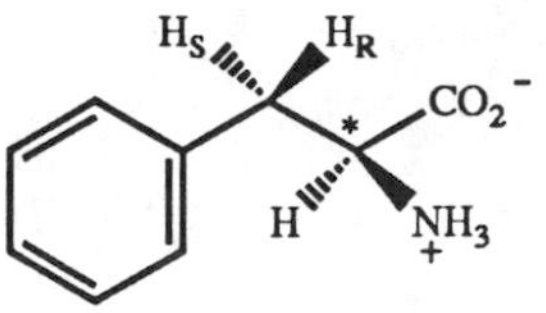

(42) 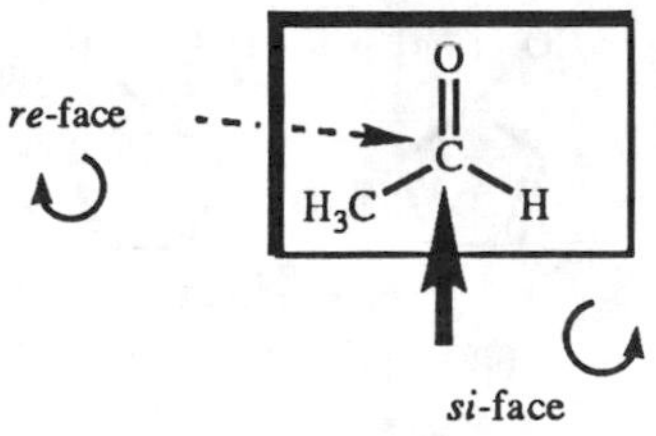   (43)

Prochirality can also be considered in terms of faces. The two faces of a molecule such as acetaldehyde (43) may be classified as *re* or *si* depending on whether the substituents viewed from a particular face appear clockwise (rectus, *re*) or anticlockwise (sinister, *si*) respectively in order of decreasing priority. Achiral reagents such as sodium borohydride will attack the carbonyl group from either of the two faces indiscriminately but a chiral reagent can attack preferentially from one of the two faces.

1.7   Selectivity in Organic Synthesis

There are various aspects of selectivity with which organic chemists are concerned:

1.7.1   Chemoselectivity

Considerations involve the intrinsic differences in reactivity of different groups in a molecule without involving special activating or blocking groups. If two functional groups in a molecule are substantially different then chemoselectivity, i.e. reaction with one group in preference to the other, may be achieved relatively easily. For instance if a ketone and ester groups are present in a molecule it is possible to reduce the ketonic carbonyl without affecting the ester carbonyl group by employing $NaBH_4$ under mild conditions. In other cases, the chemoselectivity may be more difficult to attain . For instance, reduction of cyclopentenone (44) with $NaBH_4$ leads to a mixture of compounds (45) and (46) in one of which (45) only the ketonic carbonyl is reduced while in the other (46) the double bond is also reduced. The chemoselectivity of the reduction can be enhanced by addition of $CeCl_3$, which leads to a preferential reduction of the ketonic carbonyl to the corresponding alcohol without appreciable reduction of the olefin [17].

(44)   **(45)**   **(46)**

Catalysts have played a key role in enhancing chemoselectivity. For instance if geraniol (**47**) is subjected to epoxidation with *m*-chloroperbenzoic acid the 6,7-double bond is oxidised with some preference to the the 2,3-double bond affording the products (**48**) and (**49**) with 2:1 selectivity but if the allylic OH group is coordinated with vanadium, then the 2,3-olefin is oxidised selectively [18,19].

(47)

(48)   **(49)**

(**48** : **49** = 2:1)

(47)   **(49)**

There are numerous other examples of enhancement or reversal of reactivity. For instance nucleophiles attack a double bond which is conjugated to an electron withdrawing group such as a carbonyl but they do not attack an isolated double bond. In the presence of palladium salts, however, this reactivity can be reversed [20].

1.7.2   *Regioselectivity*

Another important consideration is the tendency of a reagent to attack one functional group in preference to another in a molecule. Selective attack at one site rather than another can be achieved by suitable choice of reagent or by altering the pathway of the reaction by modifying the reaction conditions. The effect of the choice of reagent is illustrated by the addition of the elements of water to olefins in a Markownikov sense by oxomercuration/reduction, and in a anti-Markownikov sense by hydroboration/oxidation.

The change of reaction conditions can also alter the orientation of attack by changing the mechanistic pathways of the reactions. For instance thiophenol (50) attacks olefins e.g. (51) in an acid catalysed ionic reaction to give the more substituted product (52) but by a radical pathway it affords the less substituted product (53) [21,22].

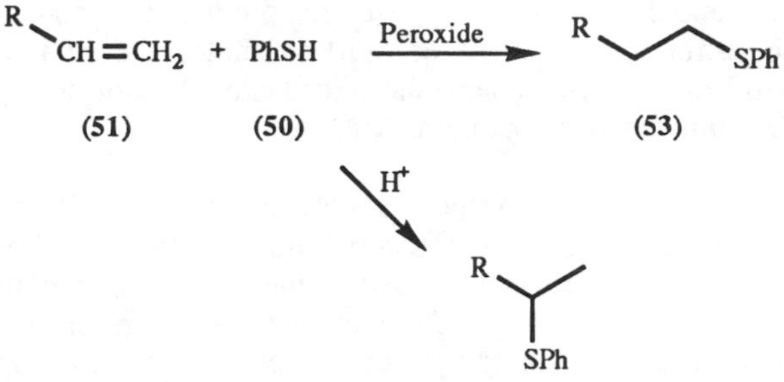

A judicious choice of directing groups can also be employed to attain desired regioselectivity. For instance acylation of the cyclohexene derivative (54) under Friedel-Crafts conditions proceeds at either end of the double bond to give (55) and (56). However the desired control can be achieved by utilising the known ability of silicon to stabilise a positive charge at the carbon β- to the silicon atom (and a negative change α- to it). Hence in silylated olefins, the acylation occurs exclusively at the silicon bearing olefinic carbon (Figure 1.1) [23].

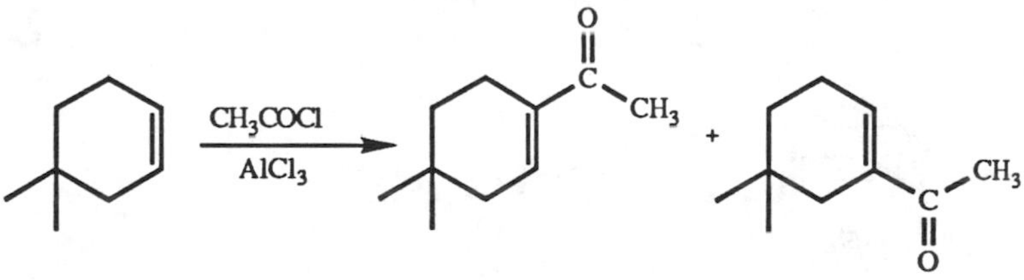

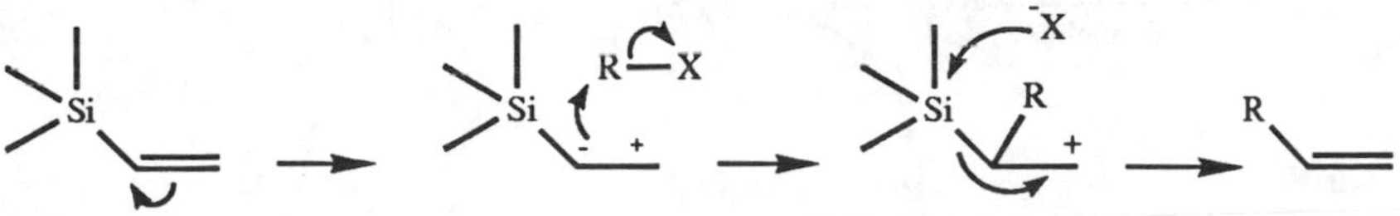

Figure 1.1. Acylation of silylated olefins.

1.7.3   *Diastereoselectivity*

Diastereoselectivity considerations in organic reactions relate to the control of *relative* stereochemistry. Diastereoselectivity may be considered with reference to the starting material or with reference to the product. One needs to consider the "stereoselectivity" or "stereospecificity" aspects in such reactions. *Stereoselective reactions* are those which involve the preferential formation of one stereoisomer when more than one can be formed. *Stereospecific reactions* on the other hand involve preferential formation of one stereoisomer of the product which is dictated by the stereochemistry of the starting materials, so that different stereoisomers of the starting materials afford stereoisomerically different products under the same reaction conditions.

It is important to consider the kinetic aspects involved when considering stereoselective reactions. Different diastereomeric substrates can either lead to the same product at different rates, or different diastereomers can give rise to different products at different rates. For example the axial alcohol (57) reacts at a rate 3.2 times greater than the corresponding equatorial alcohol (58) on oxidation with CrO_3 in aqueous acetic acid to give the product (59). Typical examples of stereoselective reactions include addition reactions to alkenes e.g. phenoxycarbene addition to cyclohexene (60) to give the products (61) and (62), or the reduction of cyclic ketones e.g., (63) with LAH affording (64) and (65) in a ratio of 9:1 respectively.

(63)   (64) 90%   (65) 10%

In the above examples the same starting material gives rise to the formation of one stereoisomer of the product in partial preference to the other stereoisomer.

Examples of *stereospecific reactions* include addition reactions to olefins such as epoxidation of alkenes or the elimination of halides . For instance the (Z) and (E) alkenes (66) and (67) afford different products (68) and (69) respectively. Similarly the *threo* (70) and *erythro* (71) halides undergo elimination to give products (72) and (73) respectively.

(66)   (68)

(67)   (69)

(70) *threo*   (72)

(71) *erythro*   (73)

The stereochemistry of attack at a reaction site can be influenced by the presence of an existing asymmetric center in the vicinity of the new center being generated by the reactions. In general the nearer the exisiting center to the new chiral center being generated the greater the control. In conformationally rigid systems, the knowledge of conformations of the substrate molecules can help to solve stereochemical problems. In non-rigid molecules however the problem of predictable stereocontrol is more difficult and it may be necessary to impose conformational rigidity, for instance by metal chelation. For example in the nucleophilic addition to carbonyl groups in a flexible non-cyclic molecule (74) Still and co-workers succeeded in introducing sufficient conformational rigidity so as to allow attack of the carbonyl group from one face of the molecule with Grignard reagent affording the product (75) [24].

(74)   (75)

Transition metal templates have also been employed to impose conformational rigidity in acyclic systems. Thus a Pd(0) complex with phosphine ligands has been used to induce ionisation of the lactone (76) in the conformation of the intermediate (77) shown. The intermediate is attacked by malonate anion faster than stereo-randomisation can occur so that the stereochemistry of the starting material is transmitted to the product (78) [25]. Thus the creation of metallocycles or complexation with transition metals allows the imposition of rigidity in otherwise conformationally mobile systems to achieve the desired stereoselectivity.

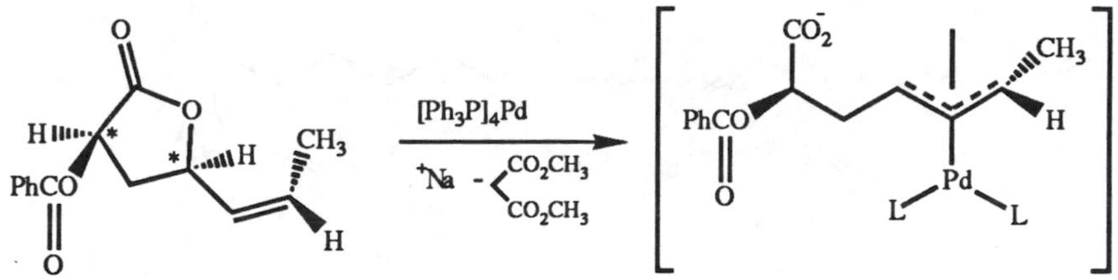

(76)

(77)

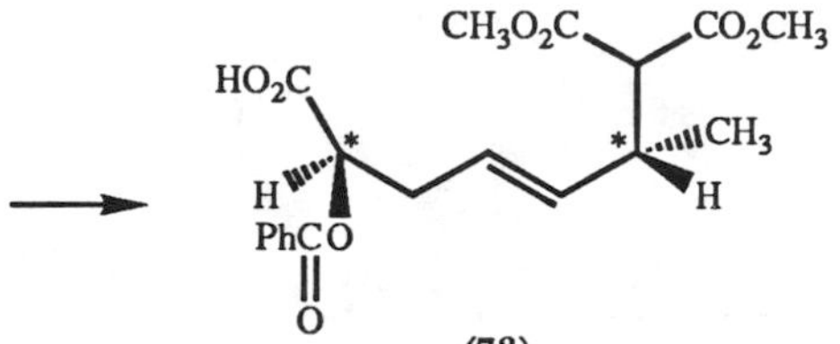

(78)

The introduction of diastereoselectivity can be particularly difficult when the new chiral center to be introduced is remote from the directing effects of an existing chiral center, and when the molecule is non-rigid. Such problems may be tackled by employing reactions in which the transition states are generated in a predictable manner or in which the intermediates are formed predictably [26]. For instance, substrate (79) affords the product (81) via the predictable transition state (80). Similarly the I_2-induced cyclization of the homoallylic phosphate (82) proceeds through the transition states (83) and (84) to give the product (85) in 63% yield [27]. This approach was utilised by Bartlett et al in the synthesis of nonactic acid.

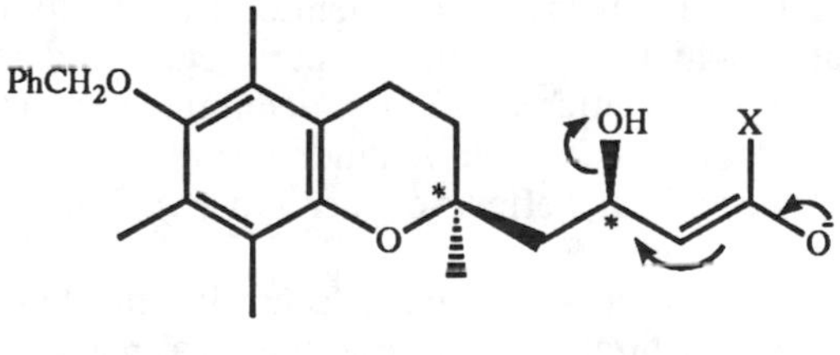

(79)

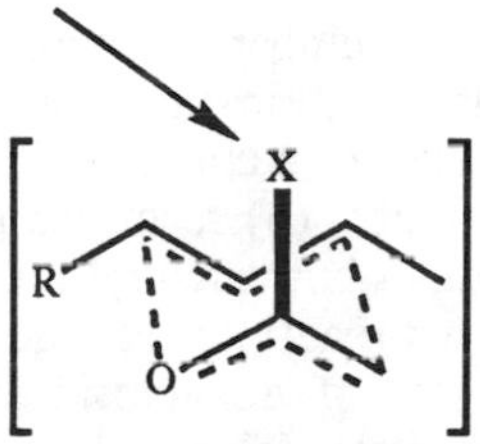

(80)

Predictable transition state

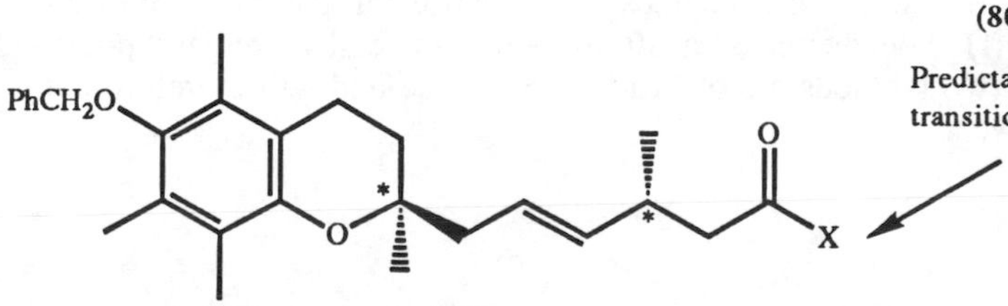

(81)

(82)   (83)

(84)   (85) *erythro*

1.7.4   *Enantioselectivity*

While in diastereoselectivity we were concerned with the *relative* configuration of the products, in enantioselectivity it is the control of the *absolute configuration* of the molecule which is involved. Enantio-differentiation relies on the chiral environment in which the reaction occurs (eg. chiral reagent, solvent, catalyst) whereas diastereo-differentiation is determined by structural elements within the substrate molecule (e.g. steric hindrance of groups near the reaction center, or the electronic and other effects of such groups) [28].

A number of approaches have been developed for achieving enantioselective transformations: (a) reactions of achiral substrates with achiral reagents in the presence of a chiral additive which does not become covalently bonded with the substrate; (b) reactions in which the substrate initially may be achiral but it becomes convalently linked with a chiral auxiliary which is removed after the desired reaction; (c) reactions in which the reagents are chiral and involve the transfer of an achiral species to the enantiotopic groups or faces of the substrates; (d) reactions with enzymes ; (e) reactions in the presence of chiral solvents, and (f) photochemical transformations induced by circularly polarised light. The last two methods are of little practical use and will therefore not be discussed below.

1.7.4.1   Reactions in Presence of Chiral Additives

Chiral additives have been used to induce chirality in reactions in which both the substrate and the reagent are achiral. There are many examples of such reactions including the heterogeneous catalytic hydrogenation of ketones in the presence of a chiral acid, homogeneous hydrogenations with rhodium or ruthenium catalysts prepared from chiral phosphines, intramolecular aldol condensations in the presence of a chiral base etc. Examples include stereoselective reactions of the substrates (86), (88) and (90) to give the products (87), (89) and (91) respectively.

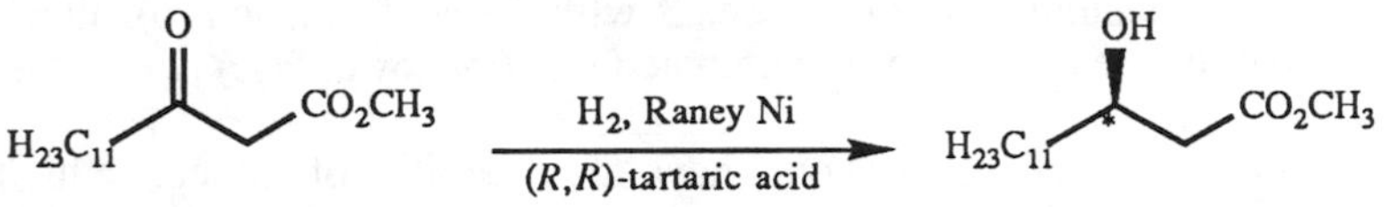

(86)

H_2, Raney Ni
(R,R)-tartaric acid

(87) 85% e.e.

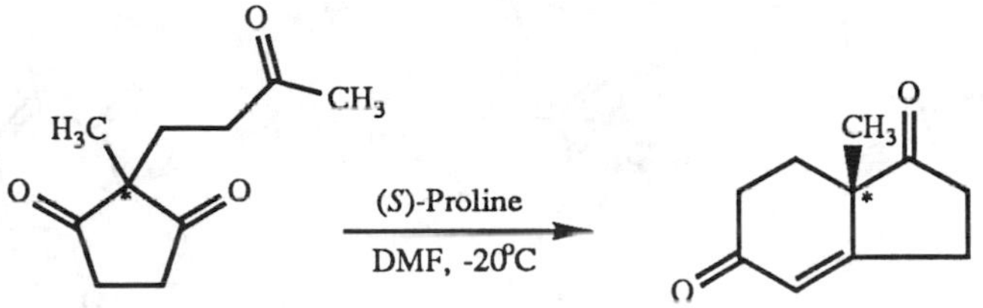

(88)

(S)-Proline
DMF, -20°C

(89) 86% e.e.

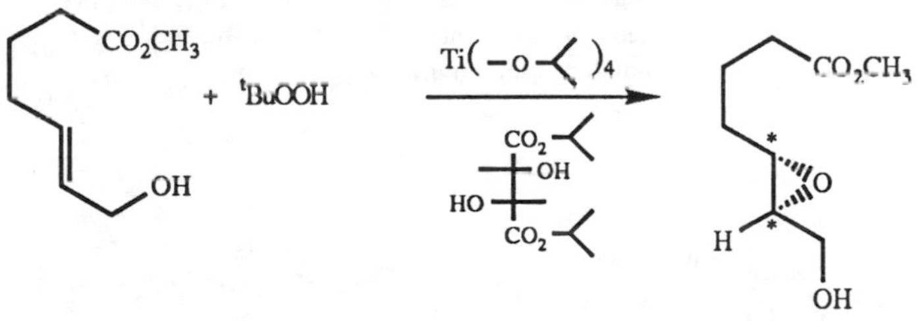

(90)

+ ^{t}BuOOH

(91) 95% e.e.

1.7.4.2 Reactions Involving Covalent Linkages of Chiral Auxiliary Groups with Substrates

The more successful approach to achieving enantioselectivity involves covalent bonding of the chirality inducing group with the substrate and the cleavage of the auxiliary groups on completion of the reaction. For example a cleavable chiral group (which is the chirality inducing agent) may be incorporated in the diene in the Diels-Alder reaction for inducing chirality in the product. The diene (92) can react with the dienophile (93) to form (94) in two different conformations (a) and (b) of which (b) is the more stable one due to π-stacking interactions. The reaction therefore proceeds with the conformation (b), attack occuring from only one of the two enantiotopic faces (the lower face) [29].

A chiral auxiliary substance should have the capability of inducing a high level of chirality, it should be recoverable and it should be available economically in both enantiomeric forms.

(92)

(a) large group (OCH₃) projects towards the diene causing severe non-bonded interactions.

(b) large group (OCH₃) points away from the diene.

(92) (b)   (93)   (94)

1.7.4.3   Reactions with Chiral Reagents

Asymmetric induction can be achieved by using chiral reagents to preferentially attack one of the two enantiotopic faces of an achiral substrate [30]. Thus reduction of the ketone (95) in the presence of NB-enantrane (96) afforded the products (97) and (98) in high optical yield. The reaction proceeds in 97% enantiomeric excess (ee) (98.5% - 1.5% = 97%).

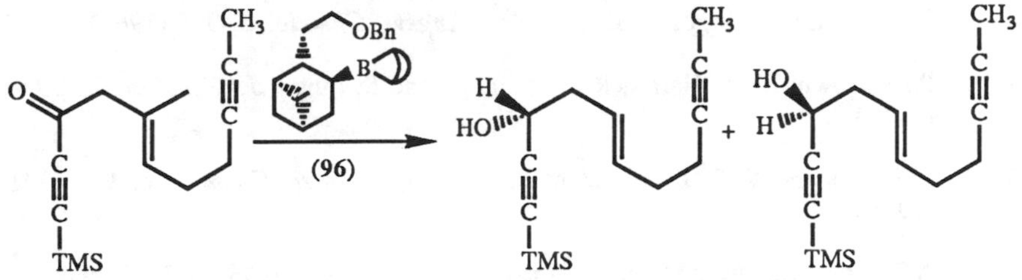

(95)   (97)   (98)

(97 :98 = 98.5 : 1.5)

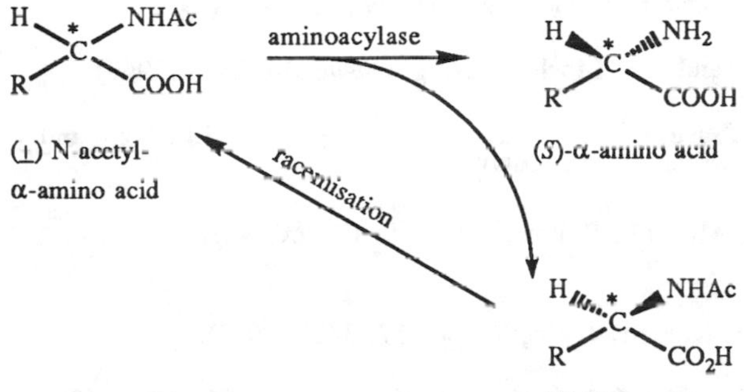

Figure 1.2. Reaction of aminoacylase with (±)-N-acetyl-α-amino acid.

1.7.4.4   Reactions with Enzymes

Microbes and enzymes can carry out asymmetric transformations which can result either in optical resolutions or in asymmetric synthesis. For instance racemic N-acetyl-α-amino acids can undergo asymmetric hydrolysis with amino acylase derived from *Aspergillus* to afford (S)-α-amino acids and the unchanged (R)-N-acetyl-α-amino acid as shown in Fig. 1.2 [31].   The latter can be recovered, racemised and recycled.

1.8 References

1.   Nomenclature of "Organic Chemistry", Part E. *Pure Appl. Chem.*, **45**, 11 (1976).

2.   K. Mislow and M. Raban, *Top. Stereochem.*, **1** (1967).

3.   J.K. O'loane, *Chem. Rev.*, **80**, 41 (1980).

4.   R.S. Cahn, C.K. Ingold and V. Prelog, *Angew. Chemie.*, **5**, 385 (1966).

5.   P.A. Brown, M.M. Harris, R.Z. Mazengo and S. Singh, *J.Chem.Soc.C.*, 3990 (1971).

6.   W.L. Waters, W.S. Linn and M.E. Caserio, *J. Amer. Chem. Soc.*, **90**, 6741 (1968).

7.   A.C. Cope and A.S. Mehta, *J. Amer. Chem. Soc.*, **96**, 3906 (1974).

8.   S.G. Know, *Top. Stereochem.*, **5**, 31 (1969).

9.   S.H. Wilen, *Top. Stereochem.*, **6**, 107 (1971).

10.   S.H. Wilen, A. Collet and J. Jacques, *Tetrahedron,* **33**, 2725 (1977).

11.   J.A. Dale, D.L. Dull and H.S. Mosher, *J. Org. Chem.*, **34**, 2543 (1969).

12.   C.H. Heathcock, C.T. Buse, W.A. Kelschick, M.C. Pissuing, J.E. Sohn and J. Lampe, *J. Org. Chem.,* **45**, 1066 (1980).

13.   S. Masamune, S.A. Ali, D.L. Snitman and D.S. Garrey, *Angew. Chem.,* **19**, 557 (1980).

14.   F.A. Carey and M.A. Kuehne, *J. Org. Chem.*, **17**, 3811 (1982).

15.   R. Noyori, I. Nishida and J. Sakata, *J. Amer. Chem. Soc.,* **103**, 2108 (1981).

16.   D. Seebach and V. Prelog, *Angew. Chem.,* **21**, 654 (1982).

17.   J.L. Luche, L. Rodriguez-Hahn and P. Crabbe, *J. Chem. Soc. Chem. Commun.,* 601 (1978).

18.   K.B. Sharpless and R.C. Michaelson, *J. Amer. Chem. Soc.,* **95**, 6136 (1973).

19.   K.B. Sharpless and T.R. Verhoeven, *Aldrichimica Acta,* **12**, 63 (1979).

20.   B.M. Trost, L. Weber, P.E. Strege, T.J. Fullerton and T.J. Dietsche, *J. Amer. Chem. Soc.,* **100**, 3426 (1978).

21.   C.G. Screttas and M. Micha-Screttas, *J. Org. Chem.*, **43**, 1064 (1978).

22.   C.G. Screttas and M. Micha-Screttas, *J. Org. Chem.*, **44**, 713 (1979).

23. (a) S.R. Wilson, M. S. Hague and R.N. Misra, *J.Org.Chem.*, **47**, 747 (1982)
    (b) I. Fleming, *Chem. Soc. Rev.*, **10**, 83 (1981).

24. W.C. Still and J.A. McDonald, *Tetrahedron Lett.*, 1031, 1035 (1980).

25. (a) B.M. Trost and T.P. Khun, *J. Amer Chem. Soc.*, **101**, 6756 (1979).
    (b) B.M. Trost and T.P. Khun, *J. Amer. Chem. Soc.*, **103**, 1864 (1981).

26. N. Cohen, R.J. Lopresti, C. Neukom and G. Saucy, *J. Org. Chem.*, **45**, 582 (1980).

27. P.A. Bartlett and K.K. Jernstedt, *Tetrahedron Lett.*, 1607 (1980).

28. For a latest review on enantioselective synthesis, see: *Chem.Rev.*, **92**(5), 739 (1992).

29. B.M. Trost, D.O. Krongly and J.L. Belletire, *J. Amer. Chem. Soc.*, **102**, 7595 (1980).

30. M.M. Midland and A. Kazubski, *J. Org. Chem.*, **47**, 2814 (1982).

31. I. Chibata in : "Asymmetric Reactions and Processes in Chemistry", (E.L. Eliel, and S. Otsuka, eds.), Amer. Chem. Soc., Washington, D.C., (1982).

2 Stereoselective Catalytic Reductions

Many natural products such as amino acids, carbohydrates and nucleotides exist in living organisms predominantly in one enantiomeric form. This is the result of the remarkable ability of enzymes to affect stereoselective transformations by controlling several stereochemical aspects in single step reactions, a property which has fascinated synthetic organic chemists and triggered efforts to develop corresponding laboratory procedures for enantioselective synthesis.

The high enantioselectivity of enzymes, which usually react with only one enantiomer of the substrate, is due to their ability to "recognize" one enantiomer from the other. This recognition stems from the three-dimensional structure of the reactive site in the enzyme structure which interacts differently with the two enantiomeric forms of the substrate, taking into account *all* the structural features of the substrate molecule——its chirality which would orient the various groups at the chiral centres in the two enantiomers differently with respect to the enzyme site (so that only one of the two enantiomers would usually act as the substrate), bulkiness of the side chains, hydrogen bonding etc. This limits the specificity of enzymes to a rather narrow class of reactant substrates. In the case of metal complexes used as catalysts, however, it is *mainly the reactive site of the substrate* which is of primary importance and it is the steric interactions at these reactive sites which largely determine the course of the reaction, the overall structure of the substrate being less critical. As a result, chemical catalysts have a broader substrate tolerance than enzymes. Moreover if the factors which determine the "enantiomeric excess" (e.e.) during the course of any reaction are understood it is possible to design optimal catalysts which will afford high enantiomeric excess in that reaction.

Although attempts to mimic enzyme-catalysed transformations in the laboratory with the aid of non-enzymatic heterogeneous catalysts initially met with only limited success, and the chiral products obtained in such reactions were merely of upto 15% e.e., this nonetheless spurred efforts to search for new catalytic systems which would afford higher e.e.s of the products. Metal complexes with organic ligands have played a key role in such transformations.

In general, there are two ways in which metal complexes can promote or catalyse organic reactions: (i) The metal complex may bind to a particular group in the substrate thereby enhancing the reactivity of the substrate to undergo a particular reaction. Thus Diels - Alder reactions can be promoted by coordination of the carbonyl group to the metal, thereby enhancing the electrophilic nature of the dienophile. (ii) Alternatively the metal complex may be necessary to bring about the cleavage of a covalent bond by providing the required activitation, without which the reaction may not occur, as in the epoxidation of allylic alcohols promoted by VO $(acac)_2$ (see chapter 5).

In order to achieve "asymmetric catalysis" i.e. the formation of one enantiomer at the expense of the other, a number of requirements must be fulfilled. The metal complex being employed as the catalyst may bind selectively to the reactant due to the presence of optically active ligands or substituents on the metal atom which can lead to steric preference in the mode of bonding. Alternatively, since the relative energies of the pathways leading to the enantiomeric products can be different, a larger quantity of the enantiomer which corresponds to the pathway with the lower intrinsic energy in the rate-limiting catalytic step may be formed. This enantioselectivity is represented in the form of an "enantiomeric excess" of the new stereogenic centre generated as a result of the reaction. The reactive complex needs to be efficiently regenerated and the product released at the end of the reaction so that the catalytic process can continue.

2.1   Homogeneous Catalytic Hydrogenations

2.1.1   Hydrogenation of Olefins

The discovery of *homogeneous* catalysts (i.e., catalysts which are soluble in the reaction medium) to promote hydrogenation marked the dawn of a new era both in mechanistic and synthetic fields. Tris(triphenylphosphine)chlororhodium, $(Ph_3P)_3$ RhCl, commonly known as Wilkinson's catalyst [1,2], represented the first and best-studied example of a homogeneous catalyst for hydrogenation of alkenes under mild experimental conditions. Asymmetric hydrogenation has received much attention due in large part to the fact that it is the basis for the first commercial catalytic asymmetric process, Monsanto's *L*-Dopa synthesis (Fig. 2.1), a drug used for the treatment of Parkinson's disease [3-5].

Rhodium complexes have been employed in this reaction, and as long as an asymmetric chelating biphosphine serves as a ligand and a dehydroamino acid or a close analogue is used as the substrate, the chiral products could be obtained in high enantioselective excesses (Table 2.2 entries 5-8). Chiral ligands with Rh^N (where "N" denotes neutral) and Rh(I) are now increasingly used as catalysts,

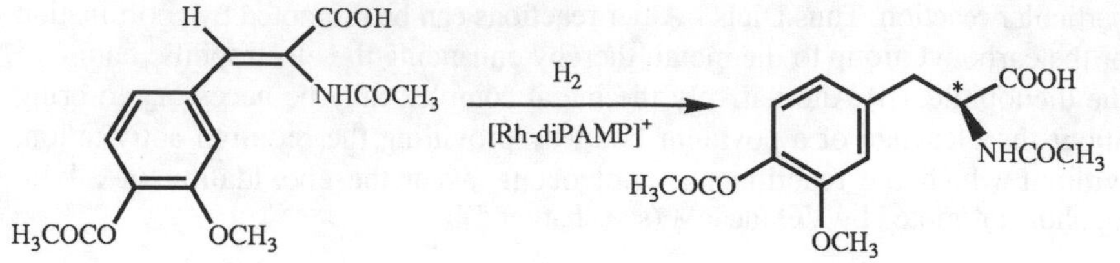

Figure 2.1. Asymmetric hydrogenation in the synthesis of L-Dopa.

giving enantioselectivity of 85-95% or even more. However, while the rhodium-based systems give higher optical yields mostly in the synthesis of amino acids and their derivatives, the recently developed ruthenium-diphosphine catalysts give good to excellent results with a much broader group of hydrogenation substrates affording useful intermediates for a variety of organic syntheses.

2.1.1.1   Hydrogenation with Rh-complexes

The earlier results with Rh(I) complexes having chiral ligands were only moderately successful since monodentate chiral phosphine ligands were usually employed which generally resulted in rather low chiral preferences. For example, when α-ethylstyrene (**1**) was hydrogenated using a chiral rhodium-phosphine complex (**4**) as catalyst[*], (*S*)-(+)-2- phenylbutane (**5**) was obtained in 7.8% e.e. [6].

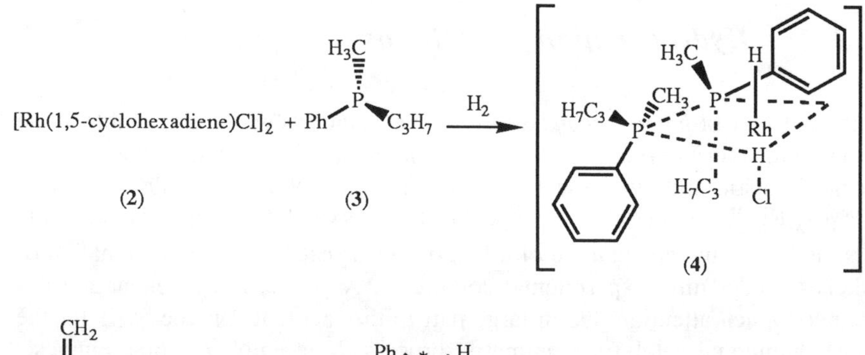

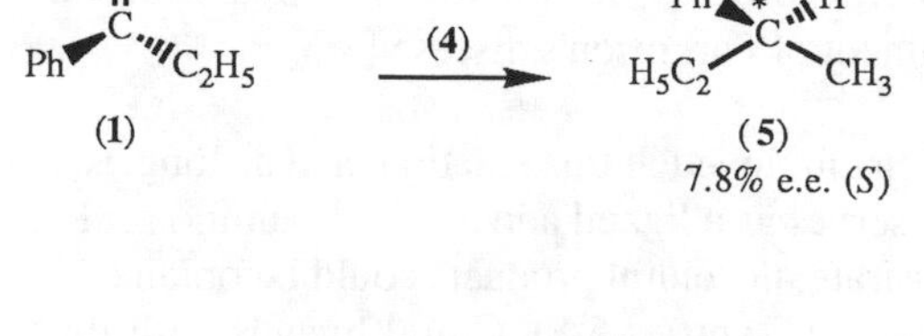

---

[*]Formed *in situ* from [Rh (1,5-cyclohexadiene) Cl]$_2$ (**2**) and (S)-(+)-methyl-phenyl-n- propylphosphine (**3**) in benzene at normal pressure and room temperature.

It was soon discovered that the most useful catalysts were Rh(I) and Rh[N] complexes with ligands chiral at phosphorus, phosphoranes chiral at carbon, or complexes in which optically active amides or ferrocenes were employed as ligands. Bi- and tridentate phosphine or phosphite ligands are now commonly used and they markedly increase the enantioselectivity by forming rigid complexes with transition metals. Asymmetric hydrogenation of didehydroamino acid derivatives (Fig. 2.1) is accepted as a reliable standard technique to test the efficiency of new optically active ligands or catalysts. This is because the hydrogenation reaction of didehydroamino acid derivatives is of primary importance and is widely used in the enantioselective synthesis of biologically active molecules. For example, the dopamine agonist (8) was obtained in a number of steps from (R)-homotyrosine (7), which was itself synthesized by hydrogenation of the corresponding acrylate derivative (6) with Rh -biphosphine catalysts [7]. Furthermore, the asymmetric hydrogenation of didehydroamino acid derivatives is used in labelling studies, such as in the synthesis of H^3 and C^{14} labelled unnatural amino acids which is achieved with rhodium catalysts having BPPM as ligands in high optical yields [8]. Table 2.1 shows some optically active ligands now used in stereoselective homogeneous catalytic hydrogenations.

(6)

(7)
80% e.e. (R)

(8)

Since the discovery of the asymmetric hydrogenation reaction used in L-Dopa synthesis (Fig 2.1), considerable effort has gone into understanding its mechanism, involving the study of reaction kinetics, X-ray crystallography, identification of reaction intermediates, synthesis and study of models of iridium complexes [30], and simulation of the reaction pathway by molecular mechanics and computer graphics [31]. A well studied reaction is the homogeneous catalytic hydrogenation of methyl-(Z)-α-acetamidocinnamate (11) with [Rh(diPAMP)]+ as catalyst to afford (S)-N-acetyl phenylalanine methyl ester (12) as the predominant product in 95% e.e. The intermediates involved in the catalytic cycle shown in Fig 2.2 were identified as (14A), (14B), (15A), (15B), (16A) and (16B) and it was shown that the dehydroamino acid substrate binds to the rhodium complex leading to the formation of two different diastereomeric complexes (**14A**) and (**14B**) both of which have the alkene and amide groups coordinated. The complex (**14A**) formed in lower quantities happens to be the more reactive one towards hydrogen and is the one involved in the catalytic cycle leading to the predominent (S) enantiomer——the major complex (**14B**) would result in the wrong (R) enantiomer (**13**). The rate determining step was shown to be the one involving the addition of hydrogen to the dehydroamino acid complex.

(**11**)   (**12**) 95% e.e. (S)

The mechanism of these reactions is believed to involve three steps: (a) hydrogen activation (b) substrate activation and (c) hydrogen transfer. Many metal complexes activate hydrogen molecules but very few are able to reduce unsaturated compounds either because the latter have not been activated on the metal complex or because transfer of activated hydrogen was energetically not feasible. Thus the interplay of the various factors involved is of paramount importance for the hydrogenation reaction to be successful.

(a) *Hydrogen Activation*: There are a number of factors responsible for the ability of a metal ion or complex to induce fission of molecular hydrogen [32]. In order to dissociate a hydrogen molecule by, say, a d^8 metal complex ML_4 (where L=ligand), the complex is oxidized by hydrogen to form a d_6 species:

$$H_2 + ML_4 \rightarrow H_2ML_4$$

The energy of formation of the metal-hydrogen bond (E_{M-H}) should be greater than the bond-dissociation energy of hydrogen (E_{H-H}) and the promotional energy P of the metal in the complex:

$$2E_{M-H} > E_{H-H} + P$$

It has been suggested that electron withdrawing ligands increase the promotional energy and reduce the ability of the metal complex to activate molecular hydrogen. The primary interaction of molecular hydrogen with a metal complex in solution is believed to involve the transfer of electron density from filled metal orbitals to an antibonding acceptor orbital of one hydrogen atom of the hydrogen molecule.The charge separation induced in the H-H bond by such electron flow will lead, in the proximity of the metal atom, to bond weakening, bond lengthening and synchronous formation of the two metal-hydrogen bonds.

This process is facilitated by the presence of electron donating ligands and a low oxidation state of the metal. Alternatively the hydrogen molecule can be activated by the metal *via* donor acceptor bonds where low-lying vacant "d" orbitals in the metal can accept electrons to form MH^2 species [1b].

Table 2.1. Some Chiral Ligands used in Homogeneous Catalytic Hydrogenations

A) Ligands chiral at carbon:

Structure		Abbreviated name	Reference
	$R^1 = c\text{-}C_6H_{11}$ $R^2 = H$	CYCPHOS	[9]
	$R^1 = C_6H_5$ $R^2 = H$	PHENPHOS	[10]
	$R^1 = CH_3$ $R^2 = CH_3$	CHIRAPHOS	[11]
	$R^1 = CH_3$ $R^2 = H$	PROPHOS	[12]
	$R = {}^iPr$	VALPHOS	[13,14]
	$R = {}^tBu$	t-LEUPHOS	[13]
	$R = C_6H_5CH_2$	PHEPHOS	[13,14]
	$Ar = C_6H_5$	DIOP	[5,15]
	$Ar = 2\text{-}C_{10}H_7$	2-Naph-DIOP	[15]
	$Ar = m\text{-}CF_3 C_6H_4$	m-CF$_3$-DIOP	[15]
	$R^1 = PPh_2,\ R^2 = H$	MDPP	[16]
	$R^1 = H,\ R^2 = PPh_2$	NMDPP	[16]
		CAMPHOS	[17]

(Table 2.1. contd.)

B) Ligands chiral at phosphorus:

Structure	Abbreviated name	Reference
	(10)	[28]
	CYCLODIOP	[29]
	diPAMP	[3]
	(9)	[28]

R = H	FPPM	[18]
R = Ph	BZPPM	[19]
R = CH$_3$	APPM	[20]
R = ^{t}Bu	PPPM	[21]
R = NHPh	Ph-CAPP	[22]
R = O^{t}Bu	BPPM	[23]

(Table 2.1. contd.)

Structure	Abbreviated name	Reference
	BPPFA	[24]
	MPFA	[24]
	BPPFOH	[25]
	BINAP	[26]
	BDPCH	[27]

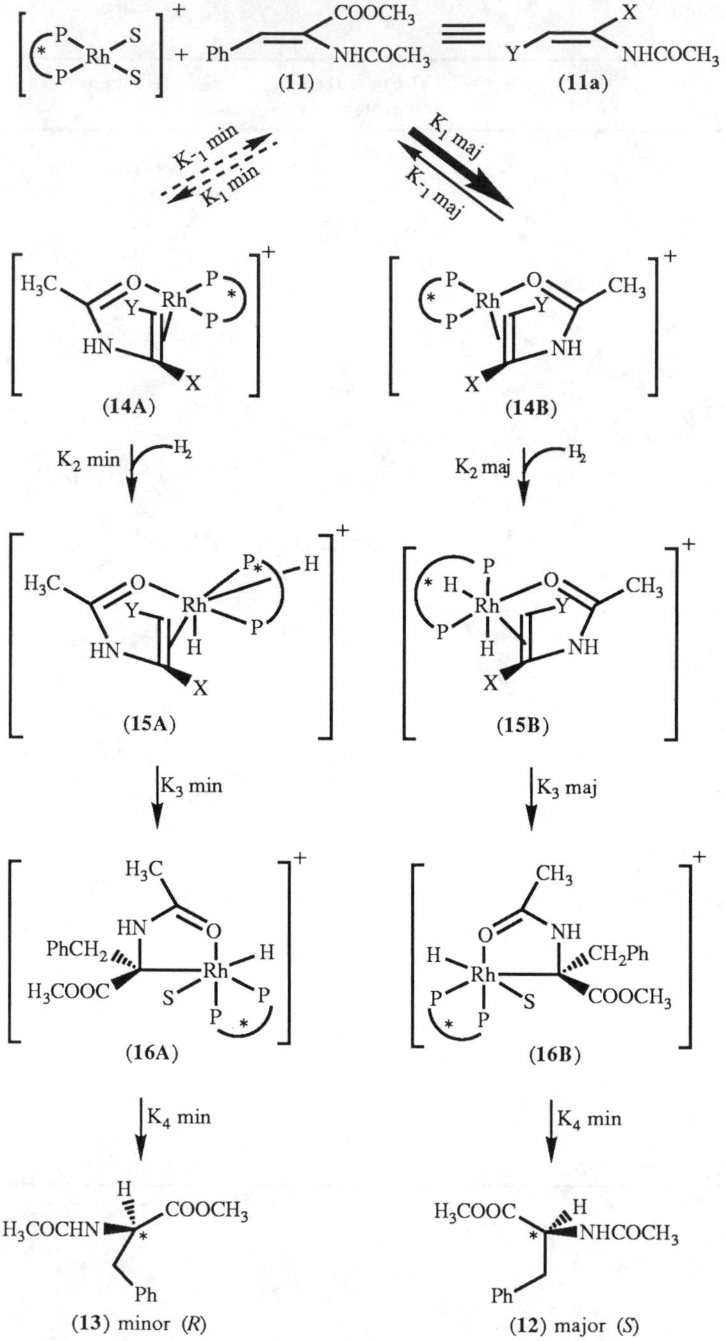

Figure 2.2. Mechanistic scheme for [Rh(diPAMP]⁺-catalyzed hydrogenation of methyl-(7)-α-acetamidocinnamate (**11**), X = COOCH₃, Y = Ph.

(b) *Substrate Activation*: The next step in the sequence of events leading to the hydrogenated product appears to be the generation of an intermediate activated complex in which both the molecular hydrogen and the olefin required to be reduced are coordinated to the metal atom. The hydrido species (H_2ML_4) activates olefins only if there is a vacant site for olefin coordination. Complexes such as $IrCl(CO)(PPh_3)_2$ etc. can activate molecular hydrogen rapidly and reversibly, but are unable to catalytically hydrogenate olefins. Such species are said to be *coordinatively saturated* as there is no vacant site for olefin coordination. Since *cis* - reduction occurs rapidly and stereospecifically, the vacant site can be assumed to be *cis* to the two M-H bonds.

(c) *Transfer of Hydrogen*: The hydrogen transfer step to the olefin commences by the competitive displacement of the bound solvent by the olefin from the dihydrometal species, H_2ML_4 (solvent), followed by the stereospecific *cis*-transfer of the bound hydrogen to the olefin. As shown in Fig 2.3 the olefin occupies a position *cis* to both Rh-H bonds in the complex (**17**) and both the hydrogens are then transferred simultaneously, each by a three centre transition state (**18**). The saturated substrate (**19**) diffuses away from the transfer site, leaving the complex again ready to activate the dissolved hydrogen molecules. Such a triangular transition state has been postulated on the basis of an overall low kinetic isotope effect shown by hydrogenation reactions.

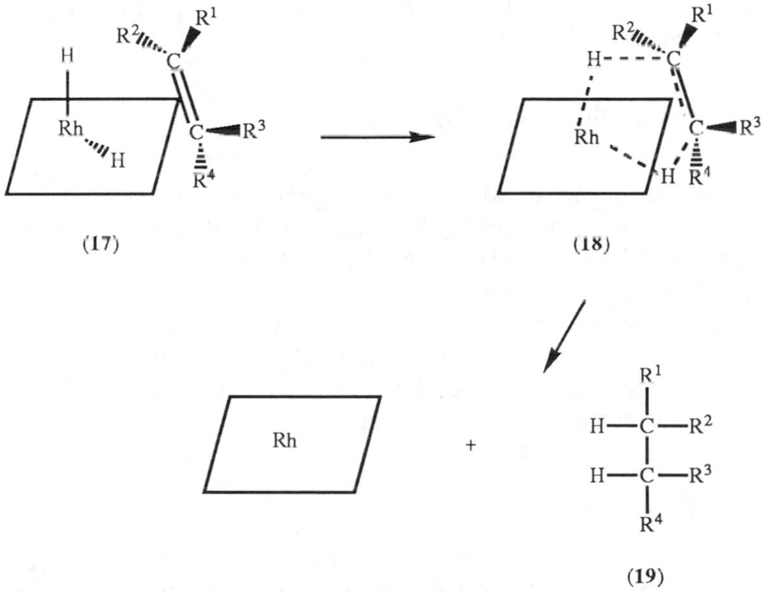

Figure 2.3. Mechanism of hydrogen transfer in the reduction of olefin.

The earlier studies suggested that the stereoselectivity, particularly of the polydentate substrate, may be determined by the binding step of the substrates with the catalyst rather than by the relative rates of hydrogenation or diastereomeric complex formation.

Recent studies on the mechanism of catalytic asymmetric hydrogenation have been carried out with α-acylamino acrylic and cinnamic acid derivatives, as shown in Fig. 2.2 for the [Rh(diPAMP)]$^+$ catalyzed hydrogenation of (11). The conclusion drawn from these studies leads to a dramatic change in the concept regarding the mechanism of asymmetric catalytic hydrogenation, at least in these reactions. It has been concluded from these studies that the diastereomeric catalyst-substrate adducts have different *rates* of hydrogenation and it is this overwhelming difference in the rates of these adducts which dictates the enantioselectivity of the asymmetric catalytic hydrogenation. The major adduct detected during the course of the reaction turned out to be an "unproductive" complex which interconverts into the minor less stable "productive" diastereomer, followed by rapid reaction of the latter with H2. This conclusion is contrary to the prevalent earlier view that it was the preferred mode of *initial* binding of the prochiral substrate to the chiral catalyst which was responsible for the enantioselectivity of the asymmetric hydrogenation reaction. For several other asymmetric catalytic reactions, e.g. hydroformylation (as discussed in Chapter 4, Section 4.1.2), however, it has been proposed that the enantioselectivity is determined by the preferred mode of binding of the prochiral substrate to the chiral catalyst [33].

2.1.1.1.1   Tetrasubstituted Olefins

Tetrasubstituted olefins can be hydrogenated using rhodium catalysts with chiral (aminoalkyl) ferrocenyl-phosphine (20) as ligands affording extremely high enantioselectivity (92-98% e.e.) [34]. Thus 2-phenyl-3-methyl-2-butenoic acid (21) was catalytically hydrogenated to give (22) in 98.4% e.e. by using [Rh-(20)] as catalyst. The asymmetric hydrogenation of (E)-2-phenyl-3-methyl-2-pentenoic acid (23) with [Rh-(20)] yielded (2S,3S)-2-phenyl-3-methyl-pentanoic acid (24) in 97.3% e.e. In this reaction two vicinal chiral centers are formed through *cis* addition at the *Re* face of the substrate*.

---

*The exceedingly high enantioselectivity achieved by the catalyst [Rh-(R,S)-(20)] can be explained by assuming that the terminal amino group in the ferrocenyl ligand [(R,S)-(20)] forms an ammonium carboxylate with the carboxyl group of the olefinic substrate and consequently attracts the substrate to the coordination sphere of the catalyst, thus facilitating hydrogenation. This assumption has been further supported by the fact that the esters of the substrates (21) and (23) did not undergo hydrogenation because the attractive interactions between the amino group of the catalyst and the esters were lacking.

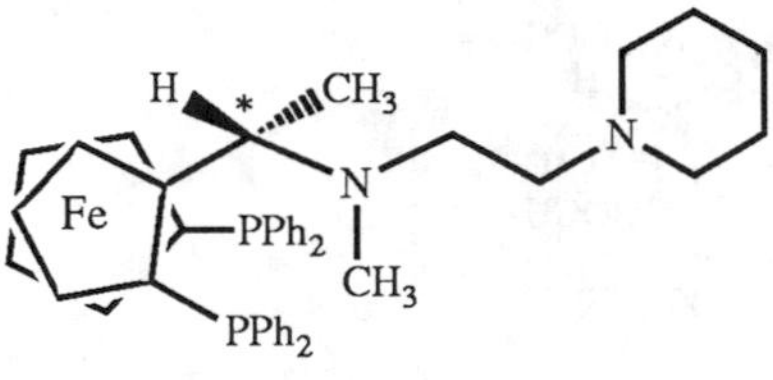

(20)

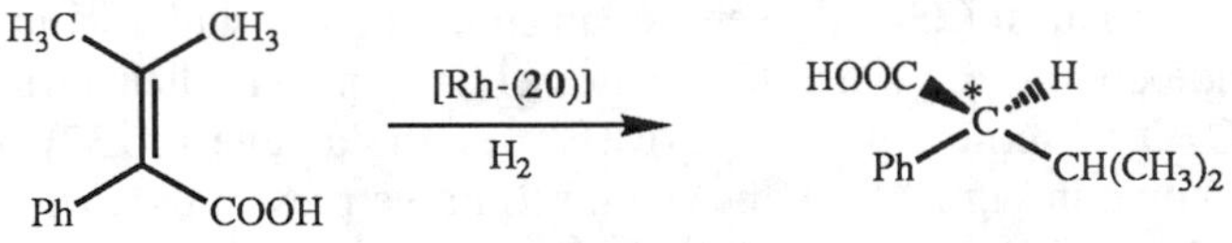

(21)   (22) 98.4 e.e. (S)

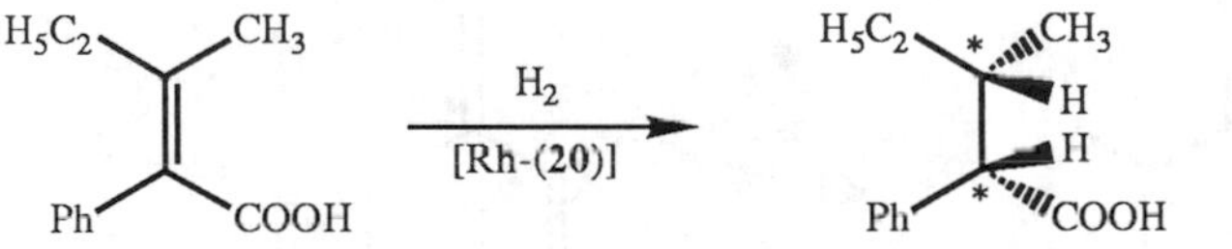

(23)   (24) 97.3 e.e. (2 S, 3 S)

2.1.1.1.2   Substituted Itaconate Esters

In the case of 3-substituted itaconate esters, the ester group at the chiral site exerts a powerful directing effect when a catalyst such as (R,R)-diPAMP-derived (25) is employed leading to very high diastereoselectivity in the reduced product [35]. The exceedingly high diastereoselectivity obtainable with catalyst (25) is evident from its use in the hydrogenation of itaconate (26) to give the reduced product (27) in 100% e.e.

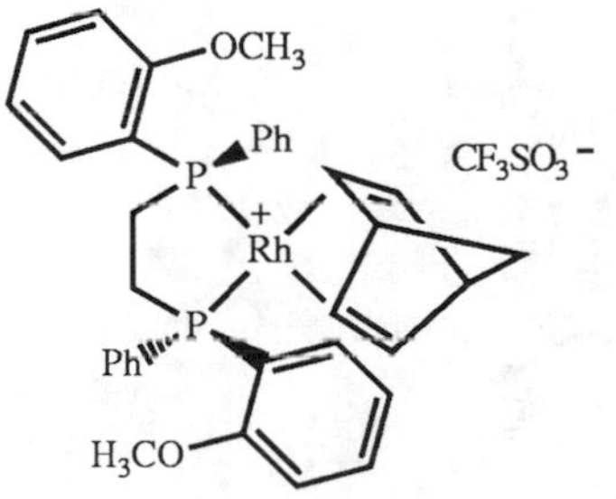

(25)

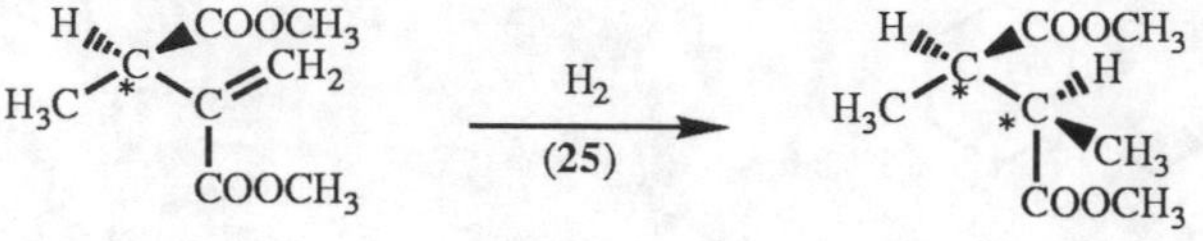

(26)   (27) 100% (R,R)

Catalyst (25) also leads to high kinetic resolution of the substrate (16:1) with preference for the (R)-configuration at the new asymmetric centre. The diastereoselection of catalyst (25) has been explained in analogy with rhodium and iridium enamide complexes which have an essentially planar chelate ring incorporating N-C=O-M moiety and the α-carbon of the alkene [36,37]. A similar model, as shown in Fig 2.4, may be employed in the present case where replacement of trigonal-NH by tetrahedral -CHCH$_3$ results in a non-planar chelate.

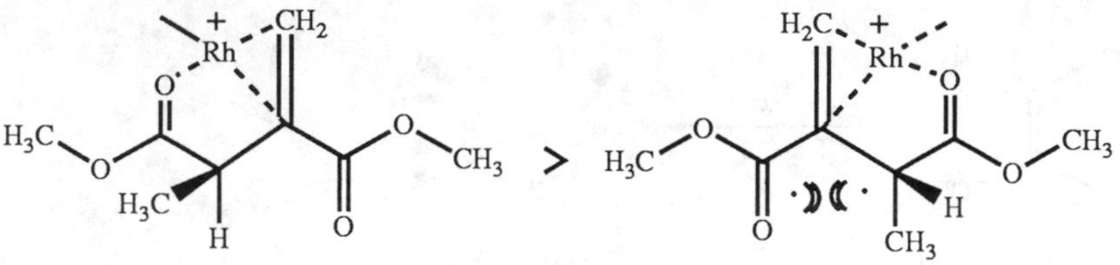

(A)   (B)

Figure 2.4. Diastereoisomeric chelates formed by coordination of (26) to rhodium. Note that (A) is more stable than (B).

In Fig. 2.4. (B), the methyl group at C-3 is *antiperiplanar* to the coordinated double bond and interacts unfavourably with the α-ester. This explains the reasons why the precursor of [(R,R)-(27)] is preferred over the precursor of [(R, S)-(27)].

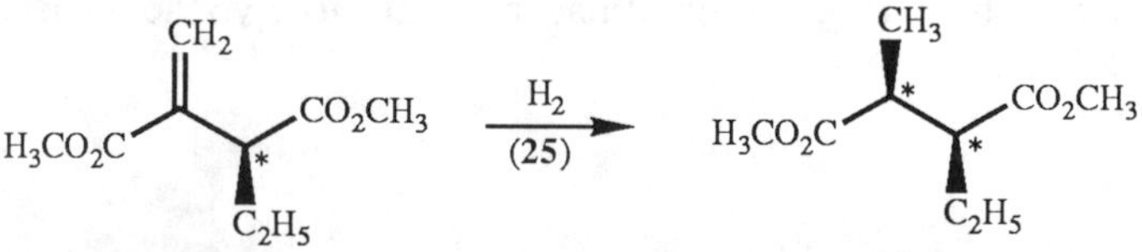

(S)-(28)   (29)   (slow)

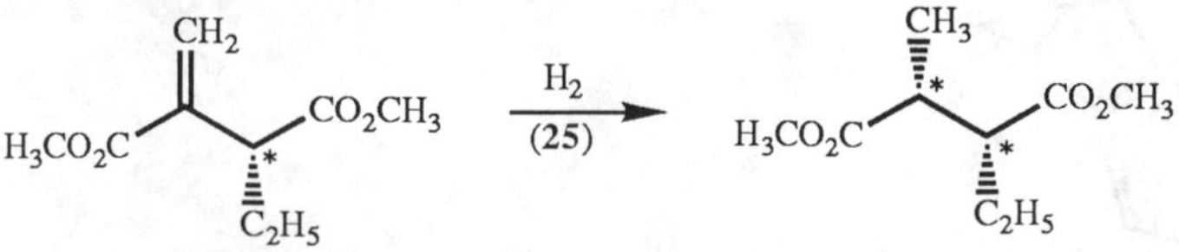

(R)-(28)   (30)   (fast)

It is interesting to note that while in the asymmetric hydrogenation of dehydroamino acids, optical yields increase with increasing temperature [38], the reverse is true in the complex (25) catalysed hydrogenation of 3-substituted itaconate esters. This is reflected in the ratio for the hydrogenation of substrates (S)-and (R)-(28) leading to the hydrogenated products (29) and (30) respectively in which the K_R/K_S is highest at lower temperatures and diminishes above 50°C. Table 2.2 shows some representative examples of asymmetric hydrogenation of olefins catalysed by Rh-complexes.

Table 2.2. Rhodium Catalyzed Asymmetric Hydrogenation of Olefins

Substrate	Ligand	Product	Optical yield (% e.e.)	Reference
Ar, H / C=C / NHCOCH$_3$, COOH	a	ArH$_2$C—*C—H / HOOC, NHCOCH$_3$	79 (R)	[5]
H, H$_5$C$_6$ / C=C / NHCOCH$_3$, COOCH$_3$	a	H$_5$C$_6$H$_2$C—*C—H / H$_3$COOC, NHCOCH$_3$	55 (R)	[5]
H, H$_5$C$_6$ / C=C / COOCH$_3$, NHCOCH$_3$	b	H$_5$C$_6$H$_2$C—*C—H / H$_3$COOC, NHCOCH$_3$	97.2 (R)	[30b]
Ph, CH$_3$ C=O, Ph—C—COOH	c	Ph—*C—CH$_3$(H) / Ph—C—COOH(H)	92.1 (2 S, 3 R)	[34]
H$_3$C, CH$_3$ C=C—COOH (4-Cl-phenyl)	c	CH(CH$_3$)$_2$ / C*—COOH (4-Cl-phenyl) H	97.4 (S)	[34]
H$_3$C, CH$_3$ C=C—COOH (4-H$_3$CO-phenyl)	c	H$_3$C, CH$_3$ CH / C*—COOH (4-H$_3$CO-phenyl) H	96.7 (S)	[34]
H$_3$C, H / C=C / NHCOCH$_3$, COOH	d	H$_3$C—H$_2$C—*C—H / HOOC, NHCOCH$_3$	84 (R)	[39]

(Table 2.2. contd.)

Substrate	Ligand	Product	Optical yield (% e.e.)	Reference
	(25)		96 (R, R)	[35]
	(25)		93 (R, R)	[35]
	e		92 (S)	[40]

a = (-)-DIOP,   b = (R, R)-diPAMP,   c = [(R, S)-(20)],   d = (-)-BINAP,   e = BPPM, NEt$_3$

2.1.1.2   Hydrogenation with Ru-complexes

An exciting development in the field of asymmetric hydrogenation has been the discovery that *ruthenium complexes* can effect asymmetric hydrogenations in good yields and high enantiomeric excesses. The catalytic potential of the use of ruthenium complexes such as HRu(PPh$_3$)Cl was demonstrated by Wilkinson but the problem in developing asymmetric analogues was the tendency to form complexes of the type (P$_2$)$_2$RuXY instead of the desired (P)$_2$RuXY which are capable of binding a substrate and hydrogen. It was later discovered that when bulky and rigid chiral ligands such as BINAP were used, the desired complexes could be generated and a number of alkenes with adjacent polar groups (e.g. allylic alcohols) could be hydrogenated with excellent enantioselectivity, often with >96% e.e. [41].

2.1.1.2.1   *Allylic and Homoallylic Alcohols*

Prochiral allylic and homoallylic alcohols can be reduced by catalytic hydrogen with high enantioselectivities by employing Ru (II) dicarboxylate complexes having (*R*)- or (*S*)-BINAP as ligands such as [(*R*)-(**31**)] and [(*S*)-(**31**)] (shown in Fig. 2.5). For instance geraniol (**32**) can be catalytically hydrogenated to citronellol (**33**) in 98% e.e. using [(*S*)-(**31b**)] as catalyst [41,42]. Employing the complex [(*R*)-(**31a**)] as catalyst, nerol (**34**) on catalytic hydrogenation yielded the same product (**33**) in 98% e.e. The (*S*) enantiomer of citronellol (**33**) can also be obtained from either geraniol (**32**) or nerol (**34**) by simply employing the ruthenium catalyst of opposite chirality. Thus using [(*S*)-(**31c**)] as catalyst, nerol (**34**) affords [(*S*)-(**33**)] in 98% e.e. The stereochemical outcome implies that Ru-BINAP complexes differentiate the C-2 enantiofaces at a certain stage of the catalysis. Thus both (*R*) and (*S*) enantiomers are accessible by either changing the handedness of the catalysts or by changing the geometry of the allylic olefins.

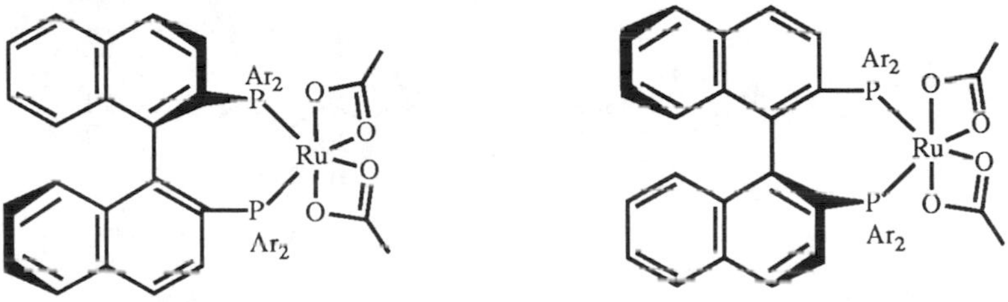

[(**31**)-(*R*)]   [(**31**)-(*S*)]

a. Ar = C_6H_5, R = CH_3
b. Ar = C_6H_5, R = $(CH_3)_3C$
c. Ar = p-CH_3 C_6H_4, R = CH_3

Figure 2.5. Ru(II)-BINAP complexes used in asymmetric hydrogenation of olefins.

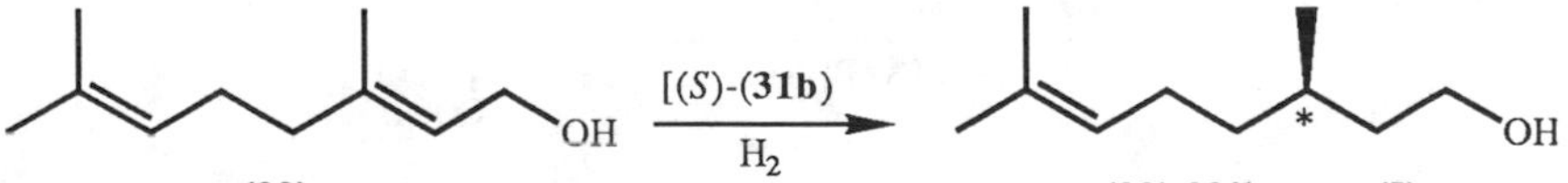

(32)   [(*S*)-(**31b**)] $\xrightarrow{H_2}$ (33) 98% e.e.   (*R*)

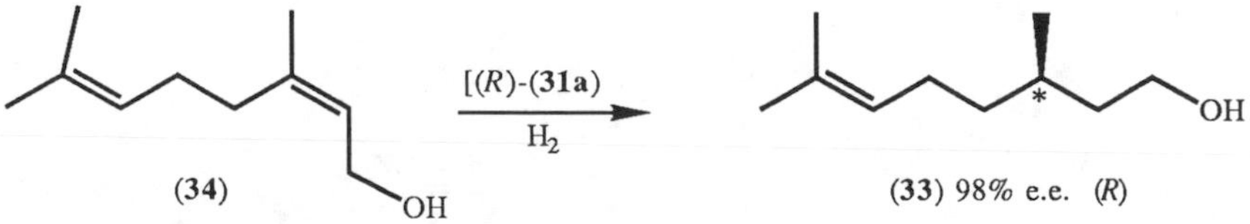

(34)   [(*R*)-(**31a**)] $\xrightarrow{H_2}$ (33) 98% e.e.   (*R*)

[Structure (34)] → [(S)-(31c)] / H$_2$ → [Structure (33)] (33) 98% e.e. (S)

2.1.1.2.2   Unsaturated Carboxylic Acids

Ru-BINAP complexes of the type (31) have also been used in the asymmetric hydrogenation of α, β or β, γ-unsaturated carboxylic acids affording the corresponding saturated products in high enantiomeric excesses and in quantitative yields [42] (Table 2.3, entries 1-3). Asymmetric hydrogenation of geranic acid (35) with [(R)-(31a)] as catalyst yielded citronellic acid (36) in 87% e.e., leaving the C$_6$-C$_7$ double bond intact.

[Structure (35)] → [(R)-(31a)] / H$_2$ → [Structure (36)] (36) 98% e.e. (S)

2.1.1.2.3   Dicarboxylic Acids

Dicarboxylic acids such as itaconic acid (37) which have structural features closely related to dehydroamino acids are reduced catalytically by ruthenium complexes in high optical yields. The complex Ru$_2$Cl$_4$[(+)-BINAP]$_2$ NEt$_3$ catalyses the asymmetric hydrogenation of itaconic acid (37) to (S)-methyl succinic acid (38) in 90% e.e [43, 44].

[Structure (37)] → H$_2$ / Ru$_2$Cl$_4$[(+)-BINAP]$_2$NEt$_3$ → [Structure (38)] (38) 90% e.e. (S)

2.1.1.2.4   *Dehydroamino Acids*

Dehydroamino acids, such as acrylic acid and its derivatives, are reduced with Ru_2Cl_4 [(-)-BINAP]$_2$ NEt$_3$ in high optical yields (Table 2.3, entries 7,8) [39]. A notable feature in such ruthenium catalysed reductions is that the chirality induced in the products by the use of Ru_2Cl_4[(-)-BINAP]$_2$ NEt$_3$ is opposite to that obtained by the [Rh-(-)-BINAP] system (Table 2.2, entry 7)*.

2.1.1.2.5   *Prochiral Ketones*

Asymmetric hydrogenation of prochiral ketones has also been successfully accomplished with ruthenium catalysts [45], which have proved to be generally superior to the corresponding rhodium complexes giving > 99% e.e.[41]. These reactions will be discussed in Section 2.1.3.2.

While the detailed mechanism of catalysis with ruthenium complexes is not known, a number of differences in comparison to rhodium complexes are discernible. The turnover rate of the ruthenium catalysts is quite slow, and the enantiomeric excess is more dependent on variations in reaction temperature, pressure or substrate concentration** . Some representative examples of Ru-catalysed hydrogenation of olefins are shown in Table 2.3.

---

*Interestingly the esters of dicarboxylic acids when hydrogenated with ruthenium complexes gave somewhat lower enantioselectivities than the free acids. This is in contrast to the Rh-diPAMP catalysed hydrogenation of itaconic acid (**37**) which is hydrogenated with low optical yields, because of dimer formation through hydrogen bonding, whereas the corresponding methyl ester gave much higher optical yields [44]. This suggests that the C-4 carboxylate group which coordinates with Ru in the transition state plays an important role in the enantioselection displayed by the Ru-BINAP catalyst. Moreover, the hydrogen bonding interference to chelate formation encountered in Rh-diPAMP catalysed system probably has little effect in enantioface discrimination in the case of the ruthenium catalyst.

** It is interesting that in the dehydroamino acid hydrogenation reaction described in Table 2.2, entry 7 and Table 2.3, entries 7, 8 employing (-)-BINAP opposite product enantiomers are obtained with Ru and Rh. This can be either due to the rate determining step being different in the two cases or due to differences in the coordination geometry. In the rhodium complex hydrogen adds to a square planar complex while in ruthenium complexes a higher coordination geometry may exist.

Table 2.3. Ruthenium Catalyzed Asymmetric Hydrogenation of Olefins

Substrate	Ligand	Product/Optical yield (% e.e.)	Reference

Substrate:

$$\underset{H_3C}{\overset{H}{\diagdown}}C=C\underset{COOH}{\overset{CH_3}{\diagup}}$$

Ligand: a

Product:

$$H_3C\cdots\overset{*}{C}\cdots H,\ H_5C_2\text{—}\overset{*}{C}\text{—}COOH$$

91 (2R)   [42]

Substrate:

$$\underset{H_3C}{\overset{H}{\diagdown}}C=C\underset{CH_2CO_2CH_3}{\overset{COOH}{\diagup}}$$

Ligand: a

Product:

$$HOOCCH_2\text{—}\overset{*}{C}\text{—}CH_2COOCH_3$$ with H_3C and H

95 (2R)   [42]

Substrate: naphthalene with $\overset{CH_2}{\underset{}{C}}\text{—}COOH$

Ligand: b

Product: naphthalene with $\overset{CH_3}{\underset{}{\overset{*}{C}}}\text{—}H,\ COOH$

97 (2S)   [42]

Substrate: $(CH_3)_2C=CH\text{-}CH_2\text{-}CH_2\text{-}C(CH_3)=CH\text{-}CH_2OH$

Ligand: c

Product: $(CH_3)_2C=CH\text{-}CH_2\text{-}CH_2\text{-}\overset{*}{C}H(CH_3)\text{-}CH_2\text{-}CH_2OH$

98 (3R)   [41]

Substrate: 6,7-dimethoxy-1-(3,4-dimethoxybenzylidene)-2-acetyl-tetrahydroisoquinoline (H_3CO, H_3CO, NCOCH$_3$, OCH$_3$, OCH$_3$)

Ligand: c

Product: 99.5 (S)   [46]

Substrate: H_3C, H_3C, NCHO, OCH$_3$ dihydropyridine benzylidene

Ligand: d

Product: 98 (R)   [47]

(Table 2.3. contd.)

Substrate	Ligand	Product/Optical yield (% e.e.)	Reference
 H₃C—CH=C(NHCOCH₃)(OCH₃) 	e	H₃COCHN—C*(H)—CH₂CH₃ with HOOC 86 (S)	[39]
 H—C(Ph)=C(NHCOPh)(COOH) 	e	PhOCHN—C*(H)—CH₂Ph with HOOC 92 (S)	[39]

a = (R)-BINAP, b = (S)-BINAP, c = [(S)-(31)], d = (R)-Tol. BINAP], e = (-)-BINAP, Et₃N

2.1.1.3   Hydrogenation with Ti-complexes

Titanium (IV) complexes catalyse the asymmetric reduction of prochiral olefins in upto 15% e.e. [48-50]. Menthyl and neomenthylcyclopentadienyl ligands have been employed with Ti to effect hydrogenation in the presence of reducing agents such as "Red-Al", Li [H$_2$(OCH$_2$CH$_2$OCH$_3$)$_2$] (39) acting as a co-catalyst [50]. Homogeneous catalytic reduction of 2-phenyl-1-butene (40) yielded 2-phenyl butane (41) in 14.9% e.e. using [η^5(-)-menthyl cyclopentadienyl]$_2$ TiCl$_2$ as catalyst in the presence of "Red-Al".

Ph(H$_5$C$_2$)C=CH$_2$ (40) →(39)/H$_2$→ Ph(H$_5$C$_2$)C*(H)—CH$_3$ (41) 14.9% e.e. (S)

Employing [(η^5-(+)-neomenthyl cyclopentadienyl) (η^5-cyclopentadienyl)] TiCl$_2$ as catalyst, (33) was obtained in 10% e.e. with the predominent (R) enantiomeric product. However these Ti-complexes have limited applicability in the asymmetric hydrogenation of olefins, since they are inactive towards olefins with donor groups such as CN, COR, COOR etc [48].

2.1.1.4   Hydrogenation with Co-complexes

Cobalt complexes have been used with modest success in the homogeneous catalytic hydrogenation of *simple and functionalized olefins* [51-55]. Earlier attempts at asymmetric reduction with Co complexes gave poor optical yields [51, 52].   Vitamin B_{12} or cyanocob(III)alamin (**42**), the deficiency of which causes megaloblastic anaemia in man, represents a rare example of a natural product which contains a metal-carbon bond. It has been used both as a catalyst precursor and as a model to prepare similar catalytic systems in the asymmetric reduction of olefins [53-55]. Cyanocob(III)alamin (**42**) is first reduced to cob(I) alamin (**43**) in the presence of an electron donor such as metallic zinc in acetic acid as shown in Fig. 2.6. The complex (**43**) reduces the carbon-carbon double bond in α,β-unsaturated esters, amides etc. stereoselectively affording optical yields ranging from 14 to 32.7% e.e., the (Z) substrates giving predominantly the (S) products. The α,β- unsaturated substrate (**44**) is catalytically reduced to (**45b**) with the complex (**43**) in 26.7% e.e. The highest optical yield achieved with the complex (**43**) was in the reduction of α,β-unsaturated methyl ketone (**46**) which afforded (**47**) in 32.7% e.e. [55].

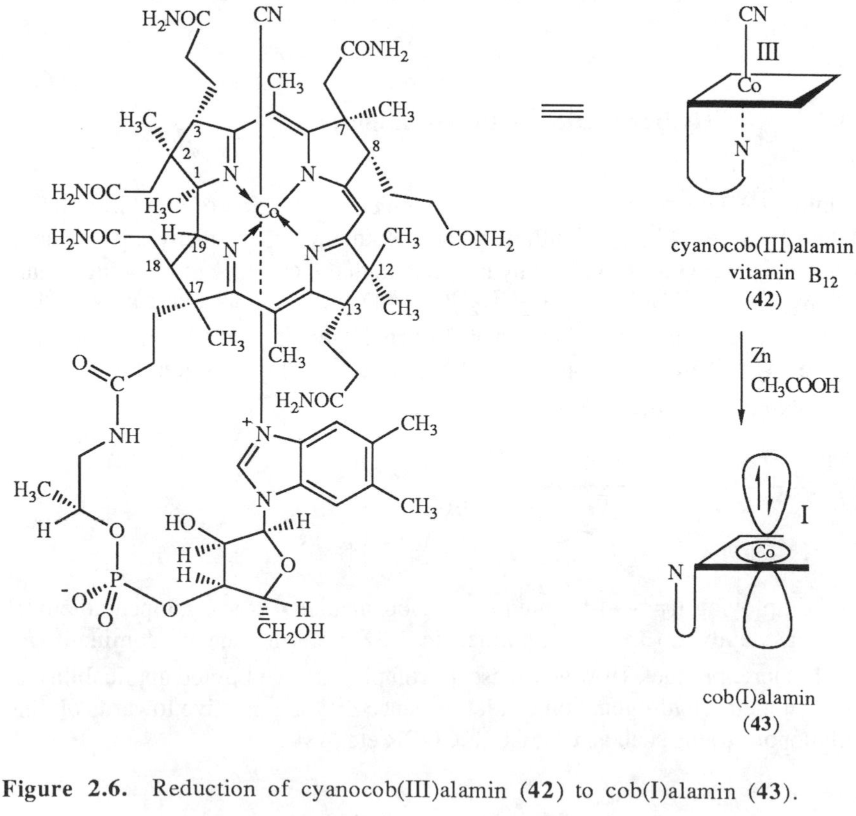

Figure 2.6.   Reduction of cyanocob(III)alamin (**42**) to cob(I)alamin (**43**).

(44) $\xrightarrow[\text{Zn,CH}_3\text{COOH/H}_2\text{O}]{(42)}$ (45b) 26.7% e.e. (S)

(46) $\xrightarrow[\text{Zn,CH}_3\text{COOH/H}_2\text{O}]{(42)}$ (47) 32.7% e.e. (S)

(Z)-(48) $\xrightarrow[\text{Zn,CH}_3\text{COOH/H}_2\text{O}]{(50)}$ (49) 20% e.e. (S)   (Z) (48) + (E)-(48)

$\xrightarrow[\text{Zn,CH}_3\text{COOH/H}_2\text{O}]{(42)}$

(49) 20% e.e. (S)

Studies on the Co- catalysed homogeneous catalytic hydrogenation of other substrates such as the reduction of the unsaturated amide (48) to (49) have been carried out employing, among others, model complexes such as the hepta-methylcob(I)yrinate complex (51) formed from heptamethyl-dicyanocob(III) yrinate complex (50) (Fig 2.7). These studies have contributed to a great extent in understanding the mechanism of Co-catalysed reductions. As shown in Fig 2.8, cob(1)alamin complex (43) attacks the electrophilic C- atom in position 3 of substrate (44) with its strongly nucleophilic d_z2-orbital on the β-side of the complex, (the face of the corrin nucleus opposite to the ribonucleotide side chain is the β-side as shown in Fig. 2.6). This attack can be directed to the si- or the re- face of the double bond in (44) and accordingly leads to the two diastereomeric alkycobalamins (50a) and (50b). These intermediates undergo reductive cleavage with retention of configuration to give (45a) and (45b) respectively [56,57]. In practice (45b) is obtained as the major product.

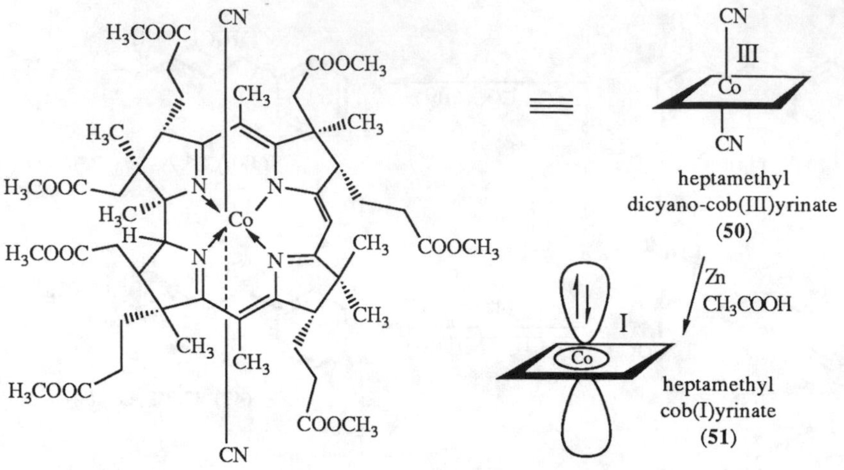

Figure 2.7. Model complex (**51**) and its formation from (**50**).

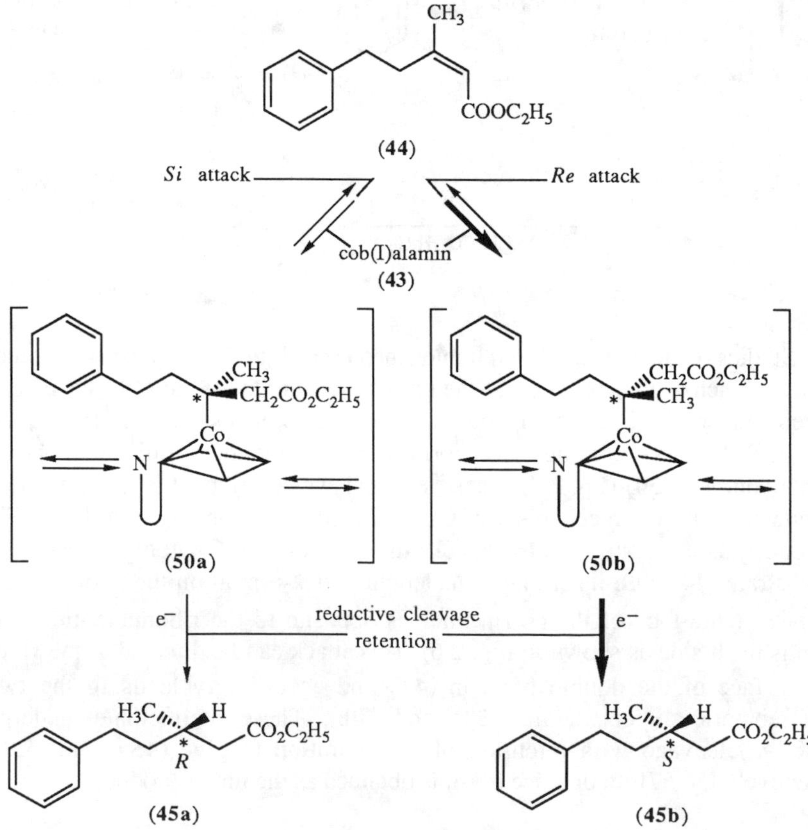

Figure 2.8. Mechanism of Co-catalysed reduction.

2.1.1.5   Hydrogenation with Heterobimetallic Complexes

Heterobimetallic complexes are metal complexes containing another chiral transition metal atom in addition to the asymmetric C (or C and N) atoms in the chelate skeleton. Stereoselective catalytic hydrogenation of olefins with such complexes has opened a new chapter in asymmetric synthesis [58, 59].

Rhodium (I) is generally employed as the central metal atom with ligands, along with another chiral transition metal atom such as Mo, V, Re etc. to form a heterobimetallic complex. These *chiral-at-metal* complexes have been used in the homogeneous asymmetric hydrogenation of olefins and in hydrosilylation reactions yielding > 98% enantioselectivities. For example, the heterobimetallic complex (51), a rhodium (I) complex containing a chiral rhenium atom in a chelating "diphosphine" backbone, catalyses the asymmetric hydrogenation of enamide precursors to α-amino acids and esters in optical yields that are among the highest observed for the rhodium system [59]. The homogeneous asymmetric hydrogenation of (Z)-α-acetamidoacrylic acid (52) yielded the reduced product (53) using [(+)-(R)-(51)] in 98% e.e. Employing ruthenium -BINAP complex as catalyst, the same reaction afforded an optical yield of 95% e.e. (R), whereas the optical yield with Rh-diPAMP catalyst was not more than 90% e.e. with the predominant (S) enantiomeric product. The higher optical yield achieved with the heterobimetallic complex, in this instance, as compared to rhodium and ruthenium complexes indicates the versatility of the heterobimetallic system. This is however not always the case. By employing other ligands in the Ru and Rh catalytic complexes, higher optical yields may be achieved.

[(+)-(R)-(51)]

2.1.2   Catalytic Hydrosilylation

2.1.2.1   Catalytic Hydrosilylation of Olefins

The catalytic reduction of prochiral olefins, imines, ketones etc. using silicon hydrides ("hydrosilylation") provides another useful method to accomplish indirect asymmetric hydrogenations. Prochiral olefins have been catalytically hydrosilylated by chiral phosphine complexes of Pt, Ni, Rh and Pd [60,61]. As shown in Fig. 2.9, the silyl group can add either at the α or β positions of the double bond, giving rise to the two adducts (54) and (55) respectively.

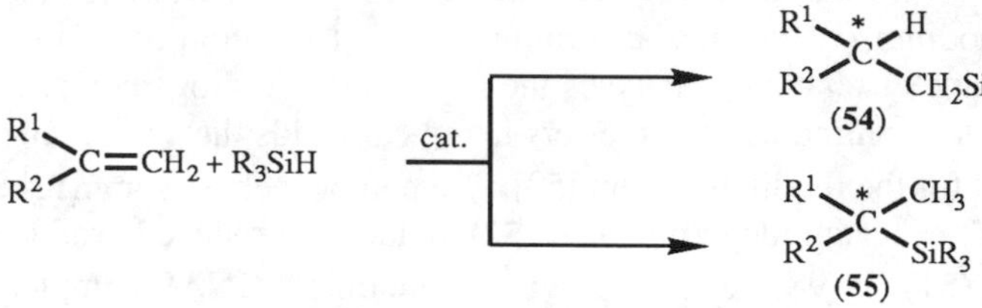

Figure 2.9. Generalized scheme of hydrosilylation giving two possible adducts.

For instance, 1,1-disubstituted olefins can be converted into the corresponding terminal adducts by methyldichlorosilane using (+)BMPP as ligand with either platinum or nickel choride. Thus α-methylstyrene (56) is hydrosilylated with [(+)BMPP (PtCl$_2$)]$_2$ or [(+)BMPP]$_2$ NiCl$_2$ as catalysts to afford (57) in 5.2% and 20.9% e.e. respectively [62,63] (entries 1,2, Table 2.4). Rhodium complexes also catalyse hydrosilylation of 1,1- disubstituted olefins. For instance (56) gives the terminal adduct (58) [62].

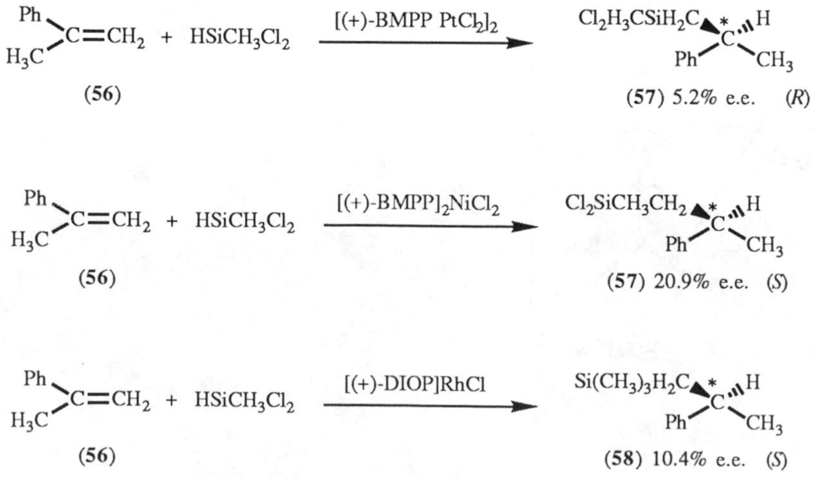

Styrene (59) represents an interesting case in homogeneous catalytic hydrosilylation. Using platinum (II) or nickel (II) complexes as catalysts, styrene gives predominantly the terminal adducts, i.e. β-adducts. With palladium (II) complexes having MDPP or NMDPP as ligands, however, styrene is hydrosilylated with trichlorosilane to give exclusively the α-adducts (60) [63].

(59)   +   HSiCl₃   →   (60) 3.3% e.e.   (R)

(61)

[(R_C, S_Fe)-PPFA] PdCl₂ (61) catalyses the hydrosilylation of styrene (59) and norbornene (62) with trichlorosilane giving high yields of the hydrosilylated products (60) and (63) respectively.

(59)   +   HSiCl₃   →   (60) 52% e.e. (S)

(62)   +   HSiCl₃   →   (63)
53% e.e.
(1R, 2S, 4S)

The silicon halides (60) and (63) can be converted to the corresponding fluoro compounds (64) and (65) on reaction with KF [64]. Oxidative cleavage of the carbon-silicon bond in such fluoro derivatives affords an efficient method for preparing optically active alcohols with *retention* of configuration. Thus treatment of (64) and (65) with m-chloroperoxybenzoic acid (MCPBA) gives the corresponding alcohols (66) and (67) in 52% and 50% e.e. respectively [64].

Catalytic asymmetric hydrosilylation of prochiral olefins provides an efficient alternative method to prepare optically active functionalized compounds. For example, bromination of the pentafluorosilicates (64) and (65) with NBS proceeds with *inversion* of configuration to give (*R*)-1-phenylethyl bromide (68) and *endo*- norbornanol (69) in 3-10% and 53% optical yields respectively.

(65)

NBS →

(69) 53% e.e.

(1R, 2 S, 4S)

Hydrosilylation of cyclopentadiene (**70**) and 1,3-cyclohexadiene (**71**) with $HSiCl_3$ in the presence of MDPP [Pd(PhCN)$_2$Cl$_2$] yielded the corresponding optically active allylsilanes (**72**) and (**73**) respectively [65].

(70)   **(72)**

(71)   **(73)**

The mechanism of hydrosilylation has been investigated in the hydrosilylation of simple olefins catalysed by chloroplatinic acid, in Pt (II)-olefin complexes such as [(C$_2$H$_4$)PtCl$_2$]$_2$ or in Ir(I)-phosphine complexes, employing optically active silanes R$_3$Si*H as well as by exchange reactions of R$_3$Si*H (D) [66,67]. As shown in Fig. 2.10, the hydrosilylation catalysed by the Pt (II)-olefin complex (**74**) involves the following stereochemical processes: (a) The platinum centre is inserted into the silicon-hydrogen bond to form (**75**) with *retention* of configuration at silicon; (b) the intermediate (**75**) is converted to (**76**) which results in *cis* addition of platinum to the double bond; (c) the product (**77**) is formed from the intermediate (**76**) with *retention* of configuration at both silicon and carbon. This final step may be regarded as proceeding by a *quasi*-cyclic (S$_{Ni}$-S$_i$) mechanism involving nucleophilic attack on silicon and electrophilic attack on carbon, both proceeding with *retention* of configuration [67]. A possible alternative to steps (b) and (c) in hydrosilylation is the direct conversion of intermediate (**75**) to product (**77**) *via* concerted addition of Si*(Si* = optically active silane) and H. However it has been shown that the intermediate (**76**) is involved in olefin isomerization, which may occur during hydrosilylation [67].

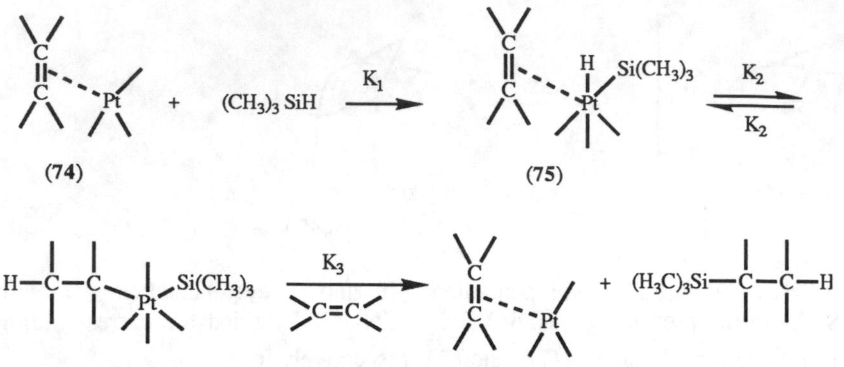

Figure 2.10. Mechanism of Pt(II) catalysed hydrosilylation of olefins.

The mechanism of hydrosilylation catalysed by palladium (II) having MDPP or NMDPP as ligands has been studied in cyclic conjugated dienes and in styrene [68]. In the hydrosilylation of a cyclic conjugated diene such as 1,3-cyclohexadiene (71) catalysed by Pd (II)-NMDPP or Pd(II)-MDPP, the product (73) has always the predominent (S) configuration. In contrast, styrene (59) gives the predominant (R) enantiomeric addition product (60) with Pd(II)-NMDPP, whereas with Pd(II)-MDPP, the product with the predominant (S) configuration is obtained (Table 2.4, entry 3). These observations have led to the suggestion that π-allylic metal intermediates are involved in these reactions.

2.1.2.2   Catalytic Hydrosilylation of Imines

Rh(PPh$_3$)$_3$Cl is an excellent catalyst for the hydrosilylation of imines giving high yields of the corresponding N-silyl amines under mild conditions. As shown in Fig. 2.11, the N-silyl amines are readily desilylated to give the corresponding amines or amides [69,70].

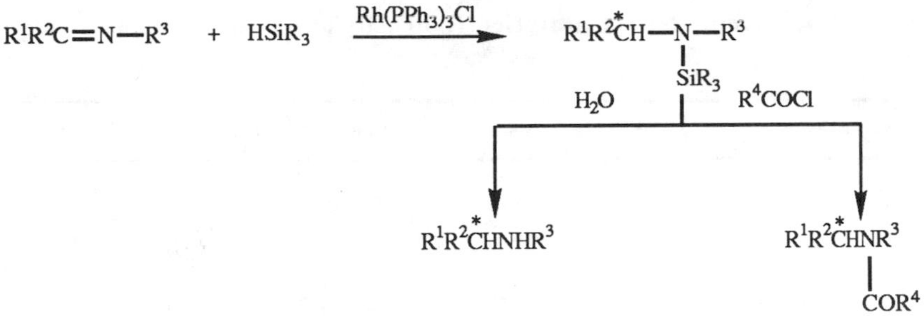

$$R^1R^2C=N-R^3 \ + \ HSiR_3 \ \xrightarrow{\text{Rh(PPh}_3)_3\text{Cl}} \ R^1R^2\overset{*}{C}H-N-R^3$$

Figure 2.11. Generalized scheme: hydrosilylation and desilylation of imines.

Hydrosilylation of imines depends not only on the type of catalyst used but is also affected by the hydrosilane employed. Dihydrosilanes have been found to react more smoothly than monohydrosilanes or trihydrosilanes. With dihydrosilanes such as $(C_2H_5)_2SiH_2$, $C_6H_5CH_3SiH_2$ and $(C_6H_5)_2SiH_2$, the activity of the catalysts decreases in the order $(Ph_3P)_3RhCl$ $(Ph_3P)_2Rh$ $(CO)Cl>$ Py_2RhCl (dmf) $BH_4 > [(1,5\text{-hexadiene}) RhCl]_2 > [(1,5\text{-cyclooctadiene})RhCl]_2 >$ $PdCl_2 > (Ph_3P)_2$ $PdCl_2$ [71].

Hydrosilylation of the ketimine (78) with diphenylsilane gives the corresponding optically active secondary amine (79) using [(+)DIOP] RhCl as catalyst. The optical yields depend significantly on temperature, the stereoselectivity increasing at lower temperatures [71,72]. Some representative hydrosilylation reactions are presented in Table 2.4.

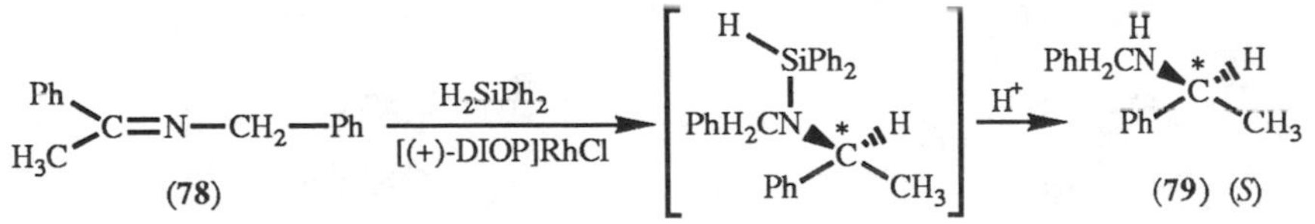

60°C = 27.5% e.e.
40°C = 39% e.e.
24°C = 50% e.e.
2°C = 65% e.e.

Table 2.4. Asymmetric Catalytic Hydrosilylation of Olefins and Imines

Substrate	Catalyst	Product / %e.e.	Reference
$\underset{H_3C}{\overset{Ph}{>}}C=CH_2$ + CH_3Cl_2SiH	$[(+)\text{-BMPP-PtCl}_2]_2$	$\underset{Ph}{\overset{H_3C}{>}}\overset{*}{\underset{}{C}}\underset{CH_2SiCH_3Cl_2}{\overset{H}{<}}$ 5.2 (R)	[62,63]
$\underset{H_3C}{\overset{Ph}{>}}C=CH_2$ + CH_3Cl_2SiH	$[(+)\text{-BNPP-PtCl}_2]_2$	$\underset{Ph}{\overset{H_3C}{>}}\overset{*}{\underset{}{C}}\underset{CH_2SiCH_3Cl_2}{\overset{H}{<}}$ 20.9 (R)	[62,63]
$\underset{H_3C}{\overset{Ph}{>}}C=CH_2$ + $SiHCl_3$	2MDPP + $Pd(PhCN)_2Cl_2$	$\underset{Ph}{\overset{SiCl_3}{>}}\overset{*}{\underset{}{C}}\underset{CH_3}{\overset{H}{<}}$ 34 (S)	[63]
$\underset{H_3C}{\overset{Ph}{>}}C=CH_2$ + $(CH_3)_3SiH$	$[(-)\text{-DIOP}]RhCl$	$\underset{Ph}{\overset{(CH_3)_3SiCH_2}{>}}\overset{*}{\underset{}{C}}\underset{CH_3}{\overset{H}{<}}$ 10.4 (S)	[62]
$\underset{H_3C}{\overset{Ph}{>}}C=NCH_2Ph$ + H_2SiPh_2	i) $[(+)\text{-DIOP}]RhCl$ ii) H^+	$\underset{Ph}{\overset{PhH_2CHN}{>}}\overset{*}{\underset{}{C}}\underset{CH_3}{\overset{H}{<}}$ 50 (S)	[71,72]
$\underset{H_3C}{\overset{Ph}{>}}C=N-Ph$ + H_2SiPh_2	i) $[(+)\text{-DIOP}]RhCl$ ii) H^+	$\underset{Ph}{\overset{PhHN}{>}}\overset{*}{\underset{}{C}}\underset{CH_3}{\overset{H}{<}}$ 40 (S)	[71,72]
+ H_2SiPh_2 i) $[(+)\text{-DIOP}]RhCl$ ii) H^+ R=		 39 (S)	[71,72]
$\underset{H}{\overset{Ph}{>}}C=CH_2$ + $HSiCl_3$	2MDPP + $Pd(PhCN)_2Cl_2$	$\underset{Ph}{\overset{H_3C}{>}}\overset{*}{\underset{}{C}}\underset{NHPh}{\overset{H}{<}}$ 22 (R)	[63]

(Table 2.4. contd.)

Substrate	Catalyst	Product / %e.e.	Reference
⬡ + HSiCl₃	Pd-NMDPP	(*R*) SiCl₃ H	[65]
⬡ + HSiCl₃	Pd-NMDPP	(*S*) H SiCl₃	[65]

2.1.3   *Catalytic Hydrogenation of Ketones*

The asymmetric catalytic hydrogenation of ketones has not been very successful and high enantioselectivity has been observed only in a few cases. The homogeneous catalysts capable of hydrogenating ketones in high enantiomeric excesses are far fewer in number than the diverse catalysts employed in olefin hydrogenation.

The mechanism of asymmetric hydrogenation of ketones has not yet been fully investigated and although some correlation exists between the specific catalyst-substrate interactions and the degree of enantioselectivity observed, very few conclusions can be drawn from the limited data available. The understanding of all the factors involved in transferring chirality to ketones is still far from complete.

The homogeneous catalytic hydrogenation of prochiral ketones can be carried out by two methods: (a) direct asymmetric hydrogenation across the double bond, and (b) derivatization of the ketone prior to asymmetric hydrogenation.

2.1.3.1   Direct Hydrogenation of Simple Ketones

Direct asymmetric hydrogenation of simple ketones is generally accomplished by using rhodium complexes with chiral phosphine as ligands, though ruthenium and iridium metals have also been successfully employed in some cases [73,74]. For example, acetophenone (**80**) is catalytically reduced using [Rh(NBD)(BMPP)$_2$]$^+$ as catalyst to afford (R)-1-phenylethanol (**81**) in 8.6% e.e. [75].

Changes in the structure of the chiral ligands attached to the metal complex markedly influence the stereoselectivity of the reaction. Introduction of a hydroxyl group in the side chain of the chiral ferrocenyl phosphine ligand (R)-(S)-BPPFA (**82**) to form (**83**) brings about a high degree of stereoselectivity in the rhodium complex-catalysed hydrogenation of carbonyl compounds.

The asymmetric hydrogenation of acetophenone described above proceeds with 43% e.e. when [Rh-(**83**)]$^+$ is used as the catalyst [25]. Interestingly, the complex [Rh-(**82**)]$^+$ catalyses the same reaction to give the product having (S) configuration in 15% e.e.

The mechanism of asymmetric hydrogenation of ketones has not attracted much attention and fully reproducible kinetic data is lacking. However in a few cases, factors affecting the rate and enantioselectivity of the catalytic hydrogenation of carbonyl compounds such as hydrogen pressure, nature of solvent, temperature, isotopic labelling and the effect of various ligands employed in catalytic systems have been studied [75,76].

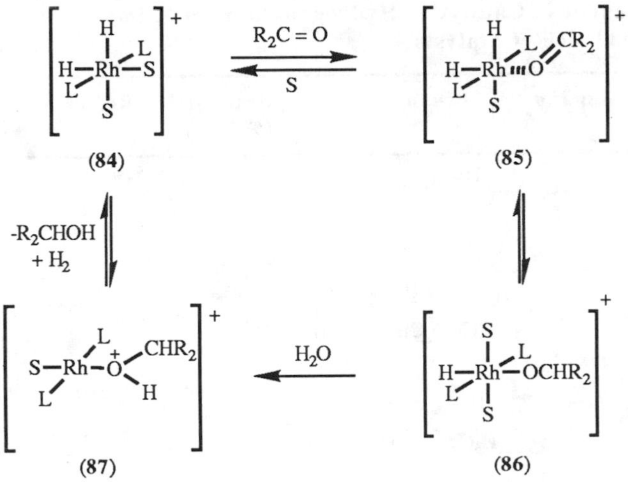

Figure 2.12. Mechanism of Rh-catalysed hydrogenation of ketones (L = ligands, S = solvent).

The results of these studies are summarized in Fig. 2.12 which shows that the first step is the electrophilic attack of the rhodium hydride complex (**84**) on the carbonyl oxygen of the substrate to form the complex (**85**) wherein the solvent has been replaced by the substrate which is coordinated to the metal atom. This is probably followed by a step-wise process involving a 1,3-hydride shift from a *cis-* site on the metal to the carbonyl carbon to form (**86**). A second proton transfer then occurs from the catalyst complex to the oxygen atom of the substrate to form (**87**) which dissociates to give the product and *re*-form (**84**). Stable analogues of all the intermediates except (**86**) have been isolated [77].Table 2.5 shows some typical examples of homogeneous asymmetric hydrogenation of simple ketones.

2.1.3.2   Direct Hydrogenation of Functionalized Ketones

Functionalized ketones or carbonyl-containing substrates having additional carbonyl, olefin, hydroxyl or ether functionalities $\alpha-$ or $\beta-$ to the carbonyl group generally afford high optical yields in the asymmetric catalytic reduction of prochiral carbonyl compounds (75-98% e.e.). This is probably due to their ability to generate a secondary interaction with the metal centre. Fig. 2.13 (A),(B) show such secondary interactions of substrates with rhodium thereby producing a chelating effect similar to that seen in olefins. This mechanism is operative in the asymmetric reduction of prochiral functionalized ketones leading

Table 2.5. Homogeneous Catalytic Hydrogenation of Simple Ketones with Chiral Rh Catalysts

Substrate	Ligand	Product	Optical yield (% e.e.)	Reference
$Ph-\overset{\overset{\displaystyle O}{\|\|}}{C}-CH_3$	a	$\underset{Ph}{\overset{H_3C}{\diagdown}}\overset{*}{C}\underset{OH}{\overset{H}{\diagup}}$	43 (R)	[25]
$H_3C-\overset{\overset{\displaystyle O}{\|\|}}{C}-CH\overset{CH_3}{\underset{CH_3}{\diagdown}}$	b	$\underset{(CH_3)_2CH}{\overset{H_3C}{\diagdown}}\overset{*}{C}\underset{OH}{\overset{H}{\diagup}}$	60 (R)	[77]
$H_3C-\overset{\overset{\displaystyle O}{\|\|}}{C}-CH_2-CH_3$	c	$\underset{H_5C_2}{\overset{H_3C}{\diagdown}}\overset{*}{C}\underset{OH}{\overset{H}{\diagup}}$	12 (R)	[78]
$H_3C-\overset{\overset{\displaystyle O}{\|\|}}{C}-CH\overset{Ph}{\underset{Ph}{\diagdown}}$	c	$\underset{(Ph)_2CH}{\overset{H_3C}{\diagdown}}\overset{*}{C}\underset{OH}{\overset{H}{\diagup}}$	32 (R)	[73]
phenyl ethyl ketone	d	1-phenyl-1-propanol (S)	14.2 (S)	[79]
4-phenyl-2-butanone	d	4-phenyl-2-butanol	8.8 (R)	[79]
acetophenone	d	1-phenylethanol	80 (R)	[73]
$H_3C-\overset{\overset{\displaystyle O}{\|\|}}{C}-\overset{CH_3}{\underset{CH_3}{\overset{\|}{C}}}-CH_3$	e	$\underset{(H_3C)_3C}{\overset{H_3C}{\diagdown}}\overset{*}{C}\underset{OH}{\overset{H}{\diagup}}$	43 (R)	[25]

(Table 2.5. contd.)

Substrate	Ligand	Product	Optical yield (% e.e.)	Reference
$\overset{O}{\underset{}{\|}}$ $C_6H_5-\overset{}{C}-CH_3$	f	$\overset{H}{\underset{OH}{C}}$ (phenyl)	15 (R)	[25]
$H_3C-\overset{O}{\underset{}{C}}-(CH_2)_5-CH_3$	b	$H_3C(CH_2)_5\overset{H_3C}{\underset{OH}{C}}H$	48 (R)	[77]

a = (R, S)-BPPFOH,   b = (structure: PPh$_2$ ... PPh$_2$, N, O=C, NHCOOC$_6$H$_5$, C$_6$H$_5$)   c = CAMP /^{i}Pr COOH,

d – (+)-DIOP,Et$_3$N (structure: H, SiCH$_3$, H) + (Ph$_3$P)$_3$ RhCl,   e = (R_C, S_{Fc})-BPPFOH,   f = (R_C, S_{Fe})-BPPFA

to high optical yields as exemplified by the asymmetric hydrogenation of ketopantoyl lactone (88) affording pantoyl lactone (89) in 86.7% e.e. using Rh(I)-BPPM as catalyst [80]* .

Figure 2.13. Complexation of functionalized ketones with rhodium. (Note the secondary interaction in (A) and (B)).

---

* The catalyst Rh(I)-BPPM serves as a biomimetic model of ketopantoyl lactone reductase which is involved in the biosynthesis of (89) from (88) enroute to pantothenic acid, a member of the B complex vitamins [81]. Employing (-)-DIOP and (-)(cyclohexyl)-DIOP as ligands, the rhodium catalysed reduction of (88) afforded (89) in 35% and 54% e.e. respectively [80,82].

(88)   (89) 86.7% e.e. (R)

2.1.3.2.1   With Rhodium-Diphosphine Catalysts

Neutral Rh-diphosphine as well as cationic Rh-diphosphine catalysts have been used in the asymmetric hydrogenation of ketoamides [83]. Chiral peralkyldiphosphine ligands such as (+)- or (-)-cyclohexyl-DIOP (cyc DIOP) have been found particularly effective for the asymmetric hydrogenation of α-ketoamides. For example, with a neutral complex, (+)-cycDIOP-Rh^N, substrate (90) was hydrogenated smoothly to give (91) in 68% e.e. Other examples are shown in Table 2.6 (entries 1,3).

(-)-cycDIOP   (+)-cycDIOP

(cyc = cyclohexyl)

(90)   (91) 68% e.e. (R)

2.1.3.2.2   With Ruthenium Complexes

Recently ruthenium complexes such as (31) (discussed under Section 2.1.2 on the asymmetric hydrogenation of olefins) as well as halogen containing complexes of the type RuX$_2$ (BINAP) have been found as excellent catalysts for the asymmetric hydrogenation of functionalized ketones. The reaction requires the simultaneous coordination of the carbonyl oxygen and hetero atom X or Y (C=sp^2 or non-stereogenic sp^3 carbon) to the ruthenium atom making a five or six-membered chelate which affords a high degree of stereodifferentiation (Fig. 2.14).

The presence of some nitrogen and oxygen containing groups such as dialkylamino, hydroxyl, alkoxyl, keto, alkoxycarbonyl, alkylthiocarbonyl, dialkylaminocarbonyl, carboxyl etc. helps to direct asymmetric induction. The oxygen-triggered asymmetric hydrogenation is achieved with halogen containing Ru complexes as catalysts whereas more basic ketones are hydrogenated equally well with the Ru-dicarboxylate complexes (31).

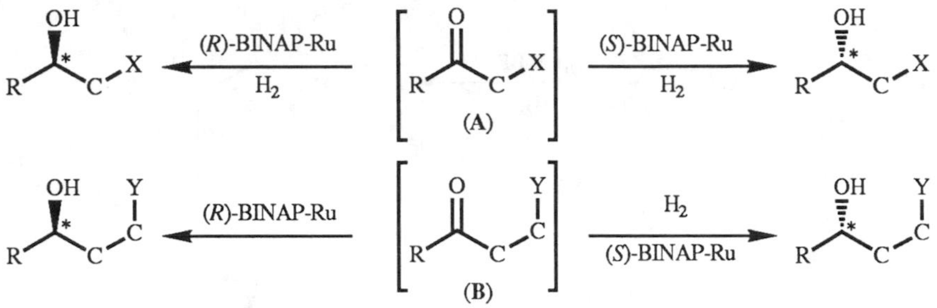

Figure 2.14. Asymmetric hydrogenation of functionalized ketones with Ru-BINAP complexes.

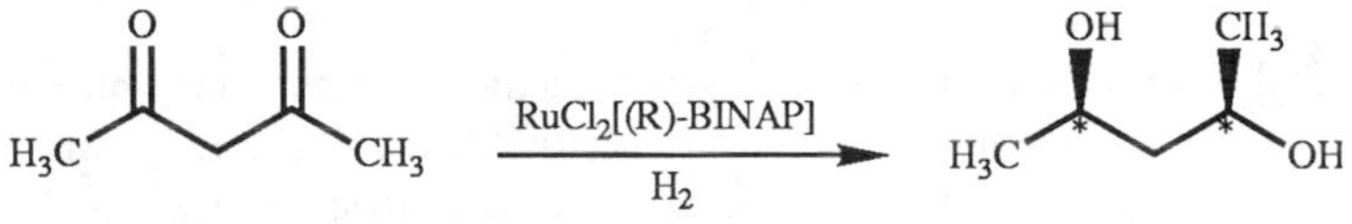

(92)   (93) 100% e.e. (R, R)

The high degree of enantioface differentiation is reflected in the asymmetric hydrogenation of pentane-2,4-dione (92) with $RuCl_2[(R)$-BINAP] as catalyst to give pentane-2,4-diol (93) in 100% e.e. Similarly hydrogenation of the unsymmetrical β-diketone (94) catalysed by $RuCl_2$ [(R)-BINAP] afforded (1S,3R)-diol (95) in 94% e.e.

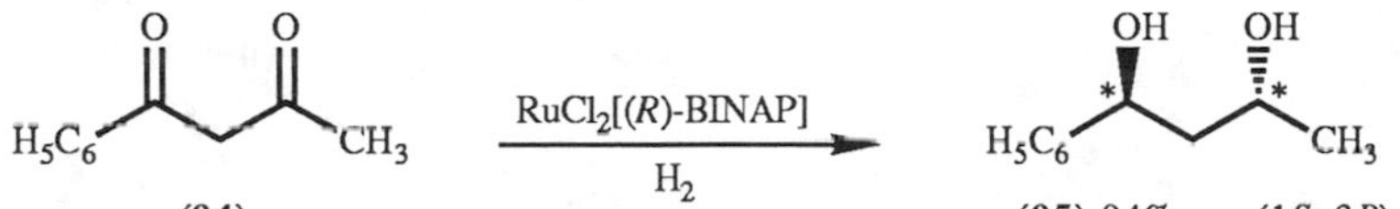

(94)   (95) 94% e.e. (1S, 3R)

The Ru-BINAP catalysed asymmetric hydrogenation of diketones described in the previous section implies that such a two-step hydrogenation is subjected to both *catalyst control* as well as *substrate control*. Both these factors, i.e. the efficacy of catalyst/carbonyl chirality transfer (catalyst control), and the structure of the stereogenic centre (substrate control) determine the overall stereochemical outcome of the reaction. Hydrogenation of acetylacetone (92) catalysed by $RuCl_2[(R)$-BINAP] produces first the (R)-hydroxyketone (96) in 98.5% e.e. at 10% conversion , as would be expected from Fig. 2.14 (B) and then results in a

99:1 mixture of (R,R)-diol (**93**) (in 100% e.e.) and the *meso*-diol (**97**). In contrast, hydrogenation of the intermediate (**96**) with the enantiomeric (S)-BINAP-based catalyst gives the isomeric diols (**93**) and (**97**) in a ratio of only 15:85. Thus the high enantiomeric purity of (**93**) obtained by the (R)-BINAP-Ru catalysed hydrogenation of (**92**) appears to be due to the result of double stereodifferentiation [84].

In the second hydrogenation step, analysis indicates that *catalyst control* (>33:1) is much more dominant over *substrate control* favouring formation of (**93**) (6:1). In the case of α-diketones such as (**98**), *substrate control* in the second hydrogenation step becomes much more important, which results in a high enantiomeric purity of the minor diol (**100**). Thus diacetyl (**98**) gave a 74:26 mixture of the *meso*-diol (**99**) and the (S,S)-diol (**100**) in 100% e.e.

The competitive ligation of the functionalities to Ru in the multifunctionalized ketones tends to decrease the enantioselectivity. The overall directivity of functional groups depends on the donicity and orientation of the nonbonding orbitals of X or Y group, bulkiness of the functional groups and kinetic properties of the resulting chelate complexes.

2.1.3.2.3   With Copper Complexes

Recently a new reagent i.e., triphenylphosphine-copper (I) hydride hexamer (**101**) has been developed which reduces a wide variety of substrates including α,β-unsaturated carbonyl compounds with high stereoselectivity and exceptional regio- and chemo-selectivity [85(a)-(d)].

The reduction of α,β-unsaturated ketones with the reagent (101) is homogeneous in the presence of excess triphenylphosphine and proceeds at room temperature under conveniently accessible hydrogen pressure.

(101)

For instance, 3,5-dimethylcyclohexanone (102) was reduced with the reagent (101) stereoselectively affording the products (103) and (104) in ratio of >100:1 in 85% yield [85(d)].

| (102) | (103) | (104) |

>100:1

The observed selectivities with the reagent (101) are superior to other methods of reduction including hydrosilylation, catalytic hydrogenation and dissolving metal reduction etc. A remarkable feature of the reagent (101) is that it does not reduce isolated double bonds, carbonyl groups etc [85(d)].

Table 2.6 shows some examples of asymmetric catalytic reduction of functionalized ketones.

2.1.3.3   Hydrogenation of Ketones via Derivatization

2.1.3.3.1   Hydrogenation of Simple Ketones via Hydrosilylation

The catalytic hydrosilylation of olefins and imines has already been dealt with in Section 2.1.2. Hydrosilylation is far by the most widely applied method in the asymmetric homogeneous catalytic hydrogenation of prochiral ketones. Fig.2.15 shows the generalized scheme for the hydrogenation of ketones via

Table 2.6. Asymmetric Catalytic Reduction of Functionalized Ketones

Substrate	Catalyst Ligand	Product/ Optical yield (% e.e.)	Reference
Ph–C(=O)–C(=O)–N(H)–(S)–CH(COOCH₃)CH₂Ph	a	Ph–CH(OH)–C(=O)–N(H)–(S)–CH(COOCH₃)CH₂Ph R,S:S,S = 14:86	[83]
H₃C–C(=O)–C(=O)–OCH₂CH₂CH₃	b	H₃C–CH(OH)–OCH₂CH₂CH₃ 76 (R)	[76a]
Ph–C(=O)–C(=O)–N(H)–*C(CH₃)(H)–CO₂CH₃	a	Ph–CH(OH)–C(=O)–N(H)–CH(CH₃)–COOCH₃ 66 (S)	[83]
H₃C–C(=O)–CH₂–N(CH₃)CH₃	c	H₃C–CH(OH)–CH₂–N(CH₃)CH₃ 96 (S)	[45]
H₃C–CH(CH₃)–C(=O)–CH₂–N(CH₃)CH₃	c	H₃C–CH(CH₃)–CH(OH)–CH₂–N(CH₃)CH₃ 95 (S)	[45]
Ph–C(=O)–CH₂–N(CH₃)CH₃	d	Ph–CH(OH)–CH₂–N(CH₃)CH₃ 95 (S)	[45]
H₃C–C(=O)–CH₂–OH	e	H₃C–CH(OH)–CH₂–OH 92 (R)	[45]
H₃C–C(=O)–CH₂–C(=O)–OEt	d	H₃C–CH(OH)–CH₂–C(=O)–OEt >99 (R)	[45]

(Table 2.6. contd.)

Substrate	Catalyst Ligand	Product/ Optical yield (% e.e.)	Reference
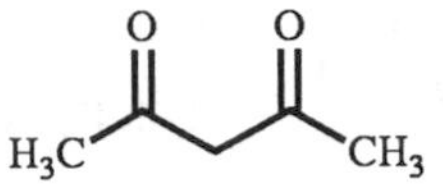 H₃C—CO—CH₂—CO—CH₃	e	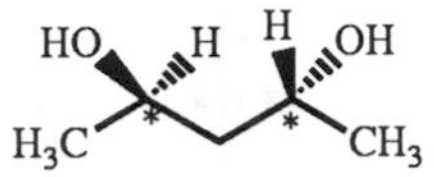 HO—H / H—OH, H₃C—*—*—CH₃ 100 (R,R)	[45]
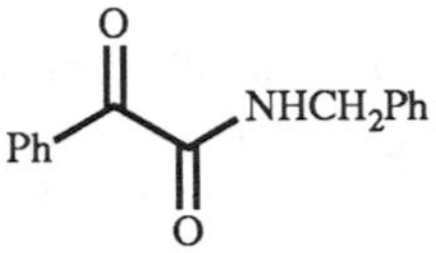 Ph—CO—CO(=O)—NHCH₂Ph	f	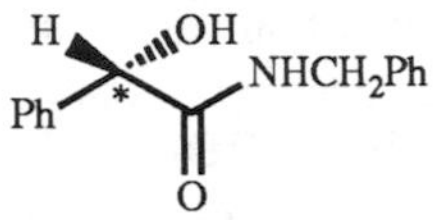 H—OH, Ph—*—CO—NHCH₂Ph 71 (S)	[84b]

a = (-)cycDIOP-RH^N,   b = BPPM- Rh,   c = [Ru(S)-BINAP],   d = RuBr₂[(S)-BINAP],
e = RuCl₂[(S)-BINAP],   f = Rh-cycDIOP

hydrosilylation. Earlier studies in hydrosilylation with simple ketones employing [(+)BMPP-PtCl₂]₂ and other chiral phosphine -Pt (II) complexes gave low optical yields of the corresponding secondary alcohols, but it was interestingly found that the mechanism of stereoselective addition of hydrosilanes to the enantiotopic faces of the ketones is different from that of similar additions of hydrosilanes to olefins [85(e,f)]. For example, acetophenone (**80**) upon hydrosilylation with HSiMeCl₂ using [(+)BMPP)PtCl₂]₂ as catalyst yielded 1-phenylethanol in 7.6% e.e. with (*S*) configuration. The same reaction, however, with α-methylstyrene (**56**) yielded the (*R*) product (**57**) in 5.2% e.e. It was later found that all the three components (ignoring the participation of solvent), i.e. the substrate, the silane and the catalyst take part in the reaction and their structures greatly influence the course of the reaction [61].

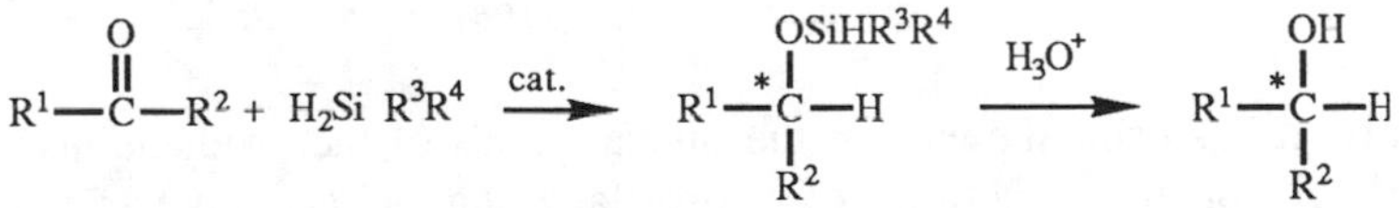

$$R^1-\overset{O}{\overset{\|}{C}}-R^2 + H_2Si\ R^3R^4 \xrightarrow{cat.} R^1-\overset{OSiHR^3R^4}{\overset{*|}{C}}-H \xrightarrow{H_3O^+} R^1-\overset{OH}{\overset{*|}{C}}-H$$

Figure 2.15. Generalized scheme for the hydrogenation of ketones *via* hydrosilylation.

(80)   +   HSiCH$_3$Cl$_2$ i) [(+)-BPPM)PtCl$_2$]$_2$ → (81) 7.6% e.e. (S)
                              ii) H$^+$

(56)   +   HSiCH$_3$Cl$_2$ i) [(+)-BPPM)PtCl$_2$]$_2$ → (57) 5.2% e.e. (R)
                              ii) H$^+$

The influence of substrate structure on the optical yield of the product in hydrosilylation is best represented by the reaction of acetophenone (80) with Ph$_2$SiH$_2$ using [(-)- DIOP)RhCl] as catalyst giving (81) in 30% e.e. On changing the substrate structure from (80) to (105) by replacing the methyl group of acetophenone with *tert.*-butyl group, the same reaction afforded the secondary alcohol (106) in 41% e.e. By keeping the structures of the substrate and the hydrosilane constant and changing the nature of ligands, variations in opticals yields and configuration have been observed. Thus acetophenone (80) yielded the (R) sec. alcohol (81) in 49% e.e. with [MPFA-RhCl] as catalyst, whereas employing ligand (107) with RhCl as catalyst, an optical yield of 79% e.e. with (S) configuration was obtained.

(80)   +   Ph$_2$SiH$_2$ i) [(-)-DIOP]RhCl → (81) 30% e.e. (R)
                              ii) H$^+$

(105)   +   Ph$_2$SiH$_2$ i) [(-)-DIOP]RhCl → (106) 41% e.e. (R)
                              ii) H$^+$

Similarly, it has been shown that the optical yields of the products also depend markedly on the structure of hydrosilanes employed and various combinations of hydrosilane-catalyst have been used. Thus with HSiEt$_3$, acetophenone (80) afforded (81) in only 3.8% e.e., whereas by changing the hydrosilane to H$_2$SiPh1-Naph, the same reaction gave the product in 58% e.e. [86-88].

(107)

(MPFA)

$$
\text{(80)} \quad Ph\text{-CO-}CH_3 + Ph_2SiH_2 \xrightarrow[\text{ii) H+}]{\text{i) (107)}} \quad \text{(81) 79\% e.e. } (S)
$$

$$
\text{(80)} \quad Ph\text{-CO-}CH_3 + HSiEt_3 \xrightarrow[\text{ii) H}^+]{\text{i) [(+)-DIOP]RhCl}} \quad \text{(81) 3.8\% e.e. } (S)
$$

$$
\text{(80)} \quad Ph\text{-CO-}CH_3 + HSiPh1\text{-Naph} \xrightarrow[\text{ii) II+}]{\text{i) [(+)-DIOP]RhCl}} \quad \text{(81) 58\% e.e. } (S)
$$

Optically active nitrogen ligands have also been used with rhodium and platinum complexes in the hydrosilylation of simple prochiral ketones [89,90]. Some of these ligands and their complexes are shown in Fig. 2.16. It has been shown that the stoichiometric complexes (111-116), though catalytically active, give only slight optical induction of between 0-2% e.e. [89,90]. If the hydrosilylation catalysts are synthesized *in situ* from [(COD)RhCl]$_2$ or K[PtCl$_3$(C$_2$H$_4$)] H$_2$O with a large excess of the ligand, the optical induction is increased manyfold. In the hydrosilylation of acetophenone using diphenylsilane with the complex [(COD)RhCl]$_2$ (108) having a 1:13 ratio of Rh/ligands, 95% hydrosilylation was achieved with an optical induction of 57% (R). In the system K[PtCl$_3$(C$_2$H$_4$)] H$_2$O (110), with Pt/ligand ratio of 1:36, hydrosilylation occurs in 74% yield with 5% e.e. of the compound with (S) configuration. With the corresponding complex (115) with a Pt/ligand ratio of 1:1, hydrosilylation occurs in 10% yield without any enantioselectivity.

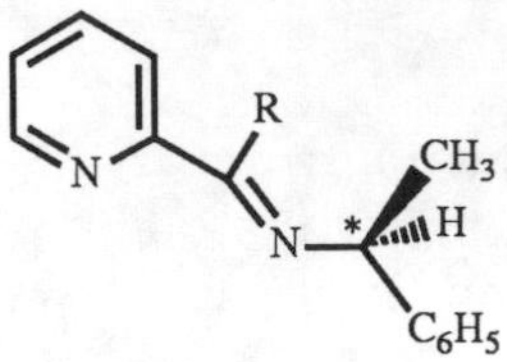

(108) R = H

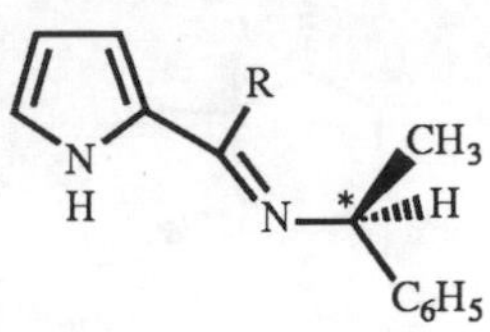

(109) R = CH$_3$

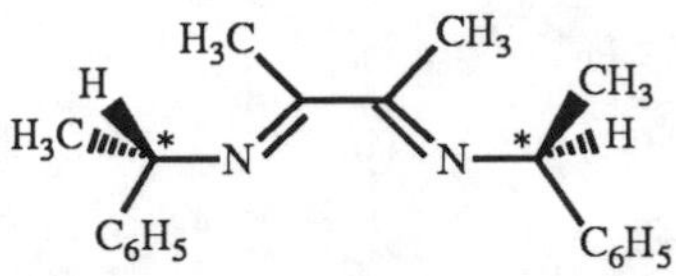

(110)

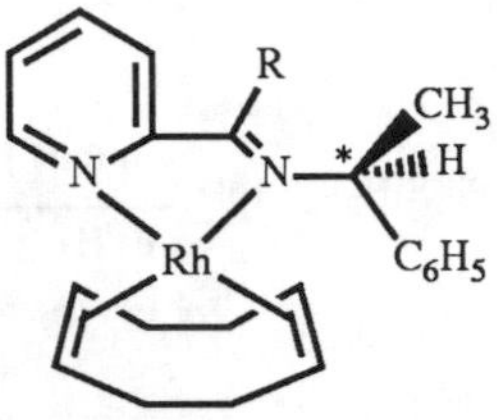

(111) R = CH$_3$
(112) R = C$_6$H$_5$

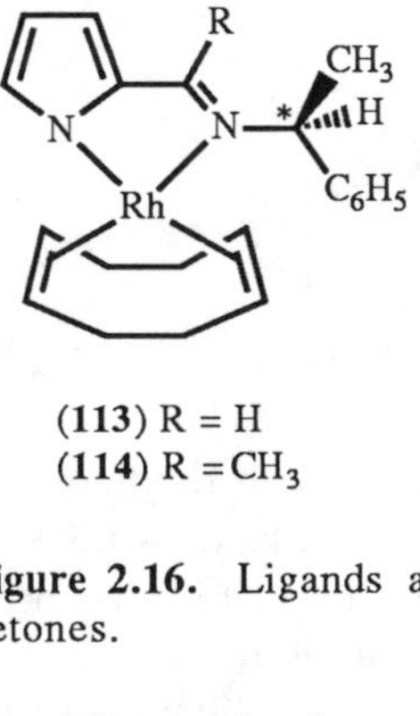

(113) R = H
(114) R = CH$_3$

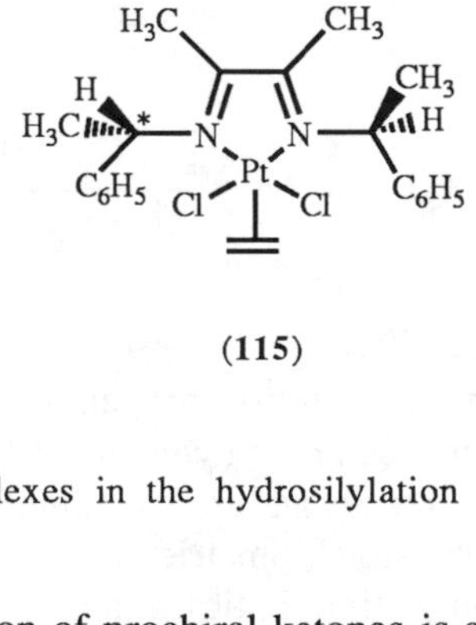

(115)

Figure 2.16. Ligands and their complexes in the hydrosilylation of simple ketones.

The mechanism of hydrosilylation of prochiral ketones is shown in Fig. 2.17 [86-88]. The various steps involved are: (a) oxidative addition of the hydrosilane to the rhodium (I) complex (**116**) to give (**117**), (b) insertion of the carbonyl of the substrate into the Si-Rh bond of (**117**) to form the diastereomeric α-siloxylalkylrhodium hydride intermediate (**118**) (this step determines the predominant configuration and the extent of the enantiomeric excess of the product) and (c) reductive elimination to afford the optically active silyl ether of the secondary alcohol. Table 2.7 shows some representative examples of asymmetric hydrogenation of simple ketones *via* hydrosilylation.

Table 2.7.   Asymmetric Hydrogenation of Simple Ketones via Hydrosilylation

Substrate	Hydro-silane	Cata-lyst	Product	Optical yield (% e.e.)	Reference
Ph—C(O)—CH₃	a	A	HO, H / Ph—*—CH₃	13 (S)	[91]
Ph—C(O)—C₂H₅	b	A	HO, H / Ph—*—C₂H₅	56 (S)	[92]
Ph—C(O)—ⁱPr	c	B	HO, H / Ph—*—ⁱPr	56.3 (S)	[93]
ᵗBu—C(O)—CH₃	d	C	HO, H / H₃C—*—ᵗBu	41 (R)	[94]
ⁿBu—C(O)—CH₃	e	D	HO, H / ⁿBu—*—CH₃	30 (S)	[86-88]
H₅C₂—C(O)—CH₃	b	B	HO, H / H₃C—*—C₂H₅	42 (R)	[92]
PhCH₂—C(O)—CH₃	f	A	HO, H / PhCH₂—*—CH₃	5.3 (S)	[91]
Ph—C(O)—ᵗBu	e	E	HO, H / PhCH₂—*—CH₃	25 (S)	[93]

a = PhCH₃SiH₂, b = H₂SiPh 1-Naph, c = Ph(CH₃)₂SiH, d = Ph₂SiH₂ e = (C₂H₅)₂SiH₂,
f = (C₂H₅O)₃SiH.
A = (+)-DIOP-RhᴺN, B = [(+)-BMPP]₂ - Rhᴺ, C = (MPFA)₂ - Rhᴺ, D = [(-)-BMPP]₂ - Rhᴺ,
E = [(-)-DIOP]-Rhᴺ

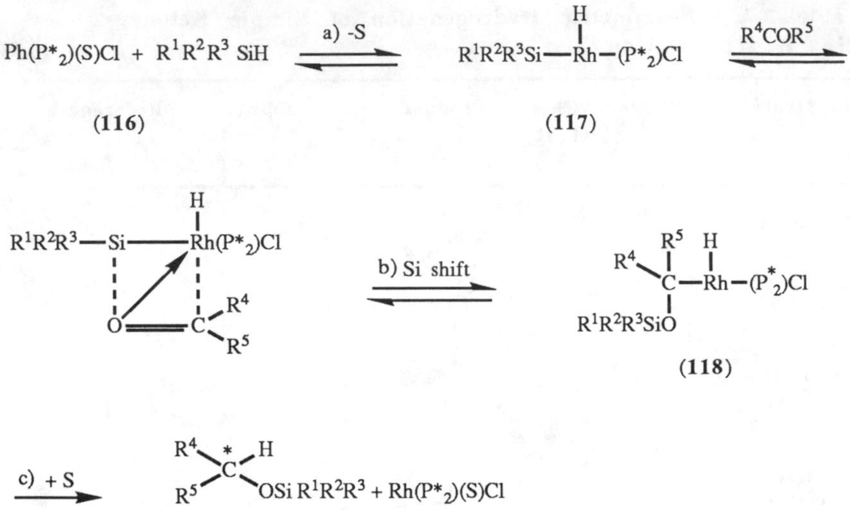

$$\text{Ph(P*}_2\text{)(S)Cl} + \text{R}^1\text{R}^2\text{R}^3 \text{ SiH} \xrightleftharpoons{\text{a) -S}} \text{R}^1\text{R}^2\text{R}^3\text{Si}-\overset{\overset{\text{H}}{|}}{\underset{|}{\text{Rh}}}-\text{(P*}_2\text{)Cl} \xrightleftharpoons{\text{R}^4\text{COR}^5}$$

(116)   **(117)**

$$\text{R}^1\text{R}^2\text{R}^3-\text{Si} \cdots \overset{\overset{\text{H}}{|}}{\text{Rh(P*}_2\text{)Cl}} \xrightleftharpoons{\text{b) Si shift}} \overset{\text{R}^5 \ \ \text{H}}{\underset{\text{R}^1\text{R}^2\text{R}^3\text{SiO}}{\text{R}^4 \diagdown \underset{|}{\overset{|}{\text{C}}}-\text{Rh}-\text{(P}^*_2\text{)Cl}}}$$

(118)

$$\xrightarrow{\text{c) +S}} \ \ \overset{\text{R}^4 \diagdown \ \ * \diagup \text{H}}{\underset{\text{R}^5 \diagup \overset{|}{\text{C}} \diagdown \text{OSi R}^1\text{R}^2\text{R}^3}{}} + \text{Rh(P*}_2\text{)(S)Cl}$$

Figure 2.17. Mechanism of hydrosilylation of ketones with Rh $(P_2)(S)Cl$, $(S = \text{solvent}, P^* = \text{chiral phosphine})$.

2.1.3.3.2   *Hydrogenation of Functionalized Ketones via Hydrosilylation*

The asymmetric reduction of α-, β- and γ-ketoesters, typically pyruvates, levulinates and phenylglyoxylates as well as ketoamides has been achieved by chiral rhodium complex-catalysed hydrosilylation. Generally the optical yields of functionalized ketones are much higher than those achieved with simple prochiral ketones. This is due to the presence of the extra functionality e.g. the ester group in ketoesters as well as the type of hydrosilane used which largely determine the extent of asymmetric induction. Rhodium (I) complexes with chiral phosphines such as (+)-BMPP and (+)- or (-)-DIOP have been used in the hydrosilylation of ketoesters. The hydrosilylation of alkylpyruvates and phenylglyoxylates with 1-naphthylphenylsilane gave the highest optical yields employing (+)- or (-)-DIOP-Rh(I) as catalyst.   Thus employing 1- naphthylphenylsilane, n-propylpyruvate (**119**) afforded n-propyllactate (**120**) in 85.4% e.e. [95].

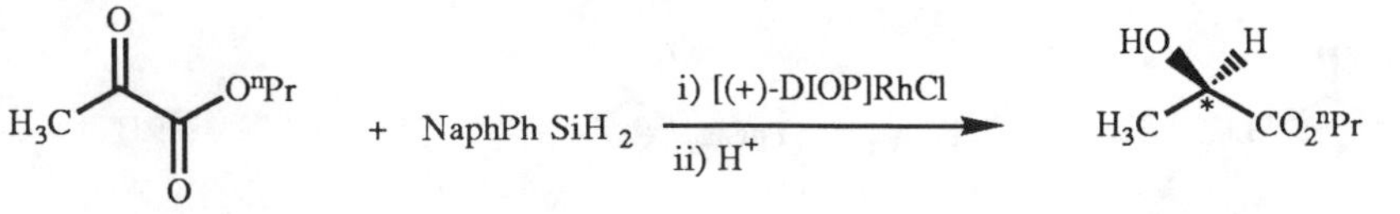

$$\underset{\text{(119)}}{\overset{\text{O}}{\underset{\text{H}_3\text{C}}{\overset{||}{\diagup}}} \overset{\text{O}}{\underset{||}{\text{C}}} \text{O}^n\text{Pr}} \ \ + \ \ \text{NaphPh SiH}_2 \ \xrightarrow[\text{ii) H}^+]{\text{i) [(+)-DIOP]RhCl}} \ \underset{\text{(120) 85.4\% e.e. }(R)}{\overset{\text{HO}\diagdown \ \diagup\text{H}}{\underset{\text{H}_3\text{C} \diagup \overset{*}{\text{C}} \diagdown \text{CO}_2{}^n\text{Pr}}{}}}$$

"Double asymmetric reduction" i.e. reduction of a chiral substrate with a chiral catalytic complex has been studied with (-)-menthyl esters of phenylglyoxylic acid and pyruvic acid using DIOP-rhodium complexes. Thus (-)-menthyl phenylglyoxylate (**121**) afforded (-)-menthyl-(S)-(+)-mandelate (**122**) in 77% e.e., using 1-napthPhSiH$_2$ as the hydrosilylating agent and (+)-DIOP- Rh(I) as catalyst. The optical yields obtainable with double asymmetric reduction were much higher than those using either achiral substrates or achiral catalysts. Thus the same reaction catalysed by $(PPh_3)_3RhCl$ resulted in the formation of (S)-mandelate in only 17% e.e. Similarly, double asymmetric reduction of (-)-menthyl pyruvate (123) afforded (124) in 85.6% e.e., using Rh-(+)-DIOP as catalyst. The high optical yields in double asymmetric reduction reflect the synergetic effect of both (+)-DIOP and (-) menthyl group, both of which favour the product with (S) configuration.

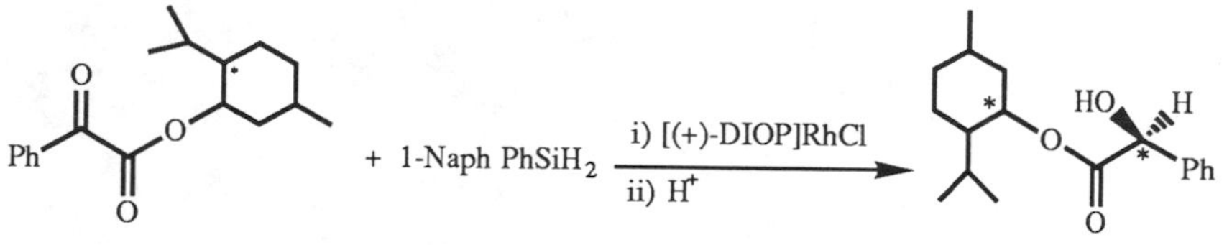

(121)   (122) 77% e.e. (S)

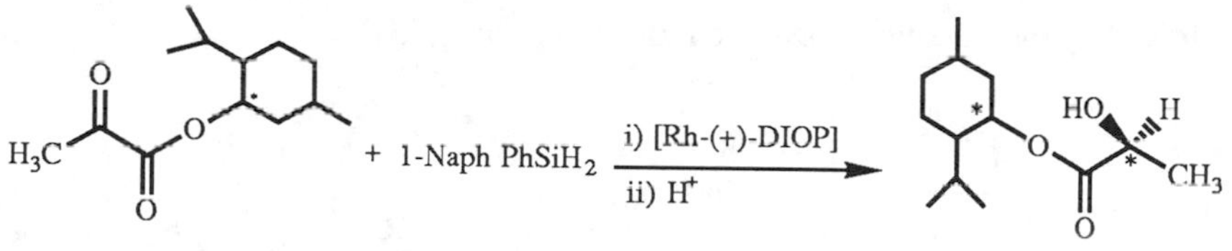

(123)   (124) 85.6% e.e. (S)

The asymmetric reduction of methylacetoacetate (125) via hydrosilylation afforded (126) in 23.5% e.e. On the other hand, asymmetric reduction via hydrosilylation of levulinates such as (127) afforded 4-methyl-γ-butyrolactone (129) through the silyl ether of 4-hydroxypentanoate (128) in 76.2% e.e.

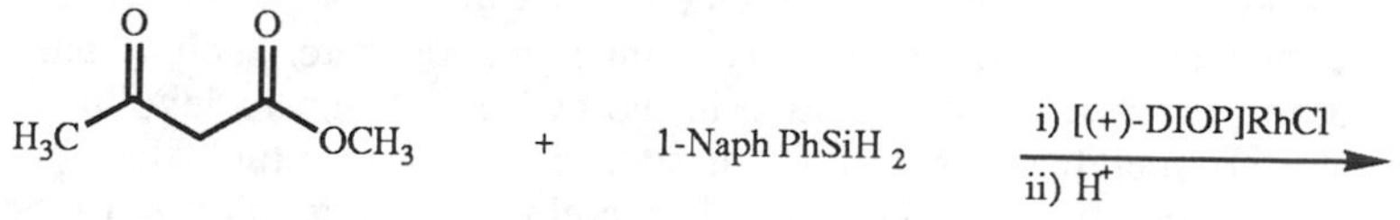

(125)

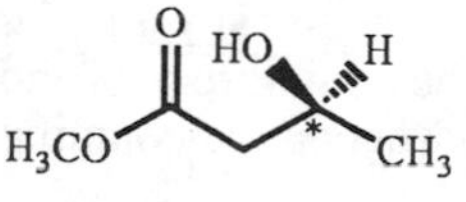

(126) 23.5% e.e. (S)

H$_3$CC(CH$_2$)$_2$COCH$_3$ + 1-Naph PhSiH$_2$ $\xrightarrow{\text{[(+)-DIOP]RhCl}}$ H$_3$CCH(CH$_2$)$_2$COCH$_3$

OSiHPh 1-Naph

(127)   (128)

(129) 76.2% e.e. (S)

The asymmetric reduction *via* hydrosilylation of α-ketoacylamino esters catalysed by chiral phosphine-Rh complexes has been achieved in high optical yields. Thus the reaction of (130) with H$_2$SiPh1-Naph using [(+)-DIOP]-Rh(I) followed by methanolysis gave the reduced product (131) in 82% e.e.

(130)   (131) 82% e.e. (S,S)

The mechanism of hydrosilylation of functionalized ketones is similar to the one proposed in Figure 2.17 (Section 2.1.3.3.1) for the hydrosilylation of simple ketones [86-88,95]. However, the marked increase in optical yields in the reduction of α-ketoesters *via* hydrosilylation may arise due to a ligand effect, i.e. an attractive interaction of the ester moiety in the transition state. Such a ligand effect has been proposed to be operative in the hydrosilylation of levulinates which form a six-membered ring chelate such as (132) in which the ester oxygen forms a strong dative bond with Rh, thereby removing chlorine to the outer sphere as a counter anion in the transition state. However such a ligand effect has not been observed in acetoacetates which may form a five-membered ring chelate such as (133) in the transition state.This may be due to the fact that the silyl ether could arise not only by hydrosilylation, but also by *in situ* transfer hydrogenation of the silyl enol ether (which would be a mixture of *E* and *Z* isomers) formed in the first step of the reaction, by dehydrogenative coupling of the enol with the hydrosilane, as shown in Fig. 2.18.

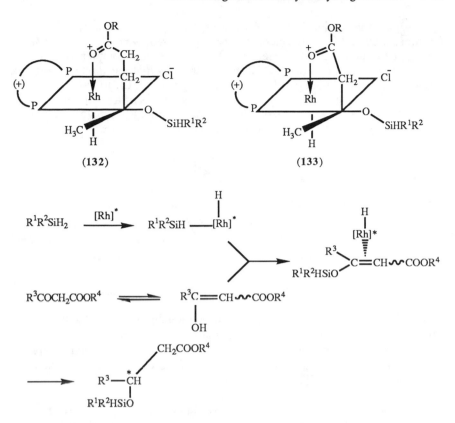

Figure 2.18. Asymmetric hydrogenation of unsymmetric ketones *via* hydrosilylation.

In the hydrosilylation of pyruvates, the preferred conformation shown for the α-silyloxyalkyl rhodium intermediate (**134**) results from the attractive interaction between the alkoxy oxygen and the silyl group. The hydride shift in this conformation should afford the (*R*)-lactate and this has indeed been found to be the case.

An alternative mechanism for the asymmetric induction in asymmetric hydrogenation and hydrosilylation has been proposed, which is based on space filling CPK-type molecular models of (+)- DIOP-rhodium complex system [96]. According to this mechanism, a diastereomeric silylhydrido-rhodium complex having octahedral (or trigonal bipyramidal) structure (135) is assumed as an intermediate, which distinguishes the enantiotopic faces of a prochiral ketone in terms of *steric approach control*. Predictions regarding the major enantiomer produced can be made by this mechanism.

(135)

However the mechanisms of hydrosilylation and hydrogenation are quite different from each other when carbonyl compounds are used as substrates, the α-silyloxyalkyl-rhodium complex (134) being formed due to *product development control* [97], so that the mechanism involving (134) seems to be more plausible.

Some representative examples of asymmetric reduction of functionalized ketones are presented in Table 2.8.

Table 2.8. Asymmetric Reduction of Functionalized Ketones via Hydrosilylation

Substrate	Hydro-silane	Cata-lyst	Product /Optical yield (% e.e.)	Reference
	a	A	76.5 (S)	[95]
	b	B	72.1 (S)	[95]

(Table 2.8. contd.)

Substrate	Hydro-silane	Cata-lyst	Product /Optical yield (% e.e.)	Reference
	b	A	47.2 (S)	[95]
	b	A	75.1 (S)	[95]
	b	B	72 (R, S)	[98]
	b	A	82 (S, S)	[98]
	b	A	77 (S)	[95]
	b	A	83.5 (S)	[95]

a = Ph$_2$SiH$_2$ b = H$_2$SiPh 1-Naph   A = (+)-DIOP-Rh^N,   B = (-)-DIOP-Rh^N

2.1.3.3.3   Hydrogenation of Ketones via Enol Phosphinates

Another indirect route to the asymmetric reduction of prochiral carbonyl compounds is the asymmetric hydrogenation of enol phopshinates prepared regio- and stereo-selectively from unsymmetric ketones [77].

As shown in Fig. 2.19 the unsymmetric ketone (136) is converted into the enol phosphinate (137) by phosphorylation of the lithium enolate generated under *kinetically* controlled conditions (LDA in THF at - 78°C).

The phosphinate (137) is hydrogenated in the presence of a chiral phosphine-Rh complex as catalyst to give the optically active secondary alkylphosphinate (138). Conversion of (138) into the corresponding alcohol (139) is effected by treatment with excess of methyl lithium in ether followed by hydrolysis.

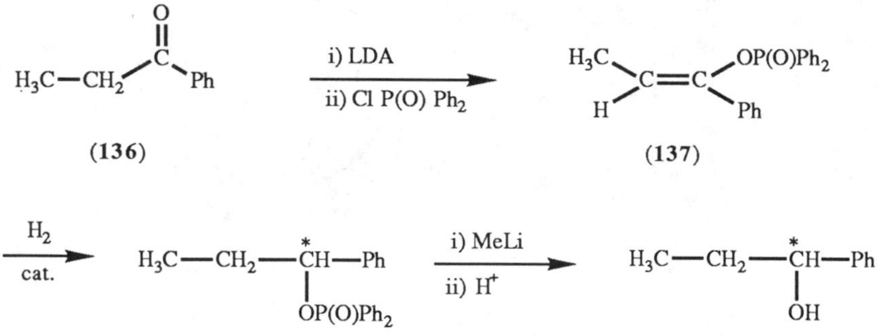

Figure 2.19. Asymmetric hydrogenation of unsymmetric ketones *via* enol phosphinates.

Table 2.9 includes some examples of asymmetric hydrogenation of prochiral ketones *via* enol phosphinates.

2.2   Heterogeneous Catalytic Hydrogenations

While the soluble catalytic reacting system used in homogeneous catalytic hydrogenations (Section 2.1.) has its merits in that it is often easier to vary the steric and electronic environment of the catalytically active site to optimize chemical and optical yields in an asymmetric synthesis, one encounters a problem of practical significance. The separation of the catalyst from the products as well as the continuous removal of the products formed requires special treatment and the catalyst is often destroyed in the process.

Table 2.9. Asymmetric Hydrogenation of Enol Diphenylphosphinates Catalyzed by Chiral Phosphine-Rhodium Complexes[†]

Substrate	Ligand	Product	Optical yield (% e.e.)
CH_2 $Ph\text{—}OP(O)Ph_2$	(R,S)-BPPFOH	$H_3C\text{—}^*C\text{—}H$ $Ph\text{—}OH$	78 (R)
CH_2 $H_3C\text{—}Ph\text{—}OP(O)Ph_2$ (H_3C)	(R,S)-BPPFOH	$H_3C\text{—}^*C\text{—}H$ $(H_3C)_2HC\text{—}OH$	60 (R)
CH_2 $H_3C\text{—}CH\text{—}OP(O)Ph_2$ (H_3C)	(-)-DIOP	$H_3C\text{—}^*C\text{—}H$ $HO\text{—}CH(CH_3)_2$	55 (S)
CH_2 $n\text{-}H_{13}C_6\text{—}OP(O)Ph_2$	(R,S)-BPPFOH	$H_3C\text{—}^*C\text{—}H$ $n\text{-}H_{13}C_6\text{—}OH$	48 (R)
CH_2 $H_3C\text{—}C\text{—}OP(O)Ph_2$ (H_3C, CH_3)	(R,S)-BPPFOH	$H_3C\text{—}^*C\text{—}H$ $HO\text{—}C(CH_3)_3$	39 (S)
CH_3 $Ph\text{—}OP(O)Ph_2$	(R,S)-BPPFOH	$H_5C_2\text{—}^*C\text{—}H$ $HO\text{—}Ph$	49 (R)
CH_2 $Ph\text{—}OP(O)Ph_2$	(-)-DIOP	$H_3C\text{—}^*C\text{—}H$ $HO\text{—}Ph$	80 (S)
CH_2 $Ph\text{—}OP(S)Ph_2$	(R,S)-BPPFOH	$H_3C\text{—}^*C\text{—}H$ $HO\text{—}Ph$	24 (S)

[†] From ref. [78]

Many attempts have been made to couple the activity and selectivity of homogeneous catalysts with the ease of recovery of the heterogeneous phase catalysts. One way to solve this problem is the recent development of water-soluble chiral catalysts [100-103]. Such catalysts have been employed in water,

water-alcohol mixtures or water-organic solvent (two-phase) systems in the hydrogenation of dehydroamino acids and itaconic acid yielding hydrogenated products in 10-35% e.e., though in some cases enantioselectivities as high as 72-88% e.e. have been achieved [102].

Another method for the facile separation of a chiral catalyst after asymmetric hydrogenation is to fix the catalyst on a solid support in such a way that it retains the advantages served in solution. Thus homogeneous catalysts have been attached to a variety of solid supports including cross-linked polymers. In this way, the catalyst is expected to acquire the property of insolubility and may retain the same reactivity in solution.

2.2.1   *Enantioselective Heterogeneous Catalytic Hydrogenations*

Enantioselective heterogeneous catalytic hydrogenations have been mostly devoted to the hydrogenation of β-dicarbonyl compounds [104-108]. Raney nickel modified with (-)-(2R,3R)-tartaric acid has been generally used in such hydrogenation reactions, while in other cases a variety of bimetallic systems have been employed [108].

The enantioselectivity of heterogeneous catalysts, such as tartaric acid - modified Raney nickel, in asymmetric hydrogenations has been shown to be greatly enhanced by the addition of various salts such as NaBr or additives such as pivalic acid [$(CH_3)_3CCOOH$]. Thus the addition of 8% NaBr to the Raney nickel catalyst modified with tartaric acid in the asymmetric heterogeneous catalytic hydrogenation of 4-hydroxy-2-butanone (**140**) enhanced the optical yield of the product (R)-1,3-butanediol (**141**), giving enantioselectivity of 70.2% e.e. [109]. Without the addition of NaBr, the optical yield dropped to 27.3% e.e. Other metals, such as Co, Cu and Pd employed as catalysts in the above reaction gave poor optical yields (< 2 % e.e.), whereas the presence of Al in the Raney nickel catalyst (leached at low temperature) resulted in a significant decrease in the optical yield (see also Section 2.2.1.1.).

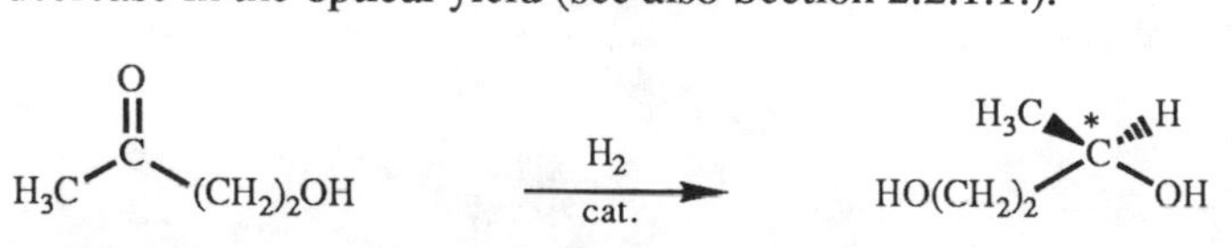

$$H_3C-\overset{\overset{O}{\|}}{C}-(CH_2)_2OH \xrightarrow[\text{cat.}]{H_2} HO(CH_2)_2-\overset{*}{C}\overset{H_3C\quad H}{\diagdown}OH$$

(**140**)   (**141**) 70% e.e. (R)

Asymmetric hydrogenation of methyl 3-oxoalkanoates $CH_3(CH_2)_n COCH_2 COOCH_3$, (where n = 0,6,8,10,12) has been carried out with (R,R)-tartaric acid/NaBr-modified Raney nickel catalytic system to afford (R)-3 hydroxyalkanoates in an average optical yield of 85% e.e. [110]. For example, methyl-3-oxohexadecanoate (**142**) yielded the corresponding hydrogenated product (**143**) in 87% e.e.

H_3C(CH_2)_{12}—C(=O)—CH_2—C(=O)—OCH_3   →[H_2 / cat.]   H_3CO_2CCH_2—C*(H)(OH)—(CH_2)_{12}CH_3

(142)   (143) 87% e.e. (R)

Addition of pivalic acid to a modified Raney nickel catalyst in the hydrogenation of simple ketones such as 2-hexanone (144) increased the optical yield giving (145) in 66% e.e. Without pivalic acid, the optical was 28% e.e [105a].

H_3C—C(=O)—(CH_2)_3CH_3   →[H_2 / cat./pivalic acid]   H_3C—C*(H)(OH)—(CH_2)_3CH_3

(144)   (145)

Recent studies on the enantioselective heterogeneous catalytic hydrogenation of prochiral olefinic substrates gave less promising results [111]. For example (E)-α -phenylcinnamic acid (146) upon hydrogenation over tartaric acid-NaBr-Raney nickel as catalyst gave the hydrogenated product (147) in only 17.03 % e.e. On the other hand, its (Z)-enantiomer gave a very poor optical yield i.e. about 0.47% e.e.

Ph(H)C=C(Ph)(COONa)   →[H_2 / cat.]   Ph—C*(H)(COOH)—CH_2Ph

(146)   (147) 17.03% e.e. (R)

The optical yields of the reactions discussed above are considered to be determined largely by two major factors: (a) the stereodifferentiating ability of the modifying reagent, and (b) ratio of enantiodifferentiating area to non-differentianting area on the catalyst surface. The stereodifferentiating ability of the modifying reagent depends on the intermolecular interaction between the modifying reagent and the substrate adsorbed on the catalyst. For instance (R,R)-tartaric acid (148) has been found to be coordinated on the nickel surface through its carboxylic groups while its hydroxyl groups are essentially free for interaction with the substrate [105b, 112]. The ratio between the enantiodifferentiating area to the non-enantiodifferentiating area on the surface of the catalyst can be altered by effective blocking of the non-enantiodifferentiating area on the surface of the catalyst which leads to an increase in the optical yield. Sodium bromide is considered to be adsorbed on the non-stereodifferentiating site thereby blocking its activity. Aluminium on the other hand, which is present in the Raney Ni prepared at low temperature, exists in a polymeric form of aluminium hydroxide [113]. The hydroxyl group of aluminium derivatives competes with the hydroxyl groups of tartaric acid in the formation of hydrogen bonding with the substrate thereby leading to the racemic product. For this reason, the use of Raney nickel leached at a low temperature results in a significant decrease in the optical yields. The interplay of these two factors as mentioned determines the extent of optical yield in the hydrogenation of ketones

discussed earlier. This is illustrated by the hydrogenation of methyl acetoacetate
(**149**) with (*R*,*R*)-tartaric acid-NaBr- modified Raney nickel as catalyst to give
methyl 3-hydroxybutyrate (**150**) in 86% e.e.

(148)

Some representative examples of enantioselective heterogeneous catalytic
hydrogenations are presented in Table 2.10.

**Table 2.10. Enantioselective Heterogeneous Catalytic Hydro-
genations over Raney Ni Modified with (2*R*, 3*R*) Tartaric Acid and
NaBr**

Substrate	Product / Optical yield (% e.e.)	Reference
	2.0 (*S*)	[105b]
	86 (*R*)	[105b]
	74 (*R*)	[107a]
	68 (*R*)	[105b]

(Table 2.10. contd.)

Substrate	Product / Optical yield (% e.e.)	Reference
H$_3$C(CH$_2$)$_8$—C(=O)—CH$_2$—COOCH$_3$	H$_3$C(H$_2$C)$_8$ C* H ; CH$_3$O C—CH$_2$ OH ; $\overset{\parallel}{O}$ 86.7 (R)	[110]
H$_3$C(CH$_2$)$_6$—C(=O)—CH$_2$—COOCH$_3$	H$_3$C(H$_2$C)$_6$ C* H ; CH$_3$O C—CH$_2$ OH ; $\overset{\parallel}{O}$ 83.3 (R)	[110]
H$_3$C—CH$_2$—C(=O)—CH$_2$—SO$_2$CH$_3$	H C* CH$_2$CH$_3$; HO CH$_2$—SO$_2$CH$_3$; 68 (R)	[106a]
H$_3$C—C(=O)—CH$_2$—CH$_2$OH	H C* CH$_3$; HO CH$_2$—CH$_2$OH ; 69 (R)	[105c]
Ph,H C=C Ph,COOH	Ph C* H ; PhH$_2$C COOH ; 17.03 (R)	[111]

2.2.2   Diastereoselective Heterogeneous Catalytic Hydrogenations

2.2.2.1   Asymmetric Hydrogenation of Carbon-Carbon Double Bonds

Olefinic substrates having diastereotopic faces have been hydrogenated stereoselectively over classical metal catalysts involving *syn* addition of H$_2$ preferentially to the less hindered face of the substrate. Various chiral auxiliaries are used to induce asymmetry, which are either removed and recycled after the reaction or integrated in the products. These will be discussed subsequently.

2.2.2.1.1   Hydrogenation of N-Acyl-α, β-Dehydroamino Acids

Heterogeneous catalytic asymmetric hydrogenation of N-acyl-α, β- dehydroamino acids has been carried out over various metal catalysts in which optically active ligands are coordinated to metals. In such reactions chiral ligands effective in homogeneous catalytic hydrogenation have been employed after attachment to various polymeric matrices. Thus the activity and selectivity of soluble catalysts can be combined with the ease of recovery of heterogeneous phase catalysts. Some of the commonly used polymer supported chiral ligands are shown in Table 2.11.

Table 2.11. Polymer Supported Chiral Ligands

Ligand	Structure	Reference
(151)		[116]
(152)		[91]
(153)		[116-118]
(154)		[114]
(155)		[114]

(Table 2.11. contd.)

Ligand	Structure	Reference
(156)		[91,115]
(157)		[116-118]
(158)		[116-118]
(159)		[119]
(160)		[116]

Earlier results with heterogeneous catalytic systems in the hydrogenation of dehydro amino acids were disappointing, though for other substrates better results were achieved in some cases [116 and references therein]. However structural modification of the chiral ligands resulted in a dramatic increase in enantioselectivity. For instance α-acetamidocinnamic acid (161) was hydrogenated in the presence of Rh(I) catalyst having (160) as ligand attached to a cross-linked polymer to afford (162) in 86% e.e [116]. Interestingly, employing ligand (156) with Rh(I), no hydrogenation product was isolated [91].

Catalytic asymmetric hydrogenation of (163) with [Rh(I)-(160)] as catalyst afforded (164) in 52-60% e.e., in contrast to the higher optical yield (73%) obtained with the soluble Rh(I) DIOP catalyst. When the catalyst [Rh(I)-(160)] was recycled in an atmosphere of nitrogen, an optical yield of 52.5% was obtained for the same reaction. Further recycling the catalyst twice afforded optical yields of 44.4% and 34.6% respectively, the efficiency of the catalyst decreasing with each additional cycle. The ligand (157) has also been used in the preparation of Rh(I) bound polymer catalysts such as (165) and (166) to achieve asymmetric hydrogenation of (161) in enantioselectivity of 91% (R) and 90%(S) respectively [120].

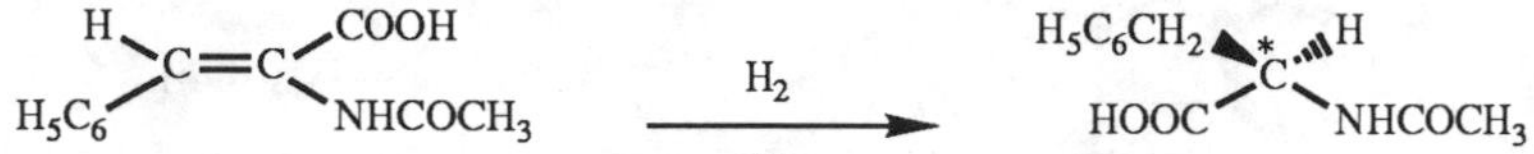

(161) $\xrightarrow[\text{cat.}]{H_2}$ (162) 86% e.e. (R)

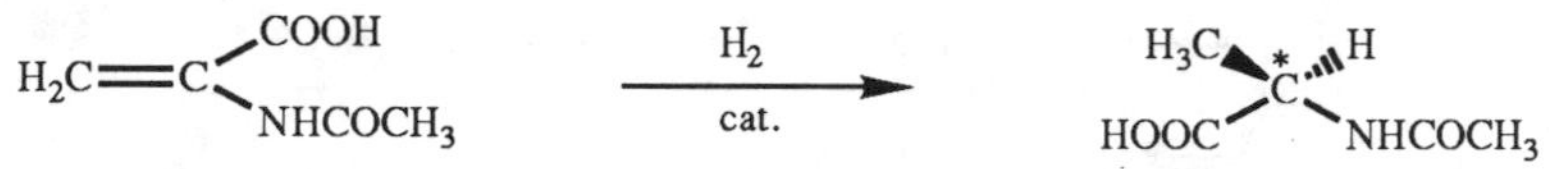

(163) $\xrightarrow[\text{cat.}]{H_2}$ (164) 52-60% e.e. (R)

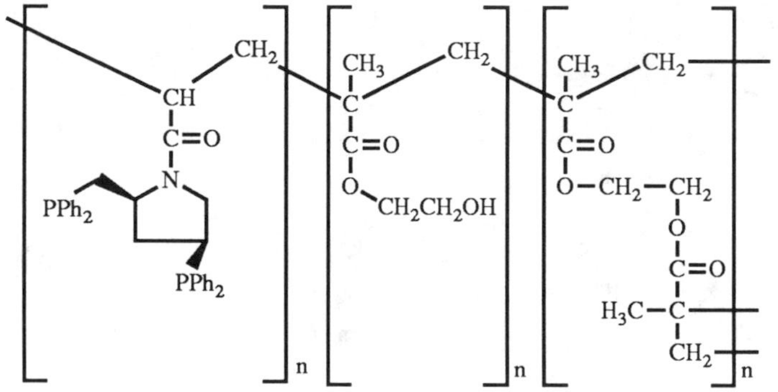

(165)

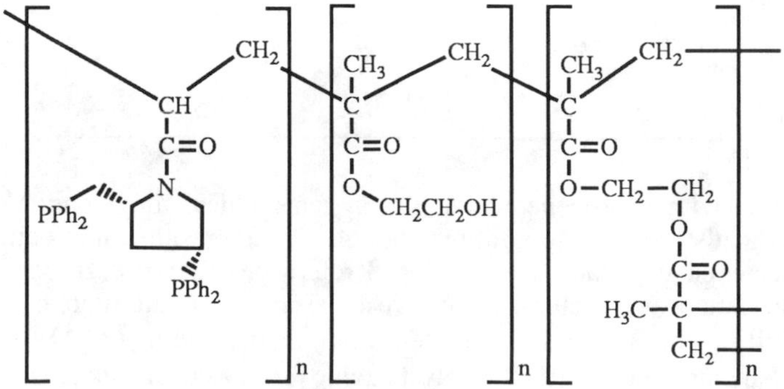

(166)

Much work has been done on the synthesis of optically active amino acids involving asymmetric heterogeneous catalytic hydrogenation of N-acyl-α,β-dehydroamino acid [121-126]. For example catalytic hydrogenation of (167) over Pd-Al_2O_3 yielded p-aminophenylalanine (168) after hydrolysis in 34% e.e [121]. A similar hydrogenation and subsequent hydrolysis of substrate (169) having L-α-methylbenzylamine as chiral auxiliary yielded D-valine (170) in 18% e.e [125]. Interestingly, D-α-methylbenzylamine in the substrate (169) gave L-valine.

(167)   (168) 34% e.e. (S)

(169)   (170) 18% e.e. (S)

Another interesting method of preparing optically active amino acids involves asymmetric hydrogenation of substrate (173) prepared from erythro-(+)-1,2-diphenylethanolamine (171) and acetylene dicarboxylic acid (172) to give (174) which after hydrogenolysis with Pd(OH)$_2$ afforded (S)-(+)-aspartic acid methyl ester (175) in 98% e.e [126].

(171)   (172)   (173)

$$\xrightarrow{\text{H}_2, \text{Ni}}$$

(174)

$$\xrightarrow[\text{H}_2]{\text{Pd(OH)}_2}$$

(175) 98% e.e. (S)

Dehydrodi- and tripeptides have been hydrogenated with high chiral induction (98%) using Raney nickel or Pd as catalysts [122,127-129]. Asymmetric hydrogenation of the dipeptide (176) yielded the hydrogenated product (177) with a *DL:LL* isomeric ratio of 75:25 [127]. A similar hydrogenation of the dehydrotripeptide (178) gave *D*- alanine (179) in 93% e.e. [122].

(176)

$$\xrightarrow[\text{H}_2]{\text{Raney Ni}}$$

(177)

(*DL:LL*, 75:25)

(178)

$$\xrightarrow[\text{Pd/C}]{\text{H}_2}$$

(179) (*D*)-alanine
18% e.e. (*R*)

The mechanism of heterogeneous catalytic hydrogenation is exceedingly complex, since the conformation of a substrate adsorbed on the catalyst surface may be substantially different from that of the most stable conformation of the same substrate in solution. Nevertheless, the classical empirical stereocorrelation rules proposed for homogeneous reactions [130-131] are still useful for arbitrary rationalization of the steric course of heterogeneous catalytic hydrogenations. As discussed in the previous section, the inability of the catalyst [Rh(I)-(156)] to hydrogenate α-acetamido cinnamic acid (161) has been explained by assuming that the catalyst contracts strongly in ethanol because of the highly hydrophobic nature of the styrene support whereas the substrate acylaminoacrylic acid (161) is practically insoluble in pure benzene. In polar solutions of the substrate, the cross-linked polystyrene beads collapse, preventing entry of the acylaminoacrylic acid to the catalytic site.

The steric course of asymmetric induction in substrates such as (167) and (169) can be explained by assuming a preferred Prelog's conformation in the solvent as shown for (169), its adsorption on the metal and subsequent hydrogenation occurring from the less hindered face (dashed arrows).

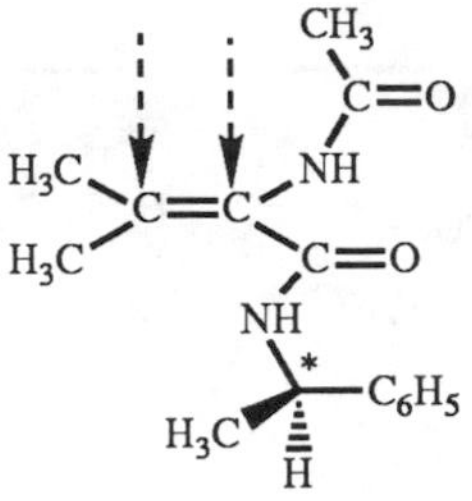

(169)

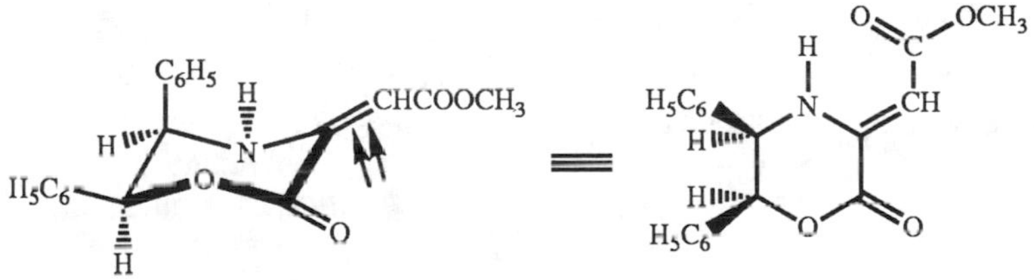

(173)

A similar reasoning can explain the almost stereospecific hydrogenation of (173) by assuming hydrogenation from the face of the double bond away from the phenyl groups in the molecule (shown by arrows). That the disposition of the two phenyl groups in (173) is an important factor has been shown by the low chiral induction (12-16%) exhibited by a similar 1,4-oxazine derivative (180) upon hydrogenation under similar reaction conditions [126].

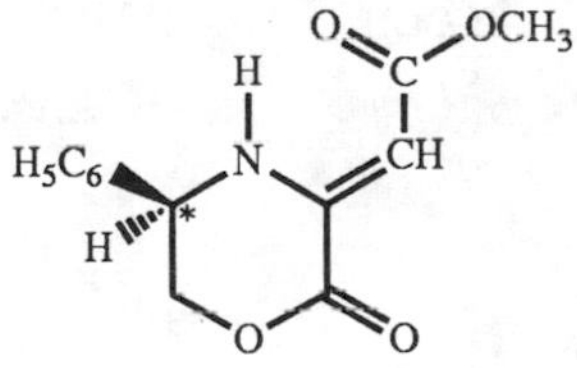

(180)

Some representative examples of asymmetric catalytic hydrogenation of N-acyl-α,β-dehydroamino acids and dehydrodi- and tripeptides are shown in Table 2.12.

Table 2.12. Asymmetric Hydrogenation of N-Acetyl α, β-Dehydro-amino Acids

Substrate	Catalyst	Product / Optical yield (% e.e.)	Reference
$\begin{array}{c}H_3C\\H_3C\end{array}C{=}C\begin{array}{c}NHCOPh\\CONHCHCH_3Ph\end{array}$	a	$(CH_3)_2CH\overset{*}{C}\begin{array}{c}H\\COOH\end{array}$ H_2N 18 (L)	[125]
$\begin{array}{c}Ph\\H\end{array}C{=}C\begin{array}{c}NH-\overset{O}{\overset{\|}{C}}-CH_3\\COOH\end{array}$	b	$PhCH_2\overset{*}{C}\begin{array}{c}H\\NHCOCH_3\end{array}$ $HOOC$ 91 (R)	[120]
CH_3O_2C OCH₃ $C{=}C\begin{array}{c}NHCOCH_3\\COOH\end{array}$ H	b	CH_3O_2C OCH₃ $H_2C\overset{*}{C}\begin{array}{c}H\\NHCOCH_3\end{array}$ $HOOC$ 88 (R)	[120]
O_2N—⬡—$CH{=}C\begin{array}{c}COOmenthyl\\NHCOPh\end{array}$	c	O_2N—⬡—$CH_2\overset{COO\ menthyl}{\underset{HN}{C}}-H$ $\begin{array}{c}C{=}O\\Ph\end{array}$ 68:32 (R, S)	[121]
Ph $\begin{array}{c}H\\H\end{array}$N $\overset{O}{\overset{\|}{C}}{-}OCH_3$ CH ... O ... O	d	$HOOCH_2C\overset{*}{C}\begin{array}{c}H\\COOH\end{array}$ H_2N 17 (S)	[123]
$\begin{array}{c}H\\C\\\|\|\\C\end{array}$ $\begin{array}{c}H\end{array}$ BOC(S)ileuNH CO(S)ProNHᵗBu	e	$H_3C\overset{*}{C}\begin{array}{c}H\\NH_2\end{array}$ $HOOC$ 90 (R)	[122]

(Table 2.12. contd.)

Substrate	Catalyst	Product / Optical yield (% e.e.)	Reference
BOC(S)(Val)NH—C(=CH$_2$)—CO(S)Pro—N⟨pyrrolidine⟩	e	H$_3$C$\overset{*}{C}$H HOOC NH$_2$ 74 (R)	[122]
Ph—C(=O)—C(CH$_3$)$_2$—CONH—CH(CH$_2$—PhOCOPh)—COOCH$_3$	a	(CH$_3$)$_2$HC CH$_2$—PhOCOPh PhCONH—$\overset{*}{C}$HCONH$\overset{*}{C}$H—COOCH$_3$ 77:23 (DL : LL)	[127]

a = i) Raney Ni,   b = i) [Rh(I)-Polymer(165)]   c = Al-Hg,   d = Pd/Al$_2$O$_3$,   e = Pd
    ii) H$_3^+$O   ii) EtOH/Et$_3$N

2.2.2.1.2   Asymmetric Hydrogenation of Cyclic Dehydropeptides

Asymmetric heterogeneous catalytic hydrogenation is an important tool to study the structure-function relationship of phytotoxins, such as the AM-Toxin, tentoxin etc., which are cyclic peptides having dehydroamino acid residues [133-135].

For example, the asymmetric hydrogenation of dehydroalanine residue in the cyclic dehydrodipeptide (181) yielded the hydrogenated cyclic dipeptide (182) in 98.4% optical yield [134]. The presence of L-valine during the hydrogenation of (181) resulted in high optical yield (>98%) whereas L-proline afforded somewhat lower chiral induction (ca. 85%). In earlier studies on the hydrogenation of diketopiperazines derivatives, L-proline was considered to be the only amino acid essential for high chiral induction [136,137]. However it was later found that the asymmetric hydrogenation of other cyclodehydrodipeptides containing dehydrophenylalanine, dehydrotryptophan, dehydroleucine and dehydrovaline moieties also afforded high chiral induction (upto 99%) [134-137].

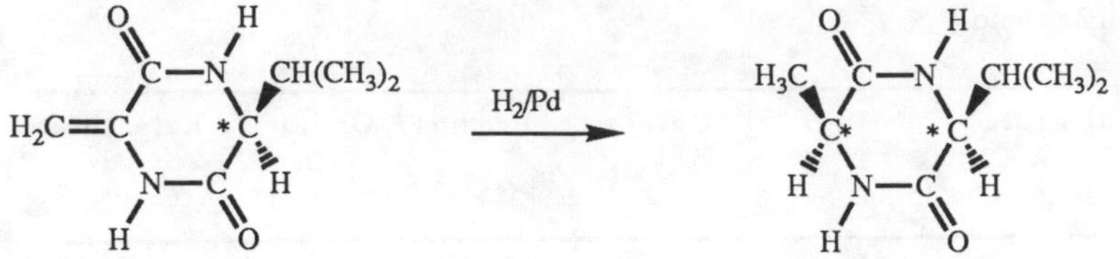

(181), (L)   (182), 98.4% e.e. (L, L)

The mechanism of asymmetric hydrogenation of cyclic dehydropeptides can be explained by assuming a partially adsorbed or anchored stage of the substrate over the metal surface prior to its complete adsorption [134,135,138,139]. The diketopiperazine ring, for instance in (181) (Fig. 2.20A), is assumed to be nearly planar and it becomes anchored vertically on the catalyst surface through its CO or NH group such that the methylene group of the dehydroalanine residue is coplanar with the ring, while the isopropyl group sticks out of the plane (Fig. 2.20 B,C). The anchored substrate is then adsorbed from its less bulky face (Fig. 2.20 D) while the isopropyl group determines whether the re- or si- face of the unsaturated moiety will be preferentially adsorbed.

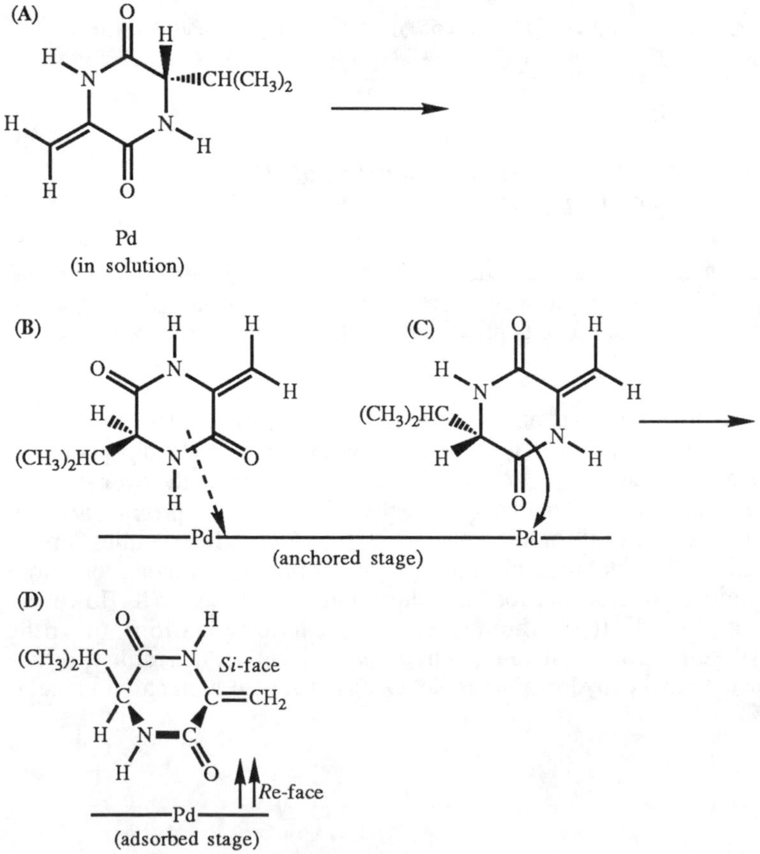

Figure 2.20. Mechanism of asymmetric hydrogenation of substrate (181).

Some examples of asymmetric hydrogenation of cyclic dehydropeptides are given in Table 2.13.

Table 2.13. Asymmetric Hydrogenation of Cyclic Dehydropeptides

Substrate	Product	Optical purity (%)	Reference
C(L-Ala-Dha)	C(L-Ala-Ala)	94.6 (S)	[134]
C(L-Val-Dha)	C(L-Val-Ala)	98.4 (S)	[134]
C(L-Leu-Dha)	C(L-Leu-Ala)	95.8 (S)	[134]
C(L-Phe-Dha)	C(L-Phe-Ala)	94.6 (S)	[134]
C(L-Pro-Dha)	C(L-Pro-Ala)	84.8 (S)	[134]
C(L-Lys-(ε-Ac)-Dha)	C(L-Lys(ε-Ac)-Ala)	91.8 (S)	[134]
C(L-Leu-Z-Dhb)	C(L-Leu-Aba)	95.6 (S)	[134]
C(L-Pro-Dhv)	Val*	>90 (S)	[137]

*after hydrolysis

Ala = alanine, Dha = dehydroalanine, Val = Valine, Leu = Leucine, Phe = Phenylalanine, Pro = Proline, Lys = Lysine, Dhb = dehydrobutyrine, Dba = 2 aminobutanoic acid, Dhv = dehydrovaline

2.2.2.2 Asymmetric Hydrogenation of other Carbonyl Compounds

There are only a few reports of diastereoselective heterogeneous catalytic hydrogenation of prochiral carbonyl compounds. These are limited to a few classes of such compounds as benzoylformic acid esters or amides and pyruvic acid menthyl esters and amides.

2.2.2.2.1 Asymmetric Hydrogenation of Benzoylformic Acid Esters

Menthyl and bornyl esters of benzoylformic acid have been catalytically hydrogenated using various catalysts such as Raney nickel, Pd, PtO$_2$ etc. and the reaction appears to depend on the pH of the medium. For instance, the catalytic hydrogenation of (-)- menthylbenzoylformate (**183**) using alkali-treated Pd on charcoal yielded (-)-menthyl-(-)mandelate (**184**) in 10% optical yield [140,141] whereas in acidic medium, the configuration of the mandelate was reversed.

(183)   (184) 10% d.e.

By modifying the substrate to (+)-bornylbenzoyl formate (185), the product obtained on hydrogenation under basic conditions was found to be (+)-bornyl-(+)-mandelate (186). However in acidic medium, (+)-bornyl-(-)-mandelate was obtained [140,141]. Similar results were obtained with sugar esters of benzoylformic acid upon asymmetric hydrogenation with various catalysts [142]. Thus the asymmetric reduction of (187) over Raney nickel or alkali-treated Pd yielded (R)-mandelic acid (188) in 44.6 and 22.8% e.e. respectively upon hydrolysis [142]. On the other hand, similar hydrogenation over acid-treated Pd gave (S)-mandelic acid in 24.1% e.e.

(185)   (186)

(187)   (188) 44.6% e.e. (R)

Cy = Cyclohexylidene

The phenomenon of inversion of configuration displayed by the hydrogenated products under acidic or basic catalytic conditions has been explained by taking into consideration the differences in substrate conformation. It is known that a solution of α-ketoesters and an optically active alcohol exhibits mutarotation [143], which takes place through hemiketal formation (Fig. 2.21) [144-146]. The (S)-trans conformation (189) is thus converted via a hemiketal (190) to the (S)-cis conformation (191).

It is evident that hydrogenation of the (S)-cis conformation using an acidic catalyst could result in the formation of the α-hydroxy acid with the opposite configuration to that obtained with the basic catalyst system.

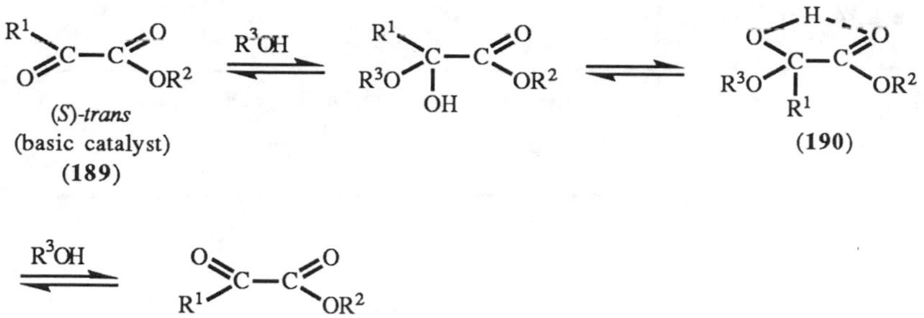

(*S*)-*trans* (basic catalyst) (**189**)

(**190**)

(*S*)-*cis* (acidic catalyst) (**191**)

Figure 2.21. Mutarotation of (**189**) to (**191**).

Some examples of asymmetric hydrogenation of phenylglyoxylic acid esters with various catalysts are shown in Table 2.14.

Table 2.14. Asymmetric Hydrogenation of Phenylglyoxylic acid Esters with Various Catalysts

Substrate	Catalyst	Optical yield (% e.e.)	Reference
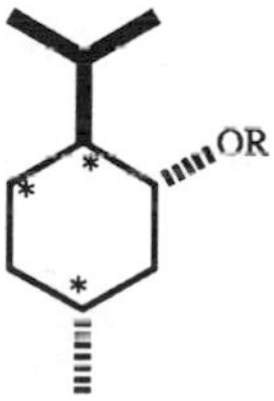	Raney Ni	16.1	[140,141]
"	5% Pd-C (basic)	10.0	"
"	5% Pd-C (acidic)	13.4	"
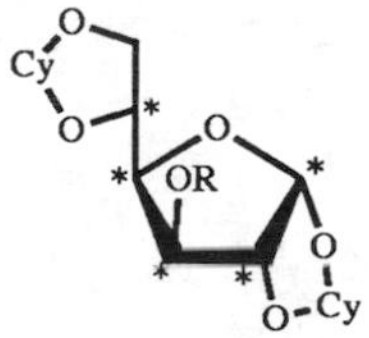	Raney Ni	44.6	[142]
"	5% Pd-C (basic)	22.8	"
"	5% Pd-C (acidic)	24.1	"

(Table 2.14. contd.)

Substrate	Catalyst	Optical yield (% e.e.)	Reference
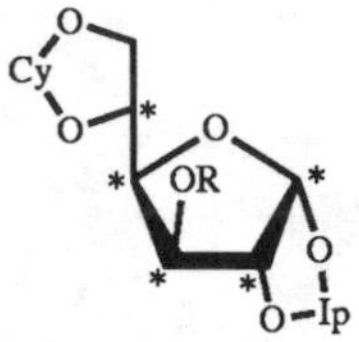	Raney Ni	36.7	"
"	5% Pd-C (basic)	15.4	"
"	5% Pd-C (acidic)	20.3	"
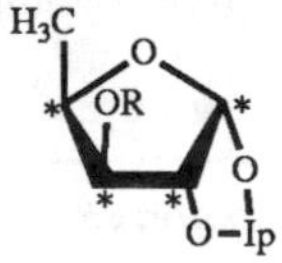	Raney Ni	14.0	[142]
"	5% Pd-C (basic)	12.7	"
"	5% Pd-C (acidic)	12.8	"

R = -COCOC$_6$H$_5$, Cy = cyclohexylidene, Ip = isopropylidene

2.2.2.2.2   Asymmetric Hydrogenation of a-Keto Amides

Asymmetric catalytic hydrogenation of α-keto amides such as pyruvamides, prepared by the coupling of pyruvic acid and optically active amines, can afford lactamides in high optical yields [147-150]. Thus, the asymmetric reduction of (S)-pyruvamide (192) in methanol over Pd-C afforded (S,S)-lactamide (193) in 59% diastereoisomeric purity (d.p.) [150].

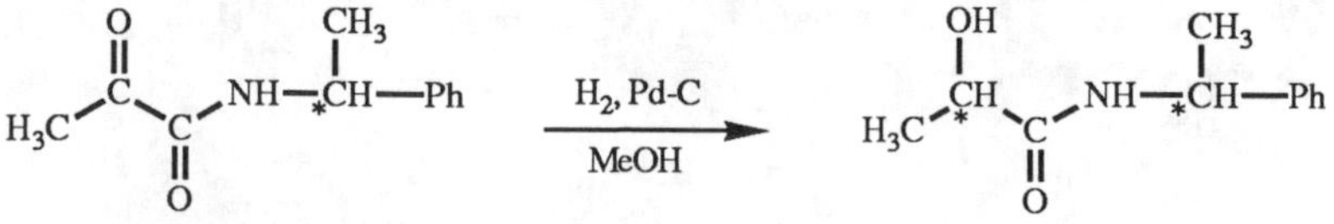

(192)

(193) 59% d.p. (S, S)

The steric course of the catalytic hydrogenation of α-ketoamides has been explained on the basis of a five-membered substrate-catalyst complex, commonly referred to as the *chelation mechanism* [150]. This involves a two-step adsorption of the substrate on the catalyst surface during the catalytic hydrogenation. An analogous mechanism to that found in the reduction of diketopiperazine derivatives also appears to be operative here .

As shown for the substrate (192) in the *S-cis* conformation (Fig. 2.22), the first stage of hydrogenation is the anchoring stage of the substrate on the metal surface through the two carbonyl groups to form a five-membered chelated

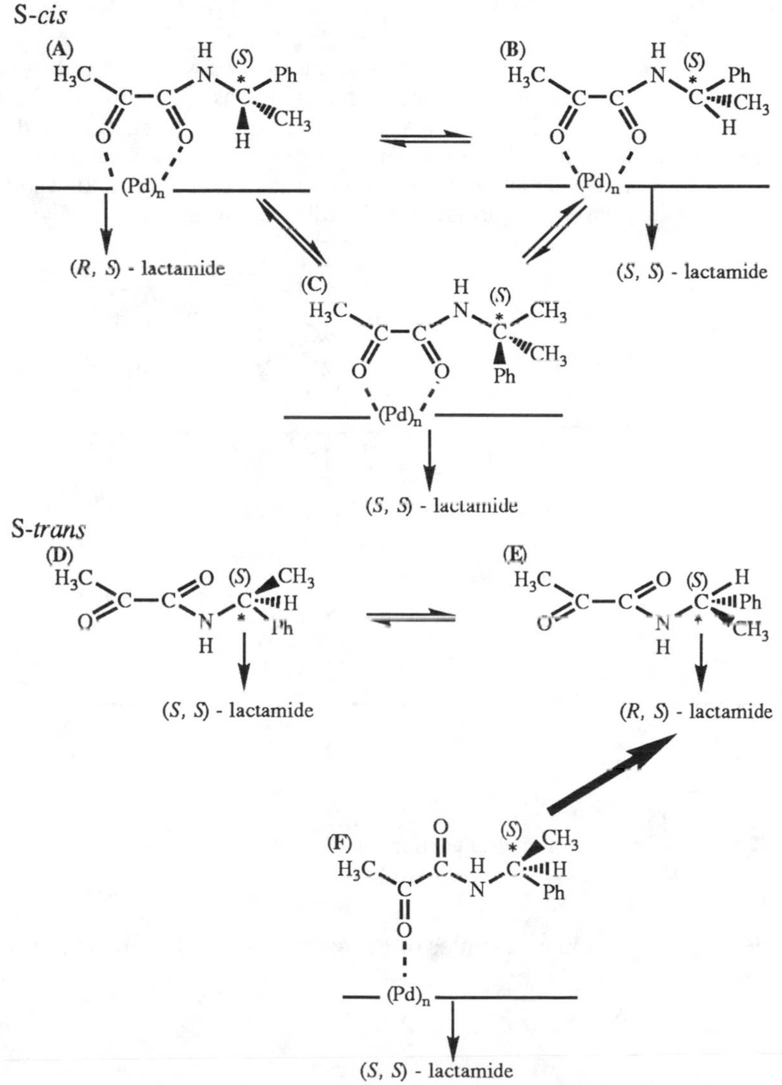

Figure 2.22. Possible steric course in the hydrogenation of substrate (192).

intermediate with the catalyst (Fig. 2.22A). The second stage involves adsorption of the chelated intermediate on the catalyst from the less bulky side of the substrate to form either conformer B or C which is then hydrogenated.

In the *S-trans* conformation (Fig. 2.22 D,E) a one-step adsorption process of the substrate on the catalyst would be involved. The five-membered chelated intermediate formed by the oxygen and nitrogen on the catalyst will be opposed by the carbonyl *vicinal* to the nitrogen as shown in Fig. 2.22 F. This would have a destabilizing effect on the catalyst and the *S-trans* adsorption state would not effect the asymmetric induction in any significant way. Conformer A therefore forms (R,S)-lactamide whereas conformers B and C produce (S,S)-lactamides.

Such interactions between solvent and substrate, and between substrate and catalyst can also be assumed in substrates such as (**194**) as shown in Fig. 2.23 Hydrogenation of (**194**) in polar solvents leads to increased interaction of the isobutyl ester group with the solvent (conformations A and B) thus decreasing the optical yield. In non-polar solvents, the isobutyl ester group tends to adsorb on the catalyst (as shown in conformation C) leading to enhanced optical yield [149].

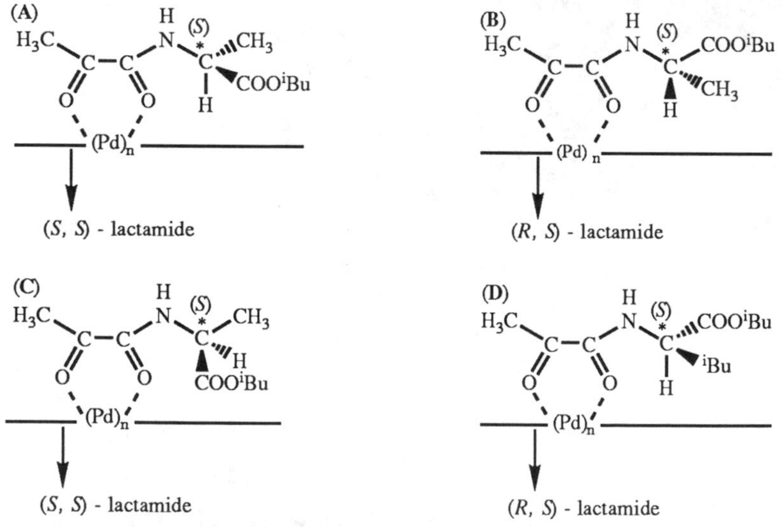

Figure 2.23. Steric course of hydrogenation of N-pyruvoyl-(S)-amino acid isobutyl ester (**194**).

Table 2.15. shows some examples of the asymmetric hydrogenation of α-keto amides.

Table 2.15.   Asymmetric Hydrogenation of α-keto Amides

Substrate	R^1	R^2	Config[a].	D.P.[b](%)	Config.[c]	Reference

	CH_3	C_6H_5	(S)	61	(S)	[150]
	C_2H_5	C_6H_5	(S)	54	(R)	[150]
	CH_3	Naph	(S)	48	(R)	[150]
	C_6H_5	COO^iBu	(S)	55	(R)	[150]
	COO^iBu	CH_3	(S)	34	(R)	[149]
	COO^iBu	$CH(CH_3)_2$	(S)	21	(R)	[149]
	COO^iBu	^iBu	(S)	21	(R)	[150]

a) Configuration of the chiral amine.
b) Diastereomeric purity.
c) Configuration of the newly formed chiral centre.

2.2.2.3   Asymmetric Hydrogenation of Carbon-Nitrogen Double Bonds

2.2.2.3.1   Hydrogenation of Imines, Oximes and Hydrazones

Imines may be hydrogenated in moderate to good enantioselectivities to the corresponding amines in the presence of suitable catalysts, and the procedure has been used in the synthesis of optically active amino acids. For instance, catalytic hydrogenation of *imines* prepared from α-keto acids and optically active amines over Pd/C resulted in imino acids which were hydrogenolysed with Pd(OH)$_2$ on charcoal to yield amino acids in 12-91% optical purities [151-155]. Similarly, catalytic hydrogenation of the imine (195) prepared from pyruvic acid and D-(+)-α-methylbenzylamine using Pd/C as catalyst yielded the hydrogenated product (196) which upon further hydrogenolysis over Pd(OH)$_2$-/C afforded D-(-)-alanine in 91.1% optical yield [154].

The configuration of the α-amino acid produced depends on the configuration of the α-methylbenzylamine employed, i.e. the (L)- amine gives the (*L*)-amino acid, while the (*D*)-amine gives the (*D*)-amino acid. The optical yield decreases as the methyl group in pyruvic acid is replaced by a bulkier group. For example, replacement of the methyl group by an isopropyl group resulted in decreased optical yield of valine formed (28% as compared to 91.1% for alanine). A number of other examples of asymmetric reduction of imines have been reported [153,155,156].

(195)

(196)

(*D*)-(-)-alanine
91.1% e.e.

The transamination reaction in amino acid metabolism has been attempted non-enzymatically in the presence of pyridoxal phosphate to prepare optically active amino acids [156-164]. Pyridoxamine analogues such as (197) with a pyridinophane structure have been used with various α-keto acids in the transamination reaction employing Zn(II) as catalyst to yield optically active amino acids in upto 94% e.e. [161-163].

(197) A : R = CH₃
       B : R = H

Similar transamination reaction between phenylpyruvic acid and pyridoxamine (198) in the presence of Cu (II) complex with ligands (199) yielded a Cu-ketimine complex (200) by its isomerization into a Cu-aldimine complex (201), which afforded phenylalanine in upto 80% e.e. [164].

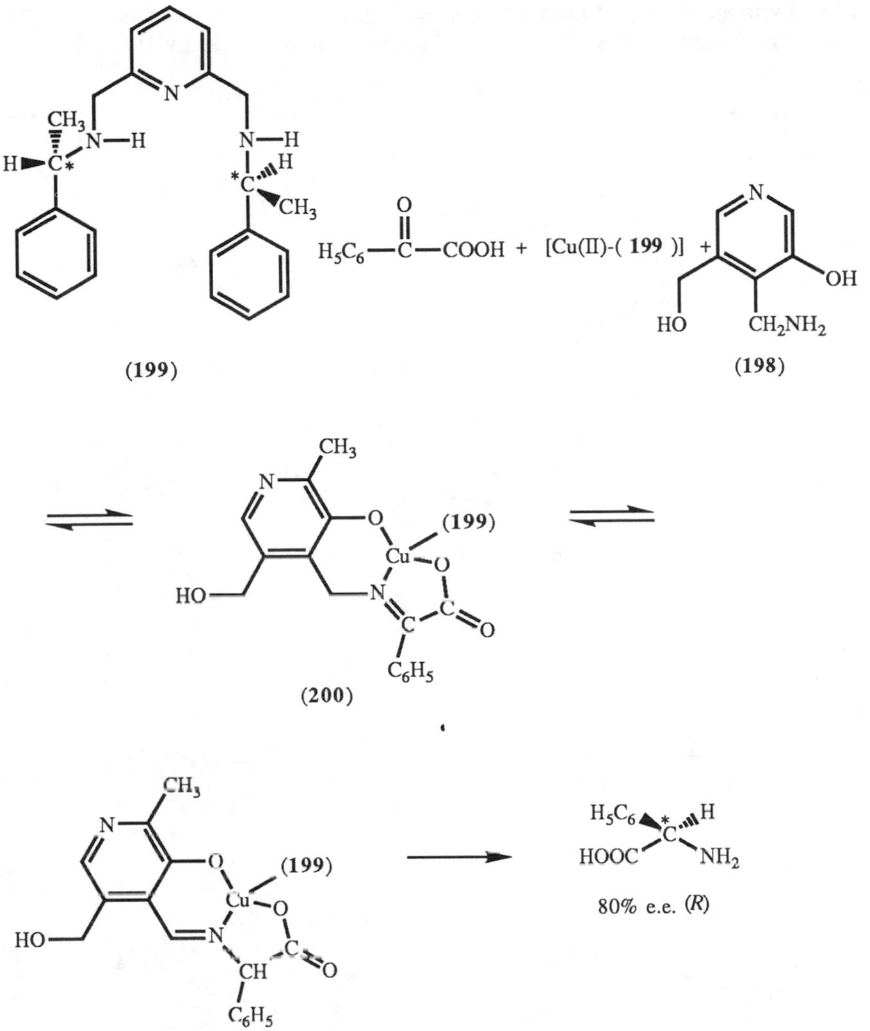

(199)

H_5C_6—$\overset{\overset{\displaystyle O}{\|}}{C}$—COOH + [Cu(II)-(**199**)] +

(198)

(200)

(201)

H_5C_6 $\overset{*}{C}$ H / HOOC NH$_2$

80% e.e. (R)

Catalytic hydrogenation of *oximes* derived from (-)-menthyl esters of pyruvic acid, benzyl formic acid and menthyl ester of α-ketobutyric acid afford, after hydrolysis, the corresponding amino acids in low to moderate optical yields [165-168]. Thus the catalytic hydrogenation of oxime (**203**), prepared from menthyl pyruvate (**202**) and hydroxylamine over 10% Pd/C yielded the hydrogenated product (**204**), which gave on hydroysis with 10% NaOH, (D)-alanine in 11.5% optical yield [165].

Asymmetric hydrogenation of *hydrazones* provides an alternative method to prepare optically active amino acids [169]. For example, the asymmetric hydrogenation of hydrazone (**205**), prepared from (S)-bornylanine derivative and ethyl pyruvate over platinum oxide yielded the hydrogenated derivative (206) which was hydrogenolysed to give (L)-alanine in 46.5% optical yield [169].

(202)   (203)

(204)

(205) (S)   (206) (S)

46.5% (L)

The mechanism of asymmetric hydrogenation of compounds having carbon-nitrogen double bond has been extensively investigated [151-176]. The steric course of the reaction in most cases has been explained by the *chelation mechanism* (discussed in the asymmetric hydrogenation of α-keto amides). It has been shown that the stereoselectivity in the catalytic hydrogenation of the Schiff's base derived from amino acid esters and α-keto esters depends largely on the bulkiness of the ester group of the amino acid ester and the alkyl group attached to the α-carbon of the α-keto esters [160]. Temperature, in most cases, has a marked effect on the stereoselectivity of the hydrogenation reaction. Lower temperatures generally favour chelate formation leading to high stereoselectivity whereas at higher temperatures the substrate is more labile and may be liberated from the catalyst surface resulting in a decrease in stereoselectivity [160,168,172]. This has been shown for the Schiff's base (**207**) in Figure 2.24.

Solvent has, in some cases, a marked effect on the stereoselectivity of the reaction [151, 166, 174]. In less polar solvents the electrostatic attraction between substrate and catalyst is stronger than in polar solvents and the solvation of the substrate in less polar solvents is also weak, leading to increased interaction of the substrate with the catalyst. In polar solvents, chelation is hindered because of solvation of the substrate, and the unchelated species leads to inversion of configuration of the product as compared to that of the chelated species.

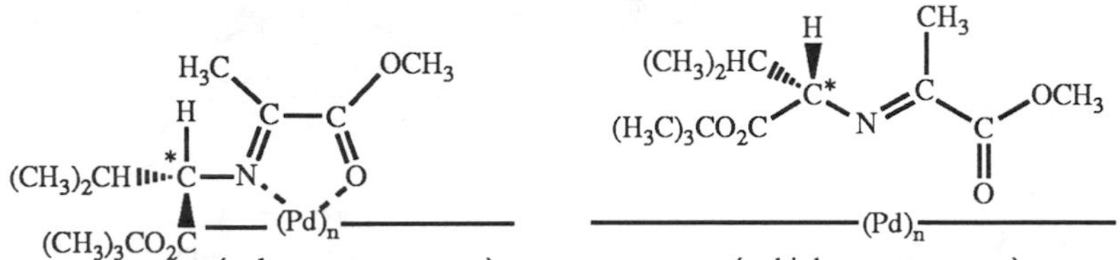

(at lower temperature) (at higher temperature)

Figure 2.24. Conformations of substrate (**207**) at lower and higher temperatures.

The steric course of catalytic hydrogenation in oximes such as (**203**) has been recently explained by a *cisoidal* conformation of the oxime-chelate intermediate (**208**) [166]. The menthyl group in the conformation shown lies apart from the catalyst surface because of steric hinderance between the menthyl group and the catalyst. Hydrogenation of (**208**) after its adsorption would lead to (*R*)- amino acids which agrees with the experimental results [165].

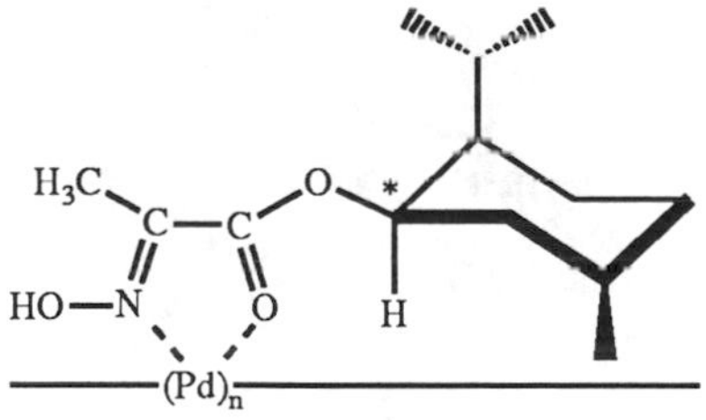

(**208**)

Table 2.16. presents some representative examples of catalytic hydrogenation of imines, oximes and hydrazones.

Table 2.16. Asymmetric Hydrogenation of Carbon-nitrogen Double Bond

Substrate	Derived from	Product/ % e.e.	Reference

Row 1:

Substrate:
$$CH_3$$
$$\underset{CO_2H}{}C=N-\overset{H}{\underset{{}^iPr}{\overset{*}{C}}}-CO_2{}^tBu$$

Derived from:
$$\overset{O}{\overset{\|}{H_3C-C}}-CO_2CH_3$$
$+$
$$(CH_3)_2HC-\overset{NH_2}{\underset{COO^tBu}{CH}}$$

Product: alanine, 71 (L)   [156]

Row 2:

Substrate:
$$CH_3$$
$$\underset{CH_2Ph}{}C=N-\overset{H}{\underset{CH_3}{\overset{*}{C}}}-CO_2{}^tBu$$

Derived from:
$$\overset{O}{\overset{\|}{PhCH_2-C}}-CH_3$$
$+$
$$H_3C-\overset{NH_2}{\overset{*}{C}}OO^tBu$$

Product:
$$H_2N-\overset{H_3C}{\overset{*}{C}}\cdots H$$
$$CH_2-Ph$$ [157]
85 (S)

Row 3:

Substrate:
$$CH_3$$
$$\underset{COR}{}C=N-\overset{H}{\underset{CH_3}{\overset{*}{C}}}-Ph$$

Derived from:
$$\overset{O\ \ O}{\overset{\|\ \ \|}{H_3C-C-C}}-R$$
$+$
$$C_6H_5\overset{*}{C}HCH_3NH_2$$

Product:
$$\overset{NH_2}{\underset{H\ \ O}{H_3C-\overset{*}{C}-C}}-R$$ [155]
58 (D)

Row 4:

Substrate:
$$Ph$$
$$\underset{CH_3}{}C=N-\overset{H}{\underset{{}^iPr}{\overset{*}{C}}}-O^tBu$$

Derived from:
$$\overset{O}{\overset{\|}{Ph-C}}-CH_3$$
$+$
$$H_2N-\overset{{}^iPr}{\underset{O^tBu}{\overset{*}{C}H}}$$

Product:
$$H_2N-\overset{H_3C}{\overset{*}{C}}\cdots H$$
$$Ph$$ [155]
85 (S)

Row 5:

Substrate:
$$CO_2H$$
$$\underset{CH_3}{}C=N-\overset{H}{\underset{CH_3}{\overset{*}{C}}}-C_6H_5$$

Derived from:
$$\overset{O}{\overset{\|}{H_3C-C}}-COOH$$
$+$
$$H_5C_6-\overset{NH_2}{\underset{CH_3}{\overset{*}{C}}}-H$$

Product: alanine, 90.7 (L)   [157]

(Table 2.16. contd.)

Substrate	Derived from	Product/ % e.e.	Reference

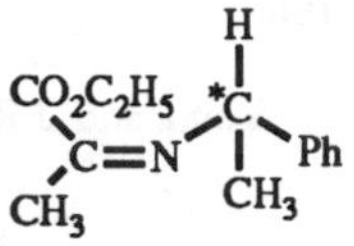

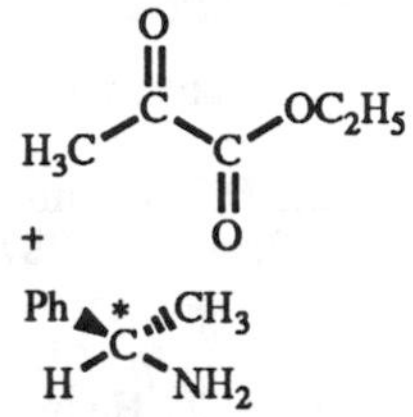

First substrate (CO₂C₂H₅, C=N, CH₃, H-C-Ph-CH₃) derived from methyl pyruvate ester + Ph–CH(CH₃)–NH₂: alanine, 62 (S) [154]

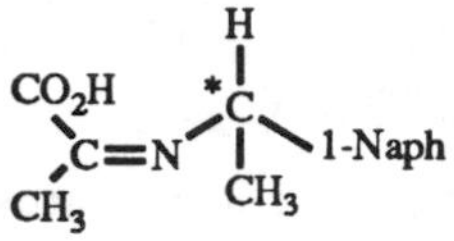

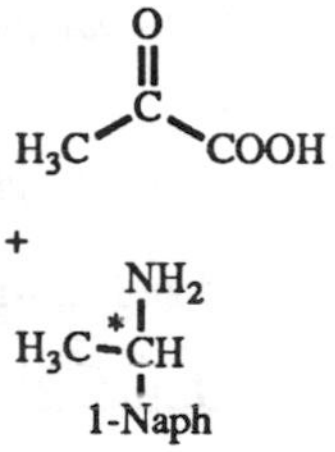

Second substrate (CO₂H, C=N, CH₃, H-C-1-Naph-CH₃) derived from pyruvic acid + H₃C–CH(NH₂)–1-Naph: alanine, 83 (R) [165]

Third substrate (CO₂C₂H₅, C=N, CH₃, N(CH₃)-C-CH₃-ⁱPr) derived from methyl pyruvate ester + H₃C–CH(CH₃)–N–NH₂ with CH(CH₃)₂: alanine, 31.8 (D) [165]

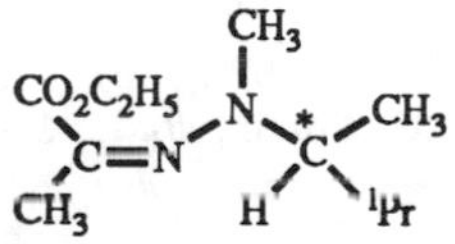

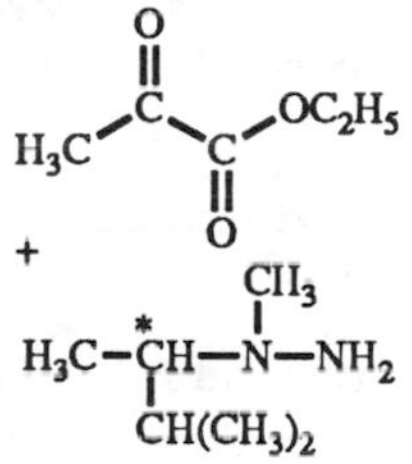

R = NHCH₂COOH

2.3 References

1. (a). J.F. Young, J.A. Osborn, F.H. Jardine and G. Wilkinson, *J.Chem. Soc.Chem.Commun.*, 131 (1965). (b). J.A. Osborn, F.H. Jardine and G. Wilkinson, *J.Chem.Soc.* A, 1711 (1966). (c). J.A. Osborn and G. Wilkinson, *Inorg.Synth.*, **10**, 67 (1967).

2. For a review of Wilkinson's catalyst, see: F.H. Jardine, *Prog.Inorg.Chem.*, **28**, 63 (1981).

3. T.P. Dang and H.B. Kagan, *J.Chem. Soc.Chem.Commun.*, 481 (1971).

4.   (a). B.D. Vineyard, W.S. Knowles, M.J. Sabacky, G.L. Bachman and D.J. Weinkauff, *J.Amer.Chem.Soc.*, **99**, 5946 (1977); (b). W. Vocke, R. Hänel and F.U. Flöther, *Chem.Tech.*, **39**, 123 (1987); (c). G. Parshall and W. Nugent, *Chem.Tech.*, **18**, 194 (1988); (d) G. Parshall and W. Nugent, *ibid*, **18**, 376 (1988).

5.   T.P. Dang and H.B. Kagan, *J.Amer.Chem.Soc.*, **94**, 6429 (1972).

6.   L. Horner, H. Seigel and H. Buthe, *Angew.Chem.*, **7**, 9420 (1968).

7.   G.D. Melillo, R.D. Larsen, D.J. Mathre, W.F. Shukis, A.W. Wood and J.R. Colleluori, *J.Org.Chem.* **52**, 5143 (1987).

8.   (a). H. Parnes and E.J. Shelton, *Int.J.Pept.Protein Res.*, **27**, 239 (1986); (b). H. Parnes, E.J. Shelton and G.T. Huang, *ibid*, **28**, 403 (1986). (c). H. Parnes and E.J. Shelton, *Synth.Appl.Isot.Labeled Comp.Proc. Int. Symp. 2nd.* 1985, P 159 (pub. 1986).

9.   D.P. Riley and P.E. Shumate, *J.Org.Chem.* **45**, 5187 (1980).

10.   R.B. King, J. Bakos, C.D. Hoff and L. Marko, *J.Org.Chem.* **44**, 1729 (1979).

11.   M.D. Fryzuk and B. Bosnich, *J.Amer.Chem.Soc.*, **99**, 6262 (1977).

12.   (a) S.B. Wilde and N.K. Roberts, *J.Amer.Chem.Soc.*, **101**, 6254 (1979); (b). M.D. Fryzuk and B. Bosnich, *J.Amer.Chem.Soc.*, **100**, 5491 (1978).

13.   (a). T. Hayashi and M. Kumada, *Acc. Chem.Res.*, **15**, 395 (1982); (b). H.B. Kagan, J.C. Fiaud, C.H. Hoornaert, D. Meyer and J.C. Poulin, *Bull. Soc. Chim. Belg.*, **88**, 923 (1979).

14.   W. Bergstein, A. Kleemann and J. Martins, *Synthesis*, 76 (1981).

15.   C.F. Hobbs and W.S. Knowles, *J.Org.Chem.*, **46**, 4422 (1981).

16.   J.D. Morrison and W.F. Masler, *J.Org.Chem.*, **39**, 270 (1974).

17.   J.D. Morrison, W.F. Masler and M.K. Neuberg, *Adv.Catal.*, **25**, 81 (1976).

18.   R. Glaser, J. Blumenfeld and M. Twaik, *Tetrahedron Lett.*, **52**, 4639 (1977).

19.   R. Glaser, S. Geresh, J. Blumenfeld and M. Twaik, *Tetrahedron*, **34**, 2405 (1978).

20.   P. Aviron-Violet, Y. Collenille and J. Varagnal, *J.Mol.Catal.*, **5**, 41 (1979).

21.   T.P. Dang, J.C. Poulin and H.B. Kagan, *J.Organomet.Chem.*, **91**, 105 (1975).

22.   (a) D.L. Allen, V.C. Gibson, M.H.L. Green, J.F. Skinner, J. Bashkin and P.D. Grebenik, *J.Chem.Soc.Chem.Commun.* , 895 (1983); (b) I. Ojima and N. Noyoda, *Tetrahedron Lett.*, **21**, 1051 (1980).

23.   K. Achiwa, *J.Amer.Chem.Soc.*, **98**, 8265 (1976).

24.   T. Hayashi, K. Yamamoto and M. Kumada, *Tetrahedron Lett.*, 4405 (1974).

25.   T. Hayashi, T. Mise and M. Kumada, *Tetrahedron Lett.*, 4351 (1976).

26.   A. Miijashita, A. Yasuda, H. Takaya, K. Toriumi, T. Ito, T. Souchi, and R.N. Noyori, *J.Amer.Chem.Soc.*, **102**, 7932 (1980).

27.   M. Tanaka and I. Ogata, *J.Chem.Soc.Chem.Commun.*, 735 (1975).

28.   R.B. King, J. Bakos, C.D. Hoff and L. Marko, *J.Org.Chem.*, **44**, 3095 (1979).

29.   S.Y. Zhang, S. Yemul, H.B. Kagan, R.Stern, D. Commereuc and Y. Chauvin, *Tetrahedron Lett.*, 3955 (1981).

30.   (a) J.M. Brown, P.A. Chaloner and G.A. Morris, *J. Chem. Soc. Perkin, Trans.II*, 1583 (1987); (b) J. Halpern and C.R. Landis, *J.Amer.Chem.Soc.*, **109**, 1746 (1987).

31.   J.M. Brown and P.L. Evans, *Tetrahedron*, **44**, 4905 (1988).

32.   (a) J. Halpern, *Adv. in Catalysis*, **11**, 301 p.73 (1959); (b) J. Halpern and R.S. Nyholm, *Proc.3rd.Int.Cong.Catal.*, (W.M.H. Sachter, G.C.A. Schuitt and P. Zwietering, eds.) Vol.I, North Holland Pub. Co., Amsterdam, P.25, 146 (1965); (c). M.F. Sloane, A.S. Matlack and D.S. Breslow, *J.Amer.Chem. Soc.*, **85**, 4014 (1963).

33.   (a). G. Consiglio and P. Pino, *Top. Curr.Chem.*, **105**, 77 (1982); (b). G. Consiglio and P. Pino, *Adv.Chem. Ser.*, **196**, 371 (1982).

34.   T. Hayashi, N. Kawamura and Y. Ito, *J.Amer.Chem. Soc.*, **109**, 7876 (1987).

35.   J.M. Brown and A.P. James, *J.Chem. Soc.Chem.Commun.*, 181 (1987).

36.   A.S.C. Chua, J.J. Pluth and J. Halpern, *J.Amer.Chem.Soc.*, **102**, 5952 (1980).

37.   (a) A.S.C. Chan, J.J. Pluth and J. Halpern, *Inorg. Chim. Acta*, **37**, 1477 (1979); (b) N.W. Alcock, J.M. Brown, A.E. Derome and A.R. Lucy, *J.Chem.Soc.Chem.Commun.*, 575 (1985); (c) N.W. Alcock, J.M. Brown and P.J. Maddox, *ibid*, 1532 (1986).

38.   R. Selke and H. Pracejus, *J. Mol. Catal.*, **38**, 213 (1986).

39.   T. Ikariya, Y. Ishii, H. Kawano, T. Arai, M. Saburi, S. Yoshikawa, and S. Akutagawa, *J.Chem.Soc.Chem.Commun.*, 922 (1985).

40.   H. Takahashi and K. Achiwa, *Chemistry Lett.*, 1921 (1987).

41.   H. Takaya, T. Ohta, N.Sayo, H. Kumobayashi, S. Akutagawa, S. Inoue, I. Kasahara and R. Noyori, *J. Amer. Chem. Soc.*, **109**, 1596 (1987).

42.   T. Ohta, H. Takaya, M. Kitamura, K. Nagai and R. Noyori, *J.Org.Chem.*, **52**, 3174 (1987).

43.   H. Kawano, Y. Ishii, T. Ikariya, M. Saburi, S. Yoshikawa, Y.Uchida and H. Kumobayashi, *Tetrahedron Lett.*, **28**, 1905 (1987).

44.   (a) W.C. Christopfel, and B.D. Vineyard, *J.Amer.Chem.Soc.*, **101**, 4406 (1979); (b) W.S. Knowles, *Acc.Chem.Res.*, **16**, 106 (1983).

45.   M.Kitamura, T. Ohkuma, S. Inoue, N. Sayo, H. Kumobayashi, S. Akutagawa, T. Ohta, H. Takaya and R. Noyori, *J. Amer.Chem.Soc.*, **110**, 629 (1988).

46.   R. Noyori, M. Ohta, Y. Hsiao and M. Kitamura, *J.Amer.Chem.Soc.*, **108**, 7117 (1986).

47.   M. Kitamura, Y. Hsiao and R. Noyori, *Tetrahedron Lett.*, **28**, 4829 (1987).

48.   E. Cesarotti, R. Ugo and H.B. Kagan, *Angew.Chem.*, **18**, 779 (1979).

49.   E. Cesarotti, H.B. Kagan, R. Goddard and C. Krüger, *J.Organomet.Chem.*, **162**, 297 (1978).

50.   R. Stern and L.Sajas, *Tetrahedron Lett.*, 6313 (1968).

51.   (a) Y. Ohgo, S. Takeuchi and J.Yoshimura, *Bull.Chem.Soc. Jpn.*, **43**, 505 (1970); (b) Y. Ohgo, S. Takeuchi and J.Yoshimura, *ibid*, **44**, 283 (1971).

52.   Y. Ohgo, K. Kobayashi, S. Takeuchi and Y. Yoshimura, *ibid*, **45**, 933 (1972).

53.   S. Takeuchi and Y. Ohgo, *ibid*, **54**, 2136 (1981).

54.   A. Fischli and D. Süss, *Helv.Chim.Acta*, **62**, 2361 (1979).

55.   A. Fischli and J.J. Daly, *ibid*, **63**, 1628 (1980).

56.   A. Fischli and P.M. Müller, *Helv.Chim. Acta*, **63**, 529 (1980)

57.   A. Fischli and P.M. Müller, *Helv.Chim. Acta*, **63**, 1619 (1980).

58.   H. Brunner, J. Wachter, J. Schmidbauer, G.M. Sheldrick and P.G. Jones, *Angew.Chem.*, **25**, 371 (1986).

59.   B.D. Zwick, A.M. Arif, A.T. Patton and J.A. Gladysz, *ibid*, **26**, 910 (1987).

60.   I. Ojima and T. Kogure, *Rev. Silicon, Germanium, Tin Lead Compds.*, **5**, 7 (1980).

61.   I. Ojima, K. Yamamoto and M. Kumada, in : *"Aspects of Homogeneous Catalysis"*, (R. Ugo, ed.), Vol.3, PP 185, D. Reidel, Dordrecht (1977).

62.   (a) K. Yamamoto, T. Hayashi, M. Zembayashi and M. Kumada, *J.Organomet.Chem.*,**118**, 161 (1976); (b) K. Yamamoto, T Hayashi and M. Kumada, *J.Amer.Chem.Soc.*, **93**, 5301 (1971).

63.   (a). K. Yamamoto, T. Hayashi, Y. Uramoto, R. Ito and M. Kumada, *J.Organomot.Chem.*, **118**, 331 (1976); (b). K. Yamamoto, Y. Uramoto and M. Kumada, *ibid*, **31**, C9 (1971).

64.   T. Hayashi, K. Tamao, Y. Katsuro, I. Nakae and M. Kumada, *Tetrahedron Lett.*, **21**, 1871 (1980).

65.   Y.Kiso, K. Yamamoto, K. Tamao and M. Kumada, *J.Amer.Chem.Soc.*, **94**, 4373 (1972).

66.   L.H. Sommer, J.E. Lyons and H. Fujimoto, *J.Amer.Chem.Soc.*, **91**, 7051 (1969).

67.   A.J. Chalk and J.F. Harrod, *J.Amer.Chem.Soc.*, **87**, 16 (1965).

68.   For a review, see: G. Paiaro, *Organomet.Chem.Rev.Sect.A*, **6**, 319 (1970).

69.   I. Ojima, T.Kogure and Y. Nagai, *Tetrahedron Lett.*,2475 (1973).

70.   M. Fieser and L.F. Fieser, in : *"Reagents for Organic Synthesis"* Vol.5, pp.739, Wiley, New York (1975).

71.   I. Ojima and T. Kogure, *Tetrahedron Lett.*, 4865 (1973).

72.   H.B. Kagan, N. Langlois and T.-P. Dang, *J.Organomet.Chem.*, **90**, 353 (1975).

73.   G. Zassinovich, C. Del Bianco and G. Mestroni, *J.Organomet.Chem.*, **197**, 85 (1980).

74.   K. Osakada, M. Obona, T. Ihariya, M. Saburi and S. Yushikawa, *Tetrahedron Lett.*, 4297 (1981).

75.   P. Bonvicini, A. Levi, G Modena and G. Scorrano, *J.Chem.Soc.Chem.Commun.*, 1188 (1972).

76.   (a) I. Ojima and T. Kogure, *J.Chem.Soc.Chem.Commun.*, 428 (1977); (b) R.R. Shrock and J.A. Osborn, *ibid*, 567 (1970).

77.   T. Hayashi, K. Kanehira and M. Kumada, *Tetrahedron Lett.*, 1417 (1981).

78.   J. Solodar, *Chem. Tech.*, 421 (1975).

79.   D. Wang and T.H. Chau, *Tetrahedron Lett.*, **24** (15), 1573 (1983).

80.   I. Ojima, T. Kogure, T. Terasaki and K. Achiwa, *J.Org.Chem.*, **43**, 3444 (1978).

81.   (a) M. Parko, W.O. Nelson and W.A. Wood, *J. Biol. Chem.* **207**, 51 (1954); (b) G.M. Brown and J.J. Reynolds, *Ann. Rev. Biochem.*, **32**, 419 (1963).

82.   T. Tanis, in : *"Organometallics"* (H. Nocaki, H. Yamamoto, J. Tsuji and R. Noyori, eds.), *Kagakuzokan* **105**, 121 Kagakudojin (Japan) (1985).

83.   K. Tani, E. Tanigawa, Y. Tatsuro and S. Otsuka, *Chemistry Lett.*, 737 (1986).

84.   (a) S. Masamune, W. Choy, J.S. Petersen and R.L.R. Sita, *Angew.Chem.*, **24**,1 (1985); (b) K. Yamamoto and S.U. Rahman, *Chemistry Lett.*, 1603 (1984).

85.   (a) J.F. Deuble, C.McGettigan and J.M. Stryker, *Tetrahedron Lett.*, **31**, 2397 (1990);   (b) W.S. Mahoney and J.M. Stryker, *J. Amer. Chem. Soc.*, **111**, 8818 (1989);   (c) T. M. Koenig, J.F. Deuble, D.M. Brestensky and J.M. Stryker, *Tetrahedron Lett.*, **30**, 5677 (1989). (d) W.S. Mahoney, D.M. Brestensky and J. M. Stryker, *J. Amer. Chem. Soc.*, **110**, 291 (1988);   (e). K. Yamamoto, T. Hayashi and M. Kumada, *J.Organomet.Chem.*, **46**, C65 (1972); (f) T. Hayashi, K. Yamamoto and M. Kumada, *J. Organomet. Chem.*, **112**, 253 (1976).

86.   (a). T. Hayashi, K. Yamamoto, K. Kasuga, H. Omizu and M. Kumada, *J.Organomet.Chem.*, **113**, 127 (1976).

87.   K. Yamamoto, T. Hayashi and M. Kumada, *ibid*, **54**, C45 (1973).

88.   I. Ojima, T. Kogure; M. Kumagai, S. Horiuchi and T. Sato, *ibid*, **122**, 83 (1976).

89.   (a). H.Brunner and G. Riepl, *Angew.Chem.*, **21**, 377 (1982).

90.   A. Kinting, H.J. Kreuzfeld and H.P. Abicht, *J.Organomet.Chem.*, **370**, 343 (1989).

91.   W. Dumont, J.-C. Poulin, T. P. Dang and H.B. Kagan, *J.Amer.Chem.Soc.*, **95**, 8295 (1973).

92.   (a) R.J.P. Corriu and J.J.E. Moreau, *J.Organomet.Chem.*, **85**, 19 (1975). (b) R.J.P. Corriu and J.J.E. Moreau, *ibid*, **64**, C51 (1974).

93.   I. Kolb, M. Cerny and J. Hetflejs, *React. Kinet. Catal.Lett.*, **7**, 199 (1977).

94.   T. Hayashi, T. Mise, M. Fukushima, M. Kagotani, N. Nagashima, Y. Hamada, A. Matsumoto, S. Kawakami, M. Konishi, K. Yamamoto and M. Kumada, *Bull.Chem.Soc.Jpn.* **53**, 1138 (1980).

95.   I. Ojima, T. Kogure and M. Kumagai, *J.Org.Chem.*, **42**, 1671 (1977).

96.   R. Glaser, *Tetrahedron Lett.*,2127 (1975).

97.   (a) G. Vavon and A. Antonini, *C.R.Acad.Sci.*, **230**, 1870 (1950); (b) G. Vavon and A. Antonini, *ibid*, **232**, 1120 (1951).

98.   I. Ojima, T. Tanaka and T. Kogure, *Chemistry Lett.*,823 (1981).

99.   Y. Amrani and D. Sinou, *J.Mol.Catal.*, **24**, 231 (1984).

100.   D. Sinou and Y. Amrani, *J.Mol.Catal.*, **36**, 319, (1986).

101.   T.P. Dang, J. Jenck and D. Morel, (Rhone-Poulenc), *Eur.Pat.Appl.Ep.* **133**, 127A1 (1985).

102.   F. Alario, Y. Amrani, Y. Colleuille, T.P. Dang, J. Jenck, D. Morel and D. Sinou, *J.Chem.Soc. Chem. Commun.*, 203 (1986).

103. R. Benhanza, Y. Amrani and D. Sinou, *J.Organomet.Chem.*, **288**, C37 (1985).

104. (a) T. Harada and Y. Izumi, *Chemistry Lett.*, 1195 (1978); (b) A. Tai, M. Nakahata, T. Harada and Y. Izumi, *ibid*, 1125 (1980).

105. (a) T. Osawa and T. Harada, *Bull.Chem.Soc. Jpn.*, **57**, 1518 (1984); (b) A.Tai, T. Harada, Y. Hiraki and S. Murakami, *ibid*, **56**, 1414 (1983); (c) S. Murakami, T. Harada and A. Tai, *ibid*, **53**, 1356 (1980).

106. (a) Y. Hiraki, K. Ito, T. Harada and A. Tai, *Chemistry Lett.*, **131** (1981). (b).K. Ito, T. Harada, A. Tai and Y. Izumi, *ibid*, 1049 (1979).

107. (a) A. Tai, K. Ito and T. Harada, *Bull.Chem.Soc.Jpn.*, **54**, 223 (1981). (b) A. Tai, H. Watanabe and T. Harada, *ibid*, **52**, 1468 (1979).

108. E.I. Klabunovskii, *Izv. Akad.Nauk SSSR, Ser.Khim.*, 505 (1984).

109. K. Ito, T. Harada and A. Tai., *Bull.Chem.Soc. Jpn.*, **53**, 3367 (1980).

110. M. Nakahata, M. Imaida, H. Ozaki, T. Harada and A. Tai, *Bull.Chem.Soc. Jpn.*, **55**, 2186 (1982).

111. M. Bartok, Gy. Wittman, Gy. Göndös and G.V. Smith, *J.Org.Chem.*, **52**, 1139 (1987).

112. A. Hatta and W. Suetaka, *Bull.Chem.Soc. Jpn.*, **48**, 2428 (1975).

113. A. Kumobatsu and S. Komatsu, *"Raney Shokubai"*, Kyoritsushuppan Co., Tokoyo (1971).

114. H.W. Krause, *React.Kinet.Catal.Lett.*, **10**, 243 (1979).

115. (a) K. Ohkubo, H. Fujimori and K. Yoshinaga, *Inorg.Nucl.Chem.Lett.*, **15**, 231 (1979); (b) K. Ohkubo, M. Haga , K. Yoshinaga and Y. Motazato, *ibid*, **17**, 215 (1981).

116. N. Takaishi, H. Imai, C.A. Bertelo and J.K. Stille, *J.Amer.Chem. Soc.*, **100**, 264, 268 (1978).

117. K. Achiwa, *Heterocycles*, **9**, 1539 (1978).

118. K. Achiwa, *Chemistry Lett.*, 905 (1978).

119. U. Nagel, H. Menzel, P.W. Lednor, W. Beck, A. Guyot and M. Barholin, *Nature*, **36**, 578 (1981).

120. G.L. Baker, S.J. Fritschel, J.R. Stille and J.K. Stille, *J.Org.Chem.*, **46**, 2954 (1981).

121. (a) A. Pedrazzoli, *Helv.Chem.Acta*, **40**, 80 (1957); (b) A. Pedrazzoli, *Chimica*, **10**, 260 (1956).

122. M. Takasaki and K. Harada, *Chemistry Lett.* , 1745 (1984).

123. (a) K.Harada and M. Takasaki, *Bull.Chem.Soc. Jpn.*, **57**, 1427 (1984). (b) M. Tamura and K. Harada, *ibid*, **53**, 561 (1980).

124. S. Yamada, T. Shioiri and T. Fuji, *Chem.Pharm.Bull.*, **10**, 688 (1962).

125. J.C. Sheehan and R.E. Chandler, *J.Amer.Chem.Soc.*, **83**, 4795 (1961).

126. J.P. Vigneron, H.B. Kagan and H. Horeau, *Tetrahedron Lett.*, 5681 (1968).

127. J.S. Davies, M.C. Eaton and M.N. Ibrahim, *J.Heterocyc.Chem.*, **17**, 1813 (1980).

128. M. Nakayama, G. Maeda, T. Kancko and H. Katsura, *Bull.Chem.Soc.Jpn.*, **44**, 1150 (1971).

129. D. Valentine, Jr. and J.W. Scott, *Synthesis*, 329 (1978).

130. V. Prelog, *Helv. Chim. Acta,* **36**, 308 (1953).

131. V. Prelog, *Bull.Soc.Chim.France,* 987 (1956).

132. E.E. Eliel, in : "*Asymmetric Synthesis*" (J.D. Morrison, ed.), Vol.2, PP.125-155. Academic Press New York (1983).

133. T. Ueno, T. Nakashima, Y. Hayashi and H. Fukami, *Agric.Biol.Chem.*, **39**, 1115 (1975).

134. N. Izumiya, S. Lee, T. Kanmera and H. Aoyagi, *J.Amer.Chem.Soc.*, **99**, 8346 (1977).

135. W.L. Meyer, L.F. Kuyper, R.B. Lewis, G.E. Templeton and S.H. Woodhead, *Biochem.Biophys.Res.Commun.*, **56**, 234 (1974).

136. H. Poisel and U. Schmidt, *Chem.Ber.*, **106**, 3408 (1973).

137. B.W. Bycroft and G.R. Lee., *J.Chem.Soc.Chem.Commun.*, 988 (1975).

138. E.J. Corey, H.S. Sachdev, J.Z. Gougoutas and W. Saenger, *J.Amer.Chem.Soc.*, **92**, 2488 (1970).

139. S. Akabori, T. Ikenaka and K. Matsumoto, *Nippon Kagaku Zasshi*, **73**, 112 (1952).

140. S. Mitsui, T. Kamaishi, T. Imaizumi and I. Takamura, *Nippon Kagaku Zasshi*, **83**, 1115 (1962).

141. T. Kamaishi and S. Mitsui, *ibid,* **86**, 623 (1965).

142. M. Kawana and S. Emoto, *Bull.Chem.Soc. Jpn.*, **41**, 259 (1968).

143. A. Mckenzie and E.W. Christie, *Biochem.Z.*, **177**, 426 (1935).

144. M.M. Jamison and E.E. Turner, *J.Chem.Soc.*, 538 (1941).

145. E.E. Turner and M.M. Harris, *Quart.Revs.* (London). **1**, 299 (1974).

146. S. Mitsui and M. Kanai, *Nippon Kagaku Zasshi*, **87**, 179 (1966).

147. K. Harada, T. Munegumi and S. Nomoto, *Tetrahedron Lett.*, **22**, 111 (1981).

148. S. Mitsui and A. Kanei, *Nippon Kagaku Zasshi*, **86**, 627 (1965).

149. K. Harada and T. Munegumi, *Bull.Chem.Soc.Jpn.*, **56**, 2774 (1983).

150. K. Harada and T. Munegumi, *ibid*, **57**, 3203 (1984).

151. K. Harada and K. Matsumoto, *J.Org.Chem.*, **33**, 4467 (1968).

152. (a) K. Harada and K. Katsumoto, *J.Org.Chem.*, **32**, 1794 (1967); (b) K. Harada, *ibid*, **32**, 1790 (1967).

153. A. Kanai and S. Mitsui, *Nippon Kagaku Zasshi*, **87**, 183 (1966).

154. R.G. Hiskey and R.C. Northrop, *J.Amer.Chem.Soc.*, **83**, 4798 (1961).

155. R.G. Hiskey and R.C. Northrop, *ibid*, **87**, 1753 (1965).

156. S. Yamada and S. Hashimoto, *Tetrahedron Lett.*, 997 (1976).

157. S. Yamada, N. Ikota and K. Achiwa, *Tetrahedron Lett.*, 1001 (1976).

158. K. Harada and T. Yoshida, *J.Org.Chem.*, **37**, 4366 (1972).

159. I. Miyazawa, K. Takashima, T. Yamada, S. Kuwata and H. Watanabe, *Bull.Chem.Soc.Jpn.*, **55**, 341 (1982).

160. K. Harada and S. Shiono, *ibid.*, **57**, 1367 (1984).

161. K. Kuzuhara, T. Komatsu and E. Emoto, *Tetrahedron Lett.*, 3563 (1978).

162. Y. Tachibana, M. Ando and H. Kuzuhara, *Chemistry Lett.*, 1769 (1982).

163. Y. Tachibana, M. Ando and H. Kuzuhara, *Bull.Chem.Soc.Jpn.*, **56**, 3652 (1983).

164. K. Bernauer, R. Deschenaux and T. Taura, *Helv.Chim.Acta*, **66**, 2049 (1983).

165. K. Matsumoto and K. Harada, *J.Org.Chem.*, **31**, 1956 (1966).

166. K. Harada and T. Yoshida, *Bull.Chem.Soc. Jpn.*, **43**, 921 (1970).

167. K. Harada and S. Shiono, *ibid*, **57**, 1040 (1984).

168. K. Harada and T. Yoshida, *J.Chem.Soc.Chem.Commun.*, 1071 (1970).

169. S. Kiyooka, K. Takeshima, H. Yamamoto and K. Suzuki, *Bull.Chem.Soc. Jpn.*, **49**, 1897 (1976).

170. K. Harada and Y. Kataoka, *Tetrahedron Lett.*, 2103 (1978).

171.   W.H. Pirkle and J.R. Hanske, *J.Org.Chem.*, **42**, 2436 (1977).

172.   K. Harada, T. Iwasaki and T. Okawara, *Bull.Chem.Soc. Jpn.*, **46**, 1901 (1973).

173.   D.E. Nichols, C.F. Barfknecht and D.B. Rusterholz, *J.Med.Chem.*, **16**, 480 (1973).

174.   K. Harada and M. Tamura, *Bull.Chem.Soc. Jpn.*, **52**, 1227 (1979).

3 Stereoselective Non-Catalytic Reductions

The vast arena of asymmetric synthesis encompassing stereoselective transition metal-catalysed reductions and reductions involving biocatalytic systems is supplemented by stereoselective *non-catalytic* reductions [1-3]. These are generally accomplished by two methods: (i) by the use of metal hydrides having chiral ligands, and (ii) by transfer hydrogenation. High optical yields are obtained by both methods, and although the mechanistic interpretation of these reactions is complex, the simulation of enzymatic reactions in some cases, such as in NADH model reactions, has greatly helped in understanding enzymatic reaction mechanisms. Furthermore such studies have led to the development of novel reducing agents in organic synthesis.

3.1 Enantioselective Non-Catalytic Reductions

3.1.1 Chiral Metal-hydride Complexes

Various metals such as Li, Al, B, Mg, Ni, Ir, Ru etc. have been used as complexes with a number of chiral groups to effect stereoselective reductions. The basic requirement in such reductions is the attainment of high enantioselectivity which is partly dependent on the ability of the reducing reagent to resist oligomerization and redistribution of the ligands. It is also often desirable that the reducing reagent should be capable of delivering one hydride per molecule of the substrate so that mechanistic interpretation of the results is facilitated.

3.1.1.1 Lithium Aluminium Hydride Modified with Chiral Groups

When ketones are reduced with lithium aluminium hydride (LiAlH$_4$, LAH), the species Al(OR)$_m$H$_n$ are assumed to be generated, where OR represents the alkoxyl group derived from the ketone and "m", and "n" are small integers. This assumption was the basis of studies done on the asymmetric reduction of unsymmetric ketones using LAH modified with *optically active* ketones such as *d*-camphor [4], which served as a landmark in asymmetric synthesis.

116   3. Stereoselective Non-Catalytic Reductions

Thus 2-butanone (1) was claimed to give optically active 2- butanol (2) on reduction with LAH modified with *d*-camphor [4].

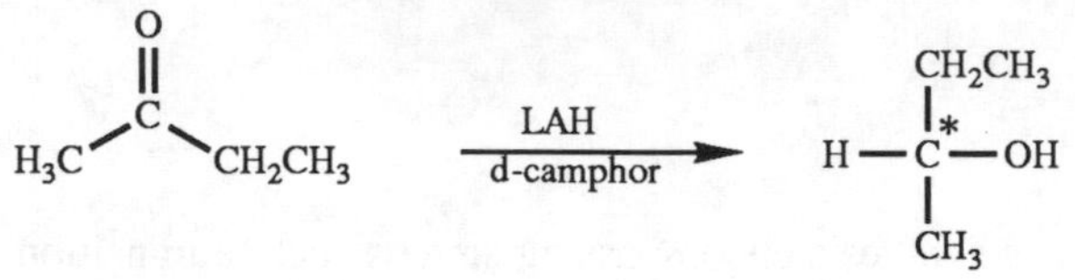

O
‖
H_3C—C—CH_2CH_3 $\xrightarrow[\text{d-camphor}]{\text{LAH}}$ $H-\underset{CH_3}{\overset{CH_2CH_3}{\underset{|}{\overset{|}{C}}}}-OH$

(1)   (2) $[\alpha]_D^{25} = +2.50$

Although these results were later challenged on the ground that they were not reproducible [5,6]*, an idea had been born. Since then tremendous progress has been made to achieve high enantioselectivities by conversion of LAH to chiral reducing agents by exchanging its hydrogens with chiral ligands and then employing such chiral modified LAH in stereoselective reductions [7,8].

$2LiAlH_3OR^*$ $\rightleftharpoons$ $LiAlH_2(OR^*)_2 + LiAlH_4$

(3)

Figure 3.1. Disproportionation of chiral LAH - alkoxy reagents.

The various modifying reagents now used with LAH are chiral alcohols, amines and amino alcohols some of which are derived from chiral terpenes, carbohydrates etc.

3.1.1.1.1   *LAH Modified with Alcohols*

(-)-Menthol has been used as a modifier with LAH (usually in a ratio of 1:3) for the asymmetric reduction of the β- aminoketone (4) to the alcohol (5) in 77% e.e. [9].

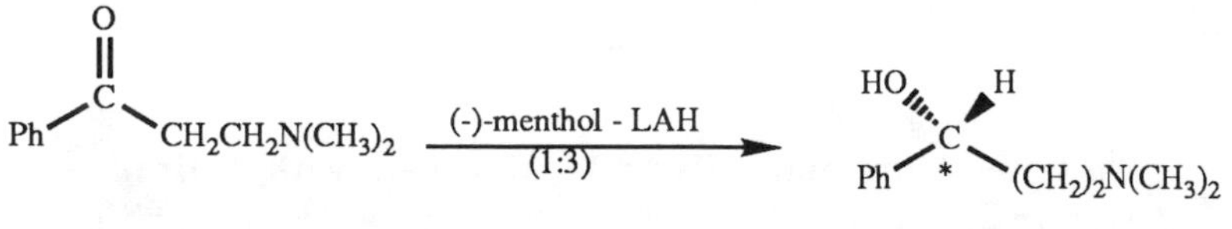

O
‖
Ph—C—$CH_2CH_2N(CH_3)_2$ $\xrightarrow[(1:3)]{\text{(-)-menthol - LAH}}$ $\underset{Ph}{\overset{HO}{\underset{*}{\overset{\diagdown}{C}}}}\overset{H}{\diagup}(CH_2)_2N(CH_3)_2$

(4)   (5) 77% e.e. (*S*)

The same system reduced α-morpholinoacetophenone (6) to give the reduced product (7) in 58.7% e.e.

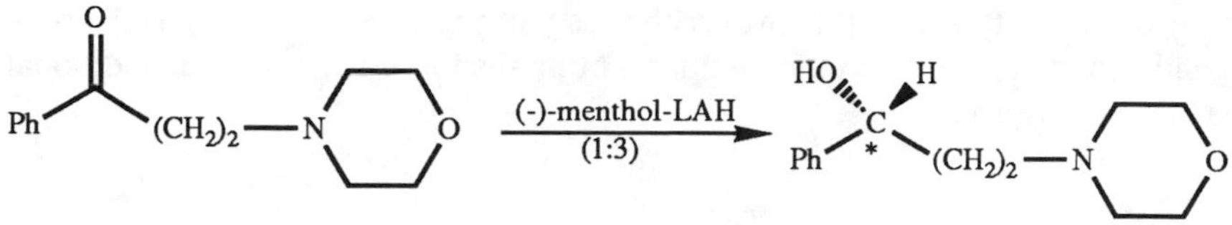

(6)   (7) 58.7% e.e. (S)

Diols and triols have been used as modifiers mainly because of their stability towards disproportionation [10-21]. For example, cis-pinanediol (8) has been employed as a modifier with LAH in the asymmetric reduction of diphenylmethylalkylketones such as (9) to yield the optically active secondary alcohol (10) in 24.85 % e.e. [18].

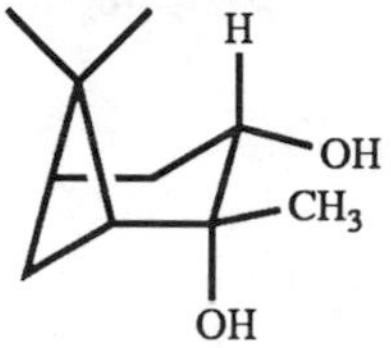

(8)

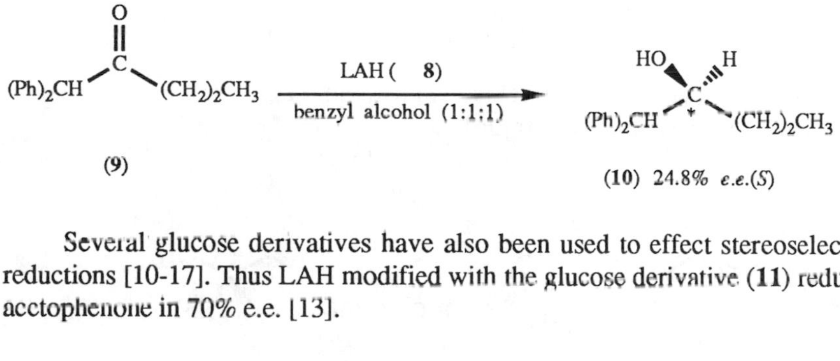

(9)

(10) 24.8% e.e.(S)

Several glucose derivatives have also been used to effect stereoselective reductions [10-17]. Thus LAH modified with the glucose derivative (11) reduced acetophenone in 70% e.e. [13].

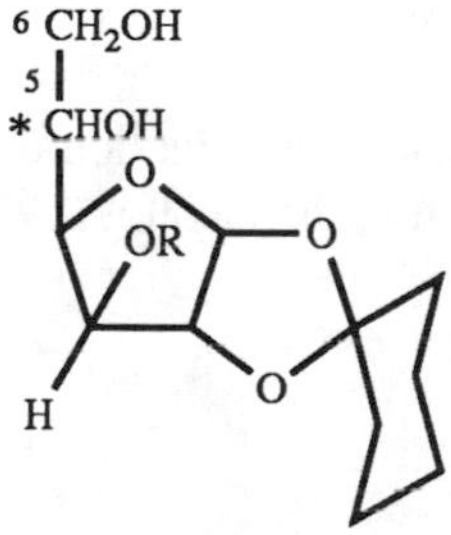

(11) (R = CH₂Ph)

Other diols, such as the binaphthol (12), surpass the optical yields achieved with the monosaccharides or other modifiers affording almost 100%

enantioselectivity [22-25]. However the reagent prepared from (12) as the only ligand is non-selective and incorporation of methanol or ethanol as an additional ligand is necessary.

(12)   (13)

Thus n-propiophenone (14) is reduced to the alcohol (15) using the complex (13) in 100% e.e.

(14)   (15) 100% e.e.(S)

Similarly, β-ionone (16) yielded (S)- β-ionol (17) in nearly 100% e.e. in the presence of the complex (13) [26].

(16)   (17) 100% e.e. (S)

The complex (13) has also been used in the enantioselective reduction of acetylenic ketones. Thus the prochiral ketone (18) is reduced to the alcohol (19) with the complex (13) in 90% e.e.

(18)   (19) 90% e.e. (S)

The difficulty in obtaining a high degree of stereoselectivity with modified $LiAlH_4$ reagents is mainly due to the disproportionation of the reagent into many reactive species which are placed in different chemical and chiral environments. Hence an overall average selectivity is obtained depending upon the concentration, reactivity, and chiral recognition ability of each species. It is of prime importance to minimize the number of reactive species and find ways to create a single highly reactive hydride possessing excellent chiral recognition ability to achieve a high degree of enantioselection.

Studies made in this regard on the reactions of ω-substituted alkylphenyl ketones, $PhCO(CH_2)_nY$ [Y=NR_2, OMe, SMe and alkyl, n = 1 - 4 = (number of methylene groups)] with $LiAlH(O\text{-menthyl})_3$, led to the following generalisations: (a) when Y is H or alkyl, hydride attacks preferentially from the si- face of the C=O group; (b) the stereoselectivity depends on the nature of the substituents on the ketone carbonyl (e.g. higher asymmetric induction is attained with $(CH_2)_n OMe$ or $(CH_2)_n NR_2$ than with $(CH_2)_n SMe$; (c) the stereoselectivity depends on the number (n) of CH_2 groups [27].

These results were explained by assuming that the reduction proceeded via a cyclic transition state such as (20) in which coordination of the carbonyl and Y groups occurs with the lithium cation [28]. In such a transition state the number of the methylene groups (n) determines the size of the chelate ring and thus affects the rigidity of the transition state, thereby affecting the stereoselectivity.

(20)

In the asymmetric reduction of prochiral ketones with the cis-pinanediol (8) - modified LAH - benzyl alcohol (1:1:1) system, the steric hindrance of the diphenylmethane group probably results in a re-attack of the hydride ion on the substrate (21) to give preferentially the product with the (S)-configuration.

re-attack   (21)   (S)-(22)

Studies on the asymmetric reduction of ketones with LAH-monosaccharide complexes have revealed that a 1:1 LAH-sugar complex affords maximum selectivities. This has been attributed to the formation of a cyclic complex such as (23) between LAH and the hydroxyl groups on C(5) and C(6) of the monosaccharide*.

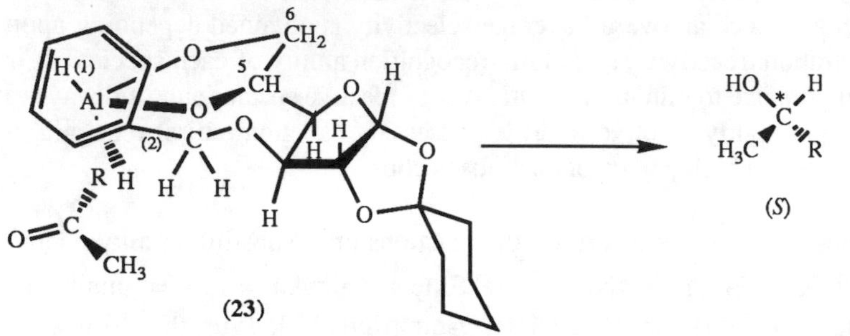

(23)

(S)

The exceptionally high stereoselectivity achieved with the binaphthol (12)-modified LAH has been proposed to proceed *via* the pathway shown in Figure 3.2.

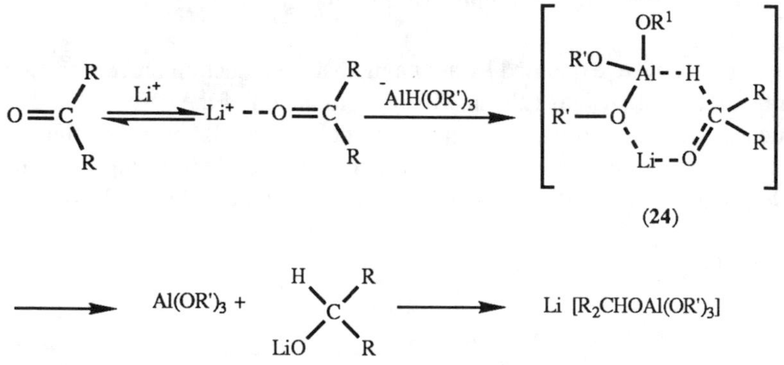

(24)

Figure 3.2. Reduction of ketones with LiAlH-(OR')$_3$ - type reagents.

The reaction is initiated by the complexation of the Lewis acidic Li$^+$ ion to the oxygen atom, which activates the carbonyl group. Hydride transfer than occurs from Al to the carbonyl carbon by way of a quasi-aromatic, six-membered ring transition state (24).

---

*As is evident from the structure of the complex (23), H(1) is more shielded by the benzyl group than the H(2). It is assumed that in the lowest energy transition state, the ketone orients itself so that the carbonyl oxygen points away from the oxygen of the aluminium complex. With this orientation of the ketone, it is predicted that alkyl-alkenyl, and phenyl-methyl methanols of predominantly S-configuration should be formed. This was indeed found to be the case [12].

3.1.1.1.2   LAH Modified with Amino Alcohols

Various amino alcohols, such as substituted ephedrines, chiral oxazolines, alkaloids etc. have been used as LAH modifiers in the asymmetric reduction of various ketones. Table-3.1 shows some amino alcohols commonly used as LAH modifiers.

Table 3.1. Amino Alcohols used as LAH - Modifiers

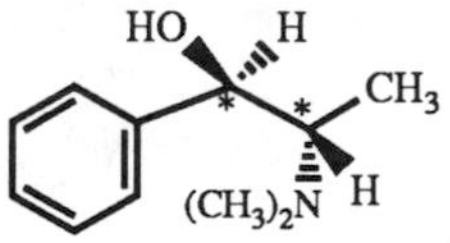

(25) 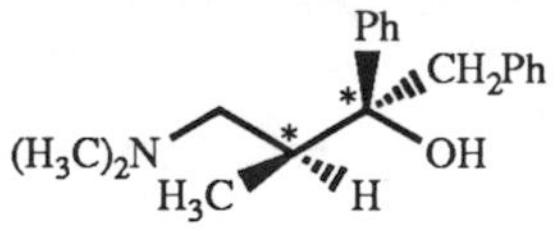   **(26)**

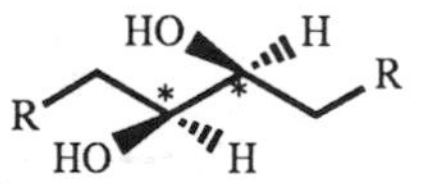

(27) R = (CH$_3$)$_2$N-,   (31) R = N(CH$_3$) Ph
(28) R = piperidino-   (30) R = OPh,
(29) R = OCH$_3$,

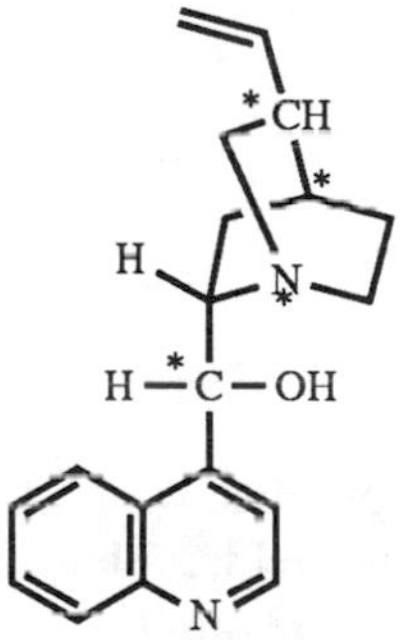

(32)

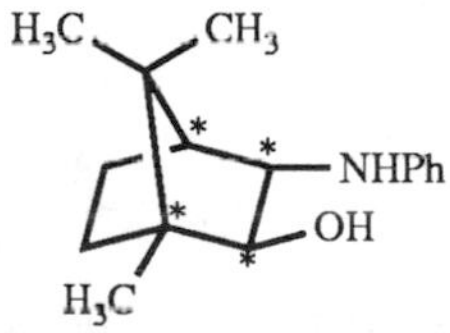

(33)

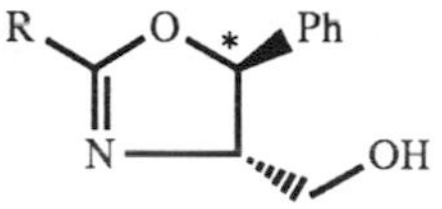

(34) R = CH$_3$ (36) R = ^{i}Pr
(35) R = C$_2$H$_5$ (37) R = PhCH$_2$

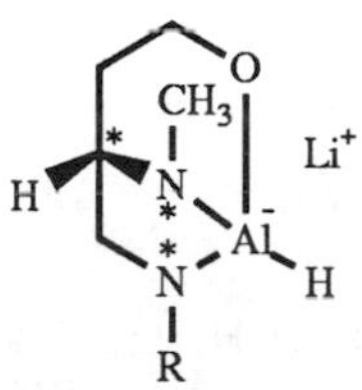

(38) R = Ph
(39) R = 2,6-dimethylphenyl   **(40)**

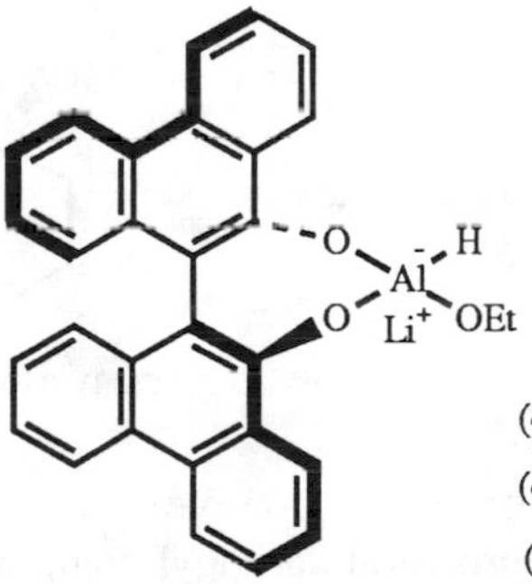

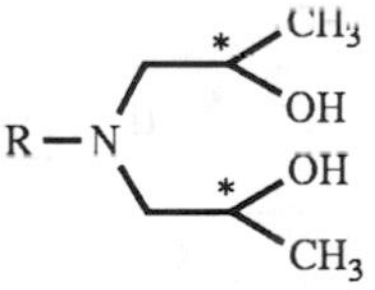

(41) R = ^{n}Bu
(42) R = $\overset{*}{C}$HCH$_3$Ph (R)
(43) R = $\overset{*}{C}$HCH$_3$ Ph (S)

These chiral ligands are used in combination with other achiral additives such as alcohols etc. as LAH modifiers in the asymmetric reduction of various substrates.

Thus N-methylephedrine (25), when employed as an LAH modifier as shown in (44) along with orcinol (3,5-dimethylphenol) in a ratio of 1:1:2 reduced α-acetylenic ketones such as (45) to (46) in 90% e.e. [29].

(45)   (46) 90% e.e. (R)

(44)

A variation of this method using (25) and 2-ethylaminopyridine with LAH (1:2:1) reduced prochiral cyclic ketones in high optical yields. Thus 2-cyclohexene-1-one (47) was reduced to 2-cyclohexene-1-ol (48) in 98% e.e. using [(25) - LAH - 2- ethylaminopyridine] (1:1:2)] as the reducing system [30].

(47)   (48) 98% e.e. (R)

The diol (27), which forms the complex (49) with LAH, has been employed in the reduction of prochiral carbonyl compounds giving upto 75% optical yields [31,32]. Thus 2,4,6-trimethylacetophenone (50) is reduced to 2,4,6- trimethylphenyl-1-ethanol (51) by the complex (49) in 75% e.e. Chiral oxazolines such as (35) have been used as modifiers in the LAH reduction of several ketones affording upto 65% e.e. of the reduced product [33].

$$(H_3C)_2N - CH_2 - \overset{*}{C}H - \overset{*}{C}H - CH_2 - N(CH_3)_2$$

(49)

(50)   (51) 75% e.e.(S)

A comparison of the reducing ability of chiral oxazolines with other asymmetric reducing agents is shown in Table 3.2.

Table 3.2.   Asymmetric Reducing Agents in Comparison with Chiral   Oxazolines

Ketone	Enantioselectivities with various reducing agents					
	(A)	(B)	(C)	(D)	(E)	(F)*
PhCOCH₃	48 (R)	71 (R)	84 (S)	6 (S)	68 (R)	65 (R)
PhCOC₂H₅	-	46 (R)	-	13 (S)	-	62 (R)
PhCOⁱPr	-	-	-	44 (S)	30 (R)	43 (R)
α-Tetralone	-	-	-	0	-	3.7 (S)
PhCH₂COCH₃	3 (R)	-	16 (S)		-	0.5 (S)
hexCOCH₃	6 (S)	25 (R)	-	-	-	4 (S)

(A) = LAH - alk [34]	(B) = LAH - monosaccharide [3,12]	(C) = LAH - amine [35]
(D) = R₃Al [36]	(E) = LAH - darvon alcohol [37]	(F) = LAH - (35)

Darvon alcohol (26) has also been used as an LAH modifier, particularly in the enantioselective reduction of α,β-acetylenic ketones, giving high optical yields of the reduced products [37-41]. It is interesting that there is a marked ageing effect with this reagent, i.e. the sense of stereoselectivity is dependent upon the length of time that the reagent has been allowed to stand before it is used for reduction. Thus when acetophenone is reduced with this reagent, either 30 sec. or 3 min. after its preparation, 68% e.e. of the reduced product is obtained having the (R)-configuration. However when the reagent is prepared and left overnight or when refluxed in ether for a few minutes, it reduced the same substrate to afford the product having predominantly (S) configuration in 66% e.e.

The chiral diaminoalcohol (52) has been used as an LAH modifier to prepare the complex (38) *in situ* which reduces ketones in high optical yields [42,43]. Thus the ketone (53) was reduced to the corresponding optically active alcohol (54) with the complex (38) in 77% e.e. [42].

(52)

(53)   (54) 77% e.e. (S)

The complex (38) has also been used in the asymmetric reduction of prochiral α,β-unsaturated ketones affording excellent optical yields [43]. Thus the asymmetric reduction of 2-cyclohexen-1-one (47) with the complex (38) afforded 2-cyclohexen-1-ol (48) with complete stereoselectivity [43].

(47)   (48) 100% e.e. (S)

Similarly the reduction of prochiral carbonyl compounds with the LAH modified complex (40) can be carried out in high optical yields [46]. The complex (40) prepared *in situ* reduced deuteriobenzaldehyde (55) to the corresponding alcohol (56) in 87% e.e.

(55)   (56) 87% e.e. (S)

Chiral amino diols of type (41-43) have also been used as LAH modifiers affording substantial optical yields of the reduced products [45]. For example, the aminodiol (43) when employed as the LAH modifier reduced propiophenone in upto 82% e.e.

Many attempts have been made to explain the observed stereoselectivities of amino alcohol-modified LAH reduction of prochiral carbonyl compounds, albeit with little success. The reversal in configuration of the reduced products with the age of the reagent further complicates any plausible mechanistic interpretation . This is because of the uncertainties involved in the conformation of the active species participating in the reduction process. The presence of the amino group in the reagent is postulated to be responsible for the enhancement in stereoselectivities due to a backside attack of the nitrogen on the aluminium, thereby assisting the transfer of hydride to the Li-coordinated substrate [31,32].

Table 3.3 shows some representative examples of the stereoselective reductions of prochiral carbonyl compounds with amino alcohol-modified LAH.

Table 3.3. Stereoselective Reduction of Prochiral Carbonyl Compounds with Amino Alcohol-modified Reagents

Substrate	Chiral Reagent	Product % e.e.	Reference
	(27)-LAH	 47 (S)	[31]
	(38)-LAH	 88 (S)	[42]

(Table 3.3. contd.)

Substrate	Chiral Reagent	Product % e.e.	Reference
![ketone] C(=O)C(CH₃)₃	(38)-LAH	HO, H C*-C(CH₃)₃ 86 (S)	[42]
cyclohexenone	(38)-LAH	H, OH 100 (S)	[42]
C(=O)CH₂-phenyl	(40)-LAH	HO, H C* benzyl 98 (S)	[44]
C(=O)C₂H₅	(40)-LAH	HO, H C*-C₂H₅ 98 (S)	[44]
indanone	[(25)-LAH- N-C₂H₅ pyridine]	HO, H 81 (R)	[30]
tetralone	(25)-LAH- N-C₂H₅ pyridine	HO, H 96 (R)	[30]
C(=O)CH₃	(35)-LAH	HO, H C*-CH₃ 65 (R)	[33]

3.1.1.2   Chiral Boranes and Borohydrides

The pioneering work carried out by H.C. Brown and co-workers on chiral boranes and borohydrides demonstrates the versatility of these reagents to prepare many types of chiral functionalities [1,46-48]. The dramatic success of newly developed organoborane reagents which reduce substrates with selectivities approaching that obtained with enzymes serves to illustrate their importance. Generally, the organoboranes used in asymmetric reductions are either boranes modified with naturally occurring chiral terpenes or they involve the borohydride anion $(BH_4)^-$ along with counter ions or chiral ligands.

3.1.1.2.1   Chiral Alkylboranes

Asymmetric hydroboration refers to "a stereochemically biased addition of a boron-hydrogen bond to a carbon-carbon double bond using chiral alkylboranes" and the process provides excellent organometallic intermediates for asymmetric organic synthesis [1].

Asymmetric hydroboration - oxidation provides a useful method for obtaining various classes of optically pure secondary alcohols [1]. Diisopinocamphenylborane (58) and monoisopinocamphenylborane (59), readily prepared from (+) - or (-)-α-pinene (57) are particularly useful as hydroborating agents but they have proved to be inefficient as chiral reducing agents [49-51]. Thus, the asymmetric reduction of the ketone (60) in the presence of reagent (58) or (59) yielded the reduced product (61) in 37% and 46% e.e. respectively [52, 53].

(57)   (58)   (59)

(60)   (61) (S)

with (58) : 36% e.e.;
with (59) : 46% e.e.

An interesting feature of these reductions is the ability of alkylboranes to reduce aliphatic ketones with ease, yielding higher optical yields than aromatic ketones. The rather low optical yields achieved with alkylboranes took a dramatic turn when 9-BBN (9-borabicyclo [3.3.1] nonane) combined with various terpenes were developed as the chiral reducing agents. For instance, Alpine-borane i.e. α-pinan-3-yl-9-borabicyclo- [3.3.1] nonane (62) reduced deuterioaldehydes to the corresponding α-deuterio- alcohols in high optical yields [52,53].

(62)

Thus deuteriobenzaldehyde (55) is reduced in the presence of the Alpine-borane (62) to the optically active α- deuteriobenzyl alcohol (56) in > 99% e.e. [52]. Various optically active terpenes have been used as alkyl groups to prepare B-alkyl-9-BBN reagents. Table 3.4, shows the preparation of these reagents from (+)-α-pinene (57), (-)-β-pinene (63), (-)-camphene (65) and (+)-3-carene (67).

(55)   (56) 87% e.e. (S)

Table 3.4. B-alkyl-9-BBN Reagents from Terpenes

Terpene	Hydroborating agent	B-alkyl-9BBN
(57)	9-BBN	(62)
(63)	9-BBN	(64)

(Table 3.4. contd.)

Terpene	Hydroborating agent	B-alkyl-9BBN

(65)   9-BBN   (66)

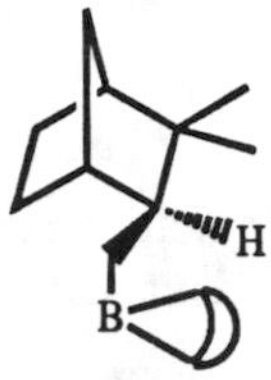

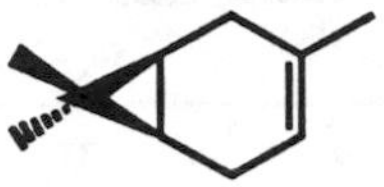

(67)   9-BBN   (68)

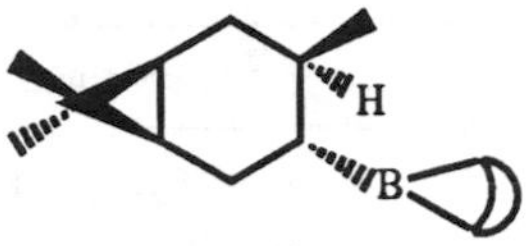

The asymmetric reduction of benzaldehyde-1-d with chiral β-alkyl-9-BBN reagents gave α- deuteriobenzyl alcohol in fair to excellent enantioselectivities, as shown in Table 3.5.

Table 3.5.   Reduction of Benzaldehyde-1-d with Chiral B-alkyl-9-BBN   Reagents

Reagent	%.e.e.	Configuration
(62)	100	(S)
(64)	47	(S)
(66)	75	(R)
(68)	61	(S)

The organoborane reagent (69) obtained by deuterioboration of α-pinene with 9-BBN-9-d quantitatively transfers deuterium to benzaldehyde. The asymmetric reduction of a variety of aldehydes can hence be achieved with incorporation of deuterium in the resulting alcohols without the necessity of prior preparation of 1-deuterated aldehydes.

(69)

Table 3.6 shows the preparation of chiral deuterated alcohols by reduction of aldehydes with the deuterated reagent.

Table 3.6.   Reduction of Aldehydes with (69)

Aldehyde	Product	% e.e.
$CH_3CH_2CH_2CHO$	$CH_3CH_2CH_2\overset{*}{C}HDOH$	83
$CH_3(CH_2)_4CHO$	$CH_3(CH_2)_4\overset{*}{C}HDOH$	64
$(CH_3)_3CCHO$	$(CH_3)_3C\overset{*}{C}HDOH$	70
$H_5C_6CH=CHCHO$	$C_6H_5CH=CH\overset{*}{C}HDOH$	60
$(CH_3)_2C=CH(CH_2)_2-\overset{CH_3}{\underset{\vert}{C}}=CHCHO$	$(CH_3)_2C=CH-(CH_2)_2-\overset{CH_3}{\underset{\vert}{C}}=CH\overset{*}{C}HDOH$	58
		70
		87
		86
		76

(Table 3.6. contd.)

Aldehyde	Product	% e.e.
![structure with CHO, H3CO]	![structure with CHDOH, H3CO]	70
![structure with CHO, (H3C)2N]	![structure with CHDOH, (H3C)2N]	61

Alpine-borane (62) also reduces α, β- acetylenic ketones affording high optical yields of the corresponding alcohols [54]. Thus 4-methyl-1-pentyn-3-one (70) when reduced with (62) yielded (71) in 99% e.e. [54].

$$HC\equiv C\overset{O}{\underset{}{\overset{\|}{C}}}CH(CH_3)_2 \xrightarrow{(62)} HC\equiv C\overset{HO\quad H}{\underset{}{\overset{}{\overset{*}{C}}}}CH(CH_3)_2$$

(70)   (71) 99% e.e. (S)

Similarly the α,β-acetylenic ketone (72) was reduced with Alpine-borane (68) stereoselectively to yield the acetylenic alcohol (73) in 100% e.e.

$$Ph\overset{O}{\underset{}{\overset{\|}{C}}}C\equiv C-COOC_2H_5 \xrightarrow{(68)} Ph\overset{HO\quad H}{\underset{}{\overset{}{\overset{*}{C}}}}C\equiv C-COOC_2H_5$$

(72)   (73) 100% e.e. (S)

A new reagent diisopinocampheylchloroborane (-)-Ipc₂ (74) has been developed which reduces aromatic ketones rapidly with excellent optical induction [55]. The reagent (74) is readily prepared from (+)-β- pinene (57) by hydroboration followed by treatment with dry HCl in diethyl ether.

$$2 \quad \text{(57)} \xrightarrow[0°C,THF]{BH_3, S(CH_3)_2} \text{(58)} \quad)_2BH \xrightarrow[\substack{0°C, diethyl \\ ether}]{HCl} \text{(74)} \quad)_2BCl$$

(57)   (58)   (74)

Prochiral α-tertiary ketones are also reduced by (74) affording high optical yields of the corresponding alcohols. Thus treatment of 3, 3-dimethyl-2-butanone (75) with (74) yielded the corresponding alcohol (76) in 95% e.e.

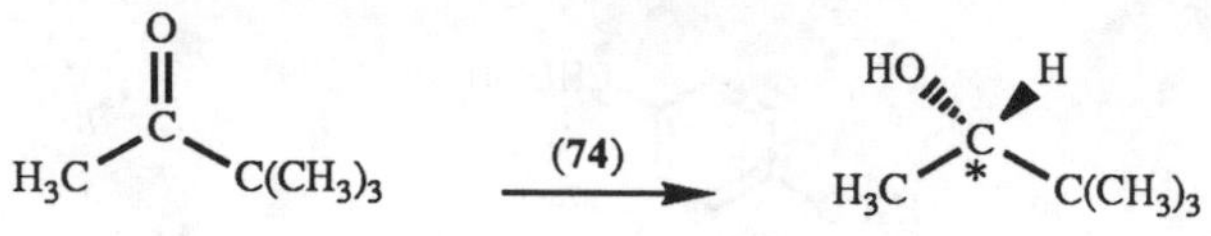

(75)

(76) 95% *e.e.*(S)

Similarly 2, 2-dimethyl acetoacetate (77) yielded the corresponding alcohol (78) on treatment with (74) in 84% e.e.

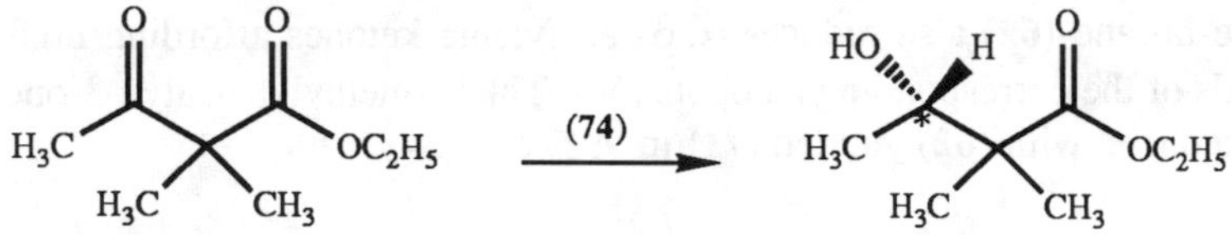

(77)

(78) 84% e.e. (S)

Reductions of hindered alicyclic derivatives with the reagent (74) proceed in high optical yields. Thus reduction of 2,2-dimethylcyclopentanone (79) yielded the corresponding alcohol (80) in 98% e.e.

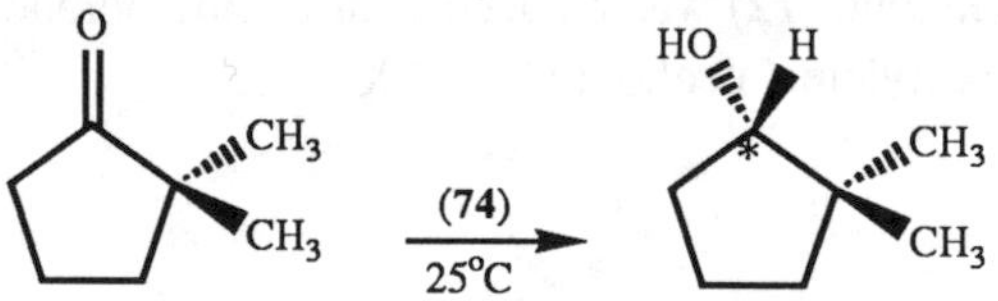

(79)

(80) 98% e.e. (S)

Two pathways have been proposed to explain the course of the reaction in the reduction by alkylborane reagents [55,56]. Pathway A involves a two step dehydroboration - reduction sequence (Fig. 3.3).

Pathway A:

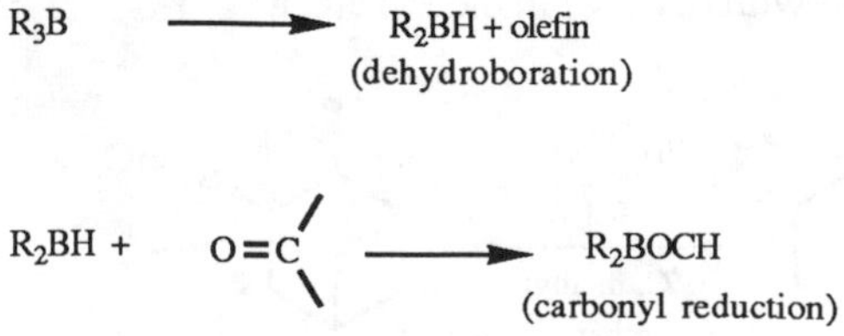

$R_3B \longrightarrow R_2BH + $ olefin
(dehydroboration)

$R_2BH + \quad O=C \longrightarrow R_2BOCH$
(carbonyl reduction)

Pathway B:

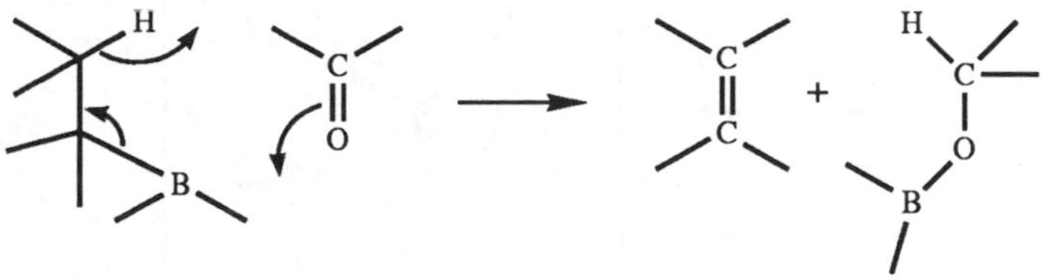

Figure 3.3. Carbonyl reduction via pathways A and B.

In the second pathway B, the reduction could proceed in a concerted manner by a cyclic mechanism in which the β-hydride of the organoborane is transferred to the carbonyl carbon, analogous to the Meerwein-Ponndorf-Verley type process. It seems more likely that both these pathways are operative depending upon the nature of the substrate. Thus with less hindered ketones, the cyclic mechanism is favoured while with hindered ketones, the mechanism switches to the dehydroboration - reduction pathway. In the case of reduction by deuterated Alpine-borane prepared from (+)-α-pinene, the aldehyde must approach the borane with the R group above the pinane ring and the aldehydic hydrogen above the methyl group of the α-pinene, as shown in Fig. 3.4.

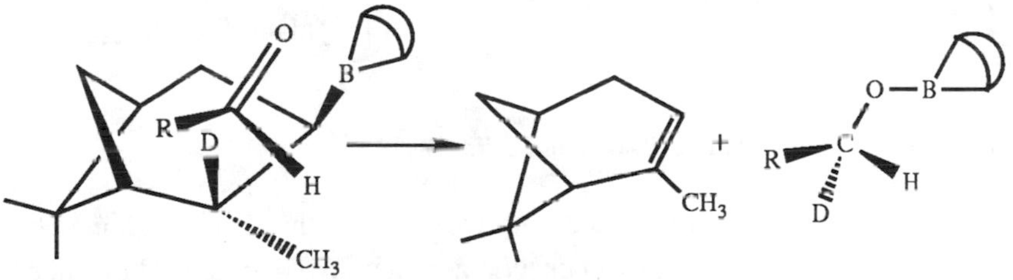

Figure 3.4. Reduction of ketones with Alpine-borane reagent.

Alternatively, the selectivity of the process may be explained on the basis of a transition state model in which it is assumed that the organoborane-carbonyl complex shown in Fig 3.5 which takes part in the reaction remains in an *anti*-configuration with the R group directed away from the pinanyl group [57,58]. Rotation of the complex occurs to form a boat cyclohexane - like transition state in which the bulky groups are away from each other.

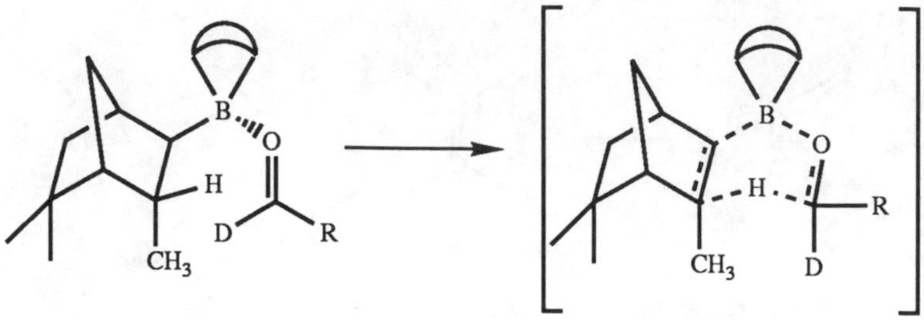

Figure 3.5. Transition state model of alkylboranes.

The boat-like transition state model shown in Fig. 3.5 explains the observation that a *syn*-planar B-C-C-H arrangement leads to a faster rate of reduction.

In the case of α, β-acetylenic ketones, it has been proposed that a bimolecular exchange mechanism is involved as shown in Fig. 3.6 [57-59].

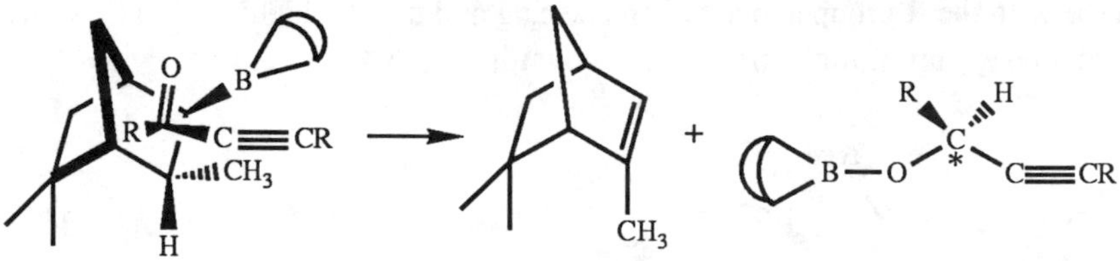

Figure 3.6. Reduction of α,β-acetylenic ketones.

The high optical yield obtained with Ipc$_2$BCl (**74**) has been explained by assuming that the reaction proceeds *via* a 6-membered cyclic "boat-like" transition state, as explained earlier [57-59]. Fig. 3.7 shows the transition states for the favoured (S) and the less favoured (R) products.

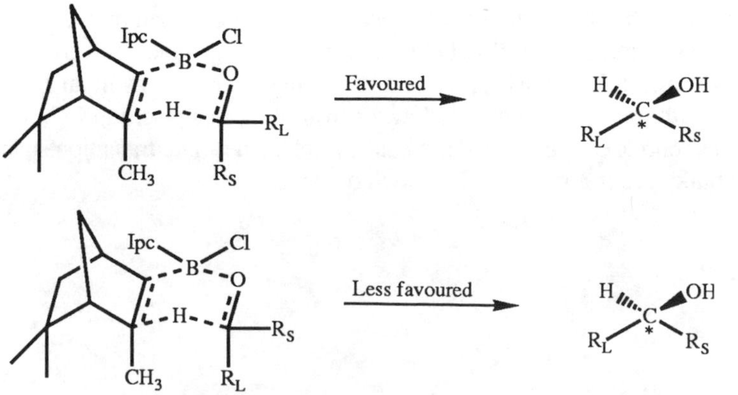

Figure 3.7. Transition states in the asymmetric reduction with (-)-Ipc$_2$BCl (**74**).

As shown in Fig. 3.7, the eliminating boron moiety and the β-hydrogen are *cis*, thereby resulting in *syn* elimination. In the preferred transition state, the bulky alkyl group (R_L) stays in an *equatorial*-like orientation, whereas the smaller alkyl group (R_S) has to face the *syn axial* methyl interaction. This explains the formation of the *S* isomer of the alcohols predominantly. Table 3.7 shows some representative examples of asymmetric reduction by alkylborane reagents.

Table 3.7. Asymmetric Reduction by Alkylborane Reagents

Substrate	Reagent	Product % e.e.	Reference
	(74)	91 (S)	[53]
	(74)	95 (S)	
^{i}Pr—C(=O)—CH$_3$	(64)	^{i}Pr—C—CH$_3$ 37 (S)	
(p-NO$_2$C$_6$H$_4$CHO)	(68)	86 (R)	
H$_5$C$_2$—C(=O)—CH$_3$	(64)	HO—C—H H$_5$C$_2$ CH$_3$ 22 (S)	[50]
PhCH$_2$O... CH$_2$COC≡CCH$_3$	(68)	PhCH$_2$O... CH$_2$CHOHC≡CCH$_3$ 88	[62]
(CH$_3$)$_2$CHCOC≡CH	(68)	(CH$_3$)$_2$CH C—C≡CH (H, OH) 99 (S)	[52]

(Table 3.7. contd.)

Substrate	Reagent	Product % e.e.	Reference
$H_5C_2OCOCOCH_3$	(68)	HO$\,$ $\,$H $H_5C_2OCO\ \overset{*}{C}-CH_3$ 89 (S)	[60]
$H_5C_6COC\equiv CCOOC_2H_5$	(68)	$H_5C_6\overset{*}{C}HOHC\equiv CCOOC_2H_5$ 100 (S)	[54]
$PhCOC\equiv C^nC_4H_9$	(68)	HO$\,$ $\,$H $Ph\overset{*}{C}\equiv C^nC_4H_9$ 89 (S)	[54]

3.1.1.2.2   Chiral Borohydride Reagents

3.1.1.2.2.1   NaBH$_4$-derived Reagents. Reagents formed from sodium borohydride modified with various monosaccharide derivatives (Table 3.8) has been employed for the asymmetric reduction of prochiral ketones [61]. A number of ketones have been reduced using such derivatives (**81-86**), affording the corresponding optically active alcohols. Upto 39% e.e. was thus obtained in the case of propiophenone, whereas for other ketones the optical yields of the alcohols were between 1.9-18.8 % e.e.

Table 3.8.   Monosaccharide Derivatives used as NaBH$_4$-Modifiers in the Reduction of Ketones

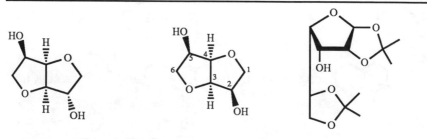

1,4:3,6-dianhydro-D-sorbitol   1,4:3,6-dianhydro-D-mannitol   1,2:5,6-di-O-isopropylidene-
        (81)   (82)   D-glucofuranose (83)

(Table 3.8. contd.)

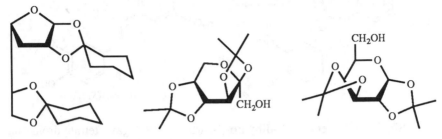

1,2:5,6-di-*O*-cyclohexylidene-   2,3:4,5-di-*O*-isopropylidene-   1,2,3,4-di-*O*-isopropylidene-
D-glucofuranose **(84)**   D-fructopyranose **(85)**   D-galactopyranose **(86)**

In contrast LAH modified with the monosaccharide derivative **(83)** reduces acetophenone in only 4% e.e. The optical yields in the monosaccharide-modified NaBH$_4$ reduction of ketones can be enhanced by using NaBH$_4$ treated with carboxylic acids such as acetic acid etc. prior to the addition of the monosaccharide derivative [62-64].

Thus the ketone **(87)** when reduced with modified sodium borohydride prepared from NaBH$_4$, (CH$_3$)$_2$CHCOOH and **(83)** yielded the alcohol **(88)** in 53% e.e. [63].

$$\text{(87)} \xrightarrow[\text{NaBH}_3\ [(CH_3)_2CHCOO)]]{\text{2x(83)}} \text{(88)}$$

(87)

(88) 53% e.e. (*R*)

Imines have also been reduced with good enantioselectivities by employing NaBH$_4$ modified with N-acyl derivatives of optically active amino acids [65,66]. Thus the acyl proline derivative **(89)** reacts with sodium borohydride to form the complex **(90)** which reduces 3,4-dihydropapaverine **(91)** to norlaudanosine **(92)** in 60% e.e. [65].

NaBH$_4$ +   3   [structure **(89)** with COPh, COOH] $\xrightarrow{\text{THF}}$ NaBH [structure **(90)** with COO, COPh]$_3$ + 3H$_2$

(89)   **(90)**

(91)   (92) 60% e.e. (S)

Similarly 1-methyl-3,4-dihydro-β-carboline (95) gave tetrahydroharman (96) on reduction with the N-acylproline - NaBH$_4$ derived from the N-acylproline derivative (93) in 79% e.e. [65]. Recently, NaBH$_4$ reduction of ketones in the solid state afforded 100% e.e. [66b].

(93)   (94)

(95)   (96) 79% e.e. (S)

3.1.1.2.2.1.1   Phase Transfer Catalyzed Reductions.   The insolubility of sodium borohydride in non-aqueous media renders it less useful for asymmetric reductions carried out in organic solvents. One way to accomplish such reductions would be to replace the sodium ion with chiral ammonium or phosphonium ions under phase transfer conditions [67-69]. Alternatively, crown ethers having two to four chiral centres have also been employed in phase transfer catalysed reductions [70]. Table 3.9 shows some ligands employed in phase transfer catalysed reductions.

Table 3.9. Chiral Ligands used in Phase Transfer Catalyzed Reductions

$\text{R} \quad \text{CH}_3$ $\text{PhCH}_2 - \overset{*}{\underset{H}{C}} - \overset{+}{\underset{C_{12}H_{25}}{N}} - \text{CH}_3 \ \bar{\text{Br}}$	$\text{OH} \quad \text{CH}_3$ $\text{Ph} - \overset{*}{\underset{H}{C}} - \text{CH}_2 - \text{CH}_2 - \overset{+}{\underset{R^1}{N}} - \text{CH}_3 \ \text{X}^-$	$\overset{*}{\text{PhCHOH}} \overset{*}{\underset{CH_3}{CH}} - \overset{+}{\underset{CH_2CH_2OH}{N}(CH_3)_2} \ \bar{\text{Cl}}$
(97) R = OH **(98)** R = H	**(99)** $R^1 = C_{12}H_{25}$, X=Br **(100)** $R^1 = CH_2Ph$, X=Cl	**(101)**

$\text{PhCH}_2 - \overset{*}{\underset{CH_3}{CH}} - \overset{+}{\underset{CH_2CH_2OH}{N}(CH_3)_2} \ \bar{\text{Cl}}$

(102)

(103)

$R^1_{\prime\prime\prime} \ \ R^2$ crown ether		

(104) $R^1 = R^2 = CON(CH_3)_2$
(105) $R^1 = R^2 = CH_2N(CH_3)_2$
(106) $R^1 = H$, $R^2 = CH_2OCH_2Ph$

$\overset{*}{\text{CH}_2}\text{CHOH CH}_2\text{OH}$

(107) $R^1 = H, R^2 = \ CON - \overset{*}{\underset{CH_3}{CH}} Ph \ (S)$ **(109)** $\overset{*}{\underset{}{C}}HOH - CH_2PPh_2Et$, $\overset{}{\underset{}{C}}HOH - CH_2PPh_2Et$

(110)

(108) $R^1 = \ CONH - \underset{Ph}{\overset{*}{C}}$ $R^2 = H$

The ligands (**97-110**) shown in Table 3.9 have been employed with NaBH$_4$ under phase transfer conditions in the asymmetric reduction of ketones [67]. Thus, the asymmetric reduction of phenyl tertiarybutyl ketone (**111**) yielded the corresponding alcohol (**112**) in 32% e.e. by using NaBH$_4$ with the ligand (**103**) under phase transfer conditions.

$$\text{(111)} \quad \xrightarrow[\text{(103)}]{\text{NaBH}_4\text{-H}_2\text{O-benzene}} \quad \text{(112)}$$

(111)   **(112)**

Employing the other ligands (97-102), the optical yields were generally lower in the reduction of ketones, ranging from 1.1-13.7% e.e. The use of crown ethers (107) as well as (108) with NaBH$_4$ afforded low optical yields with upto 8% e.e. [70]. Comparatively high optical yields are obtained by using bovine serum albumin along with NaBH$_4$ in the reduction of aromatic ketones [71]. Thus phenyl n-butyl ketone (113) is reduced to the corresponding alcohol (114) in 67% e.e. by using bovine serum albumin with NaBH$_4$.

(113)   (114) 67% e.e. (R)

The aqueous protein solution remaining in the reaction mixture can be reused without change of stereoselectivity. For example, the first reduction of propiophenone showed an asymmetric induction of 77% and the optical yield in the second repeat reduction remained almost unchanged.

Studies on asymmetric reduction under phase transfer conditions have led to the conclusion that conformationally rigid ligands result in faster reductions and high optical yields, as exemplified by the quininium salt (103). It has also been found that the hydroxy group in the ligand should be in the β- position to the onium function. This may be important for its interaction with the carbonyl group of the substrate and to favour one of the diastereomeric transition states which leads to the carbinol such as (112).

With ligands having a γ-hydroxy group, the asymmetric induction is much lower. The bulkiness of the substituents in the substrate may also play an important role in the stereochemical course of the reaction. This is shown by the lack of asymmetric induction displayed by (115) which gave no optically active products.

(115)

3.1.1.2.2.1.2   LiBH₄ Reductions. Recently LiBH$_4$ has been employed with N-benzoylcysteine (**116**) as ligand in the enatioselective reduction of alkyl arylketones in 92-93% e.e. Thus the reduction of phenyl n-propiophenone (**14**) with LiBH$_4$ having ligands (**116**) afforded the corresponding alcohol (**15**) in 93% e.e. The reaction is believed to proceed *via* the complex (**117**).

(**116**)   (**117**)

(**14**)   (**15**) *93% e.e.* (*R*)

3.1.1.2.2.2   Super Hydrides. Recently a class of very powerful reducing agents, the trialkylborohydrides or the "super hydrides" have been developed. These reagents are readily prepared from lithium or potassium hydride or lithium trimethoxyaluminohydride and trialkylboranes [72-78]. Thus a highly hindered trialkylborohydride containing an asymmetric alkyl group (Alpine-hydride) (**118**) was prepared from β-isopinocampheyl-9-BBN (**62**) in quantitative yield.

(**62**)   (**118**)

The reagent (**118**) employed in the reduction of prochiral ketones rapidly and quantitatively affords the corresponding optically active alcohols with predominant (*R*) configuration [77]. Thus the reduction of the ketone (**119**) afforded the corresponding optically active alcohol (**120**) in 37% e.e. by using the reagent (**118**) at 78°C.

(119)   (120) 37% e.e. (*R*)

The enantioselectivity observed with the borohydride reagent (118) is between 13-37% e.e. which is far less than that obtained with the trialkylborane (62) itself. This may be attributed to the fact that in the reagent (62) the hydrogen transferred is directly attached to the chiral site. Better results were obtained with thexyl-[C(CH$_3$)$_2$CH(CH$_3$)$_2$]-borane reagent (121).

(121)

Thus when the reagent (121) was employed in the reduction of the substrate (122), it afforded the alcohol (123) in 92% e.e. [73].

(122)   (123) 92% e.e.(*S*)

The high optical yield obtained in the stereoselective reduction of the ketone (122) is attributed to the van der Waals interaction between the 4-phenylphenyl carbamate moiety used as a protecting group for the C-11 OH and the enone side chain which results in a conformation in which the α-face of the substrate is effectively blocked. The bulky borohydride reagent (121) therefore preferentially attacks from the β- face to afford the corresponding alcohol (123) in high optical yield.

The reagent (124) prepared from (+)-α-pinene was employed in the reduction of prochiral ketones with generally upto 9% e.e., though in one case, i.e. in the reduction of ethyl isopropylketone, the optical yield was 46% [78,79]. The reagent (133) was also employed in the asymmetric reduction of cyclic imines with upto 25% e.e. [79,80]. Thus the cyclic imines (125) were reduced to the corresponding 2-alkyl piperidines (126) with the trialkylborohydride reagent (124, R^1=Me, n-Bu, Ph) giving optical yields in the range of 4-25% e.e.

(124)

(125) R = ^{n}Pr, Et, CH$_3$ (126) 4-25% e.e. (R)

The borohydride reagent (124) has also been used in the asymmetric reduction of 3,4-dihydropapaverine (91) to give the laudanosine methiodide (127) after methylation of the intermediate tetrahydropapaverine [79].

(91)   (127) (R)

Another useful superhydride reagent recently developed is NB-enantrane (128) [81].

(128)

α,β-Acetylenic ketones are reduced with NB-enantrane **(128)** to afford optical yields in the range of 86-96% e.e. For example, the ketone **(129)** afforded the optically active alcohol **(130)** on reduction with the reagent **(128)** in 96% e.e. [81].

Asymmetric reduction of other simple ketones, such as acetophenone, with the same reagent **(128)** afforded (S)-1-phenyl-ethanol in 70% e.e. whereas reduction with Alpine-hydride **(118)** yielded only 17% e.e. of the product. Table 3.10 presents some representative examples of asymmetric reductions by chiral borohydride reagents.

Table 3.10. Asymmetric Reductions by Chiral Borohydride Reagents

Substrate	Reagent	Product	% e.e.	Reference
	A		78 (R)	[71]
	A		66 (R)	[71]
	B		55 (R)	[63]

(Table 3.10. contd.)

Substrate	Reagent	Product	% e.e.	Reference
acetophenone ($C_6H_5COCH_3$)	B	1-phenylethanol (H, OH, CH₃)	64 (R)	[63]
propiophenone ($C_6H_5COC_2H_5$)	C	(H, OH, C_2H_5)	42 (R)	[62]
2-methyl tetrahydropyridine	D	(N–H, H, CH₃)	25 (R)	[79]
H_5C_2 CO iPr	E	HO, H, iPr, C_2H_5	46 (S)	[78]
$^nC_5H_{11}$ COC≡CH	F	H, OH, $^nC_5H_{11}$, C≡CH	95 (S)	[81]
$CH_3COC≡C_6H_5$	F	H, OH, H_3C, C≡CC_6H_5	86 (S)	[81]
$H_5C_2COC≡CC_2H_5$	F	H, OH, H_5C_2, C≡CC₂H₅	96 (S)	[81]

A = NaBH₄, bovine serum albumin,   B = NaBH₄ + (CH₃)₂CHCOOH + (83),

C = (84) + NaBH₄ + (±) PhCHEtCOOH,

D = Li⁺ [B–H, X, X] (X= (+)pin-3α-yl)   E = Li⁺ [B–H, X, X] (X=(-)-pinan-2-ene)   F = (128)

3.1.2   *Chiral Metal Alkyls and Alkoxides*

In 1901, Grignard reported the reduction of benzaldehyde with isoamylmagnesium bromide yielding mainly 1-phenyl-isohexanol along with a small amount of benzyl alcohol [82]. This laid the foundation for the development and application of a vast number of alkyl and arylmagnesium halides, known as "Grignard reagents" in organic synthesis. Vavon *et al* reported the first asymmetric reduction of phenylalkylketones using a *chiral* Grignard reagent (isobornyl-magnesium chloride) [83,84]. Since then, there have been many reports in the literature on asymmetric reductions with chiral Grignard reagents and other organometallic compounds [85-109]. However, the reaction is of limited utility since the product yield is frequently very low as a result of competing addition and enolization reactions, and the by-products formed are generally optically active and often difficult to separate from the desired product.

The basic requirement in such reductions is to use unsymmetrical substrates and chiral organometallic reagents. For example, the asymmetric reduction of methyl tert.butylketone (**75**) with the Grignard reagent (+)-2-methylbutylmagnesium chloride (**131**) afforded methyl tert. butylcarbinol (**76**) in 16% e.e. [100].

In another experiment, higher optical yields were reported in the asymmetric reduction of the phenylalkyl ketone (**53**) to the alcohol (**54**) in 91% e.e. with the reagent (**132**) [100].

Other metal alkyls such as those of Al, Be and Zn have been used in asymmetric reduction of prochiral carbonyl compounds [92- 94,96].

Chiral organoaluminium reagents reduce prochiral carbonyl compounds in upto 34% e.e. [92-94,96]. Trialkylaluminium compounds have been employed in some cases to reduce unsymmetric ketones [92]. Thus isopropylmethyl ketone (60) is reduced by (+)-tris [(S)-2- methylbutylaluminium] diethyl-etherate (133) to yield isopropylmethyl carbinol (61) in 16.9% e.e. [92].

(60)   (133)   (61) 16.9% e.e. (S)

The same reaction carried out at -60°C afforded the product in 44.4% e.e. [107]. However, the reaction was slower (at -3°C) using diethyl ether as a ligand as compared to the reagent tris[(S)-2-methylbutyl] aluminium. Recently, sterically crowded chiral organoaluminium reagents have been developed giving optical yields of upto 86% e.e. [109,110]. For instance the reagent (134) afforded high optical yields in the reduction of prochiral ketones [109]. Similarly, 2,2-dimethyl-4-nonyn-3-one (135) was reduced to the corresponding alcohol (136) with (134) giving an optical yield of 84.8% e.e. [109].

CH_2AlCl_2

(134)

(135)   (136) 84.8% e.e. (R)

Dialkylzinc compounds such as(+)-bis-[(S)-2-methylbutyl] zinc (137), (+)-bis [(S)-3- methylpentyl] zinc (138) and [(+)-bis-(S)-4-methylhexyl] zinc (139) have been employed to study the influence of the distance of the asymmetric centre from the metal atom on the stereoselectivity of the ketone reduction [93].

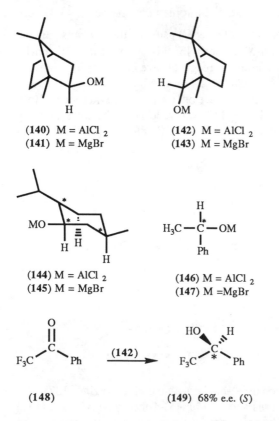

$$\left(\underset{H_5C_2}{\overset{H_3C}{}}\!\!\!\!\!\!\overset{H}{\underset{*}{C}}\!\!\!\!\!\!\overset{}{CH_2}\right)_2\!\!\!-Zn$$

(137)

$$\left(\underset{H_5C_2}{\overset{H_3C}{}}\!\!\!\!\!\!\overset{H}{\underset{*}{C}}\!\!\!\!\!\!\overset{}{(CH_2)_2}\right)_2\!\!\!-Zn$$

(138)

$$\left(\underset{H_5C_2}{\overset{H}{}}\!\!\!\!\!\!\overset{CH_3}{\underset{}{C}}\!\!\!\!\!\!\overset{}{(CH_2)_3}\right)_2\!\!\!-Zn$$

(139)

Organometallic alkoxides, such as optically active alkoxyaluminium dichlorides and alkoxymagnesium bromides, have been employed in the asymmetric reduction of ketones and aldehydes [111-118]. These alkoxides have generally been derived from bornan-2-exo-ol (isoborneol) (140,141), bornan-2-endo-ol (borneol) (142,143), (-)-p-menthane-3-ol (144,145) and (+)-1,1-phenyl ethanol (146,147). For instance, the asymmetric reduction of phenyltrifluoromethyl ketone (148) with the reagent (142) afforded trifluoro-1-phenylethanol (149) in 68% e.e. [111].

(140) M = AlCl₂
(141) M = MgBr

(142) M = AlCl₂
(143) M = MgBr

(144) M = AlCl₂
(145) M = MgBr

(146) M = AlCl₂
(147) M =MgBr

(148)

(149) 68% e.e. (S)

The optical activity is increased to 77% e.e. by using the reagent (144) in the reaction. However, by using Mg as the metal atom, the optical yield is generally low (less than 25% e.e.). This may be due to the degree of dissociation of the metal-oxygen bond, and to the degree of dissolution etc. [119].

The use of amino ketones as substrates increases the optical yields due to the presence of nitrogen as well as due to facile separation of the products [115,120,121]. For example, the asymmetric reduction of the amino ketone **(150)** with the reagent **(140)** afforded the corresponding amino alcohol **(151)** in 71% e.e [115]. The same reagent reduced the substrate **(152)** to afford the amino alcohol **(153)** in 92% e.e.

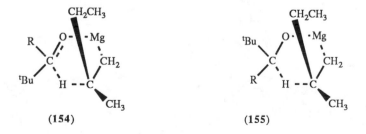

Although a number of studies have been conducted to understand the mechanism of asymmetric reduction by chiral Grignard reagents [89,90,97, 101-106], little systematic investigation has been done on the factors controlling the reduction and addition processes in relation to the nature of both the Grignard reagent and of the carbonyl substrate.

The stereospecificity of the reduction of unsymmetrical ketones such as **(75)** by the Grignard reagent derived from (+)-1-chloro-2-methylbutane **(131)** has been interpreted in terms of a six-membered transition state for the hydrogen transfer step, as shown in Fig. 3.8. In this reaction, the transition states of major importance may be postulated as **(154)** and **(155)**.

Figure 3.8. Transition states for the hydrogen transfer step in the reduction of ketones by Grignard reagents.

It has been shown that when the R group is smaller than the *tert*.butyl group, the predominant isomer of the secondary alcohol will have the configuration resulting from the transition state (154).

The reduction of prochiral carbonyl compounds by organoaluminium reagents is assumed to proceed *via* competing diastereomeric transition states as shown in Fig. 3.9.

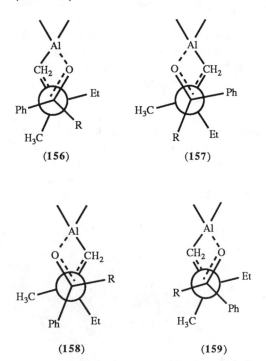

$$(+)\text{-Tris }[(S)\text{-2-methylbutyl}]\text{ Al} + \underset{R}{\overset{Ph}{\diagdown}}C=O$$

$$\xrightarrow{K_S}\ \big[(S)(S')\big] \longrightarrow (S)\text{-Carbinol}$$

$$\xrightarrow{K_R}\ \big[(S)(R')\big] \longrightarrow (R)\text{-Carbinol}$$

Figure 3.9. Reduction of ketones with organoaluminium reagents.

The mechanism of hydrogen transfer involves a cyclic transition state in which the coordinate bond between the aluminium and oxygen atoms has to be relatively loose to minimize the steric hinderance among the groups. These steric interactions may be estimated from the four conformations (156-159).

(156)   (157)

(158)   (159)

The transition states (156) and (158) lead to the S carbinol, (156) being the most stable due to steric and electronic requirements. On the other hand (157) and (159) lead to the R alcohol and although the steric interactions are similar in these two cases, the conformation (157) has the lowest energy for electronic reasons. Since conformation (156) is more favored than (157) for steric reasons, the carbinol resulting from asymmetric induction has the S configuration [92, 94].

(160)   (161)

On the basis of mechanisms proposed for the reduction of ketones by other organometallic reagents, the reduction of prochiral ketones by dialkyl zinc compounds, e.g. (138) proceeds through a six-membered cyclic transition state in which the hydrogen atom β to the zinc atom is transferred to the carbonyl carbon atom, as shown in conformations (160) and (161). Taking into account both steric and electronic factors and considering the most stable reacting conformations in which II_A is transferred to the re-face (160) or to the si-face (161) of the carbonyl compound, rationalized in terms of Newman-type projections viewed along the C---H_A–C_β axis, it appears that conformation (160) leading to the (S)-carbinol represents a lower energy transition state than (161) since in the former case the bulkiest groups, i.e. the phenyl group of the ketone and the alkyl group of the reagent (137) are in a quasi anti-position. The (S) enantiomer of the product (54) should therefore be formed prevalently, in agreement with the experimental findings.

Some representative examples of the reduction of prochiral compounds by chiral organometallic reagents are presented in Table 3.11.

3.1.3   *Chiral Dihydropyridine Reagents*

The remarkable rate acceleration as well as the regio- and stereo-selectivities achieved by enzymes has fascinated chemists since a long time. However it was not until recently that "artificial enzymes" were practically tested in the laboratory [123]. Since then a number of attempts have been made to study the practical aspects as well as mechanisms of biochemical transformations [124-127].

The dihydropyridine nucleotides NADH and NADPH are coenzymes which are closely associated in the reactions of dehydrogenases. For instance, pyruvate is stereospecifically reduced *in vivo* to D- or L-lactate by NADH in the presence of D- or L- lactate dehydrogenase respectively [128-130]. Modelled after such reactions, it has been found that 1-benzyl-1,4-dihydronicotinamide (BNAH) and its analogues reduce α-ketoesters in the presence of Mg or Zn in comparable optical yields. Thus methyl benzoylformate (162) reacts with *R*-(-)- or (*S*)-(+)-N-α-methylbenzyl-1-propyl-1,4- dihydronicotinamide (163) in the presence of magnesium perchlorate to give *R*-(-)- or S-(+)-methyl malate (164) respectively in 15% e.e. [125].

(162)   (163)

(164) 15% e.e. (*R*)

Similarly, α-ketoesters have been reduced with Hantzsch esters (dialkoxycarbonyl dihydropyridines) affording high optical yields of the reduced products [131]. For example the ester (165) is reduced to the corresponding alcohol (166) in the presence of Hantzch ester (167) and zinc in 78% e.e. [131].

(165)   (166) 78% e.e. (R)

(167)   (168)

Improved enantiospecificity was observed in the biomimetic reduction by using model compounds having both the reaction and chiral centres at the same position. It is known that only one of the two diastereotopic hydrogens at C-4 of the dihydropyridine in the natural coenzyme is transferred to and from substrates in the enzymatic reactions. Therefore the enantiospecificity in the biomimetic reduction is enhanced by employing model compounds with a chiral centre at C-4. The model compound (168) indeed showed high enantiospecificity. Thus (168) was employed in the reduction of a variety of ketones affording optical yields of upto 99% e.e. For instance the substrate (169) was reduced to the corresponding alcohol (170) in 97.6% e.e. [124].

(169)   (170) 97.6% e.e. (R)

Similarly trifluoromethyaryl ketones can be reduced to the corresponding alcohols in high optical yields. The aromatic trifluoroacetyl derivatives (171) are reduced to the corresponding alcohols (172) in over 95 %e.e. [104, 124].

(171)   (172) >95% e.e. (S)

Other model systems such as naturally derived L-amino acid or dipeptide-1,4-dihydronicotinamide derivatives (173-181) and corresponding sugar derivatives (182-184) have been developed to effect asymmetric reductions which however proceed in only moderate optical yields.

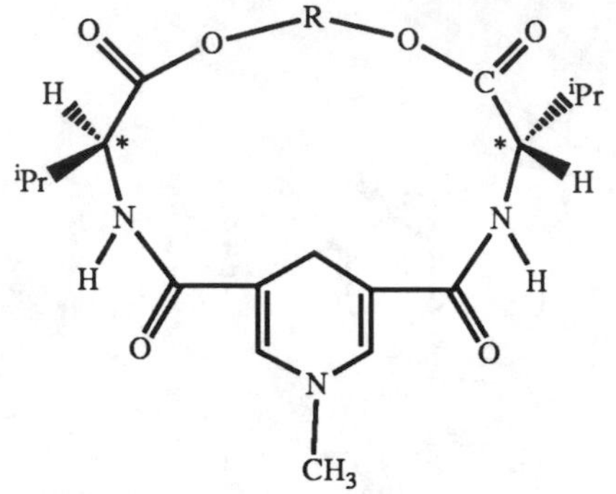

(173) R=   —(CH$_2$)$_5$—

(174) R=   —CH$_2$—⟨○⟩—CH$_2$—

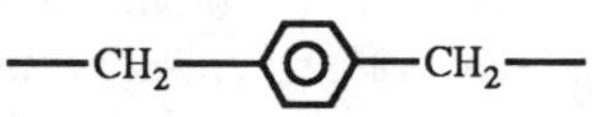

(175) R=   —(CH$_2$)$_2$—O—(CH$_2$)$_2$—

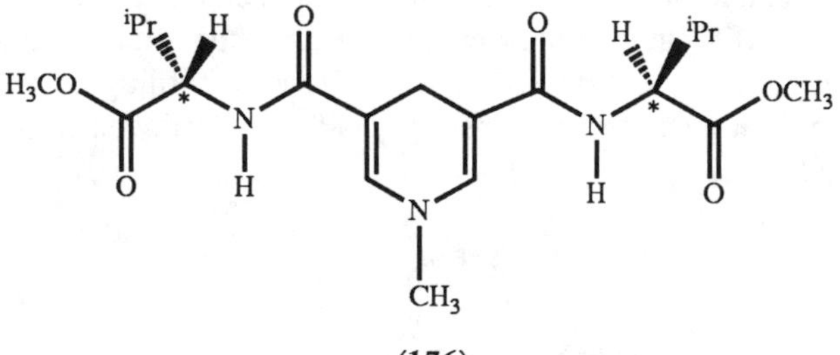

(176)

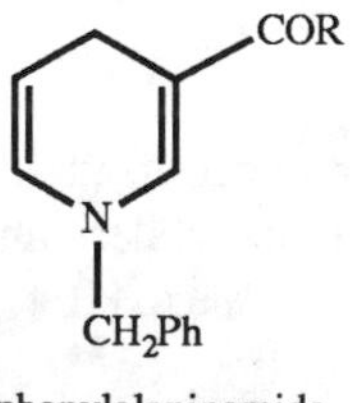

(177) R = L-phenylalaninamide
(178) R = L-alanyl-L-alaninamide
(179) R = glycyl-L-alaninamide
(180) R = L-alanyl-glycinamide
(181) R = L-prolinamide

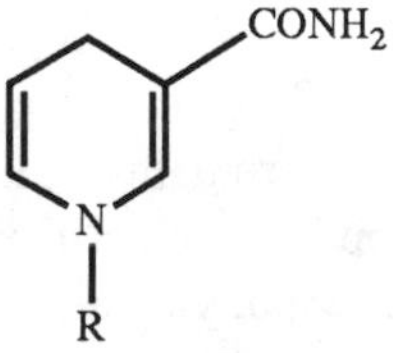

(182) R = sugars

(183) R =

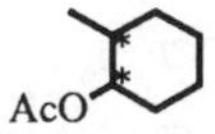

(184) R =

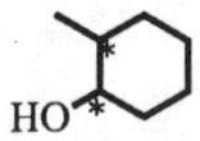

Asymmetric reduction of aryl trifluoromethyl ketones has been carried out with an achiral NADH model compound such as 1-propyl-1,4-dihydronicotinamide (185) in the presence of aqueous solution of sodium cholate micelle, β-cyclodextrin or bovine serum albumin [132]. The asymmetric induction ranges from 0.6-1.4% in sodium cholate micelle to 1-10 % in β-cyclodextrin and 16-47% in bovine serum albumin systems. The substrate (186) was reduced to the alcohol (187) in the presence of (185) and bovine serum albumin (BSA) in 22.3% e.e. [132]. The same reaction with phenyl trifluoromethyl ketone afforded the corresponding alcohol in 46.6 % e.e.

(186)   (185)   (187) 22.3% e.e. (R)

The mechanism of asymmetric reductions by chiral dihydropyridine reagents is comparable in complexity to enzymatic reactions and no generalizations have emerged although many attempts have been made to explain the experimental results [133]. It seems that the rate of the reaction as well as the stereochemistry of the product changes from substrate to substrate and from one set of experimental conditions to another. For instance, bivalent metal ions such as magnesium and zinc catalyse the reduction of certain substrates [123,126], whereas these metal ions retard the reduction of others [134,135]. With yet other substrates, the rate of reduction is affected by the metal ion in such a way that the rate initially increases, comes to a maximum and then decreases as the concentration of the metal ion is increased [136-138]. Hence the mechanism of the catalysis seems to vary from case to case. Other factors such as the presence of polar groups in the vicinity of the chiral centre of the reagent, reaction conversion (ageing) as well as additional effects of oxidized forms of the model compounds, initial and relative concentrations of the substrate, etc. greatly influence the course and stereochemistry of asymmetric reactions with chiral dihydropyridine reagents.

For instance, in the asymmetric reduction of methyl benzoylformate (169) by the reagent (168), a transition state (188) involving both the substrate and the reagent bonded to the metal atom has been proposed (Fig. 3.10). The transition state (188) has been depicted to explain the mechanism of the reaction put forward as a three-step electron-proton-electon transfer.

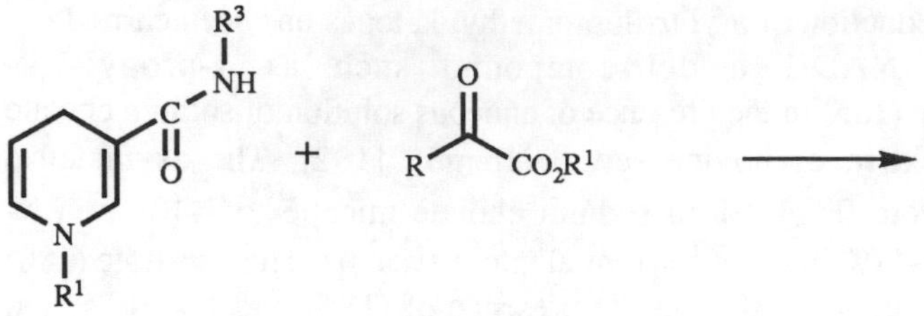

(168)   (169)

R = Ph
$R^1 = CH_3$
$R^2 = {}^iPr$
$R^3 = CH(Ph)CH_3$

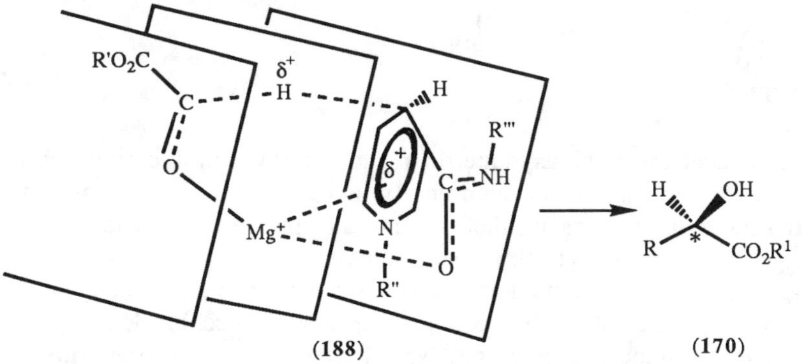

(188)   (170)

Figure 3.10. Schematic representation of the transition state (188) of the reduction of (169) by the reagent (168).

This involves an electron transfer from the reagent (168) to the substrate (169) in the transition state ternary-complex (168)- Mg^{2+}-(169), which triggers migration of a proton from the radical of (168) to the anion radical of (169). The migration of a second electron takes place spontaneously, because the pyridinyl radical can gain stabilization energy by conversion into the pyridinium ion.

The dependence of enantioselectivity on the relative concentration of substrate in an NADH model reaction has been studied [139]. The results are explained in terms of a feedback effect of the oxidized nicotinamide, which accumulates during the conversion. The oxidized form interacts ("feedback interaction") with the reductant (169) via the metal ion as described for the transition state (188). This may result in the blocking of one of the diastereotopic faces of the dihydropyridine ring, thereby facilitating easy access of the substrate to the other face of the reagent, resulting in enhanced optical purity of the product.

Chiral bis(NADH) model compounds such as (189-191) employed in the reduction of α-ketoesters lead to high optical yields (upto 98% e.e). Initial addition of the oxidized bis(NAD) form did not affect the stereochemistry of the reduction as shown by the observed constancy in the enantiomeric excess. It has been shown that the bis(NADH) reduction is a kinetically controlled process and the participation of the oxidized from *in situ* can be ruled out. The high enantioselectivity has been explained by assuming a chelation complex such as (192) in the transition state in which the reagent (189) has a C_2 conformation with the specific pro-*R* or pro-*S* hydrogens of the two juxtaposed equivalent dihydropyridine nuclei placed outside and the C_2 axis passing through the centre, connecting Mg and the *p*-xylene bridge.

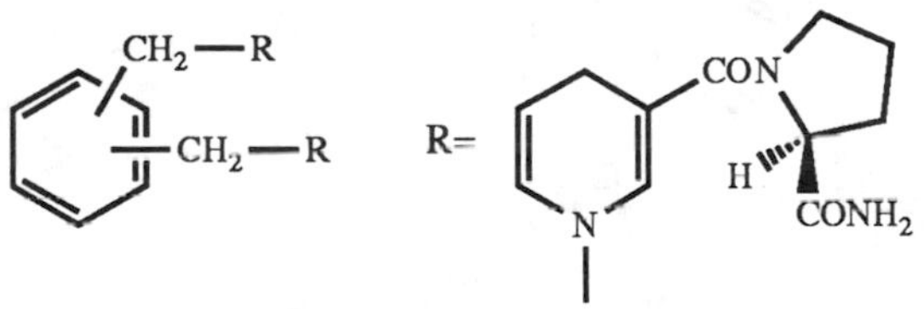

(189) R = *para*
(190) R = *meta*
(191) R = *ortho*

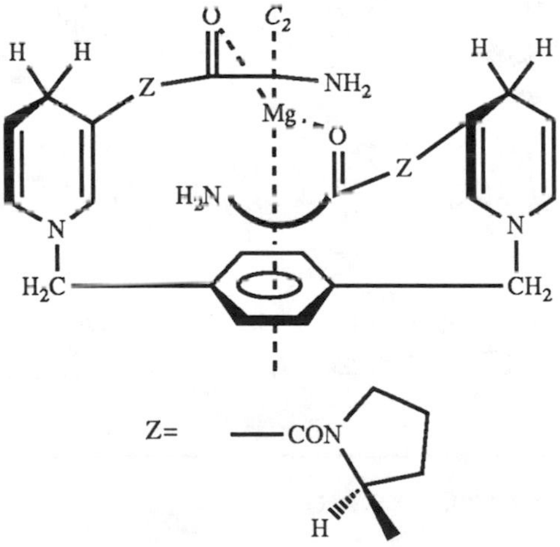

(192)

Some representative examples of asymmetric reduction by chiral dihydropyridine reagents are presented in Table 3.12.

Table 3.12.   Asymmetric Reductions with Chiral Dihydropyridine Reagents

Substrate	Reagent	Product	% e.e.	Reference
PhCOCOOEt	A	HO,,H Ph—C*—COOEt	98.1 (R)	[140]
2-acetylpyridine (C(=O)CH₃ on pyridine)	A	H,,OH pyridine—C*—CH₃	89.7 (R)	[140]
PhC(=O)CF₃	B	OH,,H Ph—C*—CF₃	46.6 (R)	[132]
PhC(=O)CO₂CH₃	C	OH,,H Ph—C*—COOCH₃	97.6 (R)	[174]
4-Br-C₆H₄—C(=O)CF₃	C	HO,,H 4-Br-C₆H₄—C*—CF₃	89.2 (R)	[124]
ᵗBu—C(=O)COOCH₃	C	HO,,H ᵗBu—C*—COOCH₃	>99 (R)	[124]
PhC(=O)COOEt	D	H,,OH Ph—C*—COOEt	90 (S)	[141]

A = (**203**),   B = (**194**)+BSA,   C = (*R,R*-**179**),   D = (**184a**)

3.2   Diastereoselective Non-Catalytic Reductions

3.2.1   Cyclic Substrates

Work on non-catalytic diasteroselective reductions is largely confined to the reduction of carbonyl compounds. Derivatives of cyclic ketones, such as those of cyclohexanone, as well as carbonyl compounds having a chiral centre at the 2,3 or 4 positions with respect to the carbonyl group (1,2-,1,3- and 1,4-asymmetric inductions) have been studied in greater detail. The basic requirement in such reductions is that the carbonyl group should have diastereotopic faces.

The stereochemical outcome of a diastereoselective non-catalytic reduction depends largely on the structure of the substrate as well as on the nature of the reducing agent. For example, reduction of the 3-keto-steroid (193) with $NaBH_4$ proceeds with 94% predominance of axial attack, affording the equatorial alcohol (194 a), whereas only 6% of the axial alcohol (194 b) is formed.

(193)

(194a) 94%   (194b) 6%

Similarly, reduction of 4-t-butylcyclohexanone (195) by the same reductant afforded trans-4-t-butyl-cyclohexanol (196 a) in 86% yield [142]. The axial alcohol (196 b) was isolated in 14% yield.

(195)   86% (196a)   (196b)14%

Introduction of axial substitutents at the C-3 and C-5 positions however dramatically changes the stereoselectivity and leads ultimately to inversion of stereoselectivity. The reduction by NaBH$_4$ of 3,3,5-trimethylcyclohexanone (197) afforded the reduced products (198) and (199) indicating that the stereoselectivity is not only eliminated, but even slightly inverted by the introduction of the axial Me group.

(197)   48% (198)   52% (199)

Introduction of a second axial methyl group pushes the stereochemistry to the opposite sense with respect to that of the unhindered ketone. Reduction of 4-(1-hydroxyethyl)-3,3,5,5-tetramethylcyclohexanone (200) by NaBH$_4$ affords a 95.5% predominance of the epimer (201c) which corresponds to the initial equatorial attack of borohydride, while the epimer (201b) possesses a severe 1,3,5-triaxial interaction between the hydroxyl group and two methyl groups. Thus the ring in (201b) flips to conformation (201c) to avoid these interactions although the hydroxyethyl group now lies in the axial position [143].

(200)   (201a) < 4.5%

(201b)   (201c) >95.5%

It has been generally found that equatorial alcohols are formed from unhindered ketones on reduction with LiAlH$_4$ or NaBH$_4$ while axial alcohols are obtained when bulky reducing agents, e.g. L-Selectride [Li(sec.Bu)$_3$ BH] are used. In the case of hindered ketones as substrates, axial alcohols are formed preferentially with any hydride donor especially those with bulky substituents, while equatorial alcohols can be prepared from hindered ketones by equilibration, e.g. over Raney nickel [144], whereby it is possible to convert axial alcohols into the more stable equatorial alcohols.

Lithium trisiamylborohydride (siamyl = 3-methyl-2-butyl), a highly hindered trialkylborohydride, containing three β-methyl substituted secondary alkyl groups reduces unhindered cyclic and bicyclic ketones with exceptional steric control. Thus 2-methyl cyclopentanone (202) is reduced to *cis*-2-methyl cyclopentanol (203) by lithium siamyl borohydride (LiSia$_3$BH) in 99.5% yield [145-148].

| (202) | (203) | (204) | (205) 91% exo |

Similarly, reduction of camphor (204) with LiAlH$_4$ proceeds by *endo* attack to give the *exo* alcohol (205) in 91% yield [145]. Using the reagent Li(2,4,6-trimethylphenyl)$_3$ BH, the same product is obtained in more than 99% yield [147,148].

Table 3.13 shows some examples of alkylcyclohexanones with various bulky hydride donors.

Table 3.13. Reduction of Alkylcyclohexanones with Bulky Hydride Donors

Substrate	Reagent	% endo	Reference
	Li(^{s}Bu)$_3$BH	93	[146]
	Li(*trans*-2-methyl-cyclopentyl) $_3$BH	>99	[148]
	"	98	[148]

(Table 3.13. contd.)

Substrate	Reagent	% endo	Reference
	Li(3-methyl-2-butyl)$_3$BH	99	[148, 149]
	Li(3-methyl-2-butyl)$_3$BH	>99	[148,149]
	Li(O^{t}Bu)$_3$AlH	93	[150]
	Li(^{s}Bu)$_2$BH	98	[151]
	Li (2,4,6-trimethylphenyl)$_3$BH	>99	[147]

3.2.2   Acyclic Substrates

A similar trend is followed in acyclic substrates, in which variation in the *syn/anti* or *erythro/threo* ratio depends on the structure as well as on the nature of the reducing agent. Thus, the ketone (**206**) on reduction with KHB$_4$ afforded the *syn* product (**207**) in 95% yield, whereas the ketone (**208**) afforded the *syn* product (**209**) in just 66% yield [152].

(206)   (207) 95% syn

(208)   (209) 66% syn

Reduction of the ketone (210) with LiAlH$_4$ afforded 98% of the *anti* product (211) [153].

(210)   (211) 98% anti

Substitution of the *tert*.butyl group by isopropyl, ethyl and methyl groups in the substrate reduces the stereoselectivity of the *anti* product to 85%, 76% and 74% respectively [153]. It has been shown that temperature has a marked influence on the stereoselectivity of the reduction, lower temperatures favouring *anti* stereoselectivity.

Similar studies have been performed on substrates involving formation of carbon-carbon bonds by attack of appropriate nucleophiles on prochiral carbonyl compounds. Thus, nucleophilic addition of phenyllithiums to the substrate (212) proceeds to afford 84% of the *syn* product (213) [146].

(212)   (213) 84% syn

3.2.2.1   1,2-Induction

Although studies on diastereoselective nucleophilic addition to carbonyl compounds have led to excellent stereoselectivities in carbon-carbon bond formations (Chapter 4), stereoselectivity in non-catalytic reductions of carbonyl compounds is generally modest. Thus, the ketone (214) is reduced to the *anti* alcohol (215) in 74% yield [153].

(214)   (215)   74% *anti*

Table 3.14. shows some examples of 1,2-induction of various substrates.

Table 3.14. Diastereoselective Non-catalytic Reduction of Ketones with 1,2 Induction

(A)   (B)

Substrate				Reagent	%Yield		Reference
R^1	R^2	R^3	R^4		A	B	
Ph	CH_3	H	CH_3	A	85	-	[159]
Ph	CH_3	H	C_2H_5	B	78	-	[153]
Ph	C_2H_5	CH_3	Ph	C	69	-	[160]
C_2H_5	CH_3	CH_3	CH_3	D	56.5	-	[159]
NH_2	CH_3	H	$C_{15}H_{31}$	C	66	-	[156]
NH_2	CH_3	H	Ph	E	95	-	[161]
NH_2	CH_2NH_2	H	Ph	E	50 : 50		[161]

(Table 3.14. contd.)

Substrate				Reagent	% Yield		Reference
^tBuNH	CH_3	H	Ph	E	100	-	[158, 162]
CH_2NHPH	CH_3	H	Ph	E	100	-	[158, 162]
NH Ac	CH_2OH	H	Ph	F	47	-	[163a]
Ph	CH_3	H	$COOCH_3$	G	>99 : <1		[163b]
Ph	C_2H_5	H	$COOCH_3$	G	>99 : <1		[163b]

A = LAH,-70°C,   B = LAH, 35°C,   C = LAH,   D = Al(O^{i}Pr)$_3$,   E = NaBH$_4$,

F = Al(O^{i}Pr)$_3$,   G = Zn(BH$_4$)$_2$

At -70°C, the yield is 85%, while using L- Selectride, i.e. Li(sec-Bu)$_3$BH as the reducing agent, the product (215) is obtained in 95% yield [155].

Aminoketones are important reagents in the pharmaceutical industry in the preparation of anaesthetics and analgesics etc. and much work has been done on their conversion to aminoalcohols by reduction and addition of organometallic reagents [156,157]. For example, reduction of the aminoketone (216) with NaBH$_4$ yielded the *anti* aminoalcohol (217) in 97% yield [158].

(216)   (217) 97% *anti*

3.2.2.2   1,3-, 1,4- and 1,6- Inductions

The stereodirecting ability of a chiral centre separated by two, three or more bonds from the carbonyl group has been fully exploited in carbon-carbon bond forming reactions (see chapter 4), but it has not been very successful in diastereoselective non-catalytic reductions [164-165].

The incorporation of an asymmetric sulfur moiety within the substrate has been used for transmission of chirality from sulfur to a carbon [166-168]. For this purpose, β-ketosulphoxides and β-ketosulfoximines have been used as substrates, which upon reduction and subsequent desulfurization yield the corresponding optically active alcohols. For instance, reduction of the sulfoxide (218) with LiAlH$_4$ afforded the β-hydroxysulfhoxide (219) in 68% e.e. Reduction with NaBH$_4$, however resulted in decreased selectivity and the same reaction yielded only 20% of the enantiomeric product.

(218)   (219) 85%, 68% e.e.

Similarly, reduction of the β-keto-sulfoximine (220) with LiAlH$_4$ afforded (221) which was hydrogenolysed to give the optically active alcohol (222) in 65% optical purity [168].

(220)   (221) 70:30 d.p.

(222) 65%o.p.

Reduction of the substrate (223) to the alcohol (224) in 40% e.e. provides an example of 1,6-induction [169].

(223)   (224) 40% e.e. (S)

Mechanistic studies on diastereoselective non-catalytic reductions and other nucleophilic addition reactions to ketones have led to the formulation of several stereochemical rationalizations to explain the selectivity of such reactions. However these interpretations of stereoselection are often gross oversimplifications, even in simple systems, and the situation becomes more complex when other factors such as the effects of solvent, temperature, molar ratio etc. are taken into consideration. Nonetheless, Cram's rule of steric control of asymmetric induction in the synthesis of acyclic systems [170] as well as Prelog's classical treatise on nucleophilic additions to α-ketoesters of chiral alcohols [171] have served as useful guidelines for the development of theoretical concepts of stereoselectivity.

As depicted in Fig. 3.11, Cram's rule is best illustrated by the nucleophilic addition of R'Z to the substrate (225) in which the groups attached to the chiral centre are (S), (M) and (L) - i.e. small, medium and large respectively.

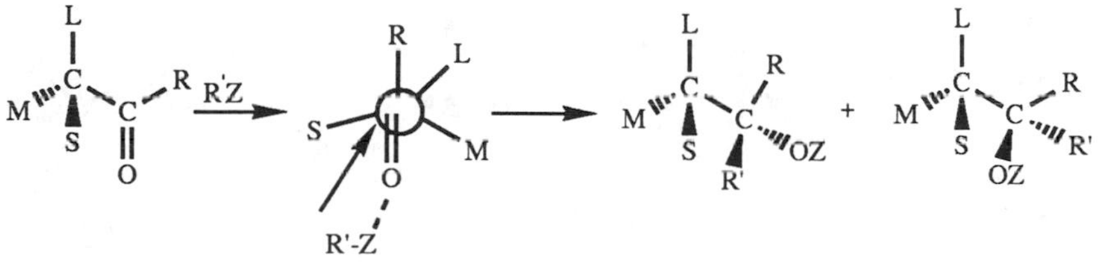

Figure 3.11.   Cram's rule

According to Cram's rule, the carbonyl group in the substrate (225) complexes with the reagent and adopts the most stable conformation (226) in which it is flanked by the two smaller groups (M) and (S). The reagent approaches the carbonyl group in the "reactant-like" transition state from the side of the smallest group (S) leading to the diastereomeric product (227). The rule strictly applies to non-catalytic reactions. It also does not apply to substrates of the type RCOCabc, where one of the substituents a, b or c such as OH, OR, NR_2 etc. is capable of chelating with the reagent. In such cases, another model, the "cyclic model", was proposed [172,173] (Fig. 3.13) and since it predicts the opposite stereochemistry substituent, such additions have been referred to as "*anti*-Cram additions". This nomenclature is inappropriate, since it was also proposed by Cram *et al* and it should more correctly be called as the Cram cyclic model, rigid model or chelate model. A comparison of predictions of the possible products by the acyclic and cyclic models is given in ref. [172].

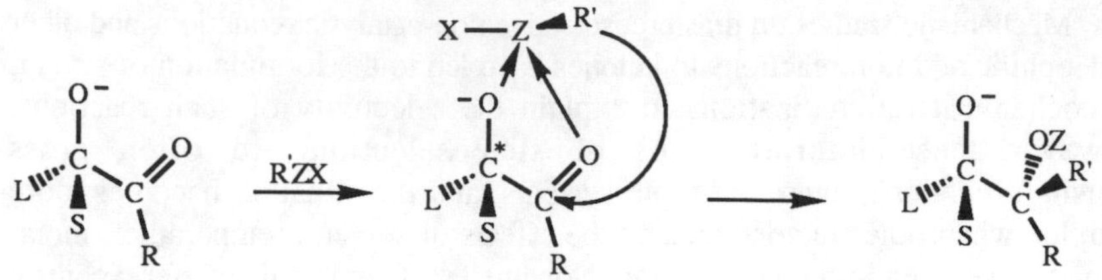

(229) (230) (231)

Figure 3.12. The cyclic model.

In cases where a substrate R-CO-C_{abX} is being reduced (X being an electronegative group), the carbonyl and C-X (e.g., C-Cl) dipoles orient themselves in an *anti*-periplanar fashion and a "dipolar" model (Fig. 3.13) has been accordingly proposed [174].

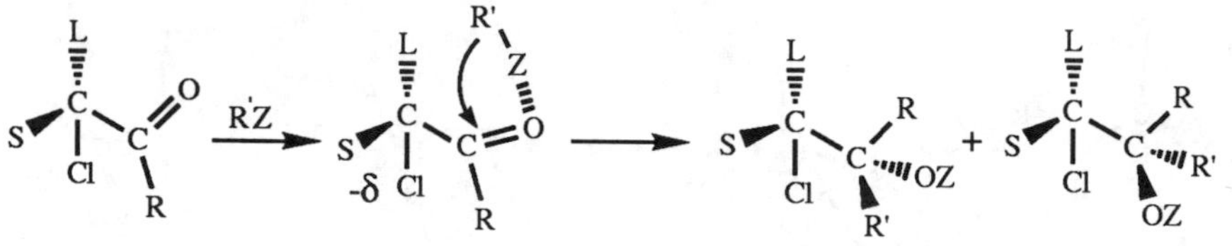

(232) (233) (234) (235)

Figure 3.13. The dipolar model.

Tables 3.15 and 3.16 show some examples of the reactions as predicted by cyclic and dipolar models respectively.

Table 3.15. Reductions and Nucleophilic Additions as Predicted by the Cyclic Model

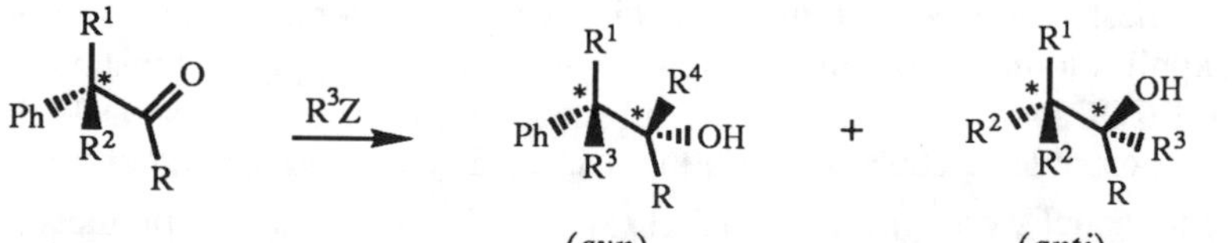

(*syn*) (*anti*)

Substrate				% *syn*
R^1	R^2	R^3	R	
OH	H	LAH	*p*-Tol	83
OH	H	PhMgBr	CH_3	96
OH	CH_3	CH_3MgI	Ph	66
OH	CH_3	PhLi	Ph	88
OH	H	PhMgBr	*p*-Tol	98

Table 3.16. Reductions and Nucleophilic Additions as Predicted by the Dipolar Model

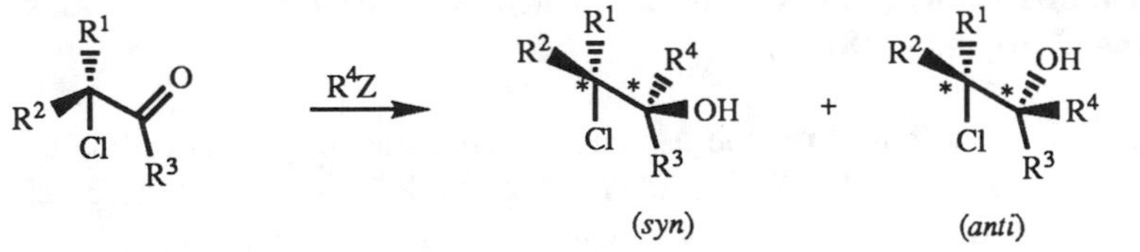

	Substrate			%syn
R^1	R^2	R^3	R^4	
H	CH_3	Ph	LAH	75
H	CH_3	Ph	$Al(O^iPr)_3$	66
H	Ph	CH_3	LAH	43
H	C_2H_5	Bu	$NaBH_4$	
H	Ph	Ph	LAH	80
				50

Other models have been proposed to explain the observed selectivities in acyclic systems, such as the Karabatsos model (Fig. 3.14) [175-177] and the Felkin's model (Fig. 3.15) [178,179].

In the Karabatsos model shown in Fig. 3.15 structures (236) and (238) are the best representations of the two minimum energy transition states leading to diastereomers (240) and (239) respectively. In both transition states, the

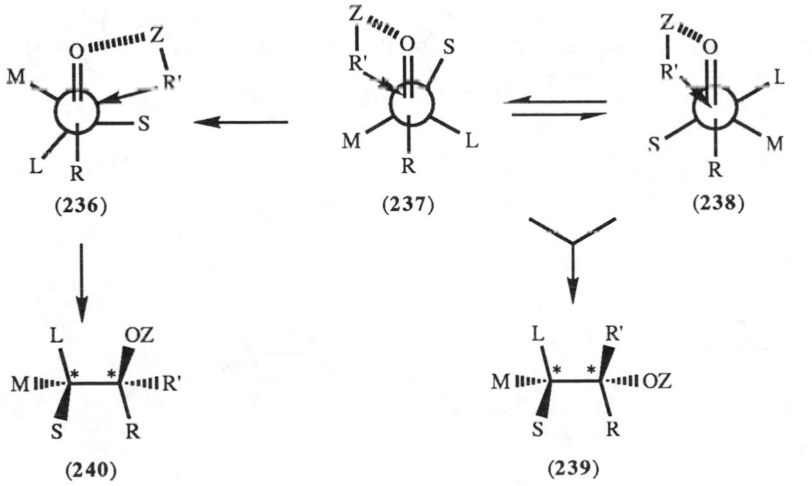

Figure 3.14. The Karabatsos model.

attacking nucleophile R' is nearest the smallest group (S). Taking R = CH₃, it has been calculated that rotamer (237) is destabilised by 3.3 kJ mol⁻¹. Thus the product ratios (240)/(239) can be correlated with the relative stabilities of structures (236) and (238).

This model is based on the assumption that the reagent coordinated to oxygen has the structure (244), provided that group R is not very large. In case, when R is very large, structure (245) would lead to the transition states (246) and (247) and the diastereomeric product ratio would obey Cram's rule.

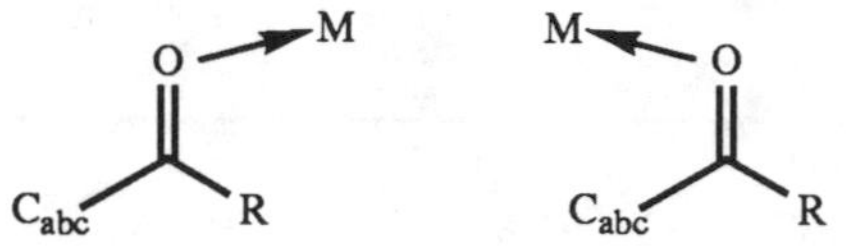

(244)   (245)

Felkin's model (Fig. 3.15) on the other hand considers not only a substrate-like transition state, but also takes into account the torsional strain generated by the existing full bonds and the partial bond R'---C---O. It also assumes that the approach of the nucleophile should be both perpendicular to the R-CO-Cα plane as well as staggered to the directing group as shown by the structures (241), (242) and (243). The minimum steric interaction occurs in structure (241) which leads to the product predicted by Cram's rule, while rotamers (242) and (243) give rise to its epimer. Moreover, the dominant interaction is between the

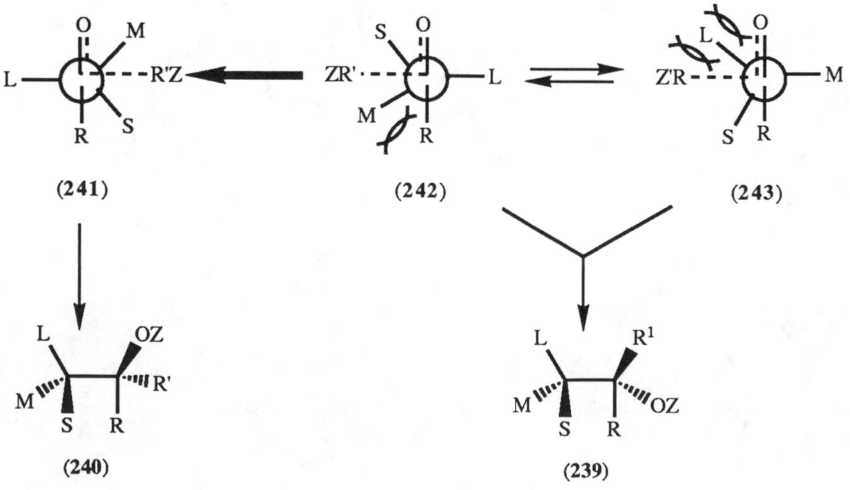

(241)   (242)   (243)

(240)   (239)

Figue 3.15. Felkin's model.

incoming nucleophile and the achiral group R attached to the carbonyl. The polar effects were also taken into account in this model and it was assumed that those transition states are stabilized in which separation between R' Z and the electronegative group at the chiral centre was the largest. This assumption was based on the observation that in the ground state, the carbonyl group and the hydrogn atom are eclipsed rather than *anti*-periplanar [180] as was assumed in the dipolar model.

(246)   (240)

(247)   (239)

Ab initio calculations [181,182] indicated that the transition state (248) according to Felkin was the most stable one, while those according to Cram, Cornforth and Karabatsos were all having higher energies (at least 11 KJ mol^{-1} more). It was also shown that the incoming nucleophile is *anti* periplanar to one of the groups attached to the *vicinal* atom, the magnitudes being Cl CH$_3$ H.

(248)   (249)

The preference of rotamer (248) over (249) followed from X-ray studies[183] and *ab initio* calculations [184] which showed that the incoming nucleophile approached the carbonyl group at an angle of 107° rather than of 90°.

3.2.2.2.1   Cyclic Ketones

In the case of cyclic ketones, a number of suggestions were made to account for the following observations:

i) preferential axial attack of the hydride reagent occurs in the reduction of unhindered cyclohexanones, and

ii) switch-over of stereoselectivity to preferential equatorial attack of the reagent is observed in the reduction of hindered cyclohexanones.

The first attempt to rationalize the variable stereoselectivity in the reduction of cyclic ketones, e.g., cyclohexanone by hydride reducing agents involved the assumption of "product development control" and "steric approach control" [185]. "Product development control "refers to the reaction going through a product-like transition state; the direction of the addition is therefore controlled by the stability of the product. This applies to unhindered ketones, e.g. (250).

In the case of hindered ketones, axial approach is hampered as shown in (251). In this case the dominating factor is the ease of attack of the reagent, i.e. "steric approach control" so that preferential equatorial attack occurs on hindered ketones resulting in the formation of axial alcohols.

(250) unhindered ketone   equatorial alcohol

(251) hindered ketone   axial alcohol

It follows that unhindered ketones appear to be reduced through a product-like transition state while hindered ketones are reduced through a reactant-like transition state. The latter statement is widely accepted, whereas for unhindered ketones various other rationalizations have also been put forward.

In the case of unhindered ketones, preference for axial attack has been attributed to repulsion by the axial hydrogens at C-2 and C-6 as shown in (252) [186]. This steric strain becomes more significant if we take into account the fact that the attack is non-perpendicular.

(252) : Steric strain

A similar conclusion can be reached by applying Felkin's model to cyclohexanones [179]. In Felkin's model, the incoming group which approaches the CO group at 90° is virtually eclipsed by the axial hydrogens at C-2 and C-6. Thus the torsional strain developing in equatorial attack between the forming C-H bond and the axial C-H bonds of C-2 and C-6 **(253)** results in a preferential attack from the axial side leading to the equatorial alcohols.

(253) : torsional strain

The Felkins model has been modified by performing *ab initio* calculations on various geometries of transition states for nucleophilic attack on CO groups [181,182] and it follows from such studies that (a) equatorial attack is not inhibited by the axial C-2 and C-6 hydrogens, and (b) that the axial attack is promoted as shown in **(254)**.

(254)

An entirely different rationalization has been put forward based on orbital interaction of the "π" orbitals of the CO group with the σ orbitals of the β C-C bonds (C$_2$-C$_3$,C$_6$-C$_5$)). Such interaction gives rise to unsymmetrical electron density on the two faces of the CO group with a higher electron density on the equatorial face **(255)**. Axial attack is thus favored [187, 188].

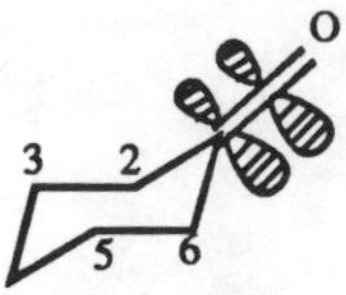

(255)

In another explanation it has been proposed that the incoming nucleophile from the equatorial side would encounter three axial hydrogens (C2, C4 and C6) whereas from the axial side, it will encounter only two axial hydrogens (C3 and C5) [189]. Quantum mechanical calculations of electrostatic potential around the carbonyl group by CNDO also predict axial attack as the favoured one as a result of this electrostatic potential [190].

Of the various possible mechanisms put forward in the case of borohydride reductions, the one shown in Fig. 3.16 is the widely accepted one. The mechanism of LiAlH₄ reduction is quite different from that of NaBH₄ reductions, the former involving a reactant-like transition state whereas the latter appears to have a product-like transition state. There is no single global explanation which would account for the stereoselectivity obtained from reductions with all reducing agents.

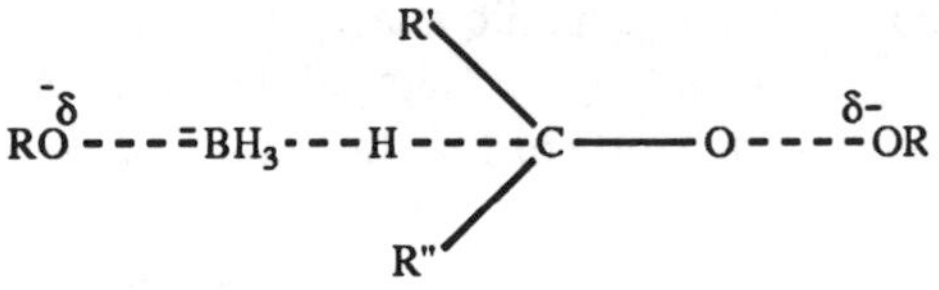

$$RO \cdots \equiv BH_3 \cdots H \cdots C \!-\!\!-\!\!-\! O \cdots OR$$

Figure 3.16. Proposed transition state for borohydride reduction of ketones.

3.3 References

1.   J.D. Morrison, ed. "Asymmetric Synthesis" Vol.2, London: Academic Press, (1983).

2.   M. Nogradi, "Stereoselective Synthesis", p. 105-141, Weinheim, VCH Verlagsgesellschaft, (1987).

3.   (a) J.W ApSimon and R.P. Seguin, *Tetrahedron* **35**, 2797 (1979). (b) J.W. ApSimon and T. Lee Collier, *Tetrahedron*, **42**, 5157 (1986).

4.   A.A. Bothner-By, *J.Amer.Chem.Soc.*, **73**, 846 (1951).

5.   P.S. Portoghese, *J.Org.Chem.*, **27**, 3359 (1962).

6.   S.R. Landor, B.J. Miller and A.R. Tatchell, *J.Chem.Soc.C*. 1922 (1966).

7.   J.D. Morrison, H.S. Mosher, "Asymmetric Organic Reactions," Englewood Cliffs, New Jersey, p.202,: Prentice Hall, (1971).

8.   For a review see: H. Houbenstock, *Top. Stereochem.* **14**, 231 (1982).

9.   R. Andrisano, S.R. Angeloni and S. Marzocchi, *Tetrahedron*, **29**, 913 (1973).

10.   S.R. Landor, B.J. Miller and A.R. Tatchell, *J.Chem.Soc.*, 227 (1964).

11.   S.R. Landor, B.J. Miller and A.R. Tatchell, *J.Chem.Soc.*, C 1822 (1966)

12.   S.R. Landor, B.J. Miller and A.R. Tatchell, *ibid*, 2280 (1966).

13.   S.R. Landor, B.J. Miller and A.R. Tatchell, *ibid*, 197 (1967).

14.   S.R. Landor, B.J. Miller and A.R. Tatchell, *ibid*, 2339 (1971).

15.   S.R. Landor, B.J. Miller and A.R. Tatchell, *J.Chem.Soc.Chem.Commun.*, 585 (1966)

16.   S.R. Landor, O.O. Sonola and A.R. Tatchell, *J.Chem.Soc.Perkin Trans. I*, 1902 (1974).

17.   S.R. Landor, O.O. Sonola and A.R. Tatchell, *ibid* 605 (1978).

18.   R. Haller and H.J. Schneider, *Chem. Ber.*, 1312 (1973).

19.   H.J. Schneider and R. Haller, *Just. Liebigs Ann.Chem.*, **743**, 187 (1971).

20.   N. Baggett and P. Stribblehill, *J.Chem.Soc.Perkin Trans.I*, 1123 (1977).

21.   E.D. Lund and P.E. Shaw, *J.Org.Chem.*, **42**, 2073 (1977).

22.   R.Noyori, I. Tomino and Y. Tanimoto, *J.Amer.Chem.Soc.* **101**, 3129 (1979).

23.   R. Noyori, I. Tomino and M. Nishizawa, *J.Amer.Chem.Soc.*, **101**, 5843 (1979).

24.   M. Nishizawa and R. Noyori, *Tetrahedron Lett.*, 2821 (1980).

25.   R. Noyori, I. Tomino, Y. Tonimoto and M. Nishizawa, *J.Amer.Chem.Soc.*, **106**, 6709 (1984).

26.   R. Noyori, I. Tomino, M. Yamada and M. Nishizawa, *J.Amer.Chem.Soc.*, **106**, 6717 (1984).

27.   S. Yamaguchi and K. Kabuto, *Bull. Chem.Soc.*, Japn., **50**(11), 3033 (1977).

28.   J.L. Pierre and H. Handel, *Tetrahedron Lett.*, 2317 (1974).

29.   J.-P. Vigneron and V. Bloy, *Tetrahedron Lett.*, **29**, 2683 (1979).

30.   M. Kawasaki, Y. Suzuki and T. Terashima, *Chemistry Lett.*, 239 (1984).

31.   D. Seebach and H. Daum, *Chem. Ber.*, **107**, 1748 (1974).

32.   M. Schmidt, P. Amstutz, G. Crass and D. Seebach, *Chem.Ber.*, **113**, 1691 (1980).

33.   A.I. Meyers and P.M. Kendall, *Tetrahedron Lett.*, 1337 (1974).

34.   O. Cervinka and O. Belovsky, *Coll.Czech.Chem.Commun.*, **30**, 2487 (1965); **32**, 3897 (1967).

35.   G.M. Giongo, F.D. Gregorio, N. Palladino and W. Marconi, *Tetrahedron Lett.*, 3195 (1973).

36.   G. Giacomelli, R. Menicagli and L. Lardicci, *J.Org.Chem.*, **38**, 2370 (1973).

37.   S. Yamaguchi and H.S. Mosher, *J.Org.Chem.*, **38**, 1870 (1973).

38.   C.J. Reich, G.R. Sullivan and H.S. Mosher, *Tetrahedron Lett.*, 1505 (1973).

39.   R.S. Brinkmeyer and V.M. Kapoor, *J.Amer.Chem.Soc.*, **99**, 8339 (1977).

40.   R.S. Brinkmeyer and V.M. Kapoor, *J.Amer.Chem.Soc.*, **94**, 9254 (1972).

41.   N. Cohen, R.J. Lopresti, C. Neukon and G. Saucy *J.Org.Chem.*, **45**, 582 (1980).

42.   T. Sato, Y. Goto and T. Fujisawa, *Tetrahedron Lett.*, 4111 (1982).

43.   T. Sato, Y. Gotoh, Y. Wakabayashi and T. Fujisawa, *Tetrahedron Lett.*, **24**(38), 4123 (1983).

44.   K. Yamamoto, H. Fukushima and M. Nakazaki, *J.Chem.Soc.Chem.Commun.*, 1490 (1984).

45.   J.D. Morrison, E.R. Grandbois, S.I. Howard and G.R. Weisman, *Tetrahedron Lett.*, **22**(28), 2619 (1981).

46.   H.C Brown, P.K. Jadhav, and B. Singaram, in "Modern Synthetic Methods" (R. Scheffold, ed.), Berlin: Spinger Verlag, (1986).

47.   M. Srebnik and P.V. Ramachandran, *Aldrichimica Acta*, **20**, 9 (1987).

48.   H.C. Brown and P.K. Jadhav, *J.Org.Chem.*, **46**, 5047 (1981).

49.   H.C. Brown, P.K. Jadhav and A.K. Mandal, *J. Org. Chem.*, **47**, 5074 (1982).

50.   H.C. Brown and A.K. Mandal, *J.Org.Chem.*, **42**, 2996 (1977).

51.   H.C. Brown and A.K. Mandal, *J.Org.Chem.*, **49**, 2558 (1984).

52.   (a) H.C. Brown, P.K. Jadhav and A.K. Mandal, *Tetrahedron,* **37**, 3547 (1981) (b) H.C. Brown and A.K. Mandal, *J.Org.Chem.*, **49**, 2558 (1984).

53.   M.M. Midland, S. Greer, A. Tramontano and S.A. Zderic, *J.Amer.Chem.Soc.*, **101**, 2352 (1979).

54.   M.M. Midland, D.C. McDowell, R.L. Hatch and A. Tramontano, *J.Amer.Chem.Soc.*, **102**, 867 (1980).

55.   H.C. Brown, J. Chandrasekharan and P.V. Ramachandran, *J.Amer.Chem.Soc.*, **110**, 1539 (1988).

56.   (a) B.M. Mikhailov, Yu.N. Bubnov and V.G. Kiselev, *J.Gen.Chem.USSR* (Engl. Trans. of *Zh.Obshch.Khim*)., **36**, 65 (1966).   (b) B.M. Mikhailov, M.E. Kuimova, and E.A. Shagova, *D.Kl.Acad.Nauk. SSR, Ser.Khim.*, **179**, 1344 (1968).

57.   M.M. Midland, A. Tramontano and S.A. Zderic *J.Organomet.Chem.*, **134**, C17 (1977).

58.   M.M. Midland, A. Tramontano and S.A. Zderic, *J.Organomet.Chem.*, **156**, 203 (1978).

59.   M.M. Midland and S.A Zderic, *J.Amer.Chem.Soc.*, **104**, 525 (1982).

60.   H.C. Brown and G.G. Pai, *J.Org.Chem.*, **47**, 1608 (1982).

61.   A. Hirao, H. Mochizuki, S. Nakahama and N.Yamazaki *J.Org.Chem.*, **44**, 1720 (1979).

178   3. Stereoselective Non-Catalytic Reductions

62.   J.D. Morrison, E.R. Grandbois and S.I. Howard, *J.Org.Chem.*, **45**, 4229 (1980).

63.   A. Hirao, S. Nakahama, H. Mochizuki, S.Itsuno and N. Yamazaki, *J.Org.Chem.*, **45**, 4231 (1980).

64.   A. Hirao, S. Nakahama, D. Mochizuki, S. Itsuno, M. Ohowa and N. Yamazaki, *J.Chem.Soc.Chem.Commun.*, 807 (1979).

65.   K. Yamada, M. Takeda and T. Iwakuma, *Tetrahedron Lett.*, **22**(39), 3869 (1981).

66.   (a) K. Yamada, M. Takeda and T. Iwakuma, *J.Chem.Soc.Perkin Trans.I*, 265 (1983).   (b) F. Toda, K. Kiyoshige and M. Yagi, *Angew.Chem.*, **28**, 320 (1989).

67.   S.Colonna and R. Fornasier, *J.Chem.Soc.Perkin Trans. I*, 371 (1978).

68.   J. Balcells, S. Colonna and R. Fornasier, *Synthesis*, 266 (1976).

69.   R. Kinishi, Y. Nakajima, I.Oda and Y. Irouze, *Agric.Biol.Chem.*, **42**, 869 (1978).

70.   Y. Shida, N. Onda, Y. Yamamoto, J.Oda and Y. Inouye, *Agric.Biol.Chem.*, **43**, 1797 (1979).

71.   T. Sugimoto, Y. Matsumura, S. Tanimoto and M. Okano, *J.Chem.Soc.Chem. Commun.*, 926 (1978).

72.   E.J. Corey, S.M. Albonico, U. Koelliker, T.K. Schaaf and R.K. Varma, *J.Chem.Soc.*, **93**, 1491 (1971)

73.   E.J. Corey, B.K. Becker and R.K. Varma, *J.Amer.Chem.Soc.*, **94**, 8616 (1972).

74.   H.C. Brown, S. Krishnamurthy and J.L. Hubbard, *J.Organomet.Chem.*, **166**, 271 (1979).

75.   C.A.Brown, *J.Amer.Chem.Soc.*, **95**, 4100 (1973).

76.   C.A. Brown and S. Krishnamurthy, *J.Organomet.Chem.*, **156**, 111 (1978).

77.   S. Krishnamurthy, F. Vogel and H.C. Brown, *J.Org.Chem.*, **42**(14), 2534 (1977).

78. M.F. Grundon, W.A. Khan, D.R. Boyd and W.R. Jackson, *J.Chem.Soc.,(C)*, 2557 (1971).

79. J.F. Archer, D.R. Boyd, W.R. Jackson, M.F. Grundon and W.A.Khan, *J.Chem.Soc.(C)*, 2560 (1971).

80. D.R. Boyd, M.F. Grundon, and W.R. Jackson, *Tetrahedron Lett.*, 2101 (1967).

81. M.M. Midland and A. Kazubski, *J.Org.Chem.*, **47**, 2814 (1982).

82. V. Grignard, *Ann.Chim.Phys.*, **24**, 468 (1901).

83. G.Vavon and B. Angelo, *Compt.rend.*, **224**, 1435 (1947).

84. G. Vavon, C. Riviere and B. Angelo, *Compt.rend.*, **222**, 959 (1946).

85. R.Macleod, F.J. Welch and H.S. Mosher, *J.Amer.Chem.Soc.*, **82**, 876 (1960)

86. E.P. Burrows, F.J. Welch and H.S. Mosher, *J.Amer.Chem.Soc.*, **82**, 880 (1960).

87. J.S. Birtwistle, K. Lee, J.D. Morrison, W.A. Sanderson and H.S. Mosher, *J.Org.Chem.*, **29**, 37 (1964).

88. J.D. Morrison, D.L. Black and R.W. Ridgway, *Tetrahedron Lett.*, 985 (1968).

89. H.S. Mosher and E. La Combe, *J.Amer.Chem.Soc.*, **72**, 4991 (1950).

90. H.S. Mosher and E.La Combe, *J.Amer.Chem.Soc.*, **72**, 3994 (1950).

91. For an earlier review on Grignard reagents see; M.S. Kharasch and S. Weinhouse, *J.Org.Chem.*, **1**, 209 (1936).

92. G. Giacomelli, R. Menicagli and L. Lardicci, *J. Org.Chem.*, **39**, 1757 (1974).

93. L. Lardicci and G. Giacomelli, *J.Chem.Soc. Perkin Trans.I*, 337 (1974).

94. G. Giacomelli, R. Menicagli and L. Lardicci, *J.Org.Chem.*, **38**, 2370 (1973).

95. L. Lardicci, G.P. Giacomelli and R. Menicagli, *Tetrahedron Lett.*, 687 (1972).

96. G.P. Giacomelli, R. Menicagli and L. Lardicci, *Tetrahedron Lett.*, 4135 (1971).

97. W.M. Foley, F.J. Welch, E. M.La Combe and H.S. Mosher, *J.Amer. Chem.Soc.*, **81**, 2779 (1959).

98.   K. Mislow and P. Newman, *J.Amer.Chem.Soc.*, **79**, 1769 (1957).

99.   A. Streitwieser, Jr., J.R. Wolfe, Jr. and W.D. Schaeffer, *Tetrahedron*, **6**, 338 (1959).

100. G. Giacomelli, L. Lardicci and A.M. Caporusso, *J.Chem.Soc.Perkin Trans. I*, 1795 (1975).

101. B. Denise, J. Ducom and J.F. Fauvarque, *Bull.Soc.Chim.France*, 990 (1972).

102. E.C. Ashby, J.Laemmle, and H.M. Neumann, *Accounts Chem.Res.*, **7**, 272 (1974).

103. W.R. Doering and I.W. Young, *J.Amer.Chem.Soc.*, **71,**, 631 (1948).

104. H.S. Mosher and E.D. Parker, *J.Amer.Chem.Soc.*, **78**, 4081 (1956).

105. H.S. Mosher, J.E. Stevenot and D.O. Kimble, *J.Amer.Chem.Soc.*, **78**, 4374 (1956).

106. H.S. Mosher and P.K. Loeffler, *J.Amer.Chem.Soc.*, **78**, 5597 (1956).

107. G.P.Giacomelli, R. Menicagli and L. Lardicci, *J.Amer.Chem.Soc.*, **97**, 4009 (1975).

108. G. Giacomelli and L. Lardicci, *J.Org.Chem.*, **47**, 4335 (1982).

109. G. Giacomelli, L. Lardicci and F. Palla, *J.Org.Chem.*, **49**, 310 (1984).

110. G. Giacomelli, R. Menicagli, A.M. Caporusso and L. Lardicci, *J.Org.Chem.*, **43**, 1790 (1978).

111. D. Nasipuri and P.K. Bhattacharya, *J.Chem.Soc.Perkin Trans.I*, 576 (1977).

112. D. Nasipuri and P.R. Mukherjee, *J.Indian Chem.Soc.*, **51**, 171 (1972).

113. D. Nasipuri and G. Sarkar, *J.Indian Chem.Soc.*, **44**, 425 (1967).

114. D. Nasipuri, C.K. Ghos and R.J.L. Martin, *J.Org.Chem.*, **35**, 657 (1970).

115. A.K. Samaddar, S.K. Konar and D. Nasipuri, *J.Chem.Soc.Perkin Trans. I*, 1449 (1983).

116. D. Nasipuri, G. Sarkar and C.K. Ghosh, *Tetrahedron Lett.*, 5189 (1967).

117. D. Nasipuri and G. Sarkar, *J.Indian Chem.Soc.*, **44**, 165 (1967).

118. D. Nasipuri and C.K. Ghosh, *J. Indian Chem.Soc.*, **44**, 556 (1967).

119. D. Cabaret and Z. Welvart, *J.Chem.Soc.Chem.Commun.*, 1064 (1970).

120. R. Andrisano, A.S. Angeloni and Merzocchi, *Tetrahedron* **29**, 913 (1973).

121. L.D. Tomina, E.I. Klabunovskii, Yu. I. Petrov, E.D. Lubuzh and E.M. Cherkasova, *Izv.Akad.Nauk, SSR, Ser. Khim.*, **11**, 2506 (1972).

122. R.A. Kretchmer, *J.Org.Chem.*, **37**, 801 (1972).

123. Y. Ohnishi, M. Kagami and A. Ohno, *J.Amer.Chem.Soc.*, **97**, 4766 (1975).

124. A. Ohno, M. Ikeguchi, T. Kimura and S. Oka, *J.Amer.Chem.Soc.*, **101**, 7036 (1979).

125. A. Ohno, T. Kimura, H. Yamamoto, S.G. Kim, S. Oka and Y. Ohnishi, *Bull.Chem.Soc.Jpn.*, **50**, 1535 (1977); *Bioorg.Chem.*, **6**, 21 (1977).

126. A. Ohno, Y. Yamamoto, T. Okamoto,S. Oka and Y. Ohnishi, *Bull.Chem. Soc. Jpn.*, **50**, 2385 (1977).

127. A. Ohno, T. Kimura, S. Oka and Y. Ohnishi, *Tetrahedron Lett.*, **8**, 757 (1978).

128. B.L. Vallee and W.E. C. Wacker, *J.Amer.Chem.Soc.*, **78**, 1771 (1956).

129. R. Bentley, : "Molecular Asymmetry in Biology" Vol. 2, p. 50, New York, : Academic Press, (1970).

130. J. Everse and N.O. Kaplan, *Adv. Enzymol. Relat. Areas Mol. Biol.* **37**, 61 (1973).

131. K. Nishiyama, N. Baba, J. Oda and Y. Inouye, *J.Chem.Soc.Chem.Commun.*, **101** (1976).

132. N. Baba, Y. Matsumura and T. Sugimoto, *Tetrahedron Lett.*, **44**, 4281 (1978).

133. Y. Inouye, J. Oda, N. Baba, : "Asymmetric Synthesis", (J.D. Morrison, ed.), Vol.2, pp.92, Academic Press Orlando, (1983) (and references cited therein).

134. D.C. Dittmer, A. Lombardo, F.H. Batzold and C.S. Greene, *J.Org.Chem.*, **41**, 2976 (1976).

135. S. Shinkai, T. Ide, H. Hamada, O. Manabe and T. Kunitake, *J.Chem.Soc. Chem.Commun.*, 848 (1977).

136. R.A. Gase, G. Boxhoorn and U.K. Pandit, *Tetrahedron Lett.*, 2889 (1976).

137. M. Huges and R.H. Prince, *J.Inorg.Nucl.Chem.*, **40**, 703 (1978).

138. A. Ohno, S. Yasui, R.A. Gase, S. Oka and U.K. Pandit, *Bioorg.Chem.*, **9**, 199 (1980).

139. N. Baba, J. Oda and Y. Inouye, *Angew.Chem.*, **21**(6), 433 (1982).

140. M. Seki, N. Baba, J. Oda and Y. Inouye, *J.Amer.Chem.Soc.*, **103**, 4613 (1981).

141. P. Jouin, C.B. Troostwijk and R.M. Kellogg, *J.Amer.Chem.Soc.*, **103**, 2091 (1981).

142. D.C. Wigfield and D.J. Phelps, *J.Org.Chem.*, **41**, 2396 (1976).

143. D.C. Wigfield, G.W. Buchanan, C.M.E. Ashley and S. Feiner, *Can.J.Chem.*, **54**, 3536 (1976).

144. E.L. Eliel and S.H. Schroeter, *J.Amer.Chem.Soc.*, **87**, 5031 (1965).

145. E.C. Ashley and J.R. Boone, *J.Amer.Chem.Soc.*, **98**, 5524 (1976).

146. H.C. Brown and S. Krishnamurthy, *J.Amer.Chem.Soc.*, **94**, 7159 (1972).

147. J. Hooz, S. Akiyama, F.J. Cedar, M.J. Bennett and R.M. Tuggle, *J.Amer. Chem.Soc.*, **96**, 274 (1974).

148. S. Krishnamurthy and H.C. Brown, *J.Amer.Chem.Soc.*, **98**, 3383 (1976).

149. H.C. Brown, J.I. Hubbade and B. Singaram, *Tetrahedron*, **37**, 2359 (1981).

150. H.C. Brown and H.R. Deck, *J.Amer.Chem.Soc.*, **87**, 5620 (1965).

151. S. Kin, K.H. Ahn and Y.W. Chung, *J.Org.Chem.*, **47**, 4581 (1982).

152. J. Canceill and J. Jacques, *Bull.Soc.Chim.France.*, 2180 (1970).

153. M. Cherest, H. Felkin and N. Prudent, *Tetrahedron Lett.*, 2199 (1968).

154. C. Zioudrou, I. Moustakali-Mavridis, P. Chrysochou and G.J. Karabatsos, *Tetrahedron*, **34**, 3181 (1978).

155. M.M. Midland and Y.C. Kwon, *J.Amer.Chem.Soc.*, **105**, 3725 (1983).

156. M. Tramontini, *Synthesis*, 605 (1982).

157. D. Hartley, *Chem.Ind.* (London) 551 (1981).

158. H.K. Mueller, E. Mueller and H. Baborowski, *J.Prakt.Chem.*, **313**, 1 (1971).

159. Y. Gault and H. Felkin, *Bull.Soc.Chim. France* 1342 (1960)

160. D.J. Cram and J. Allinger, *J.Amer.Chem.Soc.*, **76**, 4516 (1954).

161. K. Koga and S. Yamada, *Chem.Pharm.Bull.*, **20**, 526 (1972).

162. H.K. Mueller, J. Schuart, H. Baborowski and E. Mueller, *J.Prakt.Chem.*, **315**, 449 (1973).

163. H.K. Mueller, I. Jarchow and G. Rieck, *Liebigs, Ann.Chem.*, **613**, 103 (1958).

164. M. Brienne, C. Ouannis and J. Jacques, *Bull.Soc.Chim.France*, 1036 (1968).

165. A.McKenzie, *J.Chem.Soc.*, **85**, 1249 (1904).

166. R. Annuziata, M. Cinquini and F. Cozzi, *J.Chem.Soc.Perkin Trans.I*, 1687 (1979).

167. R. Annunziata, M. Cinquini and F. Cozzi, *J.Chem.Soc.Perkin Trans.I*, 1109 (1981).

168. C.R. Johnson and C.J. Stark Jr. *J.Org.Chem.*, **47**, 1196 (1982).

169. B.P. Giovanni, M. Fabio, P.G. Piero, S. Daniele, B.Achille and B. Simonetta, *J.Chem.Soc.Perkin Trans. I*, 2983 (1982).

170. D.J. Cram and F.A. Abd Elhafez, *J.Amer.Chem.Soc.*, **74**, 5828 (1952).

171. V. Prelog, *Helv.Chim.Acta*, **36**, 308 (1953).

172. D.J. Cram and K.R. Kopecky, *J.Amer.Chem.Soc.*, **81**, 2748 (1959).

173. D.J. Cram and D.R. Wilson, *J.Amer.Chem.Soc.*, **85**, 1249 (1963).

174. J.W. Cornforth, R.H. Cornforth and K.K.Mathew, *J.Chem.Soc.*, 112 (1959).

175. G.J. Karabatsos and R.A. Taller, *Tetrahedron*, **24**, 3923 (1968).

176. G.J. Karabatsos and D.J. Fenoglio, *Top.Stereochem.*, **5**, 167 (1970).

177. G.J. Karabatsos, *J.Amer.Chem.Soc.*, **89**, 1367 (1967).

178. M. Cherest, H. Felkin and N. Prudent, *Tetrahedron Lett.*, 2199 (1968).

179. M. Cherest and H. Felkin, *Tetrahedron Lett.*, 2205 (1968).

180. G.J. Karabatsos and D.J. Fenoglio, *J.Amer.Chem.Soc.*, **91**, 1124 (1969).

181. N.T. Anh and O. Eisenstein, *Nouv.J.Chim.*, **1**, 61 (1976).

182. N.T. Anh, *Top.Curr.Chem.*, **88**, 146 (1980).

183. H.B. Bürgi, D.J. Dunitz, J.M. Lehn and G. Wipf, *Tetrahedron*, **30**, 1563 (1974).

184. H.B. Bürgi, J.M. Lehn and G. Wipf, *J.Amer.Chem.Soc.*, **96**, 1956 (1974).

185. W.G. Dauben, G.J. Fonken and D.S. Noyce, *J.Amer.Chem.Soc.*, **78**, 2579 (1956).

186. J.C. Richer, *J.Org.Chem.*, **30**, 324 (1965).

187. J.Klein, *Tetrahedron Lett.*, 4307 (1973).

188. J. Klein, *Tetrahedron* **30**, 3349 (1974).

189. D.C. Wigfield and F.W. Gowland, *J.Org.Chem.*, 1108 (1977).

190. J. Royer, *Tetrahedron Lett.*, 1343 (1978).

4 Stereoselective Carbon-Carbon Bond Forming Reactions

One of the most challenging tasks for organic chemists today is the stereoselective synthesis of complex molecules in high yields and selectivities. This is rendered more demanding when factors such as efficiency, economy and praticability come into play. Carbon-carbon bond forming reactions form the backbone upon which the edifice of most synthetic plans and strategies must be built.

In asymmetric carbon-carbon bond forming reactions, the enantioselectivity may be achieved by using chiral substrates, chiral reagents, by conducting the reaction in a chiral medium or by employing chiral complexing agents. This chapter deals with such reactions.

4.1 Nucleophilic Additions to Aldehydes and Ketones

4.1.1 Enantioselective Addition Reactions

Generally carbon-carbon bond formations by asymmetric addition to carbonyl compounds involve organometallics such as alkyllithiums or Grignard reagents which can be added with enantioface differentiating ability. When achiral substrates and achiral organometallic reagents are used, then some chiral complexants such as optically active amines, ethers, amides or metal alkoxides need to be used to induce chirality. Table 4.1 shows some chiral complexants used in the enantioselective addition of alkyl metals to carbonyl compounds.

One of the first enantioselective carbon-carbon bond forming reactions reported was the hydrogenation of benzaldehyde (35) in the presence of emulsin or quinine to afford mandelonitrile (36) [1-4].

Table 4.1. Chiral Complexants used in Enantioselective Carbon-Carbon Bond Forming Reactions of Carbonyl Compounds

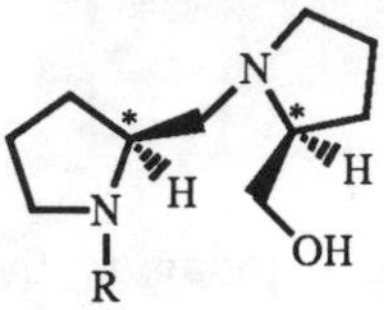

(1) R = CH$_3$
(2) R = (CH$_3$)$_3$C-CH$_2$

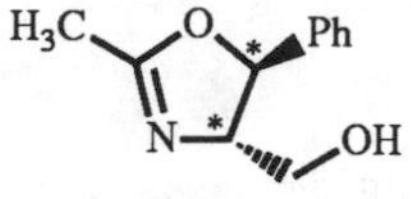

(3)

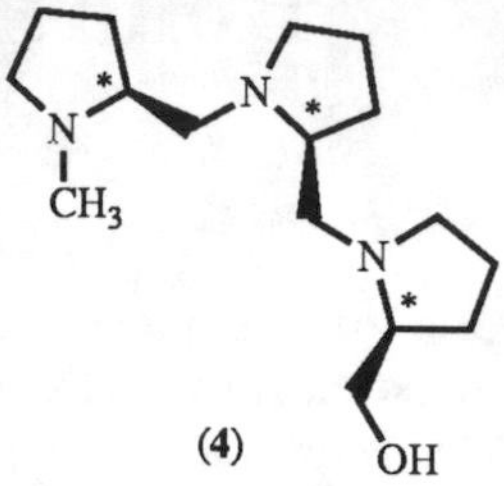

(4)

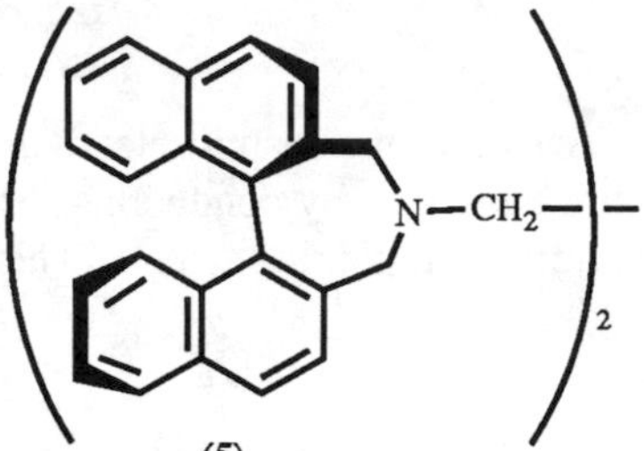

(5)

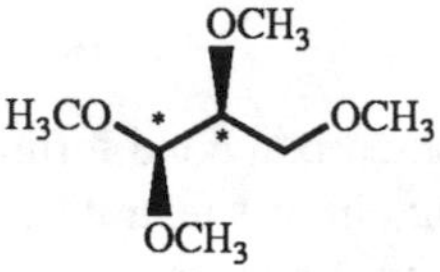

(6)

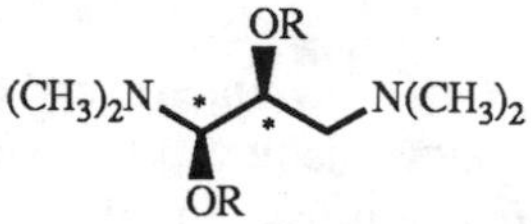

(7) R = CH$_3$
(8) R = (CH$_3$)$_2$NCH$_2$CH$_2$

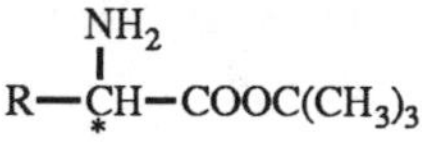

(9) R = (CH$_3$)$_2$CH
(10) R = (CH$_3$)$_3$C

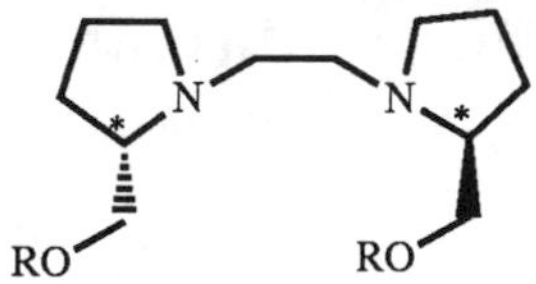

(11) R = H
(12) R = CH$_3$

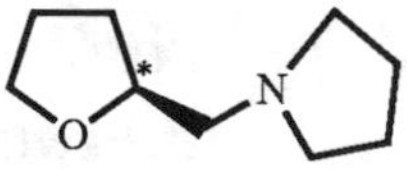

(13)

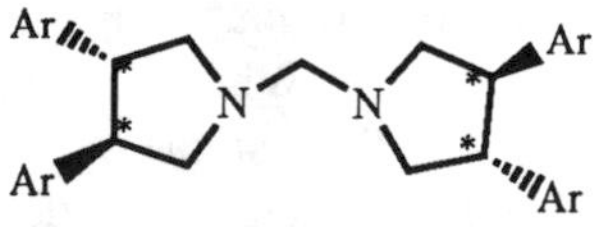

(14)

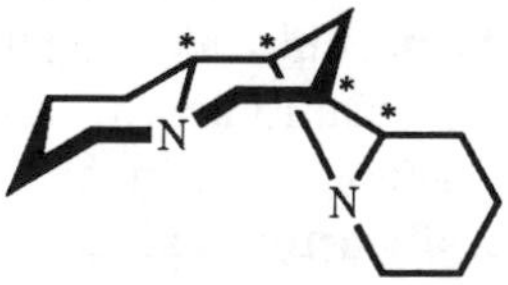

(15)

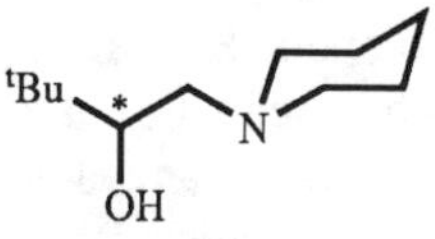

(16)

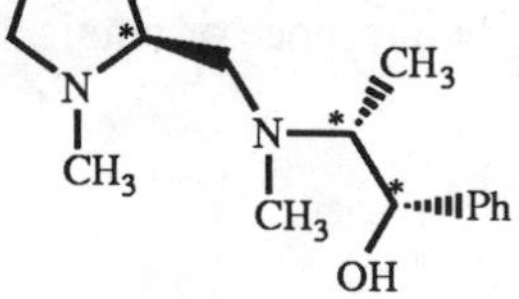

(17)

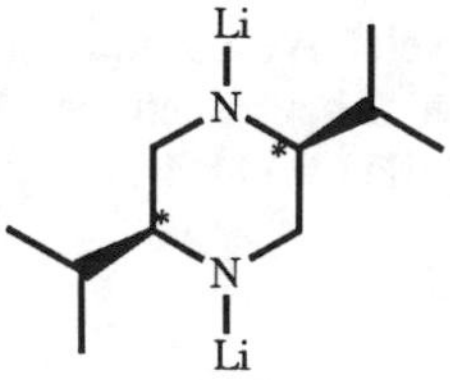

(18)

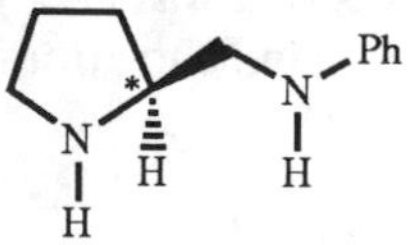

(19)

(Table 4.1. contd.)

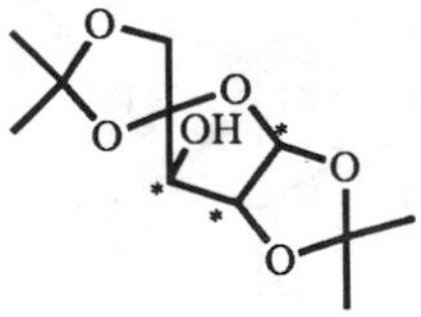

(20)

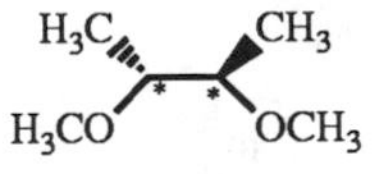

(21)

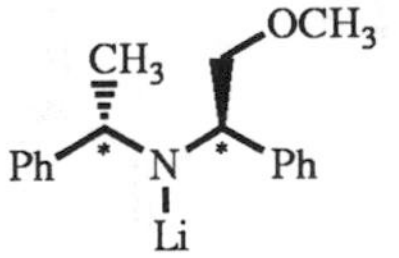

(22)

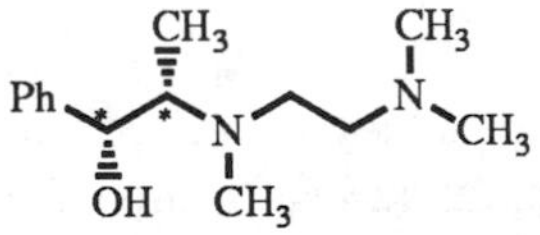

(23)

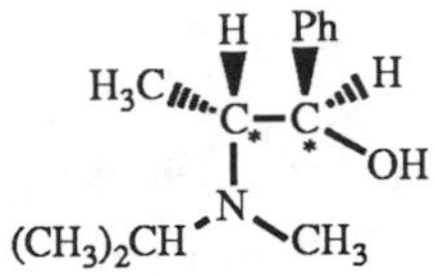

(24)

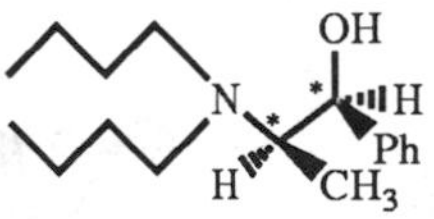

(25)

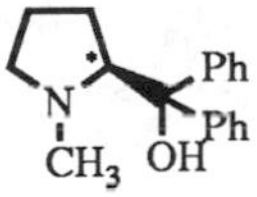

(26)

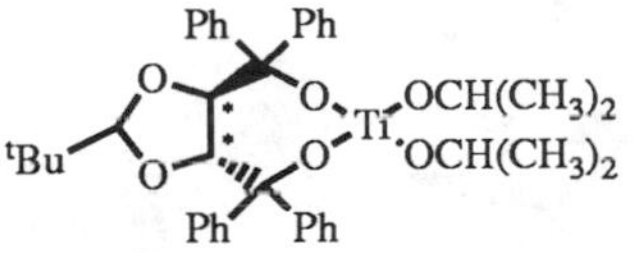

(27)

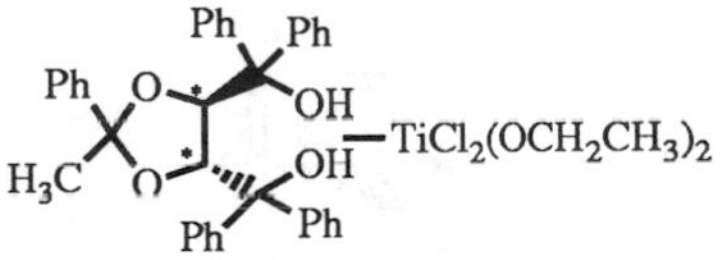

(28)

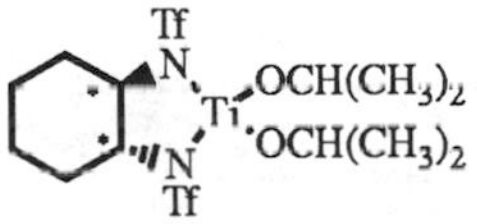

(29)

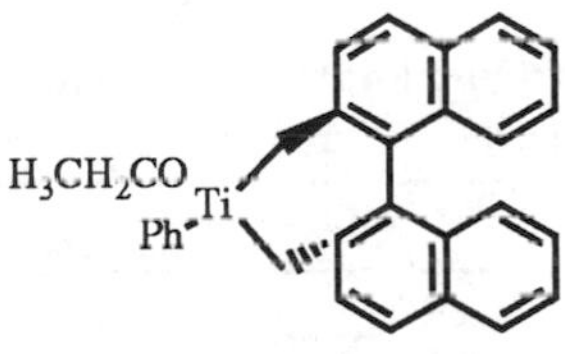

(30)

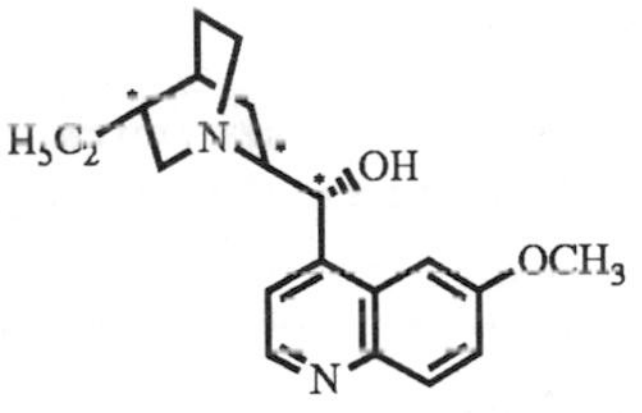

(31)

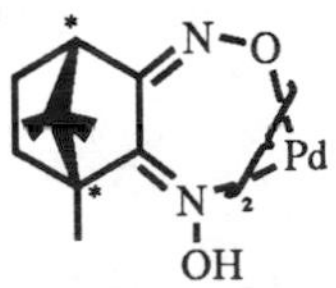

(32)

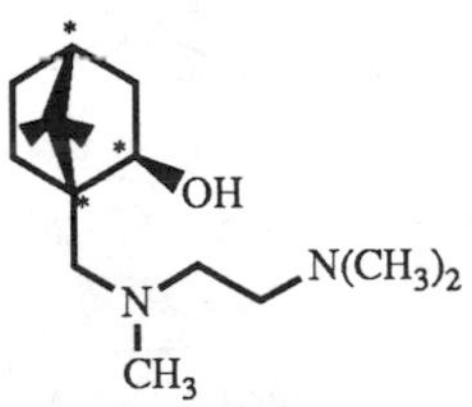

(33)

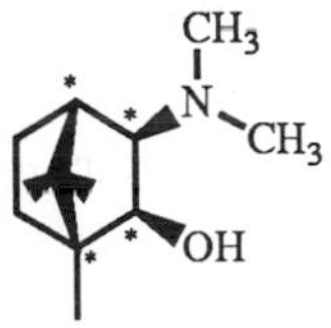

(34)

The enantioselectivity of the hydrocyanation was originally very low and the reaction was considered to be of theoretical significance only [5]. Subsequently cyclic dipeptides such as cyclo [L-phenylalanyl- L-histidines] were used as catalysts in the hydrocyanation of benzaldehyde derivatives yielding the corresponding products in upto 90% e.e [6,7].

(35)   **(36)** (*R*)

(R = H, *p*-CH$_3$, *m*-CH$_3$, OCH$_3$, *m*-OCH$_3$, etc.)

Asymmetric hydrocyanation of aldehydes has been achieved in excellent optical yields by using chiral titanium complexes such as **(28)** affording the corresponding products in upto 96% e.e. [8-10]. Thus, the asymmetric hydrocyanation of the substrate **(37)** with **(28)** in the presence of cyanotrimethylsilane yielded the corresponding alcohol **(38)** in 93% e.e. [9].

CH$_2$=CH—(CH$_2$)$_8$—C—H $\xrightarrow[\text{(28)}]{\text{Me}_3\text{SiCN,}}$ CH$_2$=CH—(CH$_2$)$_8$—C—CN

(37)   **(38)** 93% e.e. (*R*)

Table 4.2 shows some examples of asymmetric hydrocyanation of aldehydes.

Table 4.2.   Asymmetric Hydrocyanation of Aldehydes (RCHO) with Complex (30) + Me$_3$SiCN

Aldehyde RCHO	Product	% e.e.
R = n - C$_8$H$_{17}$ -	n - C$_8$H$_{17}$ —C< (HO, H, CN)	93
n - C$_9$H$_{19}$ -	n - C$_9$H$_{19}$ —C< (HO, H, CN)	85
H$_2$C=CH(CH$_2$)$_8$ -	H$_2$C=CH—(CH$_2$)$_8$ —C< (HO, H, CN)	93
PhCH$_2$ -	PhCH$_2$ —C< (HO, H, CN)	61
PhCH$_2$CH$_2$ -	PhCH$_2$CH$_2$ —C< (HO, H, CN)	91
Ph -	Ph —C< (HO, H, CN)	93

Generally asymmetric alkylation of ketones may be achieved in three ways: (i) a three-step sequence involving the preparation of a chiral intermediate (imine, hydrazone etc.) followed by its alkylation and subsequent hydrolysis of the alkylated intermediate [11-16], (ii) the reaction of an enolate with an alkylating agent having a chiral leaving group [17-19], and (iii) phase-transfer alkylation in the presence of a chiral catalyst [20- 22].

Because of their strong affinities for the heteroatoms of the chiral ligands shown in Table 4.1, various kinds of organometallic compounds have been employed. These include alkyllithiums, Grignard reagents, dialkylmagnesiums, alkylcoppers, dialkylzincs, trialkylaluminiums, etc.

The asymmetric alkylation of cyclohexanone (39) was carried out by preparing the corresponding lithioenamine using the amino acid *tert.* butyl-esters (9,10) as the chiral source. This was done by treating the imine (40), formed by reaction of cyclohexanone and the amino ester with LDA to afford the lithioenamine (41). Methylation with CH_3I then afforded α-methyl-cyclohexanone (42) in upto 97% e.e. [15].

Similarly, high optical yields (99% e.e) were obtained in the asymmetric alkylation of α-alkyl-b-keto esters in the presence of (S)-valine t-butyl ester (9) [14]. Thus the enamine (44) prepared from the substrate (43) and the amino acid ester (9) was lithiated to the lithioenamine (45) and then methylated in the presence of HMPT to give the (R) product (46) in 95% e.e. The enantiomeric (S) product could be obtained by using other additives such as THF, Me_3N or dioxolane instead of HMPT.

Addition of methyl n-propyl and n-butyllithium to benzaldehyde (35) in the presence of ligand (1) afforded the corresponding alcohols in optical yields ranging from 21% e.e. to 95 % e.e. [23]. Thus the asymmetric alkylation of benzaldehyde (35) with n-butyllithium in the presence of ligand (3) yielded 1-phenyl-1- pentanol (47) in 95 % e.e [22].

(43) → (9) → (44) → LDA, toluene →

(45) → CH₃I / HMPT → (46) 95% e.e. (R)

(35) + 2 ⁿBuLi → (3) → (47) 95% e.e. (S)

Grignard reagents also react with ketones in the presence of substituted sugars such as 1,2,5,6-di-O-isopropylidene-α-D-glucofuranose (20) to yield the corresponding alkylated products with optical yields of upto 70% e.e. [24].

Asymmetric addition of derivatives of lithium acetylides to aldehydes in the presence of chiral ligand (1) afforded optically active alkynyl alcohols in high chemical and optical yields [25]. The chiral ligand (1) has also been employed in the synthesis of corticoid hormone intermediates, such as the optically active polyene (50) [26]. Thus condensation of lithium trimethyl silylacetylide (48) with aldehyde (49) in the presence of ligand (1) afforded the optically active polyene alcohol (50) in 70% e.e. The optical yield increased to 90% e.e. on lowering the reaction temperature to - 120°C.

$$Li-C\equiv C-Si(CH_3)_3 \quad + \quad H_3C-C\equiv C-(CH_2)_2-CH=\overset{\overset{CH_3}{|}}{C}-CH_2-C\overset{H}{\underset{O}{\diagdown}}$$

(48)   **(49)**

$$\xrightarrow[-78°C]{(3)} \quad H_3C-C\equiv C-(CH_2)_2-CH=\overset{\overset{CH_3}{|}}{C}-CH_2\overset{*}{\underset{}{\diagdown}}\overset{HO}{\underset{}{C}}\overset{H}{\diagup}C\equiv C\ Si(CH_3)_3$$

(50) 70% e.e. (R)

The chiral ligand **(19)** reacts with bromobenzaldehyde **(51)** to afford a highly enatioface-selective chiral aryllithium reagent **(53)** *via* the intermediate **(52)**. The reagent **(53)** adds selectively to various aldehydes to give chiral 3-alkyl-1-hydroxy-2-oxaindanes **(55)** *via* the intermediate **(54)** in high optical yields [27].

(51)   **(52)**   **(53)**

(54)

(55) 88% e.e. (S)

Chiral organotitanium reagents have been employed in the asymmetric alkylation of aromatic aldehydes to afford the corresponding alkylated products in low to modest optical yields [28]. However, in atleast one instance, i.e. in the alkylation of p-tolylaldehyde **(56)** with the reagent **(30)**, the product **(57)** was obtained in 88% e.e. [27].

(56)   (30) →   (57) 88% e.e. (S)

Organocuprate (Gilman) reagents have also been used in the enantioselective carbon-carbon bond formation reactions with generally good enantioselectivities (75-95%) [29-32]. Thus 2-cyclohexenone (58) is alkylated with ethyllithium in the presence of complexed cuprate reagent [prepared from cuprous iodide and the aminoalcohol (23)] to yield 3-ethylcyclohexanone (59) in 92% e.e. [32]. A model such as (60) has been proposed in which lithium is complexed as a tridentate ligand to the conjugate base of (23) and associated with an alkylcopper fragment. The nucleophilic copper forms a d, π*-complex with the re-face of C-3 in (58) facilitating selective interaction whereas the carbonyl oxygen coordinates with a second lithium ion, which is held in place by the alcoholic group of (23) to form (61), leading ultimately to product (59). The alternative mechanism involving attack by copper on the si-face of C-3 in (58) is less favourable due to steric hindrance.

(58)   (23) + CuI / C₂H₅Li →   (59) 92% e.e. (R)

(60)   (61)

Chiral dithianes such as (62)-(64) have been used with success to alkylate prochiral carbonyl compounds stereoselectively [33-36]. For example, the reaction of (64) with an aldehyde in the presence of butyllithium followed by Swern oxidation (DMSO-TFAA-Et₃N) yielded 2-acyl-1,3-oxathianes (65) which reacted stereoselectively with Grignard reagents affording the corresponding tertiary alcohols (66) in > 90% diastereomeric purity [36]. Cleavage of (66) with *N*-chlorosuccinimide - $AgNO_3$ afforded tertiary α-hydroxy aldehydes which could be further reduced to glycols or oxidized to α-hydroxyacids in high enantiomeric purity.

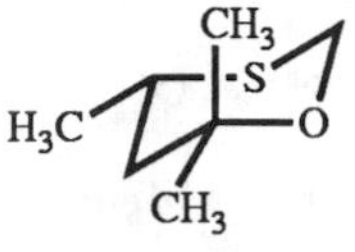

(62)

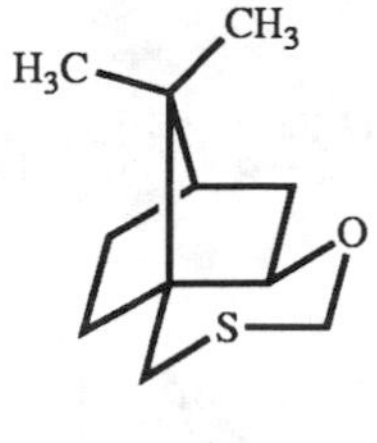

(63)

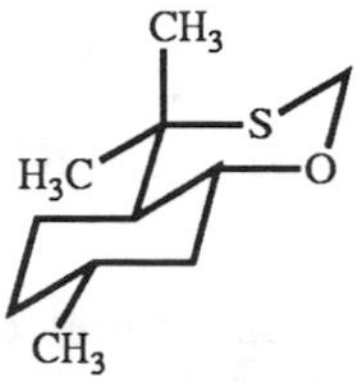

(64)

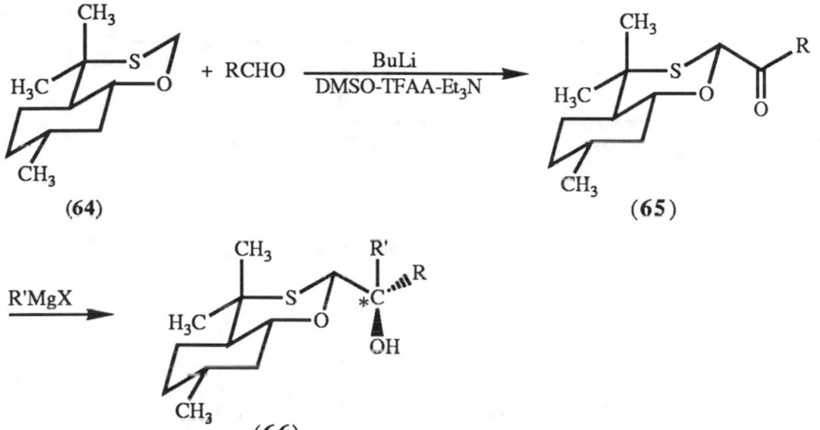
+ RCHO
BuLi
DMSO-TFAA-Et3N
R'MgX

(64) (65) (66)

Dialkylzinc reagents add enantioselectively to aliphatic, aromatic and α,β-unsaturated aldehydes as well as to formylesters in the presence of chiral catalysts such as (**23**), (**26**) etc. affording the corresponding alkylated products in high optical yields [37-43]. For example, the enantioselective alkylation of the formylester (**67**) with diethylzinc in the presence of (*1S, 2R*)- (**23**) as catalyst afforded (*S*)- (**68**) in 90% yield which upon basic hydrolysis yielded (**69**) in 92% e.e [43].

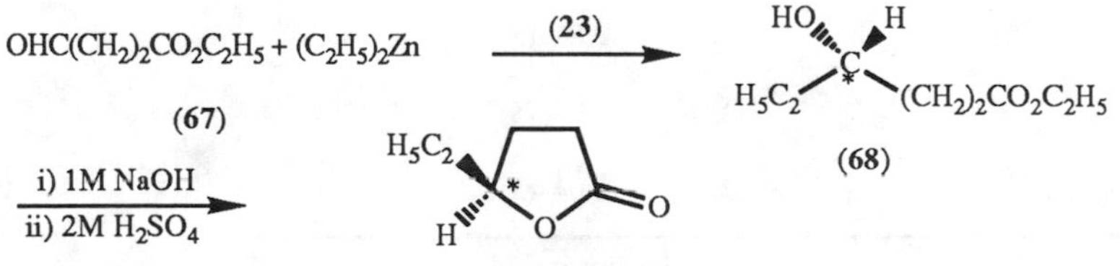
OHC(CH2)2CO2C2H5 + (C2H5)2Zn
(23)
i) 1M NaOH
ii) 2M H2SO4

(67) (68) (69) 92% *e.e.* (*S*)

Similarly, alkynylaldehydes such as (70) react with diethylzinc in the presence of chiral (+)-(26) to afford the corresponding alkynylalcohols such as (71) in 78% e.e. [42]. Table 4.3 gives some representative examples of enantioselective addition of organometallics to prochiral carbonyl compounds.

$$(H_3C)_3Si\text{—}C\equiv C\text{—}C\overset{O}{\underset{H}{\big\langle}} \quad \xrightarrow[\text{(26)}]{(C_2H_5)_2Zn} \quad (H_3C)_3Si\text{—}C\equiv C\overset{HO\;\;H}{\underset{C_2H_5}{\overset{|}{\underset{*}{C}}}}$$

(70)   (71) 78% e.e. (S)

Table 4.3. Enantioselective Addition of Alkylmetals to Aldehydes in the presence of Chiral Ligands (L*)

$$A: R^1CHO + R^2\text{-}Li \xrightarrow{(L*)} R^1\overset{OH}{\underset{|}{—CH—}}R^2$$

R^1	R^2	L*	% e.e. (Config.)	References
CH$_3$	Bu	(8)	46 (R)	[44]
Ph	Bu	(5)	89 (R)	[45]
Ph	CH$_3$	(1)	40 (R)	[23]
Ph	C$_2$H$_5$	(1)	54 (S)	[23]
Ph	Bu	(1)	95 (S)	[23]
^{i}Pr	Bu	(8)	53 (S)	[44]

$$B: R^1CHO + (R^2)_2Zn \xrightarrow{(L*)} \overset{R^1\;\;\;\;H}{\underset{HO\;\;\;\;R^2}{\overset{*}{C}}}$$

Ph	C$_2$H$_6$	(26)	92 (S)	[41]
p - ClC$_6$H$_4$—	C$_2$H$_5$	(26)	91 (S)	[41]
p - OCH$_3$ C$_6$H$_4$—	C$_2$H$_5$	(26)	100 (S)	[40]
Ph	C$_2$H$_5$	(26)	99.5(S)	[40]
^{n}C$_6$H$_{13}$	C$_2$H$_5$	(23)	88 (S)	[39]
PhCH$_2$ CH$_2$—	C$_2$H$_5$	(23)	95 (S)	[39]
^{n}BuC≡C—	C$_2$H$_5$	(26)	64 (S)	[42]

4.1.2   Diastereoselective Addition Reactions

4.1.2.1   Diastereoselective Additions to Carbonyl Compounds

The diastereoselective addition of nucleophiles to chiral carbonyl compounds and substituted cyclic ketones as well as the addition of chiral nucleophiles to prochiral carbonyl compounds has been the target of extensive studies [46,47]. Although some organometallic nucleophiles and carbanions have proved to be effective in asymmetric induction over several (n) bonds, the optical yields are generally low in cases where the induction is over three or more bonds. Such reactions involving 1,n-asymmetric induction are known as "diastereofacially selective" [44] and although this phenomenon was known almost a century ago [48,49] it was only after the work of Cram et al. that some rationalization could be made [42].

Besides 1,n-asymmetric induction, another type of diastereodifferentiation occurs in cases where a nucleophile e.g. a prochiral enolate is added to an achiral aldehyde thereby creating two new chiral centres. Such selectivity, termed as "simple diastereoselectivity" [50,51] is the basis of recent work on aldol and related reactions and will be discussed in section 4.1.3.

Diastereoselective addition of organometallic reagents to acyclic carbonyl compounds with a chiral centre adjacent to the carbonyl group (1,2-asymmetric induction) has been studied in detail and the optical yields approach as high as 100% d.p. [52-56]. For example chiral α-alkoxyketones undergo reactions with Grignard reagents in THF with high stereoselectivity [39,44]. Thus substrate (72), which is protected with methoxyethoxymethyl(MEM), when reacted with n-butylmagnesium bromide, gave (73) with complete diastereoselectivity [57].

Similarly β-alkoxyaldehydes such as (74) may be alkylated with organocuprate reagents such as dimethylcuprate to yield (75) with high threo selectivity [57].

Organotitanium and -zirconium compounds such as (76) and (77) are more selective in some cases than their lithium, magnesium and zinc analogues [58-62]. Thus the reaction of benzil (78) with the organotitanium reagent (76) yielded the product (79) in > 98% d.p. [58].

(74)   (75) (12:1 *threo*)

H_3C — Ti $(O - {}^{i}Pr)_3$

(76)

H_3C — Zr$(O - {}^{n}Pr)_3$

(77)

(78)   (79) >98% d.p.

1,3-Asymmetric induction is quite low in ketones and aldehydes devoid of additional heteroatoms and they show generally poor diastereofacial selectivity [44,59]. Reagents such as RMgX, RLi and R_2CuLi which give high 1,2-asymmetric induction do not perform well in cases involving 1,3-induction [57,63-65]. However Lewis acidic titanium reagents such as (80) give high levels of 1,3-asymmetric induction with chiral β-alkoxy derivatives [66]. For example, the reaction of (80) with the β-benzyloxyaldehyde (81) gave the diastereomeric products (82) and (83) in a ratio of 90:10.

(81)   (82)   (83)

(82 : 83 = 90:10)

Titanium reagents have also been employed with success in the stereoselective construction of the steroidal side chain [67]. Thus pregnenolone acetate (84) reacts with the organotitanium reagent (85) to give a >90:10 diastereomeric mixture of 20 *S* and 20 *R* alcohols (86) and (87) respectively.

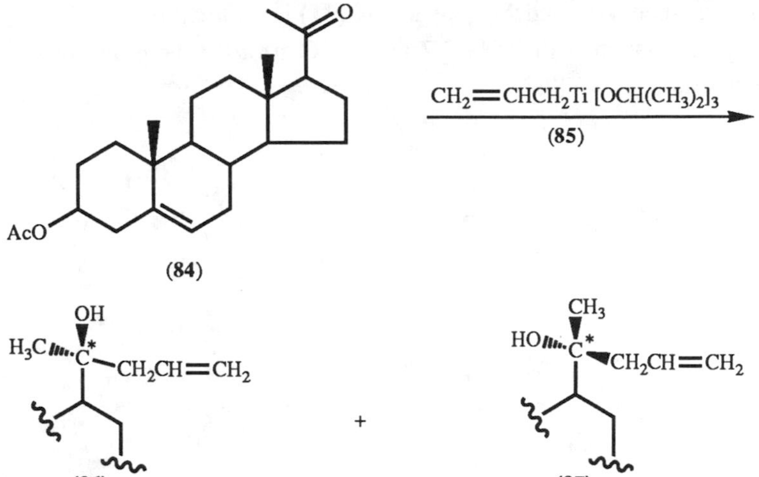

$$CH_2=CHCH_2Ti\,[OCH(CH_3)_2]_3$$

(85)

(84)

(86)   +   (87)

Stereoselective reactions involving 1,4-asymmetric induction were the first to be studied and led to theoretical formulations such as Prelog's rule [68,69]. Since the stereoselectivity in such reactions was generally low the method had little practical significance and was employed for determining the absolute configuration of chiral alcohols. Later, useful stereoselectivities were obtained with glyoxylate esters, chiral 2-oxazolidinyl ketones as well as with organotitanium reagents [70-73]. Thus glyoxylate esters of 8-phenylmenthol (88), such as (89,) on reaction with MeMgBr yielded the product (90) in a diastereomeric excess of 99.4% [71].

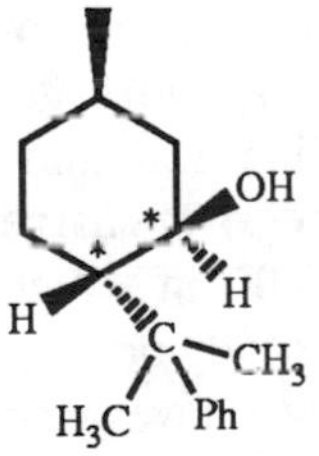

(88)

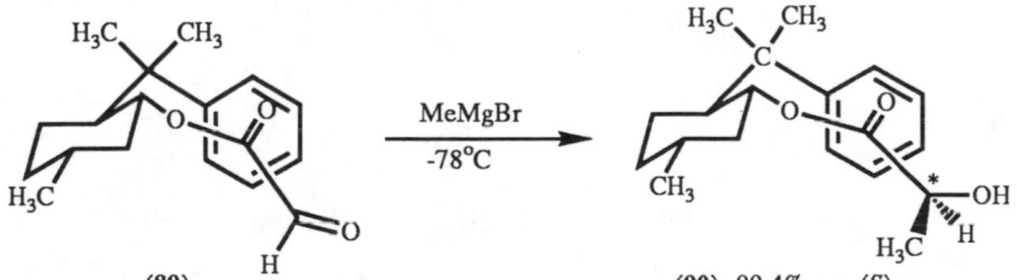

(89)

MeMgBr
-78°C

(90) 99.4% e.e. (S)

Similarly, γ-benzyloxyaldehydes such as (91) upon treatment with TiCl$_4$ followed by methylation with (CH$_3$)$_2$ Zn at 95°C afforded the products (92) and (93) in a ratio of 85:15 [73].

(91)

(92)

(93)

(92:93 = 85:15)

4.1.2.2   Diastereoselective Additions to Cyclic Ketones

Organoalkyl addition to cyclic ketones generally occurs with equatorial attack both in hindered and unhindered ketones. This is in contrast to the mode of attack in the hydride reduction of ketones (Section 3.2.2.2) where axial attack is preferred. However, atleast in one case, axial attack was preferred in the addition of ethynyl anions to conformationally defined conjugated cyclohexenones in the absence of overriding steric factors [74]. As with other reagents the selectivity of acetylide ions depends on the reaction conditions. Thus with 2-methoxycyclopentanone (94) in liquid ammonia, acetylide ion adds predominantly in a *trans* manner due to the repulsion between its negative charge and the oxygen of the methoxy group. In solvents such as THF, in which acetylide ion is not ionized, *cis* attack again predominates [75]. Recently organo-uranium, -titanium and -ytterbium reagents (e.g. (95), (96) etc.) have been employed to alkylate substituted cyclohexanones with preferred equatorial attack affording high diasteriomeric yields of axial alcohols [76,77].

(*trans* attack)
$^-$C≡CH

OCH$_3$

(94)

[{(CH$_3$)$_3$Si}$_2$N] U CH$_3$

(95)

Yb(OTf)$_3$

(96)

Bicyclic ketones show almost complete diastereoselectivity and the mode of attack of organometallic reagents depends on the substitution pattern of the substrate. Thus with ketones (97) and (98) *exo* attack is preferred, whereas in the case of (99) *endo* attack takes place.

(97)   (98)   (99)

4.1.3   Addition of Chiral Reagents

Chiral reagents such as chiral α-sulphinyl esters, lithiated oxazolines etc., have been employed to effect   carbon-carbon bond formation with good diastereoselectivities [78-80]. For example, the chiral sulphoxide (100) reacts with methyl cyclohexylketone in the presence of ^{t}BuMgBr to yield the addition product (102). The sulphoxide was subsequently liberated yielding the *tert*-butylester (103) which was converted to its methyl ester (104) of known absolute configuration.

(100)   (101)   (102)

(103)   (104)

Lithiated aryloxazolines such as (**105**) add to prochiral carbonyl compounds, e.g. (**106**) to give intermediates of the type (**107**) which readily undergo tautomerism to yield the corresponding imino lactones (**108**). In the case of (**105**) the reaction sequence proceeds in a diasteromeric ratio of 60:40. Inspite of the poor diastereoselectivity, the diastereomers were separated by crystallization which upon subsequent hydrolysis gave the phthalides (**109**) in 40-80% e.e. [77].

The product (**109**) can also be obtained by acylation of the lithiated derivative (**105**) followed by treatment with Grignard reagent to give the imino lactone (**108**) in high diastereomeric yield. With PhMgBr the acyl derivative of (**103**) yielded (**109**) in 80% e.e. [77].

Stereoselective addition of organometallic compounds to various chiral aminocarbonyl compounds has been achieved with high diastereoselectivity. Thus compound (**110**) is alkylated with CH_3MgBr to yield the products (**111**) and (**112**) in a ratio of 99:1 [81]. Other substrates such as 2-amino cyclohexanones, acylaziridines, 3-piperidones, oxobenzomorphanes, 3-pyrrolidines and several other amino carbonyl compounds give high diastereoselectivity in the addition products with various organometallic compounds [82]. For example the 1-ketoquinolizidine (**113**) reacts with Grignard reagent to afford the products (**114**) and (**115**) in a ratio of 95:1 [83].

(110)   (111) 99%   (112) 1%

(113)   (114)   (115)

(114 : 115 = >95 : <5)

Table 4.4 presents some examples of addition of organometallic compounds to various substrates.

Table 4.4.   Diastereoselective Addition of Alkylmetals to Prochiral Carbonyl Compounds

R^1	R^2	R^3	R^4	R^5M	A/B	Reference
hept	OCH_2Ph	H	CH_3	$BuMgBr$	100	[52]
CH_3	NCH_3Ph	H	CH_3	$PhMgBr$	93	[53]
C_7H_5	OMEM	H	C_4H_9	CH_3MgCl	>100:1	[52]
$^tBuSi(CH_3)_2O(CH_2)_2$	OCH_2OCH_2Ph	H	CH_3	(structure) MgBr	50:1(threo)	[52]

(Table 4.4. contd.)

ii)

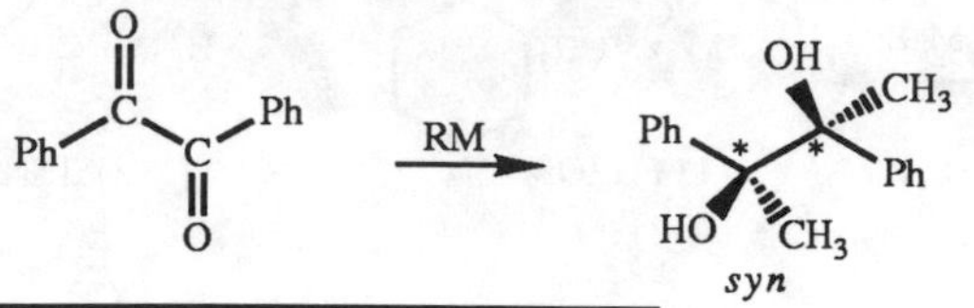

RM	syn/anti	Reference
Ti(O^{i}Pr)$_3$	>98	[58]
TiCH$_3$(O^{i}Pr)$_2$	93	[58]

iii)

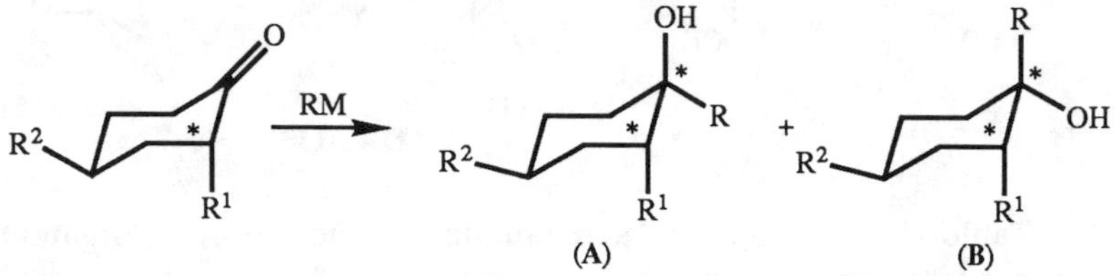

Metal Cat.	RM	R^1	R^2	A:B	Reference
(96)	^{n}BuLi	CH$_3$	H	83:17	[77]
(96)	PhLi	CH$_3$	H	>99:<1	[77]
(95)	"	H	^{t}Bu	85:15	[76]
(95)	"	H	4-methyl	85:15	[76]
(95)	"	H	^{t}Bu	90:10	[76]

4.1.4   Stereoselectivity of Nucleophilic Addition Reactions

Since the pioneering work by Emil Fischer on the addition of HCN to aldoses [1], much effort has been directed to understanding the course of asymmetric induction in nucleophilic addition to chiral carbonyl compounds [44,45]. Various models have been proposed for predicting the relative asymmetric inductions.

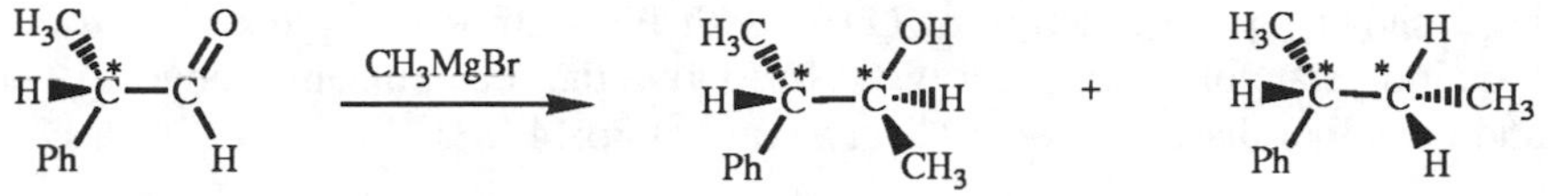

(116)   (117)   (118)

The reaction of prochiral carbonyl compounds having an a-asymmetric centre with organometallic reagents was rationalized by what is now known as Cram's rule [49]. This rule predicts that in the transition state of lowest energy, the chiral carbonyl compound such as (116) adopts a conformation in which the largest of the three α-substituents remains antiperiplanar to the carbonyl group and the nucleophilic attack therefore preferentially occurs from the least hindered side, leading to the predominant diastereomer (117) as shown in Fig. 4.1.

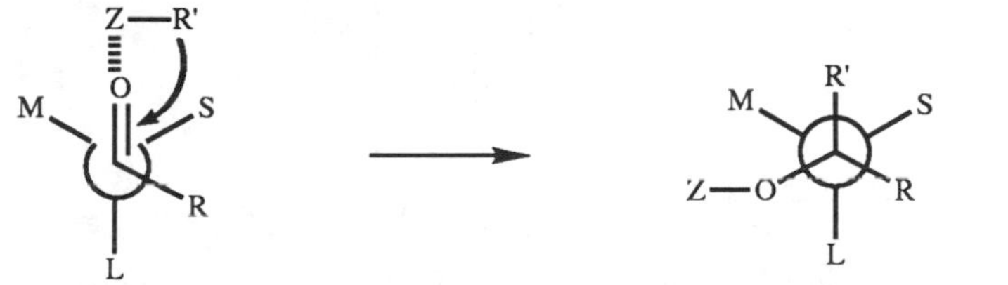

transition state of
lowest energy

predominant diastereomer

Figure 4.1. Cram's open - chain model.

In the case of CH_3MgBr, the ratio of the predominant diastereomer (117) to the minor diastereomer (118) was found to be 66:34 [49,84,85]. For carbonyl compounds having a-alkoxy, hydroxy or amino groups capable of complexing with organometallic reagents, Cram proposed a cyclic model as shown in Fig. 4.2. This involves a relatively rigid five-membered ring which fixes the conformation of the reacting species. This phenomenon has also been named as "chelation control". According to this model, chelation makes the π-face opposite to the large group on the α-carbon sterically more accessible. This has

Transition state of
lowest energy

Predominant diastereomer

Figure 4.2. Cram's cyclic model.

been shown by the reaction of (119) with PhMgBr which proceeds *via* the transition state of lowest energy (120) to give the predominant product (121) and the minor diastereomer (122) in a ratio of 86:14 [83].

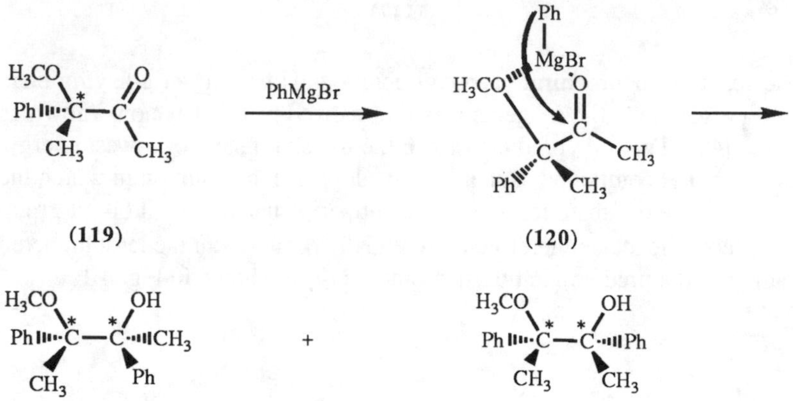

(119) $\xrightarrow{\text{PhMgBr}}$ (120) $\longrightarrow$

(121)   +   (122)

(121 : 122 = 86 : 14)

In those substrates in which the amino, alkoxy or hydroxyl groups on Cα is medium sized (with the smaller group being, for instance, hydrogen and the larger group, phenyl) the open chain and cyclic models predict the same selectivities.

Besides Cram's model, other explanations have also been put forward. For aldehydes and ketones having α-halogen substituents, Cornforth *et al.* proposed an electrostatic model, called the dipolar model, shown in Fig. 4.3 in which the halogen (X) would point away from the polar carbonyl function in the lowest energy conformer (123) and the attack of the nucleophile then proceeds from the least hindered side [84].

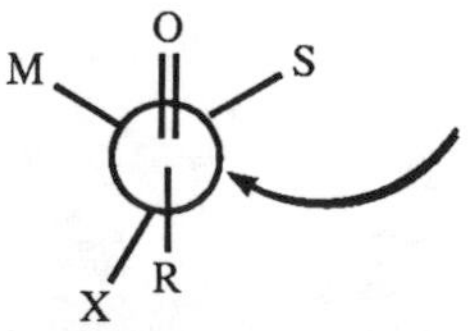

Figure 4.3. Cornforth's dipolar model (123).

Felkin *et al.* on the other hand proposed a model shown in Fig. 4.4 (124), on the assumption that polar effects stabilize those transition states in which separation between the incoming nucleophile and the electronegative group (X) is greatest [86].

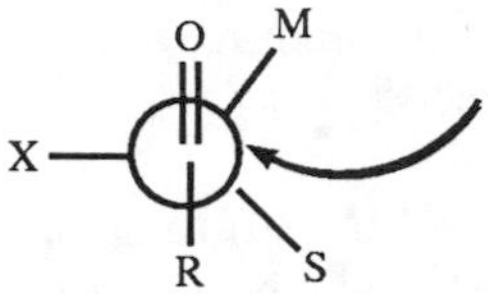

Figure 4.4. Felkin's model (**124**).

Felkin's model was later refined by Anh (Fig. 4.5) [87], who proposed that the reaction would occur through the most reactive conformation as (**125**). On the basis of molecular orbital considerations, he postulated a non-perpendicular attack of the incoming nucleophile, since an attack *anti* to the electronegative group (L=Cl) avoids certain antibonding interactions which may occur in a *syn* trajectory. These models help to explain the selectivity observed in addition reactions involving organometallic reagents, which proceed generally in two ways: i) chelation controlled additions or ii) non-chelation controlled additions.

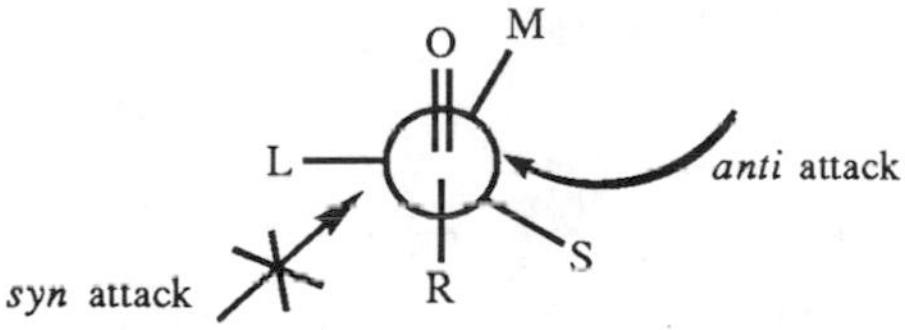

Figure 4.5. Anh model, L = Cl (**125**).

i) Chelation controlled additions involve the use of Lewis-acid reagents such as organotitanium reagents, e.g. (**80**) which form intermediate chelates with the substrates and undergo attack from the less hindered side. Thus the reaction of (**81**) with the Lewis acid reagent CH_3TiCl_3 (**80**) proceeds *via* the chelate (**126**) to give the adducts (**82**) and (**83**) in a ratio of 92:8 [66]. As shown in (**126**), attack of the methyl group bound to Ti occurs from the less hindered *si* face of the carbonyl group, leading to the predominant *S,S*-diastereomer (**82**).

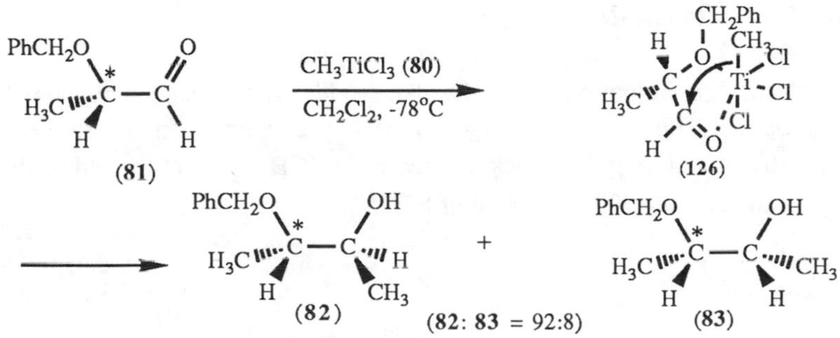

ii) Non-chelation control addition is achieved using titanium reagents of low Lewis acidity leading to the so called Felkin-Anh products.

It has been shown that by changing the ligand from Cl in (80) to alkoxy as in (76), the diastereomeric selectivity is reversed [88]. Thus by using the reagent (76), the aldehyde (81) affords predominantly (83). This behaviour can be explained by the Anh's model (125) which predicts reaction through the most reactive conformation such as (127). This conformation has the benzyloxy group positioned in such a way that the C^*-OR orbital overlaps with the pC^*-OR orbital, providing a low level LUMO. Anti-periplanar attack along the so-called Burgi-Dunitz trajectory (90°) results in the observed Felkin-Anh product (83).

(127)

Similar results have been achieved with sugars. Thus furanose (128) reacts with (76) to give 49% of the Felkin-Anh product (129) [88].

(128) : R = CH$_2$Ph   (129) : R = CH$_2$Ph, 99%   (130) : R = CH$_2$Ph, 1%

On the other hand the chelation controlled product (130) is obtained by reaction of (128) with CH$_3$MgX [89,90].

1,4 -Asymmetric induction in chiral γ-alkoxy aldehydes has been explained on similar grounds. Thus addition of TiCl$_4$ to (91) proceeds through a flexible 7-membered ring which may be attacked directly by (CH$_3$)$_2$ Zn at the aldehyde function to yield (92) and (93) in a ratio of 85:15 [88].

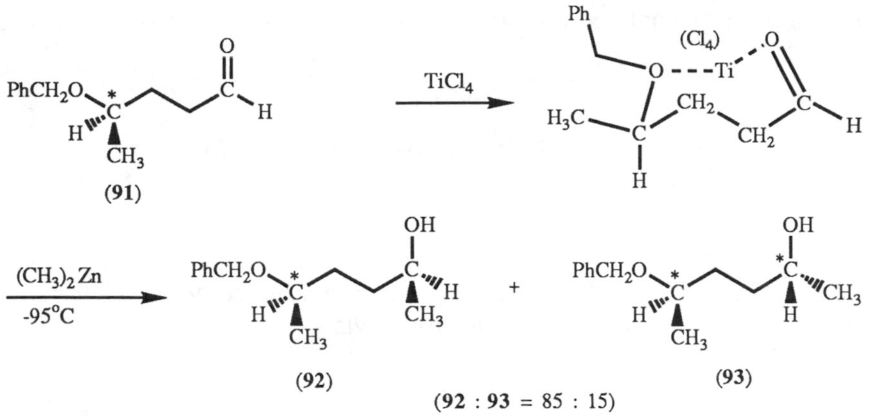

$$(92 : 93 = 85 : 15)$$

4.2 Asymmetric Catalytic Hydrocarbonylations

Hydrocarbonylation is the reaction of olefins with carbon monoxide and hydrogen in the presence of transition metal catalysts to give aldehydes. Fig. 4.6 shows a generalized scheme of hydrocarbo-nylations.

Carboxylic acids or esters can also be prepared in this way by using water or alcohol instead of hydrogen. The reaction is of considerable commercial importance and is widely used on an industrial scale. However, the asymmetric hydrocarbonylation of olefins is still in its infancy.

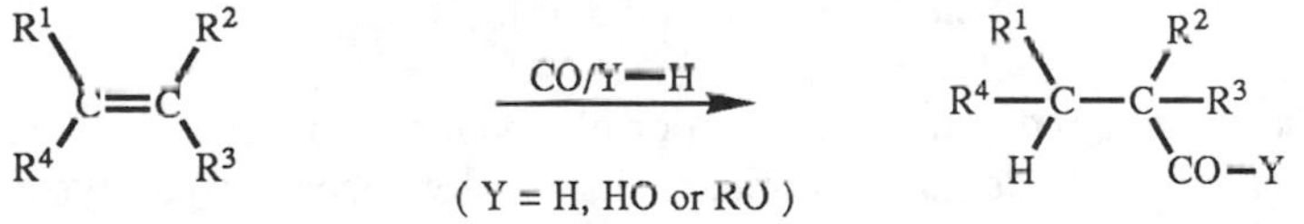

Figure 4.6. Generalized scheme for catalytic hydrocarbonylations.

4.2.1 Asymmetric Hydroformylations

One of the best studied asymmetric hydrocarbonylations is asymmetric hydroformylation, a reaction of a prochiral olefin with carbon monoxide and hydrogen in the presence of a chiral homogeneous or polymer-supported heterogeneous complexed transition metal catalyst to give the corresponding aldehyde.

4.2.1.1   Asymmetric Hydroformylation with Homogeneous Catalysts

This reaction has shown considerable promise and in some cases enontioselectivities of 80% - 96% e.e. have been achieved [91-93]. Generally, there are three types of asymmetric hydroformylations, exemplified by equations (A-C) (Fig. 4.7).

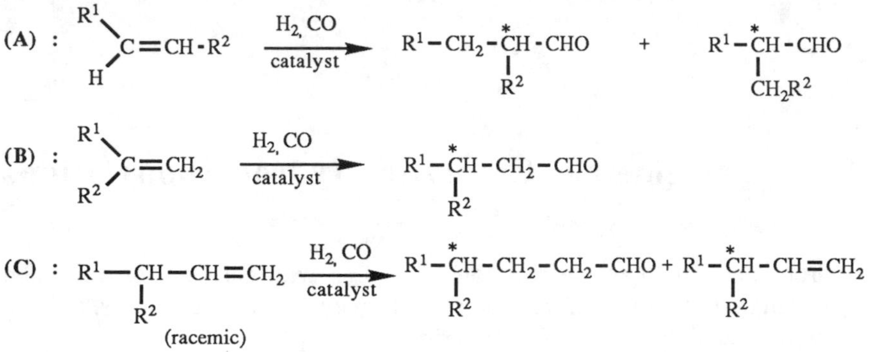

Figure 4.7. Generalized scheme for the hydroformylation of olefins.

Reactions (A) and (B) give optically active aldehydes with the chiral carbons at the α and β positions respectively, whereas reaction (C) gives optically active aldehydes by kinetic resolution [95,96].

Chiral phosphine ligands with rhodium or platinum complexes and chiral Schiff-base ligands with cobalt complexes have been used as catalysts. Transition-metal catalysts complexed to chiral phosphine ligands on polymer support have also been utilized in asymmetric hydroformylation reactions. Table 4.5 shows some chiral ligands (**131-143**) commonly used in such reactions.

Table 4.5. Ligands used in Asymmetric Hydroformylations

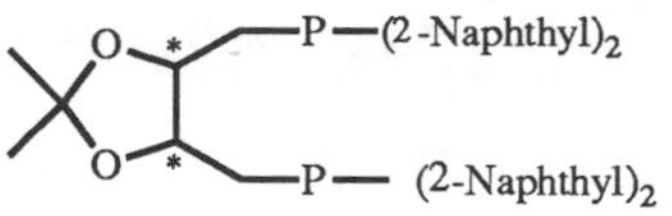

(131)

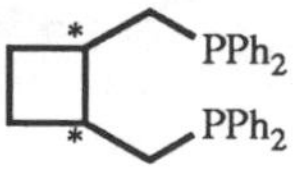

(132)

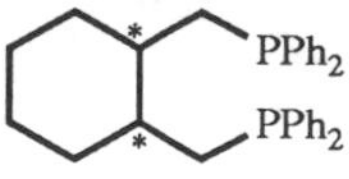

(133)

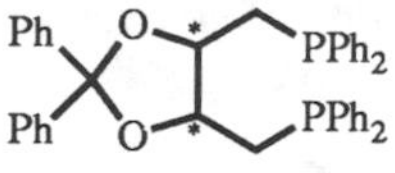

DIPH-DIOP (134)

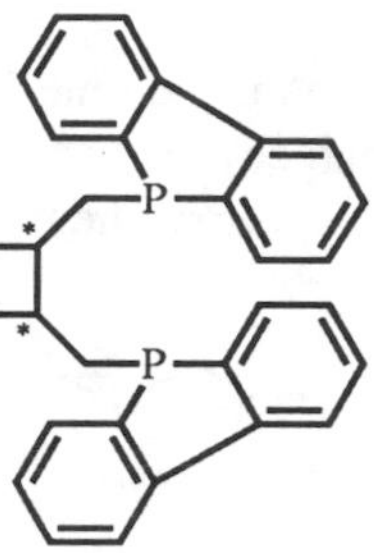

(135)

$PCH_2Ph(CH_3)Ph$ $n\text{-}C_3H_7$

(BMPP) (136)   (S) - (+) - MPPP (137)

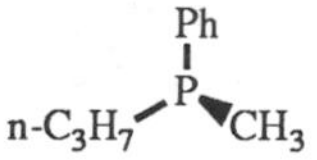

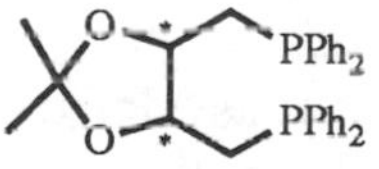

DIOP (138)

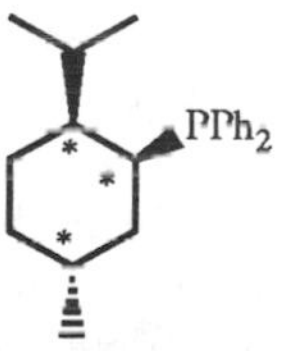

NMDPP (139)

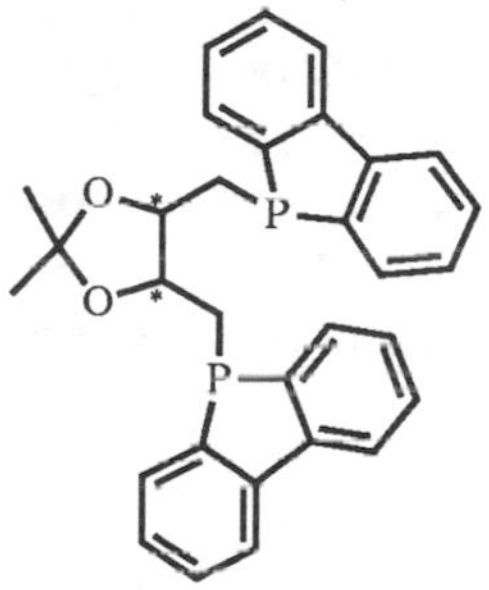

DIPHOL (140)

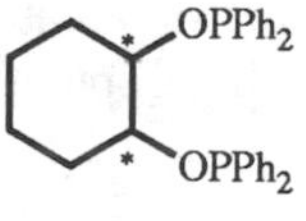

BDPCH (141)

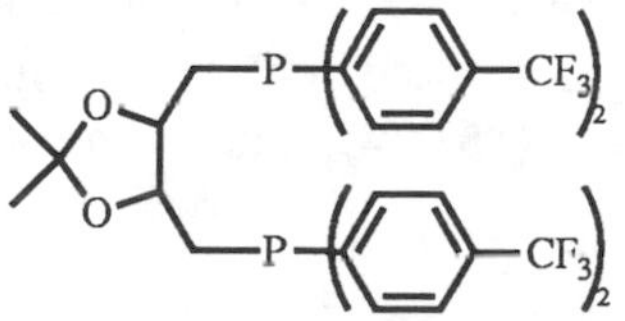

m-CF$_3$-DIOP (142)

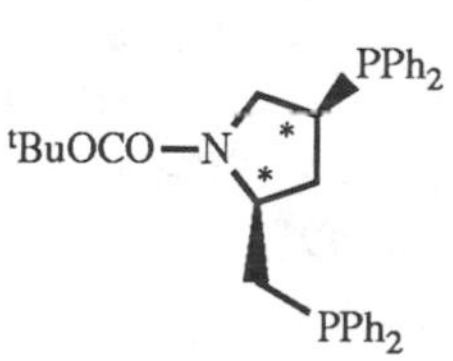

BPPM (143)

Asymmetric hydroformylation of prochiral olefins can lead to four possible products (a-d) as shown in Fig. 4.8 which can be formed from different transition states.

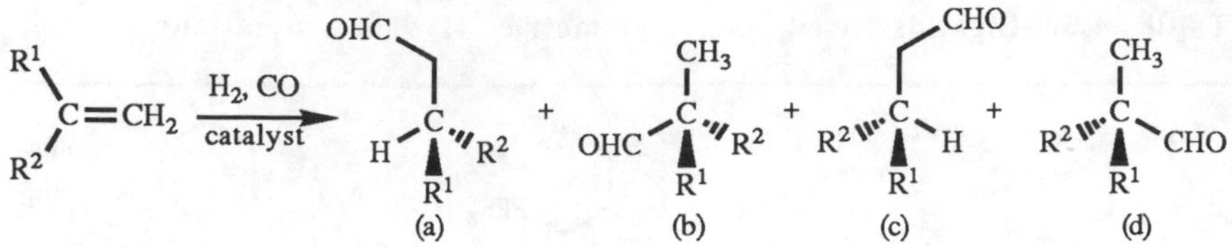

Figure 4.8. Generalized scheme for hydroformylation of olefins giving four possible adducts (a - d).

For example, the hydroformylation of styrene (**144**) gives three products namely 2-phenylpropanal (hydratropaldehyde) (**146**), 3- phenylpropanal (**147**) and ethyl benzene (**148**), the latter arising from the hydrogenation of styrene.

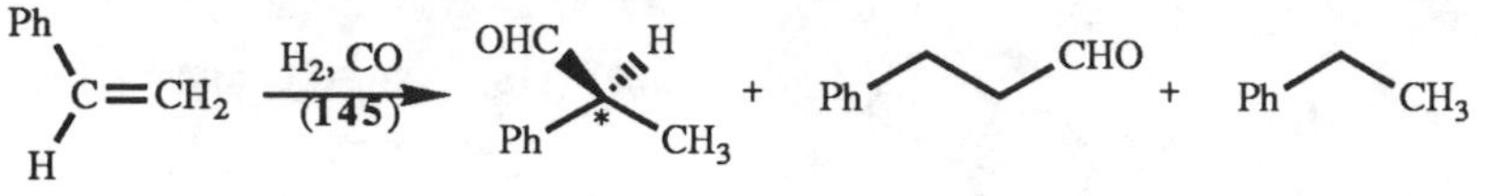

By employing the homogeneous catalytic system (**145**), i.e. [(**140**)+PtCl$_2$-SnCl$_2$], the hydroformylation of (**144**) afforded product yields of 56% for (**146**), 14% for (**147**) and ca. 20% for (**148**) with an enantiomeric excess of 73% e.e. for (**146**) [93,94]. With [(-)- DIPHOL]/[Rh(CO)$_2$Cl]$_2$ as catalyst, however the product yields were 79.8%, 21.2% and 0% respectively with a much lower enantiomeric excess of 15.6% (*S*) for (**146**) [97].

Generally, the low optical yields obtained in the asymmetric hydroformylation reaction may be attributed to the propensity of the chiral aldehydes to racemize under the reaction conditions. This can be avoided by *in situ* conversion of the chiral aldehyde into a less labile derivative, affording high enantiomeric selectivity [91]. Thus substrate (**150**) is hydroformylated in the presence of catalyst (**149**) [PtCl$_2$(**143**)-SnCl$_2$] in triethyl orthoformate to give the acetal (**151**), which is converted with pyridinium p-toluenesulphonate (PPTS) to the chiral aldehyde (**152**) without racemization and in high optical yield [91].

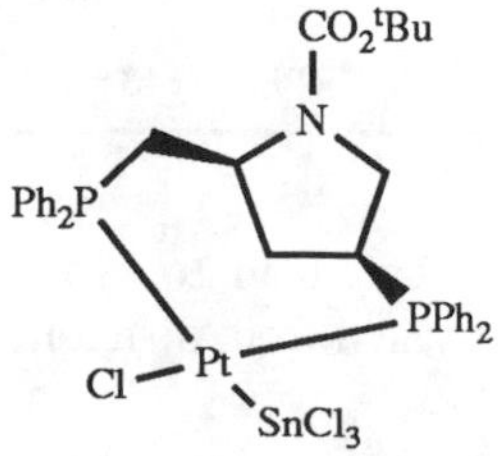

(**149**)

6 – CH₃O · Naph
$$\underset{\text{H}}{\overset{}{\diagdown}}C=CH_2$$
(150)

$$\xrightarrow[\text{(149),CH(OEt)}_3]{\text{H}_2,\text{CO}}$$

$$\underset{\text{6 – CH}_3\text{O - Naph}}{\overset{\text{H}}{\diagup}}\overset{*}{\underset{}{C}}\overset{\text{CH(OC}_2\text{H}_5)_2}{\underset{\text{CH}_3}{\diagup}}$$
(151)

$$\xrightarrow{\text{PPTS}}$$

$$\underset{\text{6 - CH}_3\text{O - Naph}}{\overset{\text{H}}{\diagup}}\overset{*}{\underset{}{C}}\overset{\text{CHO}}{\underset{\text{CH}_3}{\diagup}}$$

(152) >96% e.e. (S)

Asymmetric hydroformylation of vinyl acetate and enamides is synthetically important as their products are the precursors of chiral amino acids. The asymmetric hydroformylation of vinyl acetate (153) gives 2-acetoxy propanal (154), which is the precursor of chiral threonine, in 51% e.e. [98].

$$\underset{\text{H}}{\overset{\text{AcO}}{\diagdown}}C=CH_2$$
(153)

$$\xrightarrow[\text{[(+)-DIPHOL + Rh(COD)acac]}]{\text{H}_2,\text{CO}}$$

$$\underset{\text{H}}{\overset{\text{AcO}}{\diagdown}}\overset{}{\underset{*}{C}}\overset{\text{CHO}}{\underset{\text{CH}_3}{\diagup}}$$
(154) 51% e.e. (R)

The optical yield can be increased to as much as 98% e.e. by employing the complex (149) as catalyst in triethyl orthoformate to yield diethyl acetal (154) which can be converted to the product (155) (S-enantiomer) without racemization [91].

$$\underset{\text{H}}{\overset{\text{AcO}}{\diagdown}}C=CH_2$$
(153)

$$\xrightarrow[\text{H}_2,\text{CO}]{\text{(149), CH(OC}_2\text{H}_5)_3}$$

$$\underset{\text{AcO}}{\overset{\text{H}}{\diagup}}\overset{*}{\underset{}{C}}\overset{\text{CH(OC}_2\text{H}_5)_2}{\underset{\text{CH}_3}{\diagup}}$$
(154)

$$\xrightarrow{\text{PPTS}}$$

$$\underset{\text{AcO}}{\overset{\text{H}}{\diagup}}\overset{*}{\underset{}{C}}\overset{\text{CH}_3}{\underset{\text{CHO}}{\diagup}}$$
(155) >98% e.e. (S)

The mode of enantioface selection in the asymmetric hydroformylation of prochiral olefins has been investigated in the case of α-deuteriostyrene (156) which gives rise to the deuteriated products (157) and (158) [99].

$$\underset{\text{D}}{\overset{\text{Ph}}{\diagdown}}C=CH_2$$
(156)

$$\xrightarrow[\text{(-)-DIOP[HRhCO(PPh}_3)_3]}{\text{H}_2,\text{CO}}$$

$$\underset{\text{H}_3\text{C}}{\overset{\text{OHC}}{\diagdown}}\overset{}{\underset{*}{C}}\overset{\text{D}}{\underset{\text{Ph}}{\diagup}}$$
(157) 15% e.e. (R)

+

$$\underset{\text{Ph}}{\overset{\text{CH}_2\text{CHO}}{\diagdown}}\overset{}{\underset{*}{C}}\overset{\text{H}}{\underset{\text{D}}{\diagup}}$$
(158) 15% e.e. (S)

It was shown that both the hydrogen migration and CO insertion take place on the same (re) face of α-deuteriostyrene [99].

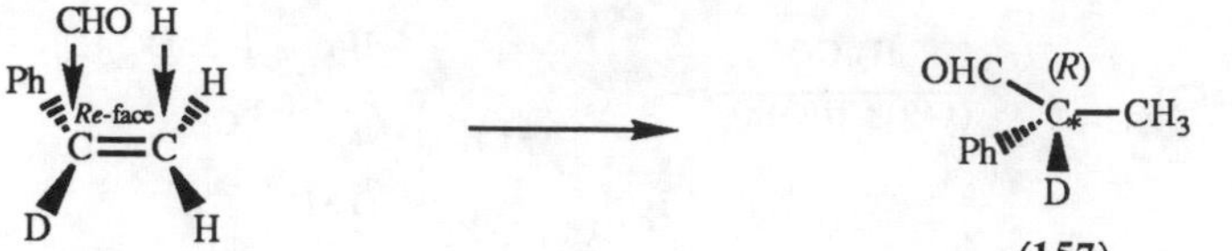

(157)

(158)

Asymmetric hydroformylation of N-substituted imides catalysed by Rh complexes affords optically active α-amino aldehydes [100]. Thus N-vinylsuccinimide (159) affords (160) which is the precursor of alanine.

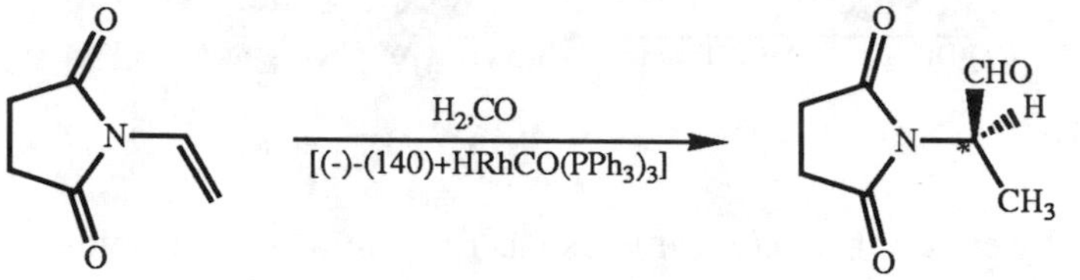

(159)

(160) 41% e.e. (R)

The asymmetric hydroformylation of acetylenes catalysed by chiral Rh complexes can occur by two possible pathways: (i) hydroformylation of acetylene to an α,β-unsaturated aldehyde followed by hydrogenation (equation A), or (ii) hydrogenation of acetylenes to olefins followed by hydroformylation, as shown by equation (B):

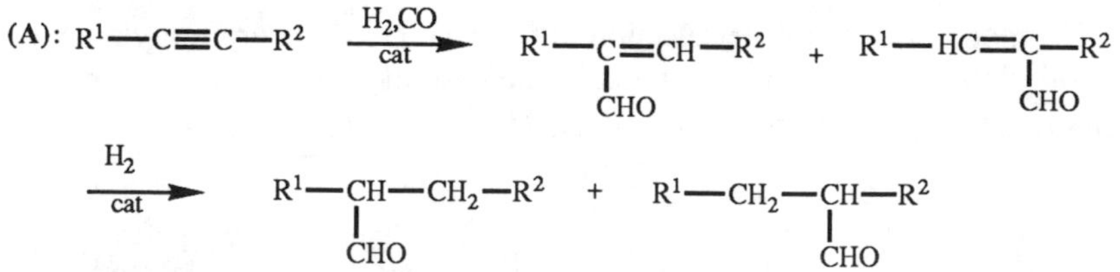

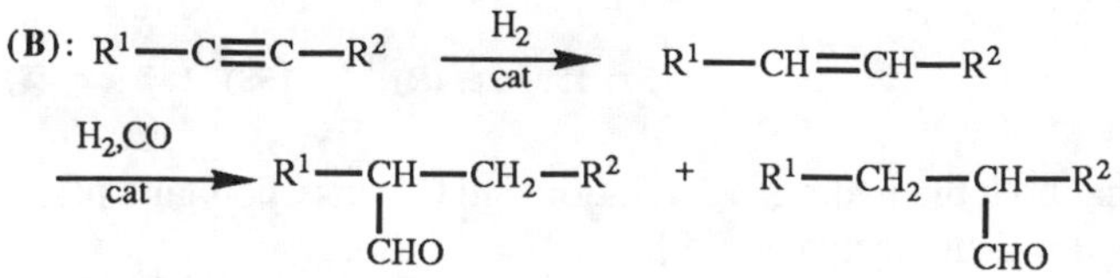

Generally, low optical yields (0.2 - 2.5 % e.e.) were obtained with the (-)-DIOP-rhodium complex catalyzed hydroformylation [101]. 1-Octyne (**161**) gives (S)- 2-methyloctanal (**162**) in 0.2% e.e, catalysed by (-)-DIOP-rhodium complex [98].

(**161**)

(**162**) 0.2% e.e. (S)

(**163**)

(**162**) 2.5% e.e. (R)

1-Octene (**163**), on the other hand, gave the same aldehyde having the (R) configuration under identical conditions [101]. These experimental evidences suggest that the path shown by equation (B) is followed, i.e. initial hydrogenation occurs followed by hydroformylation.

Other conjugated alkenes have also been hydroformylated, albeit in low optical yields [102]. However these reactions provide valuable information about the factors involved in inducing asymmetry in the catalytic cycle.

Asymmetric hydroformylation can be best explained in terms of a widely accepted mechanism proposed by Wilkinson et al. for the hydroformylation of olefins with $HRhCO(PPh_3)_3$ or $HRh(CO)_2(PPh_3)_2$ as shown in Fig. 4.9 [103].

This mechanism is based on the ability of the chiral catalyst (**164**) to differentiate between the two enantiotopic sides of the prochiral olefin. The complex (**165**) formed from the catalyst (**164**) and an olefin remains in equilibrium with the complex (**166**) which contains a rhodium - alkyl bond. The asymmetric induction takes place at this step and/or at the p-olefin - rhodium intermediate forming step (**164** → **165**) [95]. The rate determining step of the reaction is the reductive elimination of the aldehyde from the complex (**168**) whereby the tetrahedral hydrido complex (**169**) is formed which, after uptake of CO, regenerates the complex (**164**) allowing it to further add to another molecule of olefin.

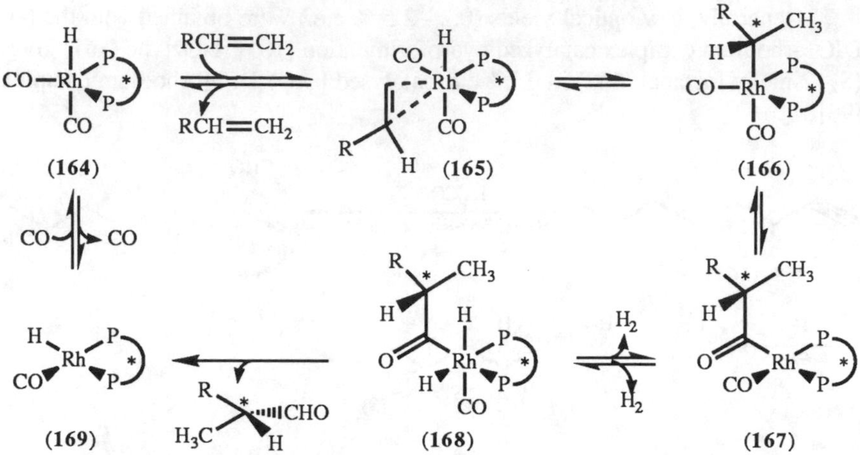

Figure 4.9. Hydroformylation of olefins catalyzed with phosphine - Rh complex.

4.2.1.2   Asymmetric Hydroformylation with Heterogeneous Catalysts

Catalytic asymmetric hydroformylations have been carried out by utilizing polymer supported chiral phosphine ligands complexed to transition metals as catalysts [104-106]. The primary advantages the heterogeneous catalytic procedure are the ease of workup and the ability to recover and recycle both the

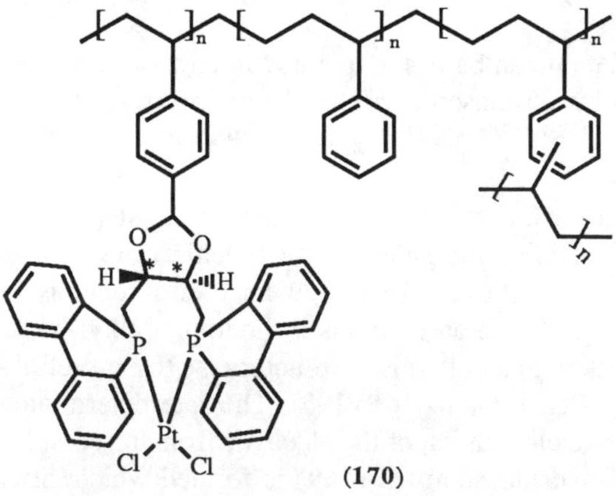

transition metal and the chiral optically active ligands [107,108]. However, most of these polymer-supported catalysts have been employed in asymmetric hydrogenations, while the asymmetric hydroformylations generally afforded low optical yields [109,110]. In some cases however, the optical yields obtained in polymer supported catalytic hydroformylations have reached as high as 98% e.e. [88,102]. For example, the platinum complexed polymer (170) was used to hydroformylate styrene (144) in 65% e.e. [105]. Table 4.6 shows some examples of asymmetric hydroformylation reactions.

Table 4.6. Asymmetric Hydroformylations

Substrate	Condition	Product	% e.e.	Reference
	a		78 (S)	[91]
	b		78 (S)	[91]
	c		28.4 (S)	[109]
	d		60	[91]
	e		>96	[91]
	f		38.3 (R)	[100]
	g		19.7 (R)	[111]
	h		33 (S)	[98]

a:SnCl$_2$, 2400 psi, H$_2$, CO   b:SnCl$_2$, 2600 psi, H$_2$, CO   c:Polymer-bound (-)(DIOP),H$_2$,CO followed by in *in situ* reduction d:SnCl$_2$, H$_2$, CO   e:(150), CH(OC$_2$H$_5$)$_3$, H$_2$, CO f:[Rh-(-)-DIPHOL], H$_2$, CO g:[Rh-(-)-DIOP], H$_2$, CO h:[Rh - (142)]

4.2.2   Asymmetric Hydroesterification

Asymmetric hydroesterification has not been as successful as its contemporary hydroformylation reaction partly due to the fact that high carbon monoxide pressures (300-700 atmospheres) are required to obtain good optical yields [112-117].

Palladium with DIOP type ligands have been widely used in the asymmetric hydroesterification of olefins. The asymmetric hydroesterification of α- methylstyrene (171), for instance, gives the optically active ester (172) in optical yields ranging from 3.5% to 58.6% e.e., depending on (a) the structure of the alcohol used, i.e. the bulkier the alcohol, the higher the optical yield, (b) the presence of solvents such as benzene or THF, which tend to improve the optical yield, (c) the magnitude of the pressure, i.e. the higher the pressure the better the optical yield,and (d) the (-)-DIOP/Pd ratio, a ratio of 0.4 in one set of reactions gave better results than a ratio of 1 or 2. On the other hand, using DIPHOL (140) as ligand, it has been shown that the ligand/Pd ratio has no effect on the stereoselectivity [113]. Thus α-methylstyrene (171) gave the product (173) in 40% e.e. at comparatively lower pressures [113].

Some representative examples of hydroesterification of olefins are given in Table 4.7.

Table 4.7.   Asymmetric Hydroesterification of Olefins

Substrate	Condition	Product	% e.e.	Reference
	^{t}BuOH, benzene, (138)/PdCl$_2$,	COO^{t}Bu, H	20.9 (R)	[114]
	^{t}BuOH, (138)/PdCl$_2$,	COO^{t}Bu, H	20.2 (S)	[114]
	CH$_3$OH, (138)/PdCl$_2$,	COOCH$_3$, H	10.3 (S)	[112]
	^{t}BuOH, (138)/PdCl$_2$,	CO$_2$^{t}Bu, H, CH$_3$	10 (S)	[114]
	^{i}PrOH, (138)PdCl$_2$(PhCN)$_2$,	CH$_2$CO$_2$^{i}Pr, H, CH$_3$	44 (S)	[116]
	PdCl$_2$, (138),^{t}BuOH,	CH$_2$COO^{t}Bu, H, CH$_3$	58.8 (S)	[112]

4.3   Asymmetric Aldol Reactions

Tremendous progress has been made during the last decade in the development of new methodologies for stereocontrolled carbon-carbon bond formations. The aldol reaction has, in particular, emerged as one of the most powerful tools for the stereoselective construction of carbon-carbon bonds [118].

The success achieved in recent years with asymmetric aldol reactions is largely due to the availability of spectroscopic and chromatographic techniques for separation and identification of complex mixtures of chiral products and due to the ever increasing demand for the asymmetric synthesis of biologically active compounds.

There are three types of aldol reactions:-

1)   Addition of a chiral enolate to a prochiral carbonyl compound.
2)   Addition of an achiral enolate to a chiral carbonyl compound.
3)   Addition of an achiral enolate to a prochiral carbonyl compounds in the presence of a chiral catalyst.

An overview of these three types of reactions is presented in the following sections.

4.3.1   *Stereochemistry of the Aldol Reaction*

In general the aldol reaction involves the addition of an enolate having homotopic double bond faces to an aldehyde or a ketone having prochiral carbonyl faces affording a pair of enantiomeric β-hydroxyaldehydes (aldol = aldehyde-alcohol) or β-hydroxy-ketones. Fig. 4.10 shows a generalized scheme of the asymmetric aldol reaction.

Employing enolates having prochiral faces, a pair of racemic diastereomers are formed. The products obtained in such type of reactions are the result of simple diastereoselection.

Double diastereoselection, on the other hand implies the reaction of a chiral enolate with a chiral aldehyde or ketone to afford a predominant diastereomer. The term "chiral enolate" is defined as an enolate which either contains a chiral centre itself or is attached to a metal which possesses a chiral ligand. The latter is most commonly employed.

As depicted in Fig. 4.10, the carbonyl compound forms the enolate which is substituted at the α-position ($R^2 \neq H$) and can react with an aldehyde, such as R_3CHO to give four diastereomeric products. Employing optimum reaction conditions and by the suitable choice of metals, it is possible to prepare either the *syn*- or the *anti*-diastereomers.

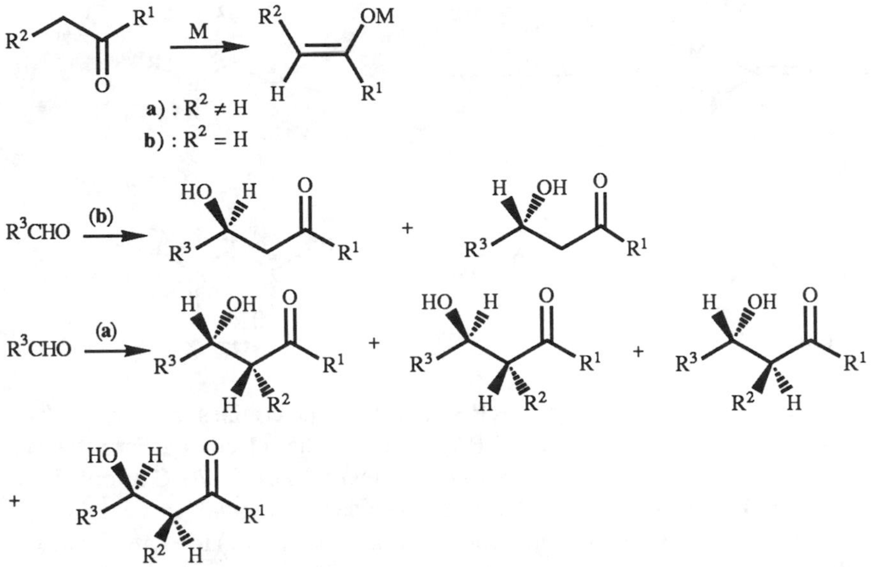

Figure 4.10. Generalized scheme for asymmetric aldol reactions (M = Li, K, Al, Mg, Ti, Zr, B, Si, Au, Zn etc.)

A much simpler picture emerges when, instead of the reagent the α-unsubstituted enolate ($R^2 = H$) is employed which can be prepared, for example, by the deprotonation of acetyl derivatives.

In such cases, only one chiral centre is formed, whereby *syn-anti* isomerism is excluded. However the enolate is unable to differentiate between the two enantiotopic sides of an aldehyde leading to a mixture of the aldol products as shown in Fig. 4.10.

4.3.1.1   Transition State Models in the Aldol Reaction

Because of the reversible nature of the aldol reaction, appropriate reaction conditions may be applied to bring the reaction under kinetic or thermodynamic control. Under conditions of kinetic control, the stereochemical outcome of the reaction can be predicted by considering the geometry of the transition state [119-121].

For the kinetically controlled aldol reaction, it has been found, in general, that the Z-enolates, e.g. (**174**) afford mainly the 2,3-*syn* aldol (**175**), while E-enolates, e.g. (**176**) lead to the 2,3-*anti* aldol (**177**) [119].

The relative size of the groups R^2 and R^3 also determines the selectivity of the reaction. The larger the size of R^2 the greater the selectivity. For instance, treatment of 2,2-dimethyl-3-pentanone (178) with LDA at -70°C afforded the Z-enolate (179) which reacted with benzaldehyde to give >98% *syn* aldol (180) while 3-pentanone under the same conditions afforded a (3:7) mixture of *anti:syn* aldols.

Under thermodynamic conditions, the *E*- or *Z*-enolates afforded mainly the 2,3-*anti* aldol irrespective of the geometry of the enolate concerned [120]. Thus thermodynomic conditions can be used to achieve *anti* stereoselection.

These reactions may proceed through a chair-like six-membered cyclic transition state in which the metal atom is bonded, besides the other ligand, to the oxygen atoms of the aldehyde and the enolate. In the case of the reaction of Z-enolates e.g. (174) with aldehyde R^1CHO, two transition states (A) and (B) could be formulated, of which the latter is disfavored by 1,3-diaxial interaction between the substituents R^1 and R^3. The reaction thus proceeds through (A) to afford the 2,3 *syn* aldol (175).

Similarly, *E*-enolates e.g. (176) react with R^1CHO *via* the transition states (C) and (D) of which the former is disfavored due to 1,3-diaxial interaction and the reaction proceeds preferentially through (D) affording the 2,3-*anti* aldol (177).

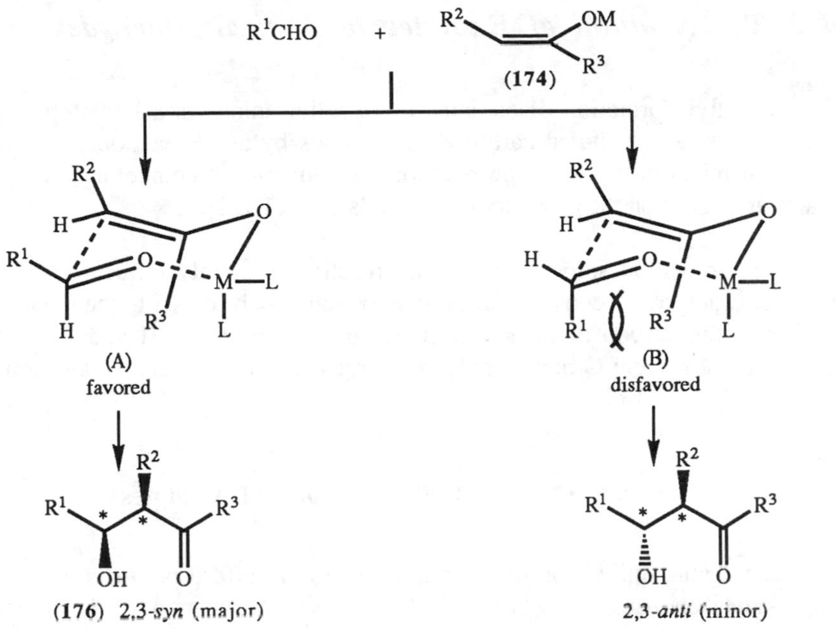

R^1CHO +   (174)

(A) favored

(B) disfavored

(176) 2,3-*syn* (major)

2,3-*anti* (minor)

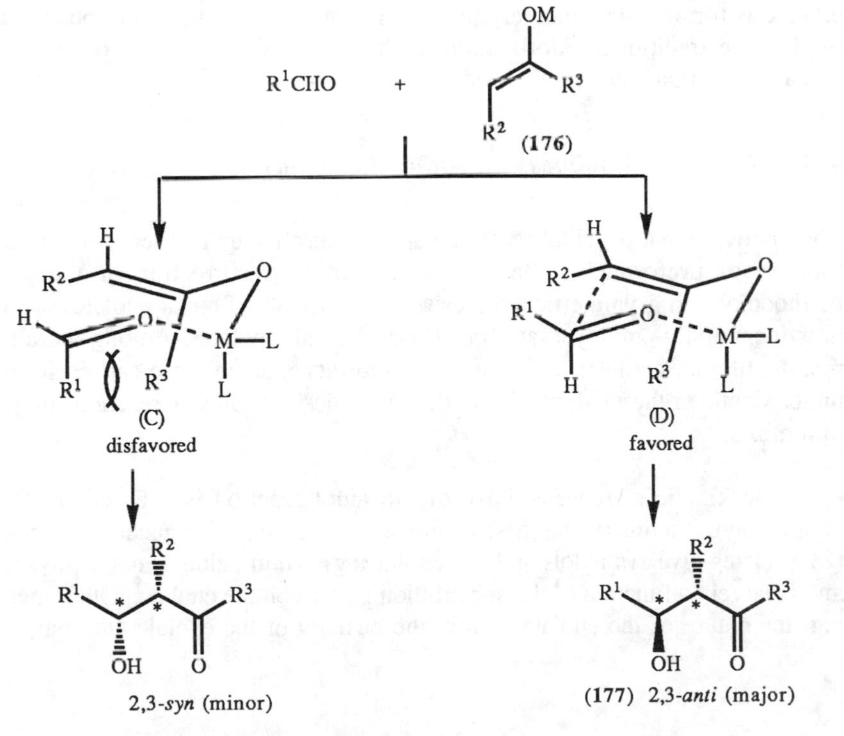

R^1CHO +   (176)

(C) disfavored

(D) favored

2,3-*syn* (minor)

(177) 2,3-*anti* (major)

4.3.2   Addition of Enolates to Achiral Aldehydes

The selective formation of enolates or oxyallyl anions is a key step in the synthesis of α-substituted carbonyl compounds by aldol reactions, since the stereochemical outcome of the reaction depends on the competing transition states and hence on the geometry of the enolates [121,122].

The transition state of an enolate reaction is largely influenced by the counter cation of the enolate. The counter cation may be closer to the oxygen or closer to the carbon atom. Metals belonging to groups I, II and III of the Periodic Table form O-metal enolates, whereas transition metals may form C-metal enolates as well.

4.3.2.1   Generation and Aldol Reactions of Enolates

The successful application of the aldol reaction to afford products with high regio- and stereo-selectivity largely is due to the development of methods for the formation of preformed enolates i.e., by adding sufficiently strong base, the enolate is formed quantitatively prior to the addition of the electrophile. This renders the traditional aldol reaction, which normally give rise to reversible enolate formation under protic conditions, irreversible.

4.3.2.1.1   Li Enolates in Aldol Reactions

The discovery that the aldol reaction can be controlled quite effectively through the use of preformed enolates resulted in the resurrection of the aldol methodology in asymmetric synthesis. Generally alkali metal enolates are the most commonly employed and most useful of all enolates. Among the alkali metals, lithium enolates are of prime importance, since these can be formed under kinetic or thermodynamic conditions as desired and side reactions may be minimized.

The kinetic stereoselectivity of the aldol reaction is a function of the enolate stereochemistry and its structure. The over simplified generalization that (Z)-enolates give *syn* aldols and (E)-enolates give *anti* aldols is not always true and is largely a function of the substitution pattern on the enolate., Other factors e.g. the nature of the enolate cation, the basicity of the enolate etc. may also exert profound effects on the stereoselectivity of the reaction. Tables 4.8A and 4.8B show some examples of the stereoselectivity of various (Z)-enolates [121].

Table 4.8A. Stereoselectivity of (Z)-Enolates

R	syn:anti
^tBu	98.7:1.3
^iPr	90:10
Et	90:10
H	50:50

† From reference [121].

Table 4.8B. Stereoselectivity of (Z)-Enolates

R^1	syn:anti
CH$_3$	100:0
C$_2$H$_5$	100:0
^nPr	98:2
^iBu	97:3
^iPr	29:71
^tBu	0:100

† From reference [120].

As is evident from Table 4.8A, the presence of a large group R in (Z)-enolates affords *syn* aldols whereas when R = H, the *syn:anti* ratio becomes 50:50.

Another structural effect which can be seen in Table 4.8B is that when R^1 is smaller in size, e.g. CH$_3$, C$_2$H$_5$, ^nPr, the selectivity is almost exclusively *syn*. However as the size of R^1 increases, e.g. from C$_2$H$_5$ to ^tBu, the stereoselectivity switches dramatically from 100% *syn* to 100% *anti* respectively. In the case of (*E*)-enolates, the R groups in the enolate exerts a similar effect on the stereoselectivity of the aldol reaction provided that the R group is large [121]. For medium sized and small R group, the enolates are stereorandom [121]. As shown in Table 4.9, the enolate with the larger R group gives rise to high *anti* selectivity.

Table 4.9. Stereoselectivity of (*E*)-enolates[†]

R	anti:syn
(di-tBu aryl ether)	>98:<2
OCH_3	60:40
iPr	50:50
C_2H_5	60:40
H	60:40

[†] From reference [121]

4.3.2.1.1.1   Ketone Enolates

The most common method for the formation of lithium enolates is the deprotonation of carbonyl compounds by lithium dialkylamide bases [50,118]. Sterically hindered amide bases are preferred in order to avoid side reactions due to nucleophilic attack on the carbonyl group. The most common bases used for deprotonation are i) lithium diisopropylamide, LDA; ii) lithium isopropylcyclohexylamide, LICA; iii) lithium 2,2,6,6-tetramethylpiperidide, LITMP, iv) lithium hexamethyldisilylamide (commonly known as lithium hexamethyldisilazane), LHMDS and v) lithium tetramethyldiphenyldisilylamide, LTDDS, etc. Other bases commonly used for the generation of Li-enolates are Ph_3CLi, nBuLi etc.

LDA   LICA   LITMP   LHMDS   LTDDS

Table 4.10 shows the stereochemistry of enolates generated with some common bases i.e. LDA, LHMDS and LITMP.

Table 4.10. Regiospecific Enolate Formation with Various Bases

(Z) (E)

R	% (Z) Isomer		
	LDA	LHMDS	LITMP
CH_3O	5	-	-
C_2H_5	30	66	16
^iPr	56	>97	32
$N-(^iPr)_2$	81	-	52
Ph	>97	>97	>97
^tBu	>97	>97	>97

Ketones with bulky groups afford (Z)-enolates with high stereoselectivity. Thus ethyl *tert.* butylketone, ethyl 1-adamantyl ketone and ethyl mesityl ketone all react with benzaldehyde to give predominantly the *syn* aldol with *syn:anti* ratio of >50:1 [121].

In the case of cyclic ketones, e.g. cyclohexanone, the stereoselectivity of the aldol reaction is generally low, mainly due to the rapid equilibration of the *anti* aldol to *syn* aldol. For instance the aldol reaction of (**181**) with benzaldehyde after normal work-up afforded the products (**182**) and (**183**) in a ratio of 90:10 [122b]. The aldol (**182**) was found to be exceedingly sensitive and was rapidly converted in solution to (**183**). However if the initial aldolate was reduced prior to work-up, a single diol (**184**) was obtained suggesting the aldol reaction itself is 100% diastereoselective.

4.3.2.1.1.2   Ester and Lactone Enolates

Esters are deprotonated with lithium dialkylamides to give predominantly (E)-enolates [121,122b]. Table 4.11 shows some examples of the aldol reaction of ester lithium enolate with aldehydes. As shown in Table 4.11, alkyl esters show no *syn-anti* selectivity. However esters which contain additional oxygen in the R^1 group show reasonably high *anti* selectivity.

Table 4.11. Aldol Reactions of Ester Lithium Enolates with Aldehydes

R^1	R^2	R^3	*anti/syn*	Reference
CH_3	CH_3	Ph	38:62	[121]
CH_3	CH_3	^iPr	55:45	[122e]
CH_3OCH_2	CH_3	^iPr	90:10	[122e]
$CH_3O(CH_2)_2OCH_2$	CH_3	Ph	75:25	[121]

(Table 4.11. contd.)

R^1	R^2	R^3	anti/syn	Reference
*DMP	CH$_3$	Ph	88:12	[122f]
DMP	CH$_2$ = CHCH$_2$	Ph	91:9	[122f]
DMP	CH$_3$	^{i}Pr	>98:2	[122f]
**BHT	CH$_3$	Ph	>98:2	[122f]

* DMP=2,6-dimethylphenyl; **BHT = 2,6-di-*tert*-butyl-4-methylphenyl

Deprotonation of the ester (185) with LDA affords the (E)-enolate (186a) which reacts with aldehydes, e.g. PhCHO to give mainly the product (187) along with some (188) with high stereoselectivity [123a].

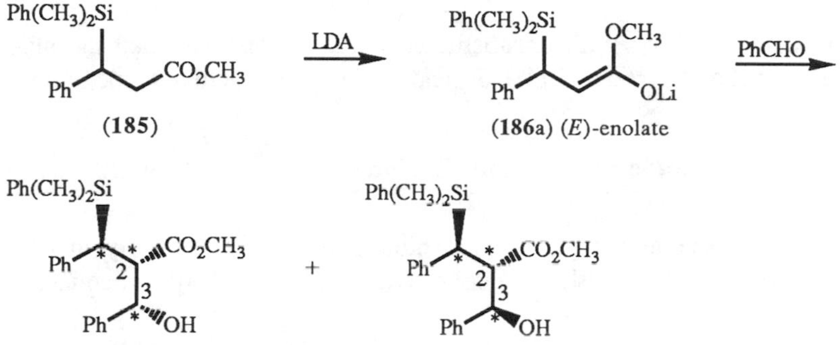

(185) → LDA → (186a) (E)-enolate → PhCHO →

(187)   +   (188)

(187:188 – 94:6)

The corresponding (Z) enolate (186b), which can be prepared by the conjugate addition of lithium bis(phenyldimethylsilyl) cuprate to methyl cinnamate, reacts with benzaldehyde to give the diastereomeric aldols (187) and (188) in a ratio of 6:94, thus favoring the formation of the 2,3 *anti* aldol.

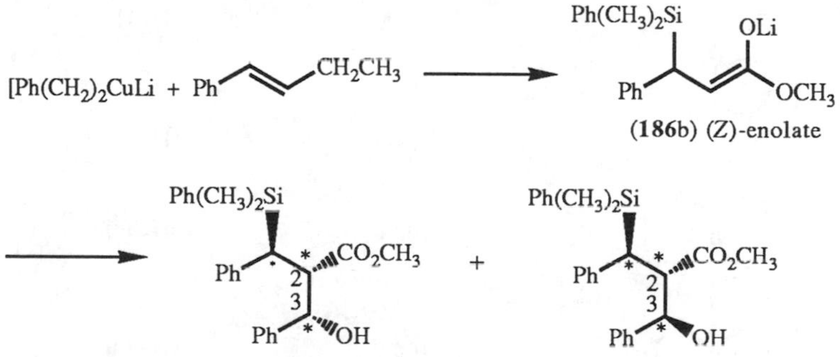

[Ph(CH$_2$)$_2$CuLi + Ph⟶⟶CH$_2$CH$_3$] → (186b) (Z)-enolate

(187)   +   (188)

(187:188 = 6:94)

Lactone enolates generally exhibit low stereoselectivity [123b]. However, substituted lactones which can form dianions show high diastereofacial selectivity [123c]. Thus the dianion of the hydroxybutyrolactone (189) reacts with aldehydes e.g. PhCHO to give the products (190) and (191) in high stereoselectivity.

(189)   (190)   (191)

(190:191 = 90:10)

The same reaction with unsubstituted butyrolactone afforded the aldol products in a ratio of 40:60 to 30:70, depending upon the reaction temperature.

4.3.2.1.1.3   Amide and Lactam Enolates

Simple N,N-dialkylamides generally exhibit poor stereoselectivity in aldol reactions [123d]. Table 4.12 shows aldol reactions of some substituted amides.

Table 4.12. Stereoselectivity in Aldol Reactions of Amides

R^1	R^2	syn:anti	Reference
CH_3	$N(CH_3)_2$	60:40	[121]
CH_3	N⟨pyrrolidine⟩	60:40	[121]
F	$N(^iPr)_2$	62:38	[123c]
CH_3	⟨phenothiazine-type S,N⟩	100:0	[123d]
CH_3	⟨phenothiazine S,N⟩	100:0	[123d]

The exclusive formation of *syn* aldols in the reaction of N-acyl derivatives of 2,3-dihydro-4*H*-1,4-benzothiazine and phenothiazine as shown in Table 4.12 has been explained by assuming a six-membered cyclic transition state (**192**) in which the phenyl group of the aldehyde lies in the equatorial position thereby avoiding a 1,3-*trans*-diaxial interaction with the heterocyclic ring [123d].

(**192**)

Lithium enolates of β-lactams are important in the synthesis of antibiotics, e.g. thienamycin [123f,g]. For instance, the β-lactam (**193**) afforded predominantly one of the four stereoisomers (**194**) in 86% yield [123f].

(**193**)   i) LDA, THF   ii) ArCHO   (**194**) Ar = O-ClC$_6$H$_4$

4.3.2.1.1.4   Thioester and Thioamide Enolates

Deprotonation of thioesters gives predominantly Z enolates [123h]. Table 4.13 shows some examples of aldol reaction of thioesters.

Table 4.13. Aldol Reactions of Thioesters

R^1	R^2	(Z):(E)	syn:anti	Reference
CH$_3$	Ph	75:25	57:43	[123i]
^{i}Pr	C$_2$H$_5$	81:19	82:18	[123i]
^{t}Bu	Ph	84:16	77:23	[123i]

(Table 4.13. contd.)

R^1	R^2	$(Z):(E)$	*syn:anti*	Reference
Ph	Ph	87:13	74:26	[123i]
Ph	C_2H_5	87:13	81:19	[123i]
^tBu	C_2H_5	84:16	76:24	[123i]
C_2H_5	$Ph(CH_3)CH$	a	95:5	[123j]
C_2H_5	$c\text{-}C_6H_{11}(CH_3)CH$	a	84:16	[123j]

a: Enolate ratio not determined

Table 4.13 shows the relationship between enolate geometry and aldol stereochemistry. The diastereoselectivity exhibited by 2-phenylpropanol and 2-cyclohexylpropanol is exceptional considering the fact that the enolate ratio is probably 75:25. The high stereoselectivity, apparently independent of enolate geometry, has been interpreted as indicative of an open chain transition state [123i].

Thioamides, like thioesters, give generally (Z)-enolates which react with aldehydes to give *syn*-aldols with high stereoselectivity [124-126]. Table 4.14 shows some examples of the aldol reactions of thioamides derived from primary as well as secondary amines. In the case of thioamides of primary amines, two equivalents of base are required to give dianions having the (Z)-configuration which react with aldehydes to afford predominantly *anti*-aldols [125,126]. Similarly, thioamides of secondary amines give (Z)-enolates, but the aldol reactions of these enolates afford the *syn*-aldol [124].

Table 4.14. Aldol Reactions of Thioamides

R^1	R^2	R^3	R^4	*anti:syn*	Reference
CH_3	CH_3	CH_3	Ph	13:87	[124]
CH_3	CH_3	CH_3	C_2H_5	10:90	[124]
^iPr	CH_3	CH_3	$PhCH_2CH_2$	68:32	[124]
CH_3	Ph	H	CH_3	69:31	[125]
CH_3	Ph	H	^iPr	88:12	[125]
CH_3	Ph	H	Ph	62:38	[125]

(Table 4.14. contd.)

R^1	R^2	R^3	R^4	*anti:syn*	Reference
^iPr	Ph	H	^iPr	93:7	[125]
^iPr	Ph	H	Ph	69:31	[125]
Ph	CH_3	H	^iPr	78:22	[125]
Ph	$CH_2CH_2OCH_3$	H	Ph	94:6	[125]

4.3.2.1.1.5   Carboxylic Acid Dianions

Carboxylic acids give dianions with LDA which add to aldehydes affording the *anti*-aldol products in high yields [122b, 123j, 124c]. As shown in Table 4.15, the diastereomeric ratios under both conditions A and B are essentially the same for the first four entries showing that there is no equilibration even after prolonged reaction time at higher temperature. However if the combined steric bulk of R^1 and R^2 is great enough, then there is significant equilibration to the *anti* isomer.

Table 4.15. Aldol Reactions of Carboxylic Acid Dianions

R^1	R^2	*anti:syn*(A)	*anti:syn*(B)
CH_3	^tBu	50:50	50:50
CH_3	Ph	55:45	55:45
C_2H_5	Ph	52:48	55:45
^iPr	Ph	55:45	58:42
^tBu	^iPr	79:21	90:10
^tBu	^tBu	80:20	>98:<2
^tBu	Ph	60:40	88:12
Ph	Ph	71:29	92:8
^tBu	Mesityl	64:36	93:7
Mesityl	Mesityl	91:9	>98:<2

A: -50°C, 10 min., B: 50°C, 3 days

4.3.2.1.2   Boron Enolates in Aldol Reactions

While the lithium enolates discussed in section 4.3.2.1.1. exist as dimeric, tetrameric and hexameric aggregates, the boron (boryl) enolates appear to be homogeneous in solution. Moreover the B - O and B - C bonds are significantly shorter than the Li - O and Li - C bonds and the nucleophilicity of boron enolates is generally less pronounced than that of lithium enolates. These characteristics result in boron enolates forming comparatively tight and organized transition states in the aldol reactions, affording high stereoselectivity. The majority of stereoselective aldol reactions mediated by boron enolates involve alkenyloxydialkylboranes ($R^1R^2C = CR^3OBXY$, where X and Y are alkyl groups), while alkenyloxydialkoxyboranes (X and Y = alkoxy groups) have also been utilized [127,128].

Boron enolates such as (**197,198**) prepared from chiral ethyl ketone (**195**) and dialkylboron- trifluoromethanesulphonate (**196**) have been employed in the aldol condensation of achiral aldehydes [129,130].

(**195**)   + R_2BOTf

(**196**)

(**197**) BR_2 = 9-BBN -
(**198**) R = cyclopentyl -

Thus the aldehyde (**199**) reacts with the boron enolate (**198**) to yield the *erythro* product (**200**) exclusively [129,130]. A cyclic transition state of the type (**201**) has been proposed to account for the observed stereoselectivity of the reaction, in which the substituents attached to the chiral centre are oriented so as to minimize steric interactions.

(**199**)

erythro (**200**)

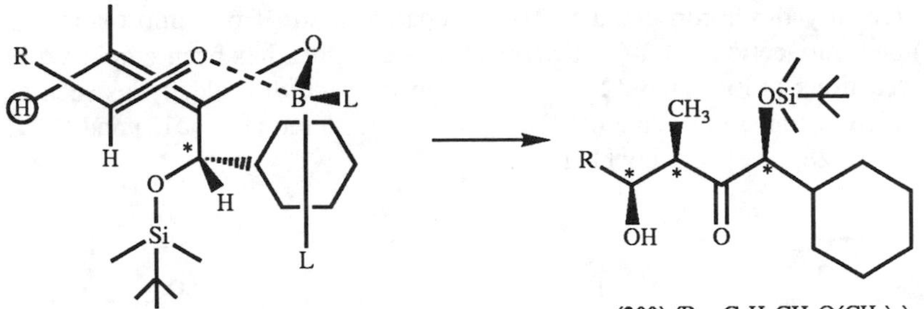

(201)

(200) (R = $C_6H_5CH_2O(CH_2)_2$)

Similarly, the boron enolate **(203)** derived from the chiral oxazolidinone **(202)** reacts with aldehydes RCHO to afford the corresponding products **(204)** in 99% optical yield [130].

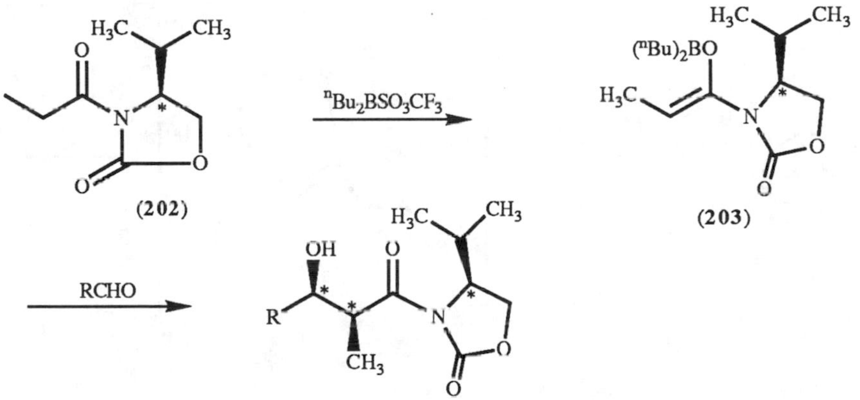

(202)

$^nBu_2BSO_3CF_3$

(203)

RCHO

(204)   (R = n-hexyl > 99.9% o.p.)

Employing the α-unsubstituted derivative of **(203)** such as the reagent **(205)** in aldol condensations with aldehydes RCHO, the adducts **(206)** and **(207)** are obtained in almost equal amounts in 52% and 48% yields respectively.

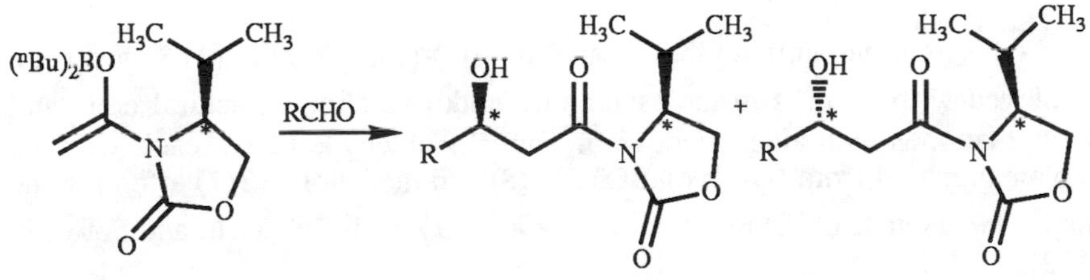

(205)

RCHO

(206)   52%

+

(207)   48%

Recently the boron enolate (210), prepared *in situ* from chloroborolane (208) and thioacetic acid 3-(3-ethyl)-pentyl ester (209), has been employed in aldol reactions [130]. Thus (210) was shown to react with aldehydes such as (211) *via* the transition state (212) to give the diastereomers (213) and (214) in a ratio of 28:1 respectively [131].

(208)   (209)   (210)

(211)   (212)

(213)   (214)

(213 : 214 = 28 : 1)

(-)-Diisopinocampheylboron triflate [(-)-(Ipc)$_2$BOTf] (215) has been employed with success in the asymmetric aldol reaction of achiral ethyl and methyl ketones with aldehydes [132]. Thus the aldehyde (216) reacts with the enolate prepared from (-)- (Ipc)$_2$BOTf (215) and the ketone (217) affording the major *anti* isomer (218) in a *syn -anti* ratio of 97:3 in 79% yield and 86% e.e. [132].

(215)

H$_2$C=C $\overset{\text{CH}_3}{\underset{\text{CHO}}{}}$ + ^{i}PrCH$_2$COEt $\xrightarrow{\text{(215)}}$ H$_3$C...

$$\text{H}_3\text{C}\overset{\overset{\text{CH}_2}{\|}}{}\overset{*}{\underset{\text{OH}}{\text{C}}}\overset{*}{\underset{\text{O}}{\text{C}}}\overset{\text{CH}_3}{}...^i\text{Pr}$$

(216)   (217)   (218)

4.3.2.1.3   *Magnesium Enolates in Aldol Reactions*

Asymmetric aldol reactions have also been carried out using magnesium enolates prepared by the reaction of substrates with Grignard reagents, e.g. EtMgBr or other reagents such as Et$_2$NMgBr [120,123j,133]. For example, the thioamide enolate (221) prepared from the substrate (219) and ^{i}PrMgBr (220) reacted with the aldehyde (222) to afford the diastereomeric aldol products (223) and (224) in a ratio of 95:5 [123j].

CH$_3$CH$_2$C—N(CH$_3$)$_2$ + ^{i}PrMgBr $\longrightarrow$ $\underset{\text{H}_3\text{C}}{\overset{\text{H}}{}}$C=C$\underset{\text{S—MgBr}}{\overset{\text{N(CH}_3)_2}{}}$ + (CH$_3$)$_2$CHCHO

$\overset{\|}{\text{S}}$

(219)   (220)   (221)   (222)

$\xrightarrow{-78°C}$

(223) *erythro*   +   (224) *threo*

The high selectivity in favour of the *erythro* product suggests that the enolate (221) has a Z-configuration [123j].

Similarly, magnesium enolates have been employed in the aldol reaction of, for instance, the substrate (225) with MeCHO (226) affording the *erythro* (227) and *threo* products (228) in a ratio of 99:1.

(225)   i) C₂H₅MgBr   (227)   (228)
          ii) CH₃CHO,
          (226)

(227 : 228 = >99 : <1)

4.3.2.1.4   Titanium Enolates in Aldol Reactions

Titanium enolates have served as the reagents of choice to achieve high diastereofacial selectivities in asymmetric aldol reactions [134-138]. High *anti* selectivity not accessible with other enolates can be achieved using titanium enolates [135,139-147]. Generally titanium is employed in the asymmetric aldol reaction either as an alkyl or alkoxytitanium reagent or as a Lewis acid in conjunction with styrenol ethers.

Thus the *E*-silylketene acetal (229) reacts with aromatic and α,β-unsaturated aldehydes e.g. benzaldehyde in the presence of TiCl₄-PPh₃ complex to give the aldol products (230) and (231) affording high *anti* selectivity [139].

(229)   (230) *anti*   (231) *syn*
          (230 : 231 = 10.5 : 1)

Similarly, the titanium enolate of the ketone (232), prepared from the preformed lithium enolate, reacts with aldehydes such as (233) to give the diastereomeric products (234) and (235) with a high level of stereocontrol [134,137].

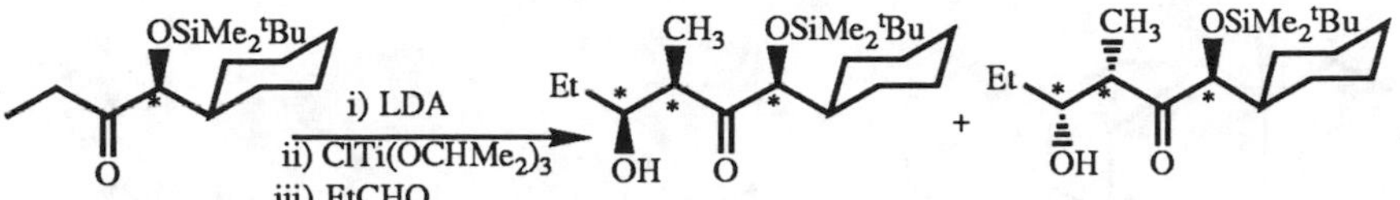

(232)   i) LDA   ii) ClTi(OCHMe₂)₃   iii) EtCHO

(233)   +   (234)

(233 : 234 = 53:1)

A non-chelated chair-like transition structure such as (235) has been proposed to account for the observed diastereofacial selectivity [134]. The bulk of the *tert*-butyldimethylsilyloxy group in the substrate (232) does not permit chelation.

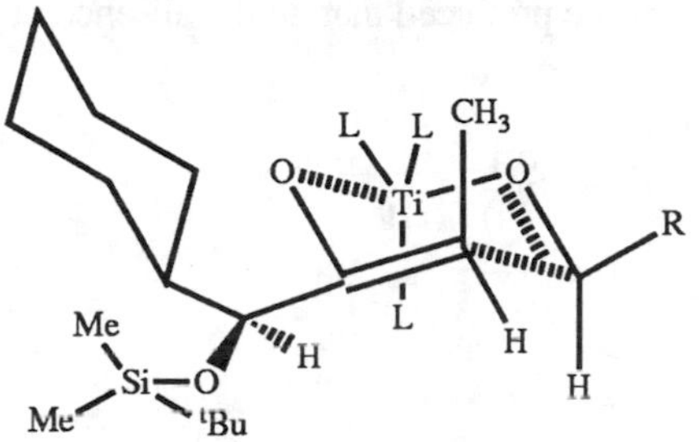

(235, L=OCHMe₂)

4.3.2.1.5   Zinc Enolates in Aldol Reactions

Zinc enolates have found application in the well known Reformatsky reaction in which the organozinc compounds formed from metallic zinc and α-haloesters add to aldehydes or ketones to give the corresponding β-hydroxyesters [148-152]. Thus benzaldehyde (35) reacts with the α- bromoester (236) in the presence of zinc and (-)-sparteine (237) to give predominantly (238) in 95% optical yield [152].

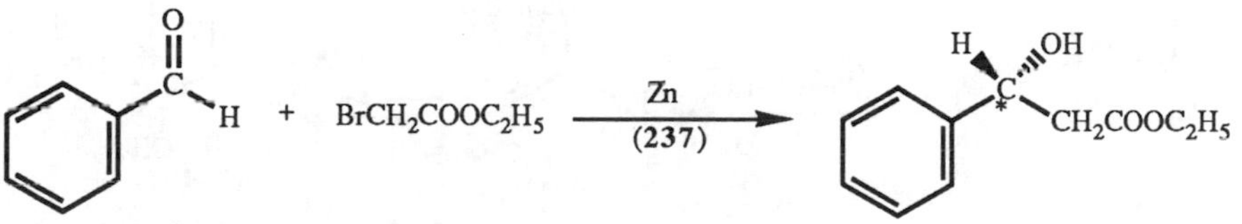

(35)   +   BrCH₂COOC₂H₅   (236) $\xrightarrow[\text{(237)}]{\text{Zn}}$ (238) 95% e.e.

(237)

Lithium enolates can be transformed to zinc enolates by the addition of zinc chloride. This increases the selectivity, yielding asymmetric aldol products in high yields [153,154]. Thus butanol (239) reacts with the lithium enolate (240) using zinc chloride as additive to give the product (241) in 86-90% yield [153]. Generally greater amounts of the *anti* isomers are produced than in the absence of $ZnCl_2$.

4.3.2.1.6   *Tin Enolates in Aldol Reactions*

Tin enolates (or enolstannanes) have been used in asymmetric aldol reactions affording high diastereoselectivity [155-157]. Thus the tin enolate obtained as both the O-and C-stannanes (244 a,b) (prepared from 1-phenyl-1-propen-1-ol acetate (242) and tri-n-butyl tin methoxide (243)) reacts with the aldehyde (245) to give the aldol products (246) and (247) in 95% yield with a *threo-erythro* ratio of 95:5 [155]. It is interesting that at comparatively higher temperatures, the selectivity is reversed and at 43°C the *threo: erythro* products obtained are in a ratio of 10:90 [156].

Similarly under kinetic control, 1-cyclohexenyltributyltin (**248**) reacts with benzaldehyde (**236**) to give the products (**249**) and (**250**) in a ratio of 80:20 [156]. Tin enolates thus exhibit unusual behaviour by giving preferentially the *threo* products at low temperature and *erythro* products at elevated temperature.

(**248**)   (**249**) *threo*   (**250**) *erythro*

(**249** : **250** = 80 : 20)

Generally, metal enolates condense with aldehydes under kinetic conditions to give the *erythro* products and under thermodymanic conditions to afford the *threo* products. It is interesting that the intermediate *threo* aldol stannane is not converted to the *erythro* aldolstannane at higher temperatures. Thus the reaction of enol stannanes with aldehydes under kinetic conditions can be classified as among the few *threo* selective reactions known so far. The reactivity of unsymmetric organotin-enolates increases in the order: $(^nBu)_3Sn\ CH_2\ COCH_3$ > $(^nBu)_3SnCH_2CO_2Et$ > $(^nBu)_3SnCH_2CN$.

4.3.2.1.7   Silicon Enolates in Aldol Reactions

Silicon enolates or enolsilanes have been extensively employed in asymmetric aldol and cross aldol reactions [158-160]. Enolsilanes are weak nucleophiles and react with carbonyl compounds in the presence of Lewis acid catalysts, e.g. $TiCl_4$, $SnCl_4$, $ZnCl_2$ etc. For instance, the (*E*)-enolsilane (**252**), prepared from ethyl propionate (**251**), undergoes condensation with the aldehyde (**253**) at -78°C in the presence of $TiCl_4$ to give the products (**254**) and (**255**) in a *threo:erythro* ratio of 94:6 [157].

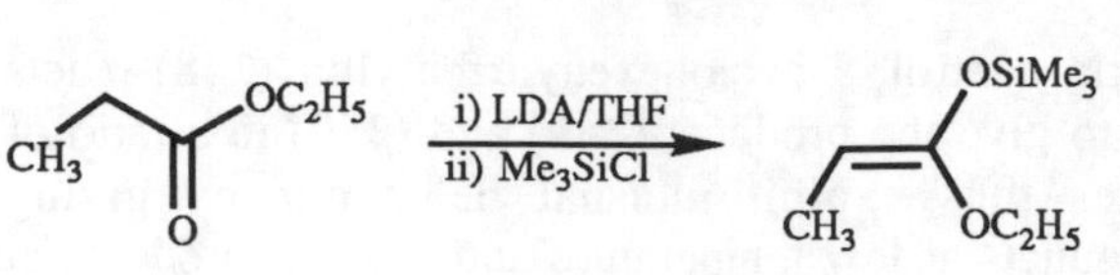

(251)

(252)

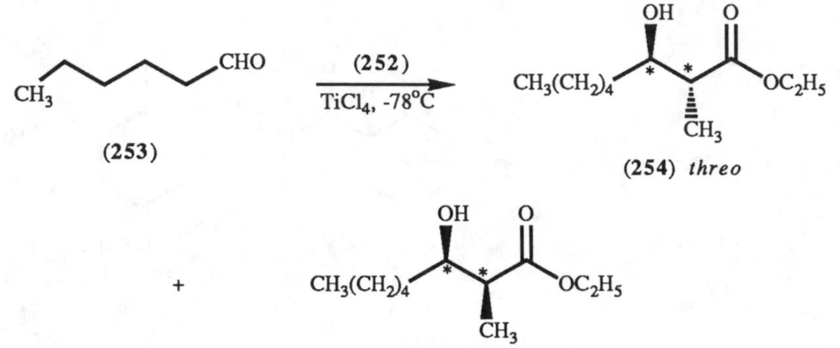

(253)

(252)
—————→
TiCl$_4$, -78°C

(254) *threo*

+

(255) *erythro*

(*threo* : *erythro* = 94 : 6)

It has been suggested that a transition state such as (256) is involved in the reaction leading preferentially to the *threo* isomer [160].

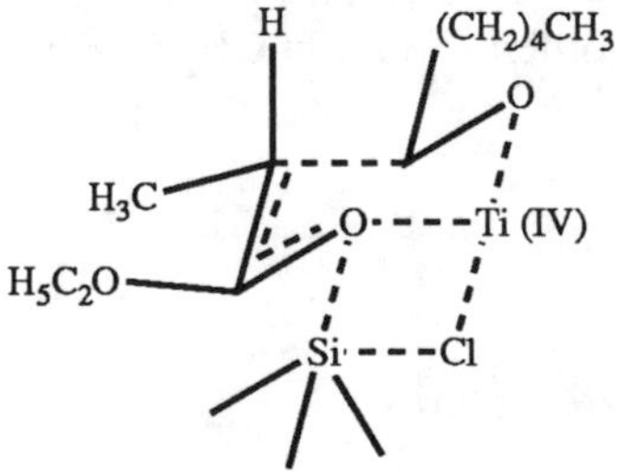

(256)

Fluoride ions derived from tetrabutylammonium fluoride (TBAF) in THF also catalyze the reaction of enolsilanes with various aldehydes affording the corresponding aldol-trimethyl silyl ethers in fair to good yield [161]. Thus, without recourse to either strongly acidic or basic reagents, asymmetric cross-aldol reaction can be carried out. For example, the reaction of 1-trimethyl-siloxycyclohexene (257) with benzaldehyde dimethylacetal (258) in the presence of trimethyl silyltrifluoromethane sulfonate (259) affords the *erythro* product (260) at -78°C in 86% yield and in a *erythro:threo* ratio of 92:8 [162].

(257)   (258)   (260)

(erythro: threo = 92 : 8)

Similarly, enolsilanes react with aldehydes in the presence of tris (diethylamino) sulphonium (TAS) difluorotrimethylsiliconate with high stereoselectivity affording *erythro* products, similar to the trimethylsilyltriflate mediated reactions [162]. Thus, the silylenolether (257) reacts with the aldehyde (222) to give the *erythro* product (261) exclusively [163].

(257)   (222)   (261)   100%

4.3.2.1.8   Zirconium Enolates in Aldol Reactions

Zirconium enolates have been developed in product selective aldol condensations, which are independent of the enolate geometry [164-166]. Zirconium enolates are generally prepared from the corresponding lithium enolates by ligand exchange with biscyclopentadienylchlorozirconium (Cp_2ZrCl_2) (262). For example, the enolate (264), prepared from the amide (263) and Cp_2ZrCl_2 (262) reacted with the aldehyde (265) to afford the products (266) and (267) in an *erythro*: *threo* ratio of 95:5 [164].

It is interesting that with amides e.g. (263) the zirconium enolate geometry is *cis*, the *trans* enolate (268) being formed in a low ratio (263:268 = 95:5). With other substrates, *trans* enolates are formed preferentially [164]. It has been suggested that *trans* zirconium enolates react preferentially *via pseudo*-boat transition states while the corresponding *cis* enolates preferentially react *via pseudo*-chair transition states [164].

Cp_2ZrCl_2 + (262)   (263)   →   (264)

$$\xrightarrow[\text{(265)}]{CH_3CH_2CH_2CHO}$$

(266) *erythro*   +   (267) *threo*

(266 : 267 = 95 : 5)

Similarly the zirconium enolate (269) reacts with aldehydes to give good *erythro* diastereoselection (96 - 98%) [165].

(268)   (269)

Other metal enolates e.g. aluminium enolates [166a-e], rhodium enolates [166f,g], cerium enolates [166k] and cobalt enolates [166l] have also been employed with verying degrees of success in asymmetric aldol reactions.

4.3.3   Addition of Chiral Enolates to Achiral Aldehydes and Unsymmetric Ketones (the Cross Aldol Reaction)

In the cross aldol reaction *chiral* enolates add to *achiral* carbonyl compounds so that the chirality is transferred to the product while the chiral moiety is ultimately removed. For instance, when the (methylthio) acetic acid derivative (270) is converted to its boron enolate and allowed to react with aldehydes, e.g. PhCHO, mainly the *syn*-diastereomer (271) is obtained. The β-hydroxycarboxylic acid (272) is obtained from (271) by desulphurization using Raney nickel followed by saponification in optical yields of upto 99% [167].

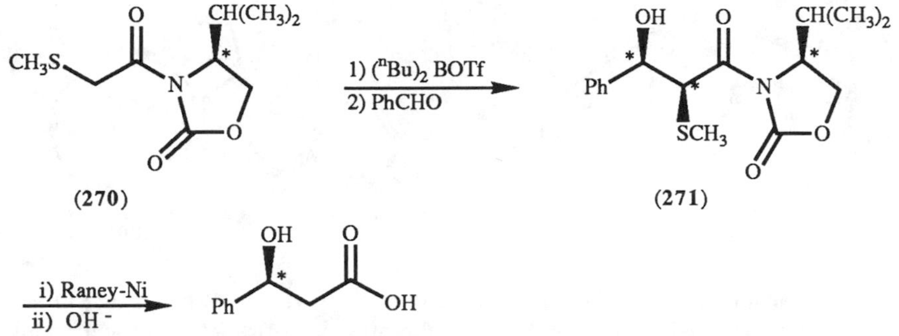

(270)

1) (^{n}Bu)$_2$ BOTf
2) PhCHO

(271)

i) Raney-Ni
ii) OH$^-$

(272) 85-99% e.e. (R)

Similarly the α-sulfinic ester (273), after deprotonation by a Grignard reagent, affords (274) which adds diastereoselectively to aldehydes, e.g. (275) yielding (276). Reductive elimination of the sulfinyl moiety from (276) affords (277) in 90% e.e. [168-171].

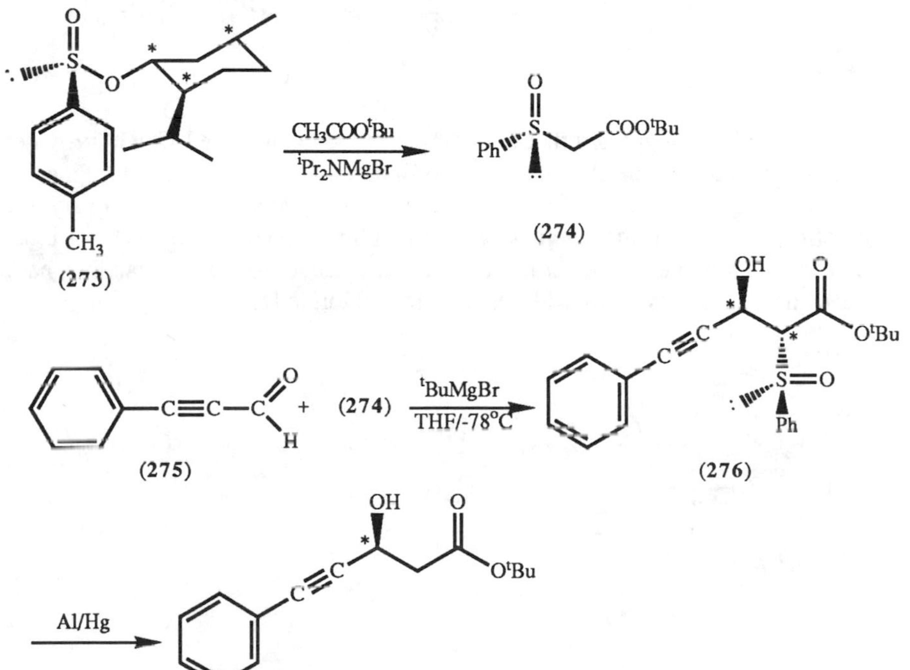

(273)

CH$_3$COO^{t}Bu
^{i}Pr$_2$NMgBr

(274)

(275)

+   (274)

^{t}BuMgBr
THF/-78°C

(276)

Al/Hg

(277)

Certain chiral diamines such as (278-280) have been employed recently as chiral promoters along with tin (II) triflate and tributyl tin fluoride in the asymmetric cross aldol reaction of silylenol ethers with aldehydes [172].

(278)   (279)   (280)

For instance, when the silylenol ether of S-ethyl ethanethioate (281) was treated with an aldehyde, e.g. PhCHO in the presence of stoichiometric amounts of tin (II) triflate and the chiral diamine (278) the aldol reaction proceeded smoothly under kinetic controlled conditions at -78°C to afford the product (282) in 78% yield with 82% e.e. [169].

PhCHO   +

(35)   (281)   (278)   (282)   82% e.e. (S)

It has been suggested that the complex (283) is involved which provides double activation to the aldehyde and the silylenol ether in the reaction. Thus the positive centre of tin (II) triflate activates the aldehyde and, simultaneously, the electronegative fluoride interacts with the silicon atom of the silylenol ether thereby making the enol ether more reactive. Similar cross-aldol reactions have been carried out previously with titanium tetrachloride [158].

(283)

4.3.3.1   Metal Atoms as Chiral Centres in Aldol Reactions

Chiral iron acyls e.g. $(\eta 5\text{-}C_5H_5)Fe(CO)(PPh_3)COMe$ (284) have been employed along with various counter ions such as $^iBu_2AlCl$, $SnCl_2$ etc. in stereoselective aldol reactions [173-175].

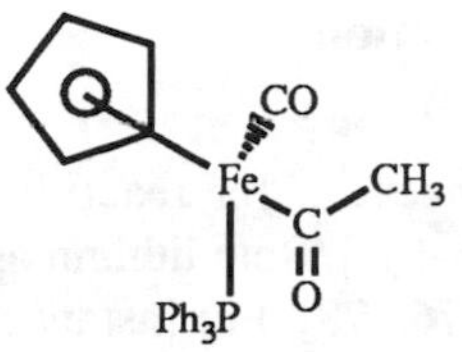

(284)

Thus **(285)** undergoes stereoselective aldol condensations with aldehydes to give the corresponding β-hydroxyacyl complexes which after decomplexation afford β- hydroxy acids.

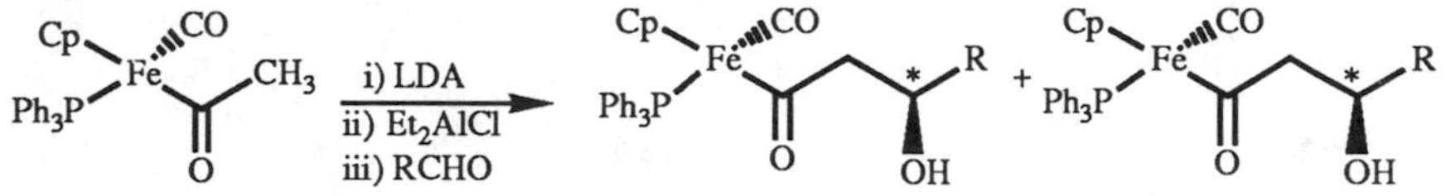

(284)   **(285)**   **(286)**

$(285 : 286 = 100 : 1)$

For instance, transmetalation of the lithium enolate of **(284)** with Et_2AlCl affords the aluminium enolate which reacts with aldehydes to afford the diastereomeric products **(285)** and **(286)** in a ratio of 100:1 [173]. Removal of the iron complex gives the corresponding β-hydroxy carboxylic acids.

It is interesting that with Sn(II) enolates, the product **(286)** is the major diastereomer, whereas with the aluminium enolate **(285)** predominates. This has been attributed to the difference in the approach of the reagents in the two transition states. In the former case a cyclic transition state such as **(287)** has been proposed to be operative whereas in the aluminium system, an extended transition state **(288)** has been postulated.

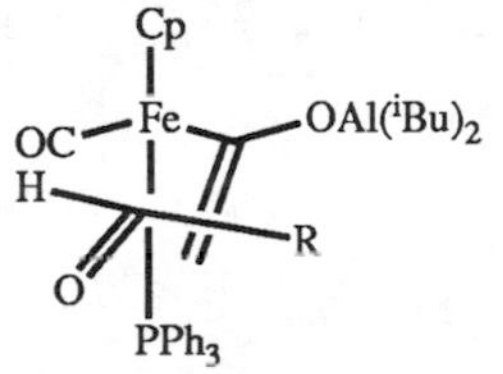

(287) 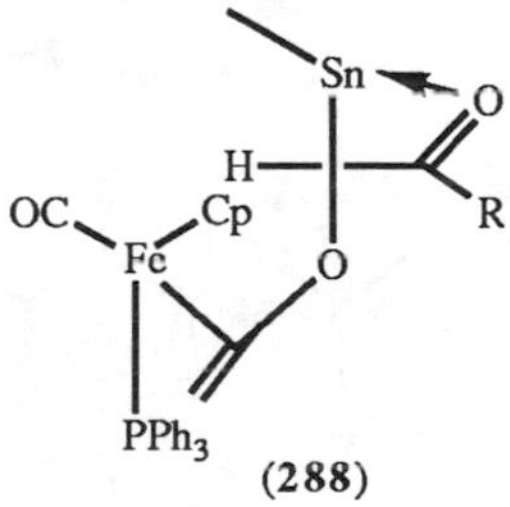   **(288)**

4.3.3.2   Chiral Ketone Enolates in Aldol Reactions

Chiral ketone enolates have been developed in asymmetric aldol reactions affording moderate to high diastereofacial selectivities [176-181]. Both lithium as well as boron enolates of ketones have been employed [176,178]. For instance, the dibutylboron enolate (290) of (±)-3-methyl-2- pentanone (289) reacts with propionaldehyde to give the diastereomeric products (291) and (292) in a ratio of (57:43) [179].

(289)   (290)

(291)   (291 : 292 = 57 : 43)   (292)

Similarly, the ketone enolate (294) prepared from the ketone (293) reacts with acetaldehyde to give the diastereomeric products (295) and (296) in a ratio of 97:3 [179].

(293)   (294)

(295)   (296)

The asymmetric induction is influenced by the metal centre, and boron enolates have been found to afford higher diastereoselection in nonpolar solvents. It has been suggested that the aldol process may proceed *via* two diastereomeric transition states (297) and (298) in which the substituents R_S and R_L are respectively designated as "small" and "large".

(297)

(299)

(298)

(300)

The transition state (297) is assumed to be preferred over (298) because of the influence of steric hinderance ($R_S \leftrightarrow B_u < R_L \leftrightarrow B_u$) at the metal centre. Thus (299) is obtained as the major postulated product in preference over the alternative product (300).

4.3.3.3   Chiral Azaenolates in Aldol Reactions

The asymmetric aldol reactions of chiral azaenolates derived from chiral oxazolines or chiral hydrazones etc . with aldehydes afford the corresponding aldol products with fairly good enantioselectivities [149,182-185].

Recently chiral oxazolines, which are useful synthetic intermediates in the synthesis of optically active β-hydroxyamino acids and their derivatives, have been prepared by using chiral ferrocenylphosphines, such as (301) and (302), and the gold (I) complex (303) as the catalyst precursor [186-192].

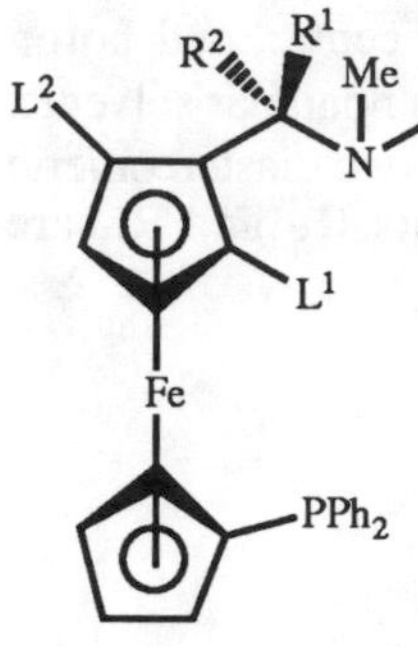

(301) R^1 = Me, R^2 = H, L^1= PPh$_2$, L^2 = H
(302) R^1 = H, R^2 = Me, L^1=H, L^2 = PPh$_2$

[Au(CN-cyclo C$_6$H$_{11}$)$_2$]BF$_4$

(303)

Thus, the aldehyde (304) reacted with ethylcyanoacetate in the presence of gold (I)-(301) as the catalyst to give the diastereomeric products (305) and (306) with 85% diastereomeric purity of the former product [186].

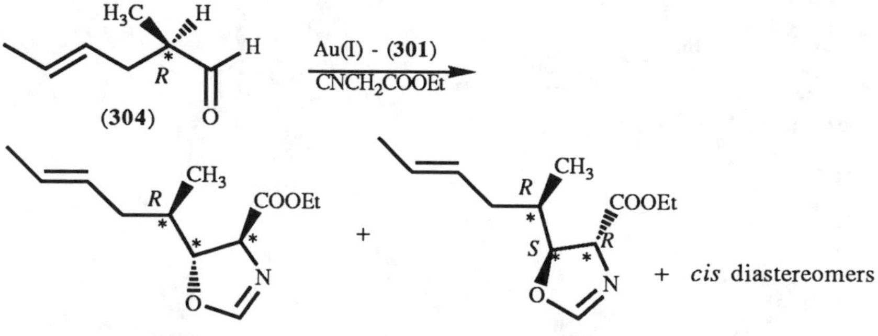

(304)

Au(I) - (301)
CNCH$_2$COOEt

(305)   +   (306)   + *cis* diastereomers

(305:306 = 85:10)

The dihydrooxazole (305) when treated with trimethyloxonium tetrafluoroborate in CH$_2$Cl$_2$ afforded the dihydroazolium tetrafluoroborate (307) which upon acidic workup with aqueous NaHCO$_3$ afforded the product (308) in 73% yield [186].

The high efficiency of the ferrocernyl phosphine ligands, e.g. (301), has been attributed to the formation of the transition state (309) in which the dialkylamino group at the end of the side chain on (301) takes part in the formation of the enolate of isocyanoacetate coordinated with gold. It may be due

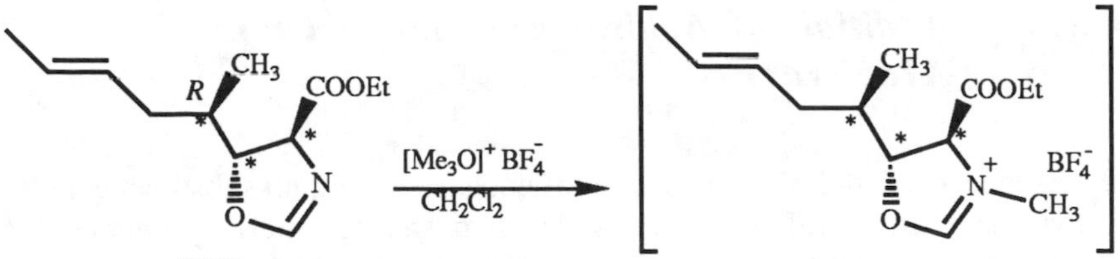

(305)   (307)

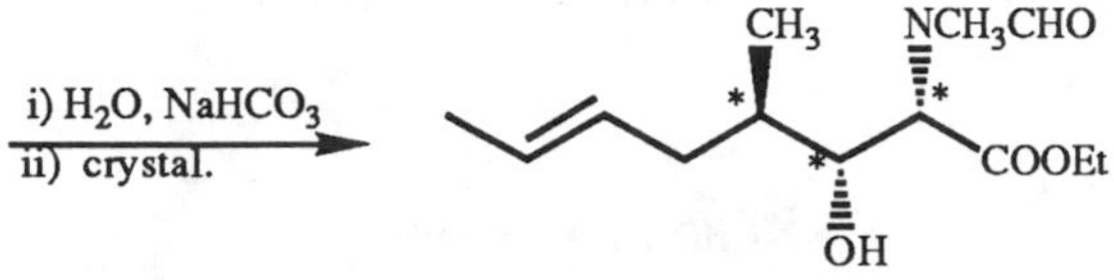

(308) 73% d.p.

to such participation of the side chain which allows a favourable orientation of the enolate and aldehyde on the gold (I) in the diastereomeric transition state which brings about high stereoselectivity.

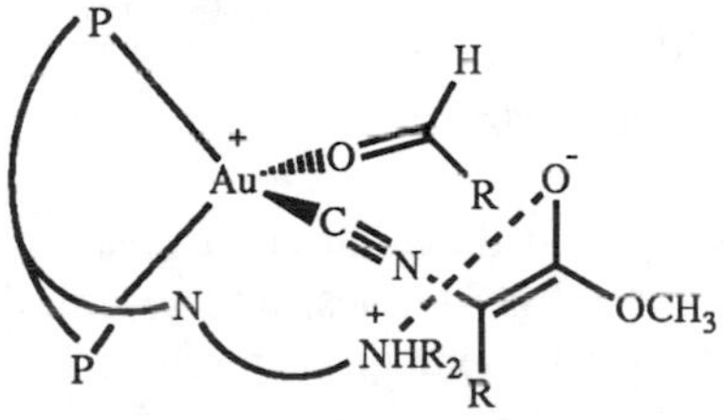

(309)

It is interesting that by varying the side chain in the phosphine ligand (301) to e.g. 3-(diethylamino) propyl side chain, the product (305) is obtained in 26% e.e. The lower optical yield reflects the larger distance between the amino group and the ferrocene moiety which is of crucial importance for selectivity. Similar chain length effects have been observed recently in the silver (I)-catalyzed asymmetric aldol reactions [191b]. Thus by increasing the side chain of the chiral ligand (301) by one methylene unit in the asymmetric aldol reaction involving chiral oxazoline formation, the optical yield decreased to as much as 44.% e.e and also led to inversion of configuration.

4.3.4   Addition of Achiral Enolates to Chiral Aldehydes

The asymmetric aldol reaction of achiral enolates having no substituents at the α-position with chiral aldehydes gives two stereoisomeric products. For instance, the reaction of the silyl enolate (310) with the aldehyde (311) afforded the diastereomeric products (312) and (313) in a ratio of 25:1 [192].

(310)   (311)

(312)   (313)

The Z- and E- enolates (314a,b respectively) substituted at the α-position, e.g. those derived from propionate esters and ethyl ketones, add to chiral aldehydes, e.g. 2-phenyl-proponal to give each four diastereomeric products i.e., (*syn*, *syn*), (*syn*, *anti*), (*anti*, *syn*) and (*anti*, *anti*).

(314a)   (314b)   (*syn*, *syn*)

(*syn*, *anti*)   (*anti*, *syn*)   (*anti*, *anti*)

The relative stereochemistry of the substituents at C-2 and C-3 in the products is controlled by the geometry of the enolate. The C-3, C-4 stereochemistry, on the other hand, depends on the direction of approach of the enolate and the carbonyl group of the aldehyde to each other. In a chiral aldehyde, the two faces of the carbonyl group are not equivalent and when an achiral enolate approaches the aldehyde it shows some preference for one face over the other with the result that one diastereomer is formed in major amount. This is designated as *diastereofacial selectivity*. This is shown in the reaction of 2-phenylpropanol with the Z-enolate of *tert*-butyl methyl ketone. The enolate approaches the α-(*re*) face of the aldehyde in the transition state (**315**) affording the product (**317**) in larger amount than the 2,3-*syn*-3,4-*anti* isomer (**318**) which arises by β-attack as shown in (**316**).

(**315**)

(**317**) 2,3-*syn*, 3,4-*syn*

(**316**)

(**318**) 2,3-*syn*, 3,4-*anti*

Generally, (Z)- enolates add to chiral aldehydes yielding syn products while E-enolates form anti- products as predicted by Cram's rule or its other subsequent modifications [82] provided that the aldehydes have no polar groups. For instance, the (Z)-lithium enolate (**319**) reacts with chiral aldehydes (**320**) to give the C-3, C-4 *syn* isomers, where R is phenyl or vinyl. In other cases, it affords *anti* products [115,119,193].

(**319**)

(**320**)

syn, syn

syn, anti

R = Ph *syn, syn / anti, syn, = 81:19,*
R = Me$_2$C = CH *syn, syn / anti, syn = 94:6,*
R = PhCH$_2$OCH$_2$ *syn, syn / anti, syn = 33:67*

Similarly enol silyl ethers such as **(320)** react with substituted cyclohexanone acetals such as **(321)** to give the equatorial adducts **(322)** in high yields [194].

(320)   (321)   (322) 95%

The enhanced equatorial attack may be due to the chair transition state **(323)** in which the bulky moiety is directed away from the ring.

(323)

Recently, an *ortho*-substituted benzaldehyde - chromium (0) complex, has been employed in asymmetric aldol reactions affording high *syn* selectivity [195b,c]. Thus the chiral chromium complex **(324)** reacted with the achiral enolate **(325)** to give the *syn* product **(326)** in >98% e.e. [195].

(324)   (325)   (326) > 98% e.e.

High *syn* selectivity was observed in substrates in which the *ortho* substituent was either TMS or C_2H_5. Changing to CH_2 lowered the *syn* selectivity.

4.3.5   Reactions of Chiral Aldehydes with Chiral Enolates

In asymmetric aldol reactions in which both the enolate and aldehyde are chiral, the inherent diastereoface selectivity of the two reactants may have reinforcing or opposing effects. The reinforcement of selectivity of the two reactants has been termed as "*consonant double stereodifferentiation*" and the reactants are termed as members of a "*matched pair*". On the other hand, the inherent stereoface preferences of the two reactants may oppose one another. This is termed as "*dissonant double stereodifferentiation*" and the reactants are regarded as a "*mismatched pair*" [196]. The degree of asymmetric induction is approximately taken as the result of (a x b) for a "matched pair" and (a ÷ b) for a "mismatched pair" where a and b are the diastereofacial selectivities of a substrate and a reagent respectively [195]. These formulations have resulted from studies on the asymmetric aldol reactions between achiral substrates and chiral reagents as well as between chiral substrates and achiral reagents. For instance the chiral enolate (S-327) attacks the α-face of the carbonyl group of the aldehyde (35) to give the diastereomeric products (328) and (329) in a ratio of 3.5 : 1. The two substituents at the C-2 and C-3 positions are *syn*-related in both the products (328) and (329) but their absolute configurations are different in that both are β in product (328) while both are α in (329). It has been shown that the absolute configurations are directly related to the stereoface selection exhibited by the chiral reagent (S- 327).

(35)   (S -327)

(328)   (329)

(328 : 329 = 3.5 : 1)

Similarly, the achiral lithium enolate (330) reacts with the chiral aldehyde (S-331) to give the diastereomeric products (332) and (333) in a ratio of 2.7:1 [122].

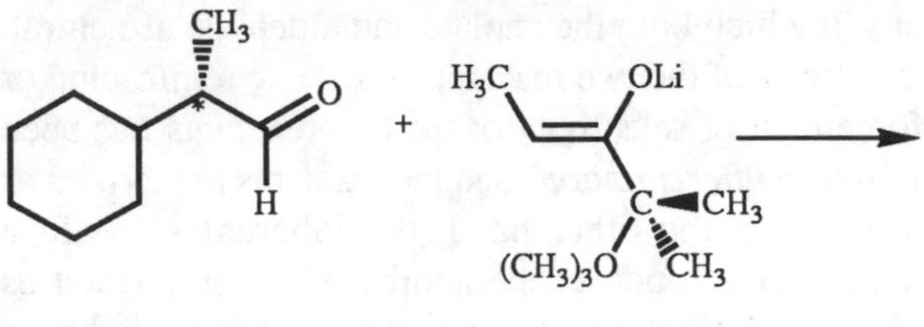

(S-331)   (330)

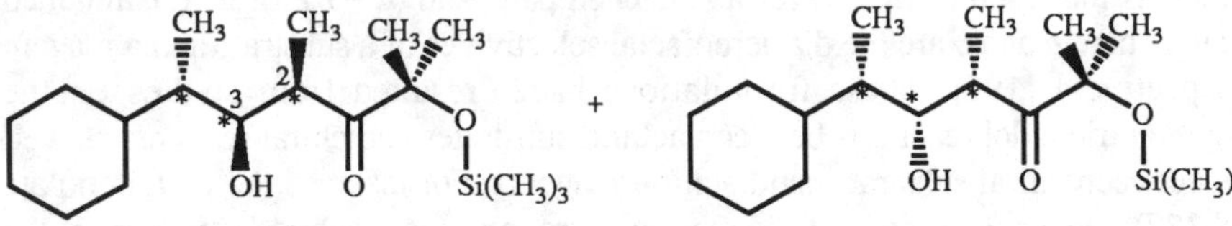

(332)   (333)

(332 : 333 = 2.7:1)

The absolute configuration of the hydroxyl group at the C-3 position in both the products (328) and (332) suggests that both the chiral enolate (S-327) and the chiral aldehyde (S-331) constitute a matched pair, i.e. their diastereoface selectivities reinforce each other. Thus the reaction of the chiral aldehyde (S-331) with the chiral enolate (S-327) would lead to enhanced stereoselection, since both favour α-attack at the carbonyl group. This is indeed found to be the case and the products (334) and (335) are obtained in a ratio of 8:1.

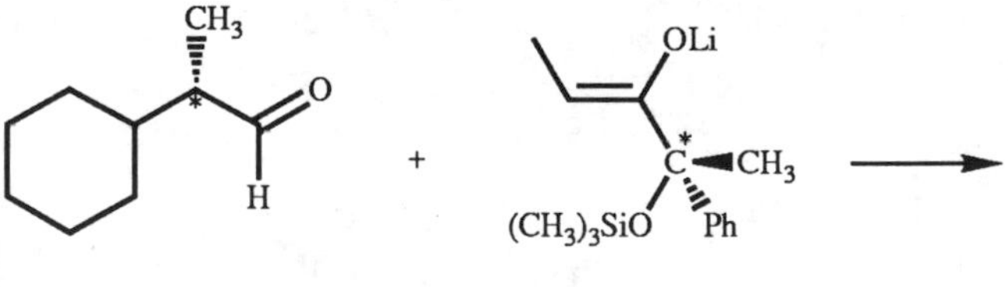

(S-331)   (S-327)

(matched pair)

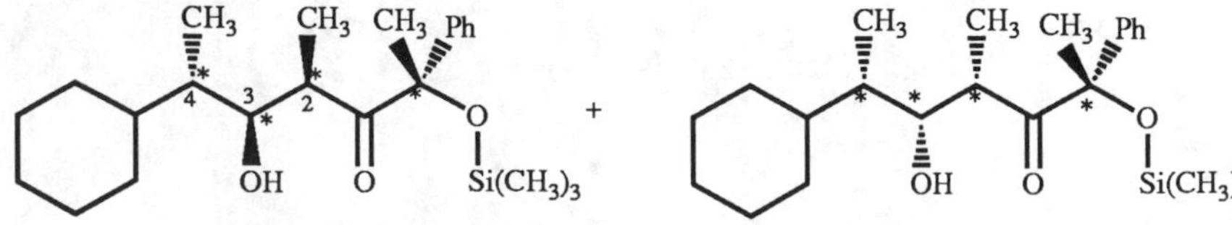

(334)   (335)

(334 : 335 = 8:1) (calcd.: 3.5 x 2.7 = 9.5)

The corresponding mismatched pair of (S-331) and (R-327) reacts with inferior stereoselection affording the products (336) and (337) in a ratio of (1:1.5) as predicted.

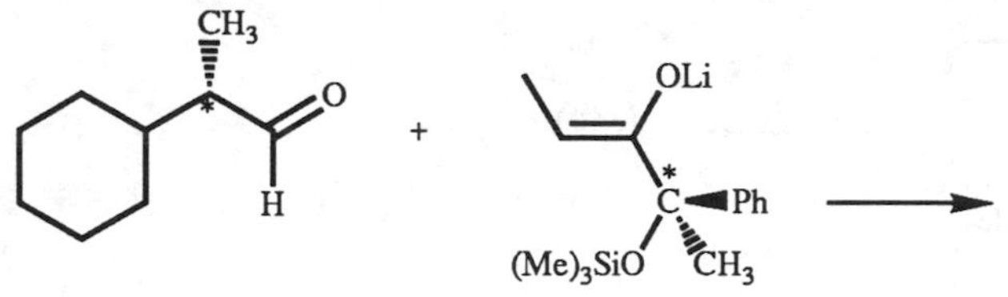

(S-331)   (R-227)

mismatched pair

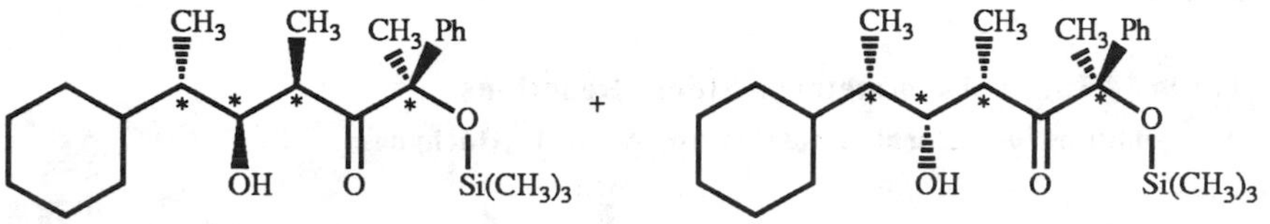

(336)   (337)

(336 : 337 = 1:1.5) (calcd.: 3.5 + 2.7 = 1.3)

The magnitude of diastereoface selectivity in the matched pair suggests a ratio of 3.5 x 2.7 = 9.5 for the diastereomers (334:335), while for the mismatched pair, the calculated value is 3.5 / 2.7 = 1.3 which is in close agreement with the experimental value

Likewise, the reaction of the chiral enolate (S-338) with the chiral aldehyde (-)-339 afforded the two diastereomers (340) and (341) in a ratio of 100:1 [197].

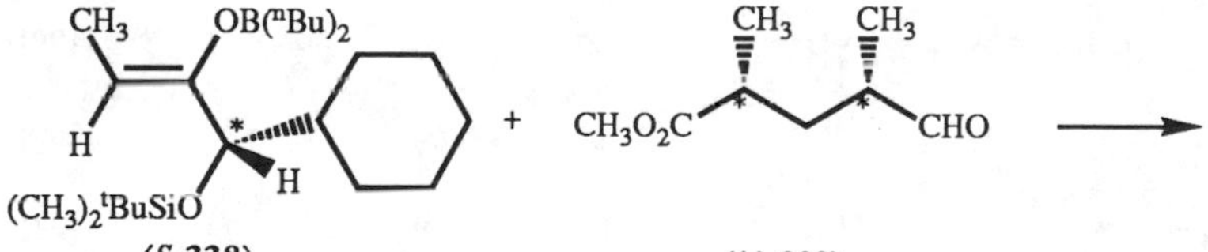

(S-338)   ((-)-339)

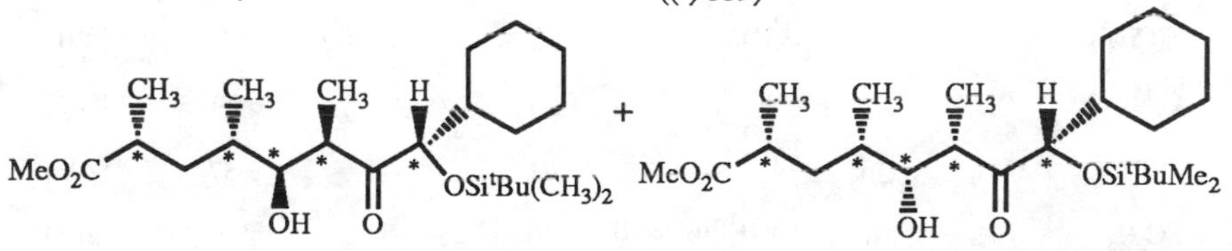

(340)   (341)

(340 : 341 = 100:1)

A change in the chirality of the enolate reagent brings about a reversal of the result. For instance the reaction of (-)-(**339**) with (*R*-**338**) leads to the formation of (**340**) and (**341**) in a ratio of 1:30.

$$CH_3, H, OB(^nBu)_2, OSi^tBuMe_2, H \quad (R\text{-}338) \xrightarrow{(-)\text{-}339} (340) + (341) \quad (1:30)$$

Table 4.16 includes some representative examples of asymmetric aldol reactions.

Table 4.16.   Asymmetric Aldol Reactions

i)   Addition of Chiral Enolates to Achiral Aldehydes:

$$R^3CHO + R^2CH=C{\Large\langle}^{OM}_{R^1} \longrightarrow R^3 \overset{OH}{\underset{R^2}{\wedge}}\overset{O}{\wedge}R^1 + R^3 \overset{OH}{\underset{R^2}{\wedge}}\overset{O}{\wedge}R^1$$

R^1	R^2	R^3	M	Enolate config.	Product config.	% o.p.	Reference
^iPr	CH_3	Ph	Li	(Z)	(S)	>98	[121]
^tBu	CH_3	^iPr	MgBr	(Z)	(S)	100	[120]
$-(CH_2)_4-$	ZnCl	^tBu	Cl	(E)	-	80	[151]
Et	CH_3	^tBu	$ZrCp_2Cl$	(E)	(S)	88	[198]
Et	CH_3	Ph	$TiCl_3$	(Z)	(S)	81	[199]
c-hex	-	Ph	9-BBN	(Z)	(S)	>97	[200]
$-(CH_2)_4-$	-	Ph	$SnPh_3$	(E)	(S)	71	[157]
$N(CH_3)_2$	CH_3	Ph	Si^tBu	(Z)	(S)	95	[201]
C_2H_5	CH_3	Ph	$-B\overset{O}{\underset{O}{\diagup}}$	(Z)	(S)	>99	[202]
OCH_3	CH_3	CH_3	Li	(E)	-	57	[203]
SC_2H_5	CH_3	^iBu	$SnO+^tBu_2OSnTf$	(Z)	-	98:2 (94)	[204]
SEt	CH_3	Ph	$Sn(OTf)_2+^nBu_3SnF$	(Z)	syn : anti	(100:0)%e.e. >98 (2R, 3S)	[205]

(Table 4.16. contd.)

ii) Addition of Achiral Enolates to Chiral Aldehydes

R^1	R^2	R^3	R^4	M	A:B:C	% o.p.	Reference
Ph	CH_3	H	CH_3	Li	3:1:0	89	[192]
Ph	CH_3	H	^tBu	$Bu(CH_3)_2Si$	24:1:0	74	[192]
CH_3	OBu	CH_3	Ph	$Si^tBu+TiCl_4$	syn:syn	97	[206a]
CH_3	OBu	CH_3	Ph	$Si^tBu+TiCl_4$	syn:anti	92	[206a]
CH_3	OBu	H	^tBu	Si^tBu+BF_3	anti	90	[66, 206a]

iii) Addition of Chiral Aldehydes to Chiral Enolates

Aldehyde	Enolate	Product/% d.p.	Reference

(Table 4.16. contd.)

Aldehyde	Enolate	Product/% d.p.	Reference
		(400 : 1)	[207]
		(660 : 1)	[207]
		(94 : 6)	[197]
		(9 : 91)	[197]

(Table 4.16. contd.)

Aldehyde	Enolate	Product/% d.p.	Reference

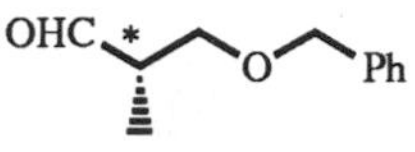

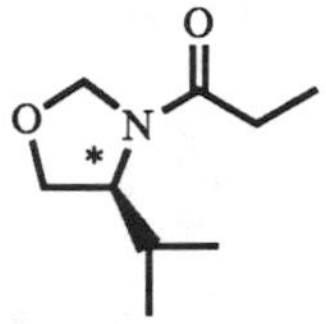

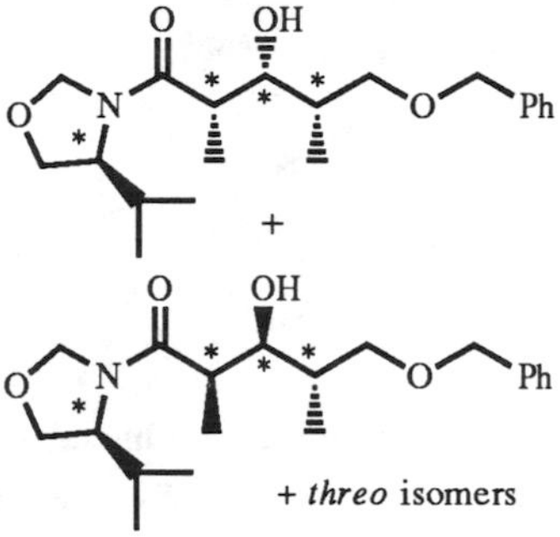

(98.7 : 0.9 : 0.4)

[165]

+ *threo* isomers

I *threo* isomers

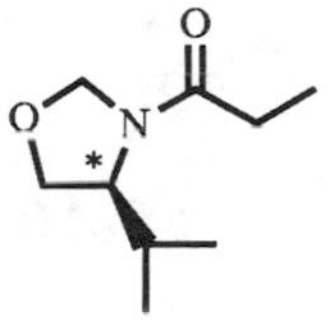

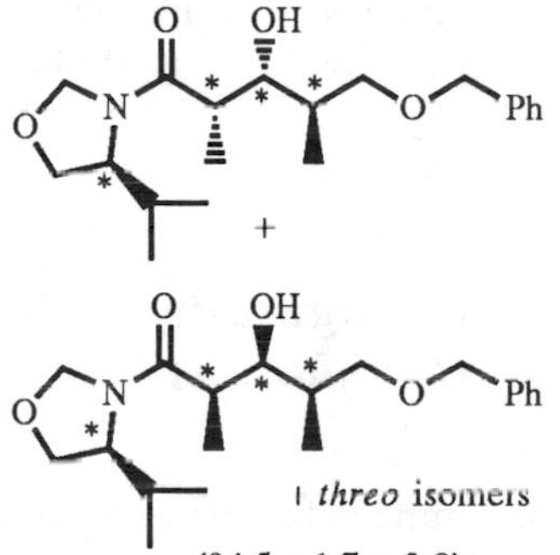

(94.5 : 1.7 : 3.8)

[165]

4.4   Allylmetal and Allylboron Additions

The addition of allyl metal derivatives to carbonyl substrates is structurally and perhaps mechanistically analogous to the stereoselective aldol additions discussed in section 4.3. These additions have many advantages over enolate - derived products since the newly formed alkenes may be readily derivatized and the operation repeated. Fig. 4.11 shows a generalized scheme for the reaction of the (E)- and (Z)-allylmetal compounds (342) and (343) with aldehydes to afford the corresponding products (344) and (345) respectively.

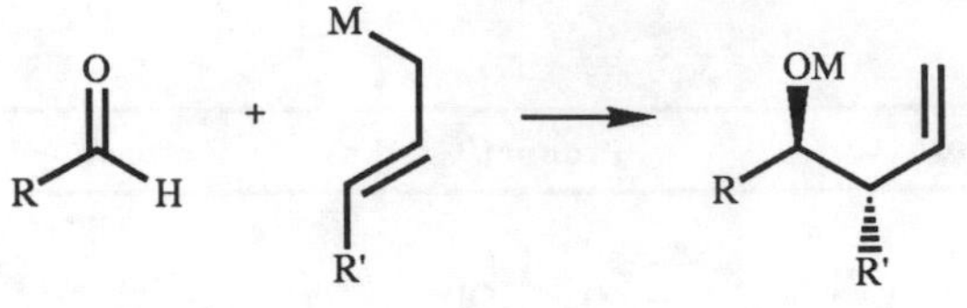

(342)   (344)

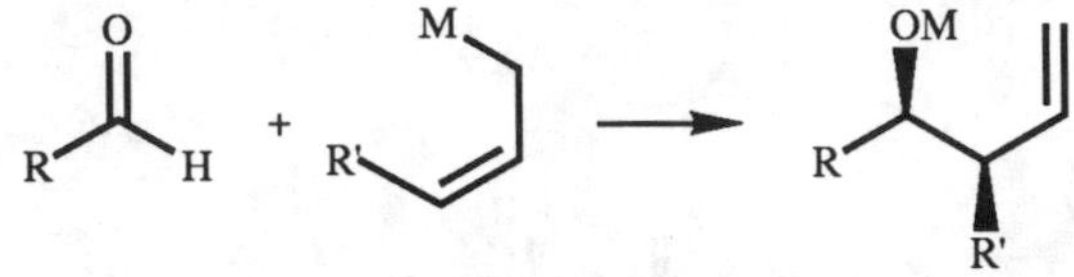

(343)   (345)

Figure 4.11. Addition of allylmetal compounds to aldehydes (R^1 = alkyl, or a heterobonded group; M = metal).

These additions are equivalent to aldol additions since the adducts (**344**) and (**345**), being homoallylic alcohols, can be readily transformed into aldol derivatives.

4.4.1   Configurational Stability of Allylmetal Compounds

Allylmetal compounds exist as either monohapto (η^1-) or trihapto (η^3-) bonded substances depending upon whether the metal atom is attached either to one carbon (η^1-) or to all the three carbons (η^3-) of the allyl group. Both monohapto as well as trihaptocompunds undergo (E)/(Z) isomerizations.

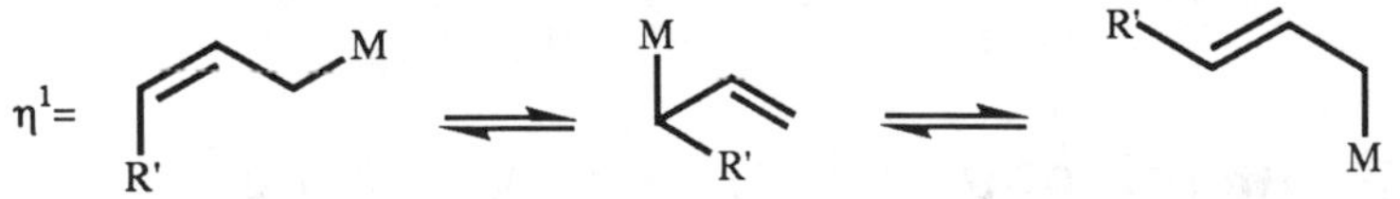

(342)   (346)   (343)

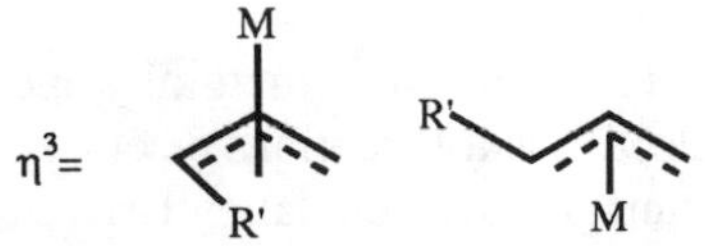

(347)   (348)

Figure 4.12. Metallotropic rearrangements in allylic system.

As shown in Fig. 4.12, the $(E)/(Z)$ isomerization of monohapto compounds (342) and (343) proceeds through metallotropic rearrangement to (346) [208,209]. Similar E/Z isomerization occurs in trihapto compounds (347) and (348) *via* the monohapto- structure (346) [210-213]. In general high diastereoselectivities in allylic additions can be obtained only with those allylmetal compounds with energy barriers of >20 Kcal/mole for the mutual interconversion of monohapto- and trihapto- forms. Among the monohapto-allylmetal compounds fulfilling this requirement are the silanes which have the lowest tendency to metallotropic isomerization [214], whereas the configurational stability decreases on going from germanium to tin compounds [215,216] which undergo $(E)/(Z)$ isomerization below 100°C or in the presence of Lewis bases. Metallotropic rearrangement in allylboron compounds can be controlled through proper choice of the substituents. For example dialkyl (allyl) boron compounds undergo borotropic rearrangement at room temperature [208,209] and can be handled without isomerization only at below - 78°C [217]. Electron-donating substituents at boron raise the energy of the vacant boron orbital to such an extent that rearrangement to the trihapto-structure is hindered. Consequently allylboron compounds substituted at boron with oxygen or nitrogen substituents are particularly resistant to borotropic rearrangement [218,219].

Among the trihapto-allylmetal compounds which are stable towards $(E)/(Z)$ isomerization are included nickel complexes [220], dicyclopentadimethyltitanium compounds [221,222] allylpotassium compounds [211,223] and dicarbonyl (cyclopentadienyl) iron derivatives [224].

4.4.2   Stereochemistry of Allylmetal Additions

The metal forms "ate" complexes involving coordination with the carbonyl of the aldehyde due to the presence of a low-lying vacant orbital. It therefore acts as a Lewis acid and a six-membered cyclic transition state has been proposed for the addition reactions [225,226] while open chain transition states have been proposed in certain other reactions (see later). As in the adol additions discussed in section 4.3. the diastereoselectivity of the allylmetal additions is explained by considering the transition metal geometries (in analogy to cyclohexane geometry) i.e. whether chair or boat forms exist, depending on the relative energies of the diastereomeric transition states. These can be influenced by factors such as the bulk of the substituents, and their axial or equatorial disposition etc. as in cyclohexane. This is illustrated in Fig. 4.13 for the addition of a (Z)-allylmetal derivative to an aldehyde [227].

The chair-like and boat-like transition states (349) and (350) respectively give rise to *syn* products whereas the transition states (351) and (352) afford *anti* products. It has been suggested that steric requirements of substituents and of the ligands are of decisive importance for the diastereoselectivity. Of the chair-like transition states, (349) is considered more stable, since it avoids 1,3-diaxial interaction between the group R and ligand L whereas in the boat-like transition states (352) is more stable than (350) for the same reason. Because of the eclipsing groups boat transition states are inherently disfavoured. However, studies have shown that the boat-like transition states are only 2-3 Kcal/mole less stable then the corresponding chair-like transition states [228]. Consequently, boat-like transition states should also be considered in allylmetal additions.

Certain allymetal additions can also take place through open chain transition states. These are catalyzed by Lewis acids e.g. boron trifluoride etherate [229-235]. Thus the addition of the allylmetals (342) and (343) to aldehydes proceeds through open-chain transition states such as (353) and (354) respectively, since the aldehyde group is already complexed with the Lewis acid catalyst and is not available for coordination with the allyl group.

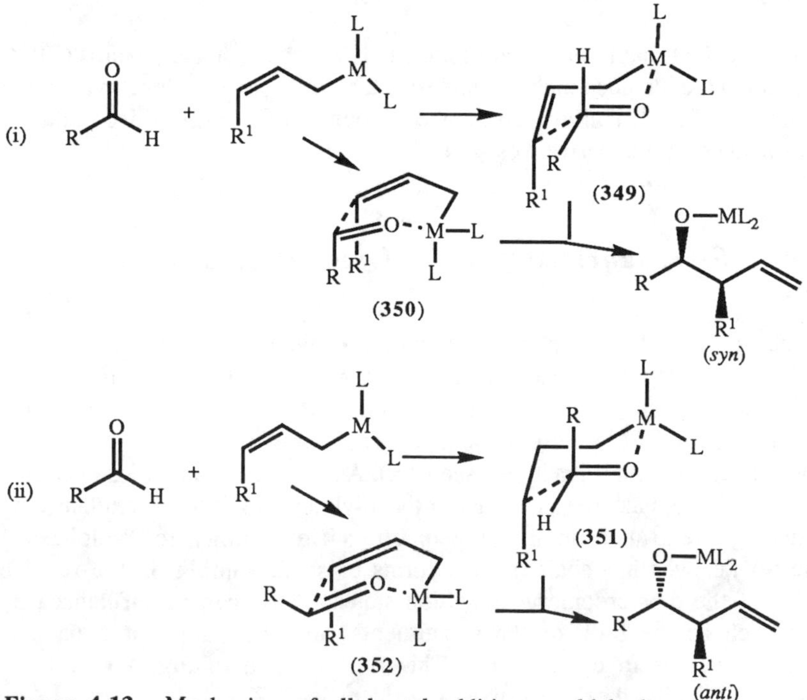

Figure 4.13. Mechanism of allylmetal addition to aldehydes.

(342)   (353)

(343)   (354)   (355)

Of the four possible transition states for each of the allylmetal compounds (342) and (343), the most favourable transition states (353) and (354) are shown in which the least steric interaction occurs along the newly formed bond resulting in the predominance of only one diastereomeric product (355) from the two stereoisomeric allyl-metal compounds (342) and (343) [236].

4.4.3   Addition of Allylboron Compounds

Chiral allylboron compounds, such as β-allyldiisopinocompheylboron (356), have been conveniently prepared and successfully employed in asymmetric carbon-carbon bond forming reactions. Thus hydroboration of α-pinene with chloroborane etherate followed by treatment with allylmagnesium bromide afforded (356) which underwent condensation with a variety of aldehydes to furnish secondary homoallylic alcohols via the borinate (357) in 83-96% e.e. [237]. Similarly, but-2-enyl-BBN (358) adds to pyruvates (359) affording the *threo* product (360) predominantly [238]. The stereoselectivity depends on the size of the R group, which interacts in the transition state with the boron ligand as depicted in (362) and (363). By increasing the size of R, the transition state (363) is destabilized due to the interaction of the ester group with the but-2-enyl ligand, thereby favouring the transition state (362) and hence leading to a high ratio of the *threo* product (360).

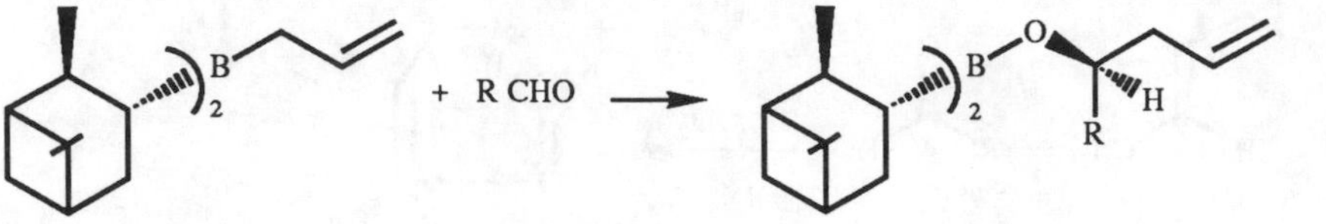

(356)   (357)

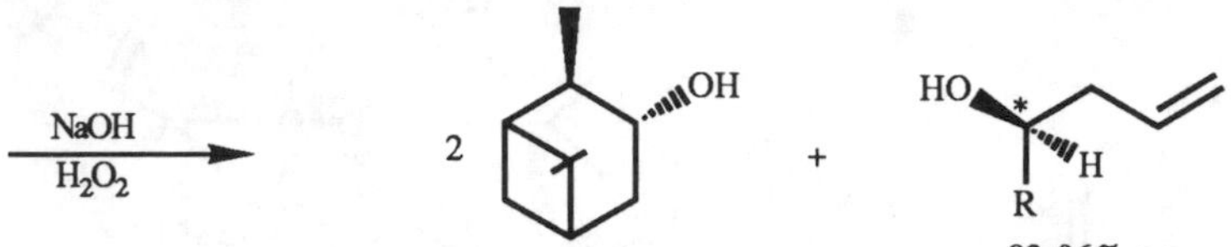

83-96% e.e.

$R = CH_3, CH_3CH_2, CH_3CH_2CH_2,$
$(CH_3)_2CH, (CH_3)_3C, Ph$

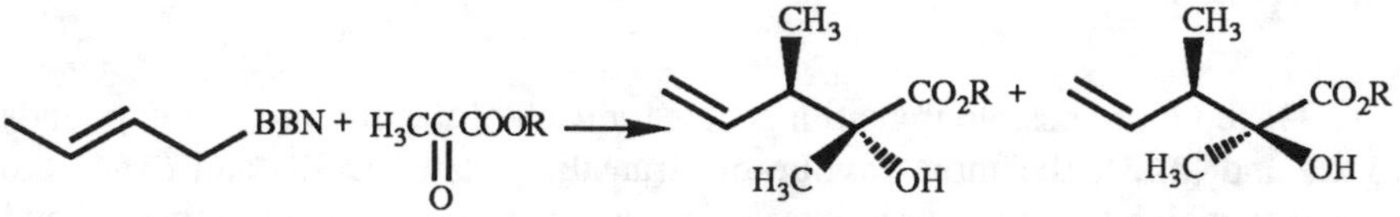

(358)   (359)   (360) *threo*   (361) *erythro*

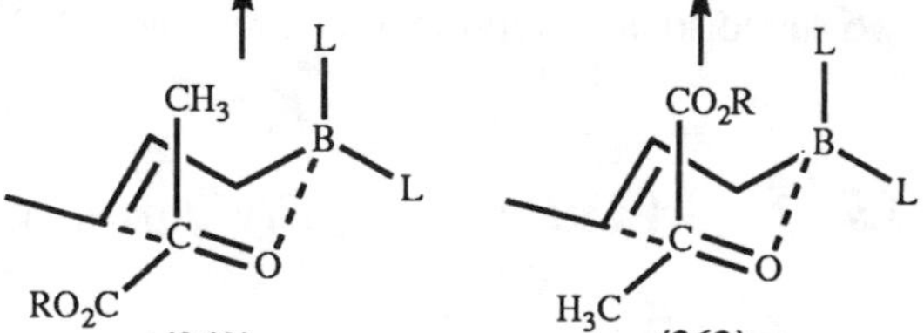

(362)   (363)

For example with R = PhCH$_2$, Ph and 2,6-*tert*-Bu$_2$-4-CH$_3$-Ph, the *threo* products are obtained in 80, 90 and 100% yields respectively. Similarly, α-trimethylsilyl-crotyl-9-BBN (364) reacts with aldehydes affording the *threo* (Z) isomer in high yields. Thus with benzaldehyde, the reagent (364) condenses to give the product (366) *via* the transition state (365) in high yield [239].

However, in order to achieve high stereoregulation, an additive such as pyridine, n-butyllithium etc. is essential, which forms an "ate" complex with the reagent (364). Without the use of such additives, an isomeric mixture of products coupled at α and γ positions of (364) along with *erythro* and *trans* isomers are obtained.

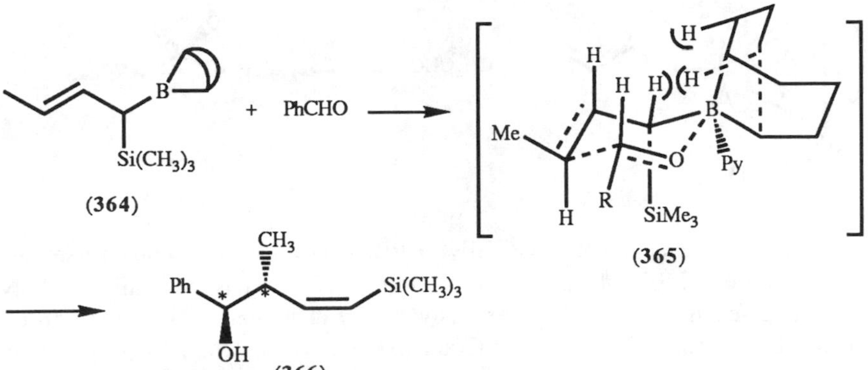

(364)

(365)

(366)

threo (Z) : *erythro* (Z) : others 92:1:6 (90% yield)

A complimentary method involving the use of allylboranes produces *threo* homoallylic alcohols in high optical yields [238b]. Thus (367) condenses with benzaldehyde to give the *trans* product (368) in 60% e.e.

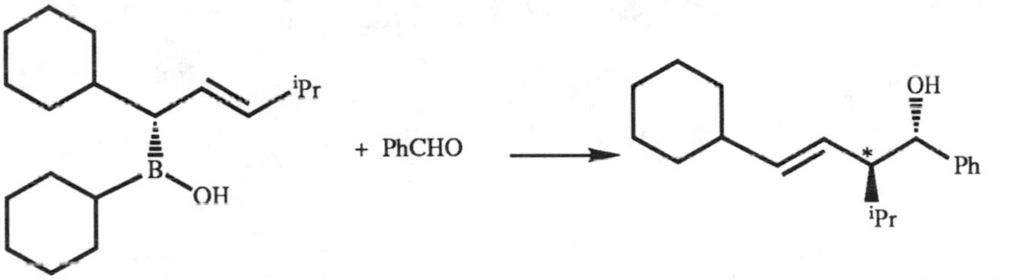

(367)

(368) *threo* : *erythro* (99:1)

The reaction is believed to proceed through the transition state (369) in which the R,R', and R" groups occupy equatorial positions.

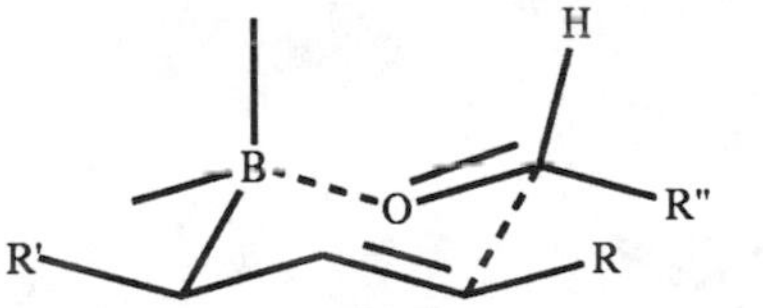

(369)

Similarly, employing B-allyl-diisocaranyl borane (370), homoallylic alcohols are produced in high optical yields [239b]. Thus acetaldehyde condenses with the allyl borane reagent (370) to yield the (R) enantiomer of 4-penten-2-ol (371) in 99% e.e.

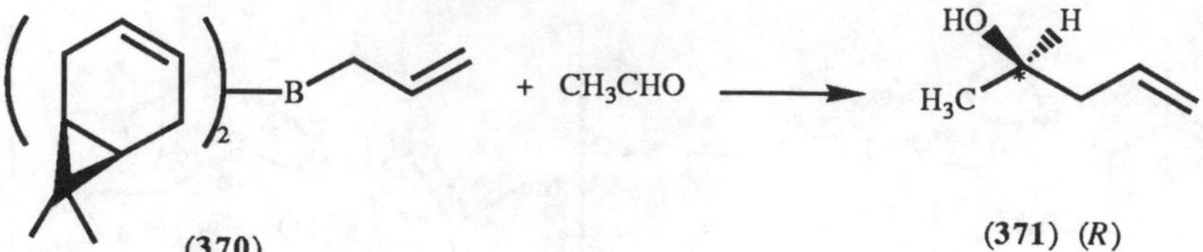

(370)

(371) (R)
>99% e.e.

Chiral imines condense with allyl-9-BBN (373) to give secondary amines in high yields [240]. Thus, the chiral amine (372) reacts with allyl-9 BBN (373) to give the product (374) exclusively. By changing the isopropyl group in (372) to *n*-propyl in (375), the Cram (376) and *anti*-Cram (377) products are obtained in a ratio of 96:4 [240]. These results have been explained by assuming the transition state (378) in which the conformation of the chiral centre is fixed in the position shown. Here the allyl group attacks preferentially from the less hindered side due to the steric influence of (L) as shown in (379).

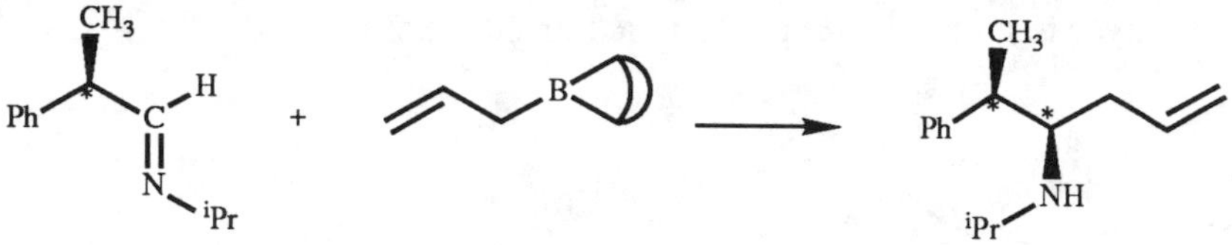

(372)   (373)   (374)   100%

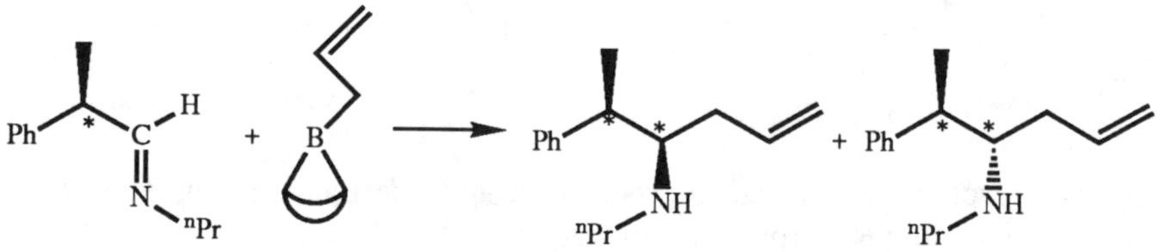

(375)   (373)   (376)   + Ph   (377)

(96:4)

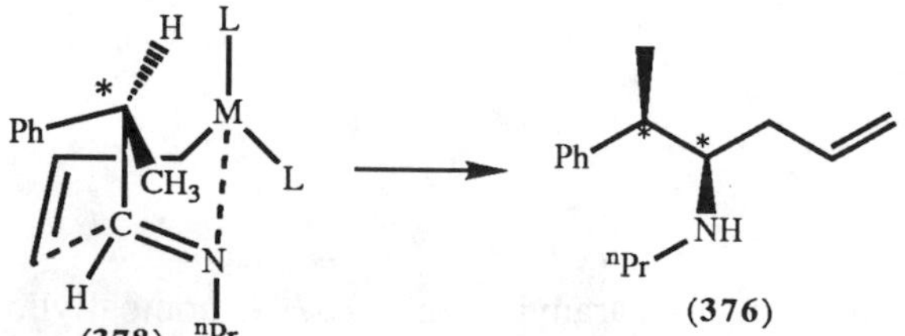

(378)   (376)

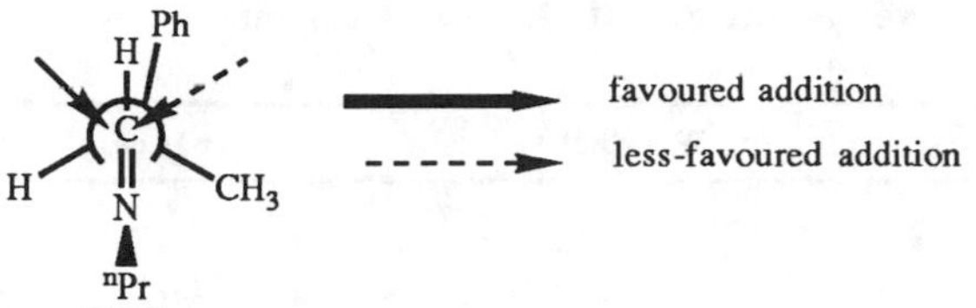

(379)

favoured addition

less-favoured addition

The alternative transition state (380) leading to the minor product (377) is not favored since the steric repulsion between the methyl group and the 9-BBN ring protons destabilizes it.

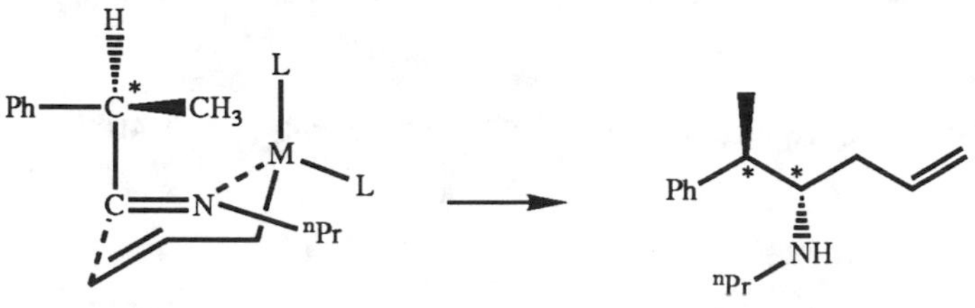

(380)   (377)

Several other allylic organoborane reagents e.g. (381-384) having different chiral auxiliaries have been developed which furnish homoallylic alcohols with variable diasteroselectivities and enantioselectivities on reaction with both achiral and chiral aldehydes.

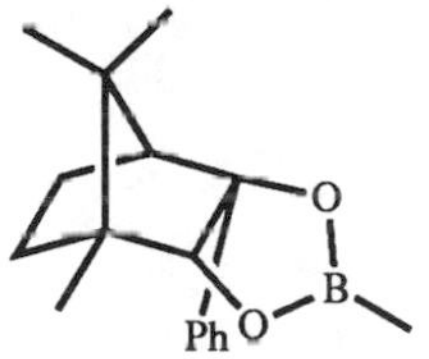

(381) 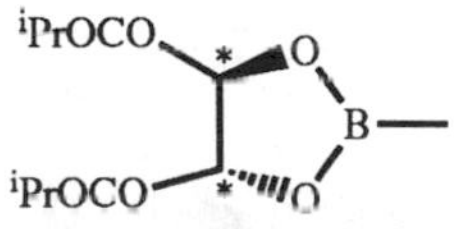   (382)

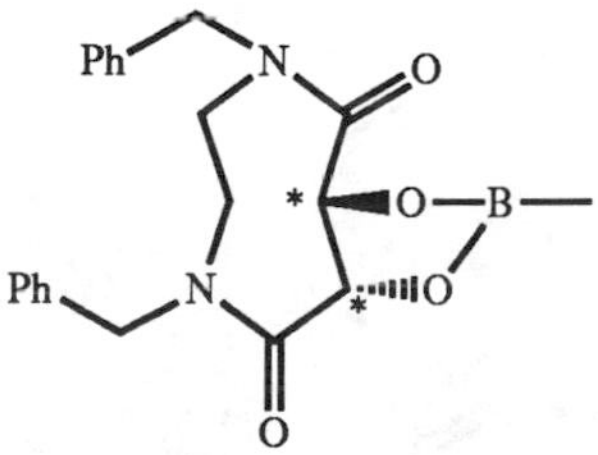

(383) 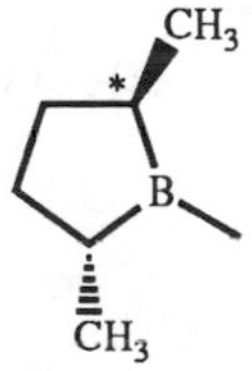   (384)

Table 4.17 presents some examples of additions of allylboron reagents to various substrates.

Table 4.17.   Stereoselective Addition of Boron Reagents to Various Substrates

Substrate	Reagent	Product/% e.e.	Reference
CH₃ CHO	(356)	93 (R)	[237]
H₃C–CH–CHO (with CH₃ substituent)	(356)	90 (S)	[237]
H₃C– CHO (with * center)	(boron reagent)	96 (R)	[241]
BzO– CHO	"	98 (R)	[241]
H₃C– CHO (OBz)	"	95 (R)	[241]
PhCHO	Ipc₂B-allyl	93%	[237]
BzO– CHO	(382)-(Z)-crotyl	95	[242]
C₂H₅CHO	Ipc₂B-(Z)-crotyl	92	[243]

(Table 4.17. contd)

Substrate	Reagent	Product/% e.e.	Reference
C_2H_5CHO	(384)-(Z)-crotyl	 96	[244]
CH_3CHO	(381)-allyl	 65 (R)	[245]
CH_3CHO	(381)-(Z)-crotyl	 *erythro* : *threo* = 98 : 2 (60% e.e.)	[246]

4.4.4   Addition of Allyltitanium Compounds

Although reactions of organometallic reagents such as Grignard reagents or organolithium reagents etc. with carbonyl compounds proceed in high yields [247], these reagents generally do not discriminate effectively between ketones and aldehydes. This problem has been surmounted by the prior treatment of the organometallic reagents with organotitanium compounds (a process called "titanation") which then results in excellent chemoselectivity [248, 249]. For instance, RMgCl and RLi react with 1:1 mixture of aldehydes and ketones to give about 1:1 adduct mixtures, whereas prior treatment with organyltitanium compounds results in 99% aldehyde selectivity. As shown in Fig. 4.14, treatment of allyllithium, allylmagnesium or allylzinc halides with various titanium compounds affords allyltitanium compounds (385-387), which allows for either aldehyde- or ketone-selectivity, depending upon the nature of the ligand at the metal [250-252].

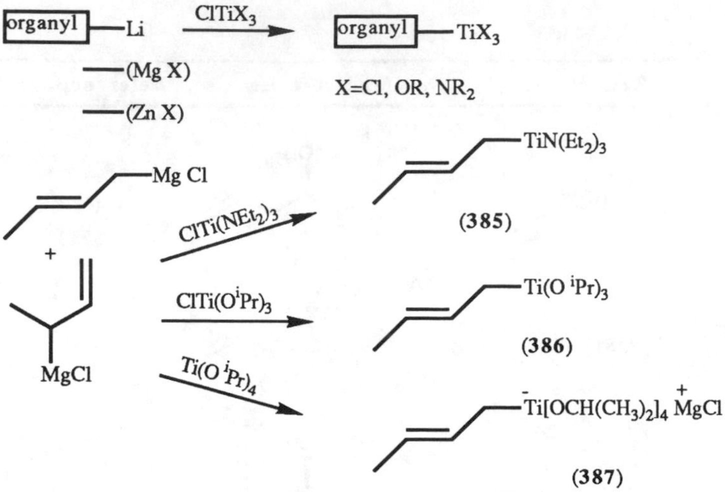

Figure 4.14. Titanation of organometallic reagents.

The use of organotitanium reagents offers an inexpensive, easily accessible and ecologically unobjectionable method for diastereoselective carbon-carbon bond forming reactions. For instance, the reaction of the ate complex (**388**) with benzaldehyde afforded the *threo* and *erythro* products (**389**) and (**390**) in a diastereomeric ratio of 99: 1 respectively [250].

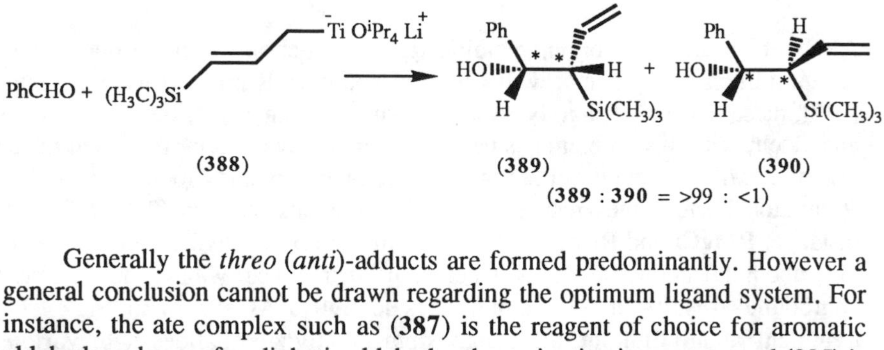

Generally the *threo* (*anti*)-adducts are formed predominantly. However a general conclusion cannot be drawn regarding the optimum ligand system. For instance, the ate complex such as (**387**) is the reagent of choice for aromatic aldehydes whereas for aliphatic aldehydes the aminotitanium compound (**385**) is more selective. The nature of the N-alkyl groups in (**385**) affects stereoselectivity, the less bulky tris(dimethylamine) analog of (**385**) being less diastereoselective. High *threo* selectivity is also achieved with but-2-enyl-titanium compounds e.g. (**391**) with both aromatic as well as aliphatic aldehydes [253]. The halogen ligand affects the diastereoselectivity of the reaction as shown in Table 4.18.

Table 4.18.   Reaction of Complexes (391) with Aldehydes RCHO

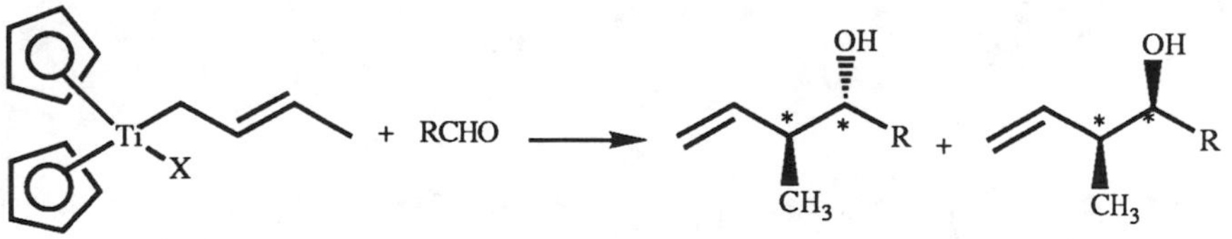

| | | | (391) | | threo | | erythro |

X in (391)	R	threo : erythro	% yield
Cl	Et	64 : 36	99
Cl	Ph	60 : 40	96
Br	Et	96 : 4	90
Br	^{i}Pr	99 : 1	87
Br	Ph	100 : 0	92
I	Et	93 : 7	86
I	Ph	94 : 6	96

Fig. 4.15 shows the formation of the allyltitanium reagent (392) which react with aliphatic as well as aromatic aldehydes giving high diastereoselectivity [254].

$$Cl—Ti(O^iPr)_3 \quad + \quad 3\,R'OH \quad \xrightarrow{C_6H_6} \quad Cl—Ti(OR')_3 \; + \; 3^iPrOH$$

$$Cl—Ti(OR')_3 \; + \; H_3C—CH{=}CH—CH_2—MgX \xrightarrow{THF} H_3C—CH{=}CH—CH_2—Ti(OR')_3$$

$$(392) \qquad + \qquad MgClX$$

Figure 4.15.   Formation of allyltitanium reagents   (392).

Thus, the reagent (392) reacts with the aldehyde (393) to give the *threo* and *erythro* products (394) and (395) in a diastereomeric ratio of 95:5 respectively.

$H_3C{-}CH{=}CH{-}CH_2{-}Ti(OC_6H_5)_3$ + $Ph(CH_2)_2CHO$ ⟶

(392) **(393)**

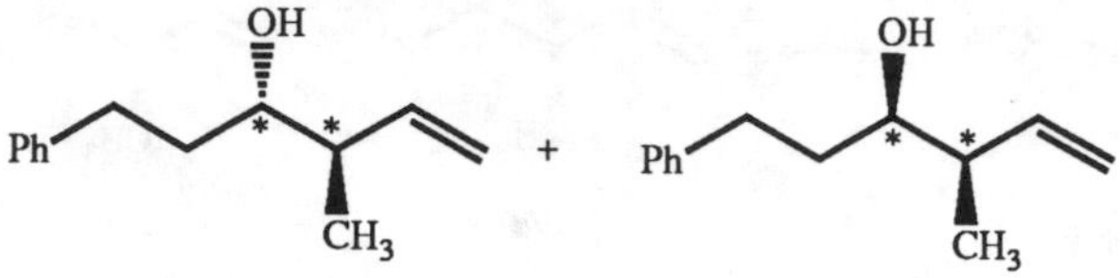

(394) **(395)**

394 : **395** = 95 : 5

It has been shown that in aromatic aldehydes, electron-donating groups at the benzene ring result in increased diastereoselectivities of up to 98% whereas electron-withdrawing groups decrease the diastereoselectivity to 80%.

Similarly, other organyltitanium reagents e.g. (**396-398**) have been successfully employed in the regio- and stereo-controlled synthesis of several compounds [255-257]. Thus reaction of the imine (**399**) with the allenic titanium reagent (**397**) affords the *threo* product (**400**) in high selectivity, whereas the *erythro* product (**401**) is not detected [255].

$H_3C{-}C{\equiv}C{-}CH{-}Ti(O^iPr)_3$

(396)

$H_3C{-}CH{=}C{=}C(Si(CH_3)_3)(Ti(O^iPr)_3)$

(397)

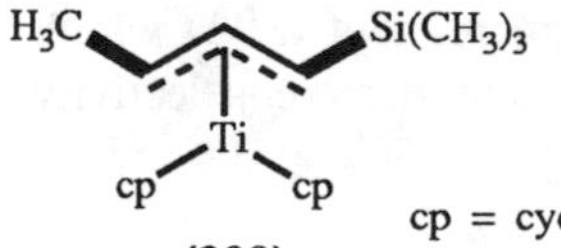

(398) cp = cyclopentadienyl

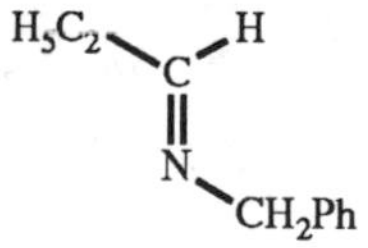

(399)

+ $H_3C{-}CH{=}C{=}C(Si(CH_3)_3)(Ti(O^iPr)_3)$ ⟶

(397)

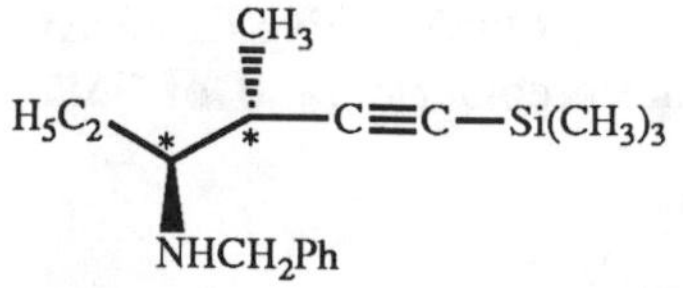

threo (**400**) >99%

+

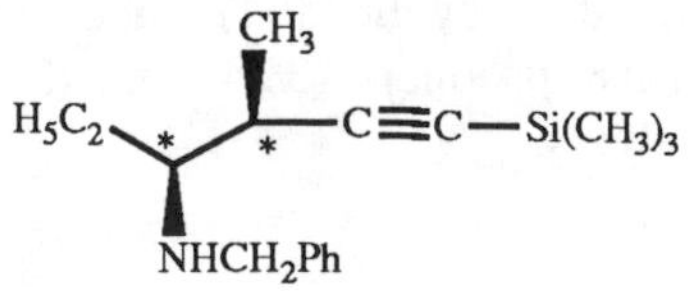

erythro (**401**) <1%

The high *threo* selectivity in this reaction has been explained by assuming that in the transition state (**402**) the R group in the substrate is placed away from the methyl group in the allenic reagent.

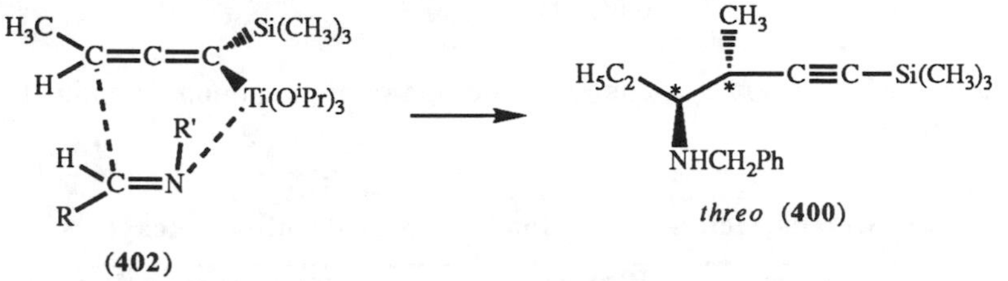

(**402**)

threo (**400**)

In the corresponding transition state (**403**), the steric interaction between the methyl group in the reagent and the R group in the substrate lowers the selectivity and hence results in the formation of (**401**).

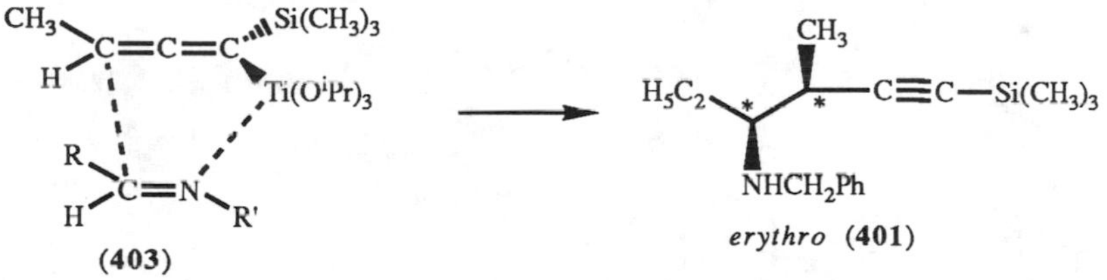

(**403**)

erythro (**401**)

Recently, a new chiral organyl titanium reagent (**404**) has been developed [258].The reagent (**404**) reacts with aldehydes, e.g. (**405**) to give the homoallylic alcohol (**406**) in good yields and with high enantiomeric purity.

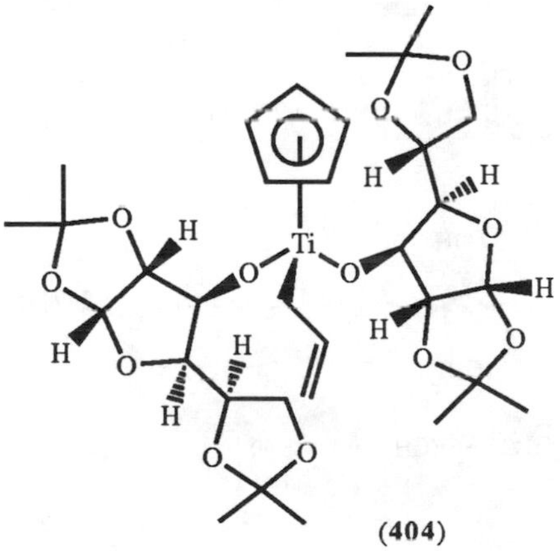

(**404**)

$$\text{CH}_3(\text{CH}_2)_8\text{CHO} \xrightarrow{\text{(404)}}$$

(405)   (406) 92% e.e. (S)

Table 4.19 shows some examples of addition of allyltitanium compounds to various substrates.

Table 4.19. Stereoselective Addition of Allyltitanium Reagents[†]

Substrate	Reagent	Product	Selectivity % (Config.)	Reference
n-nonyl-CHO	a		92 (S)	[258]
	a		91 (R)	[258]
	b		99	[256]
	c		89 (syn)	[259]
(H₅C₂)₂ CHCHO	d		99 (anti)	[254]
CH₃CH₂CHO	e		85 (trans)	[257]
	f		>99	[255]

(Table 4.19. contd.)

Substrate	Reagent	Product	Selectivity % (Config.)	Reference
PhCHO	g		100	[253]
PhCOCH$_3$	h		91	[250]
(H$_3$C)$_2$CH—CHO	i		96	[254]

† Only the major product is shown.

a:(404), b:(396), c: Ti(NEt)$_3$, d: H$_3$C Ti(OPh)$_3$, e:(398), f:(397), g:(391), h:(389), i: (392)

4.4.5 Addition of Allylstannanes

Reactions of allylstannanes with aldehydes mediated by Lewis acids exhibit an entirely different stereoselectivity. The *trans* crotylmetal compounds other than the stannanes and the silanes react with aldehydes giving the corresponding *threo* homoallylic alcohols predominantly, provided that the geometry of the double bond is retained during the reaction, whereas the *cis* derivatives afford the *erythro* isomers preferentially. The crotyltrialkylstannanes, however, react with aldehydes in the presence of BF$_3$.OEt$_2$ to give the *erythro* homoallylic adducts preferentially regardless of the geometry of the double bond. The allylic organometals containing Li, B, Al, Ti, Zr and Cd react *via* six-membered transition states, whereas the allystannanes react with aldehydes *via* acyclic transition states. Thus there are interesting parallels regarding the reactivity of allylstannanes and allylsilanes with other organometallic reagents. For instance, the crotyltin compound (407) adds diastereoselectively to benzaldehyde at 200°C to afford the *erythro* product (408) in 95% diastereomeric purity [260].

(407)   (408)   95% d.p.

(407) $\xrightarrow[\text{BF}_3.\text{OEt}_2, -78°\text{C}]{\text{PhCHO}}$ (408) >99% d.p.

The presence of Lewis acids such as $BF_3.OEt_2$ greatly modifies the reaction conditions and benzaldehyde then readily reacts with (407) at -78°C to give the product (408) in 99% diastereomeric purity. The Lewis acid-catalyzed reaction is believed to proceed through the acyclic transition states (409) and (410).

The addition of allyltri-n-butylstannane (411) to α-hydroxyaldehyde derivatives proceeds with high stereoselectivity (95:5 to 250:1) [261-263]. Thus, the addition of (411) to the *tert*-butyldimethylsiloxy substrate (412) in the presence of BF_3. Et_2O in CH_2Cl_2 at -78°C affords the products (413) and (414) in a ratio of 95:5 and in 83% yield [261]. Similarly, by protecting the α-hydroxy group as in (415) and subsequently carrying out the reaction in the presence of $MgBr_2$, the *erythro* product (416) is obtained with high selectivity.

(413:414 = 95:5)

(415 : 416 = > 250:1)

Similar high selectivity is achieved by using ZnI_2TiCl as the Lewis acid. Thus by a proper choice of Lewis acids and protecting groups, the addition of allyltri-n-butyltin compound (411) to α-hydroxyaldehyde derivatives can be controlled to give excellent stereoselectivity for the formation of *threo* or *erythro* products. Other β-substituted allylstannanes e.g. (418) and (419) also showed high diastereoselectivity in Lewis acid catalysed additions to aldehydes [264].

(418)   (419)

The addition of crotyltin compounds to aldehydes offers an alternative and superior method for the synthesis of the Prelog-Djerassi lactonic acid (422) which is an important intermediate in the synthesis of macrolide antibiotics [265,266]. The crotyltin compound (407) accordingly adds stereoselectively to the aldehyde (420) affording the *erythro* (*anti*-Cram) product (421) which is readily converted to the lactonic acid (422) in 85% yield [265].

(420)   (421) 94-97%   (422)

The high stereoselectivity has been explained by assuming a crown shaped transition state (423) in which the crotylstannane reagent (407) attacks from the direction indicated by the arrow due to steric repulsion.

(423)

Table 4.20 shows some examples of the addition of allyltin compounds to aldehydes.

Table 4.20. Stereoselective Addition of Allyltin Compounds to Aldehydes

Substrate	Reagent	Product	Diastereo-selectivity threo - erythro	Reference
$^i PrO_2C-C(=O)-H$	a	(CH_3, CO_2^iPr, OH product)	90 : 10	[267]
$^n BuO_2C-C(=O)-H$	a	(CH_3, CO_2^nBu, OH product)	80 : 20	[267]
$Ph-CH(CH_3)-C(=O)-H$	b	(Ph, CH_3, CH_3, OH product)	<1 : >99	[268]
$Ph-CH(CH_3)-C(=O)-H$	c	(Ph, OH product)	1 : 4	[268]
(cyclohexyl-CH(OCH_2Ph)-CHO)	d	(cyclohexyl-CH(OCH_2Ph), OH product)	6 : 94	[261]
$H_{17}C_8^n-C(=O)-H$	e	((S), OH, $^nC_8H_{17}$ product)	82% e.e.	[269]

(Table 4.20. contd.)

Substrate	Reagent	Product	Diastereo-selectivity threo - erythro	Reference
H_5C_6–CHO	e	HO, *, H, (S), C_6H_5	23% e.e.	[269]
Cl_3CCHO	a	Cl_3CC, CH_3, OH	0 : 100	[270]
PhCHO	f	Ph, CH_3, OH, $Sn(^tBu)_3$	90 : 10	[271]
^nBuCHO	g	^nBu, CH_3, OH, $Sn(^tBu)_3$	70 : 30	[271]
EtCHO	h	Et, OH, CH_3, OCH_2OCH_3	>99 : <1	[272]

a:(407), b:(407), BF$_3$.OEt$_2$, c:(411),BF$_3$.OEt$_2$, d.(411), MgBr$_2$, H$_2$O,

e: $\left(\begin{array}{c}Ph\\ \end{array}\right)_2 Sn \left(\begin{array}{c}\\ \end{array}\right)_2$ f: (Sn($^tBu)_3$, B)   g: (Sn(CH$_3)_3$, B)

h: H_3C — Sn(Bu)$_3$, OCH$_2$OCH$_3$

4.4.6   Addition of Allylsilanes

The stereoselective carbon-carbon bond formations with the aid of allylsilanes is analogous to the addition of allystannanes both in mechanistic and practical sense, since both additions are activated by Lewis acids. The additions of (E)- and

(Z)-allylsilanes (424) - (427) to aldehydes proceed stereospecifically to afford the *erythro* products with high stereoselectivity [273].

(424)

(425)

(426)

(427)

Thus (424) adds to the *tert*-butyraldehyde in the presence of $TiCl_4$ to give the products (428) and (429) in a ratio of 99: 1 [273].

'BuCHO + (424) $\xrightarrow[CH_2Cl_2]{TiCl_4}$

(428)

(429)

(428:429 = >99:<1)

(430)

The *erythro* selectivity of the reaction has been explained by assuming an acyclic linear transition state (430) which is favoured for steric reasons over other transition states.

Other chiral allylsilane reagents have been developed and employed in the stereoselective carbon-carbon bond formation reactions. Thus (R)-(431) adds to aldehydes e.g. pivaldehyde in the presence of $TiCl_4$ to give the product (432) in high optical yield [274]. The reaction proceeds with high enantioselectivity of upto 95% e.e. with aldehydes containing larger groups, while the enantioselectivity decreases to 70% in the reaction with acetaldehyde. It has been suggested that the reaction proceeds *via* an acyclic transition state (433) in which the aldehyde is attacked on its *re*- face leading to the alcohol. An alternative mechanism involving synclinal geometries as shown in (434) has also been suggested to account for the high stereoselectivity [274].

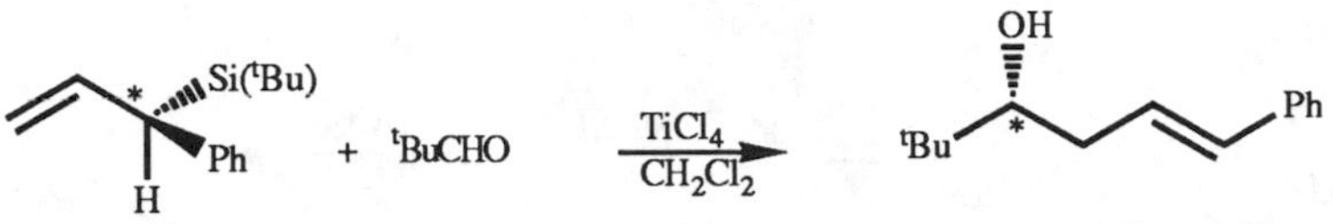

(R) - (431)

(432) 91% e.e. (R)

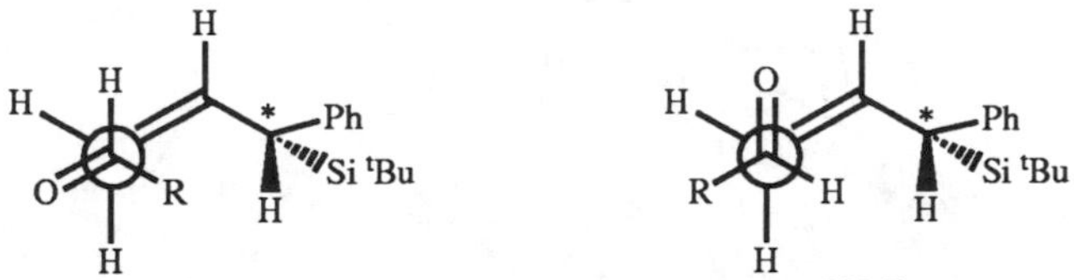

(433)

(434)

High stereoselectivity has also been achieved by using chiral substrates in the addition reaction with allylsilane [206,275]. For instance the allylsilane (435) reacts with the chiral aldehyde (436) in the presence of a Lewis acid e.g. TiCl$_4$ to afford the products (437) and (438) in a ratio of 95:5 [206].

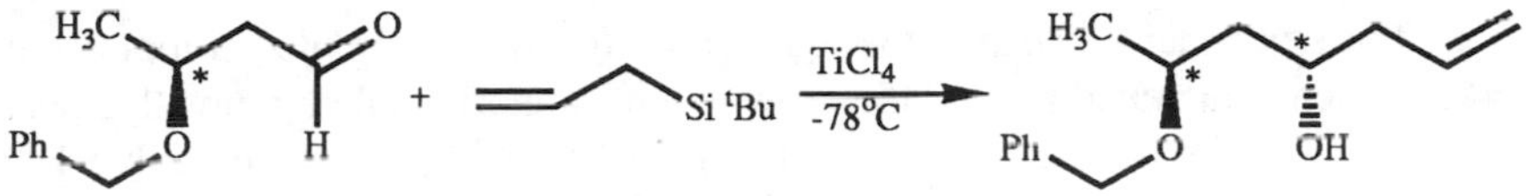

(436)

(435)

(437)

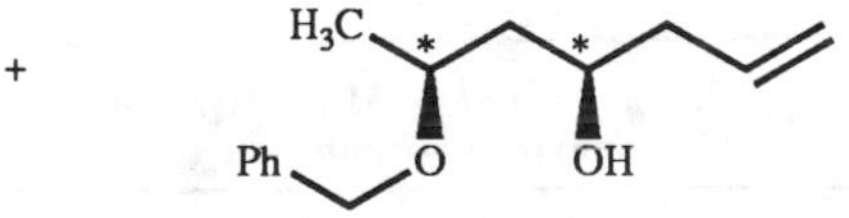

(438)

(437:438 = 95:5)

Similarly, the chiral ketone (439) reacts with the allyltrimethylsilane (435) in the presence of TiCl$_4$ to give the products (440) and (441) in a ratio of 99:1 [73].

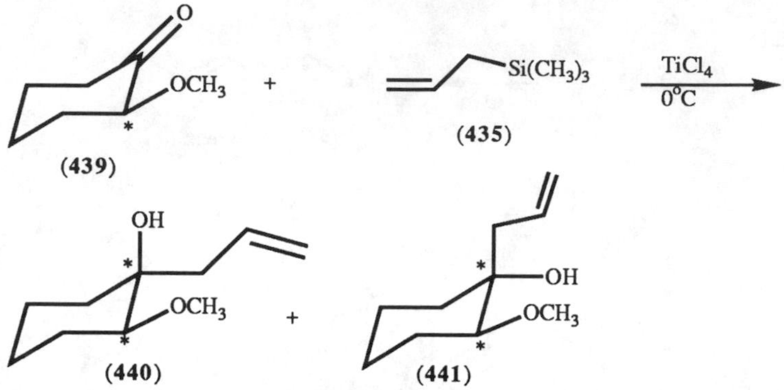

O

OCH$_3$ +   Si(CH$_3$)$_3$ $\dfrac{\text{TiCl}_4}{0°\text{C}}$

(439)   (435)

OH

OCH$_3$ +   OH   OCH$_3$

(440)   (441)

(440 : 441 = > 99 : <1)

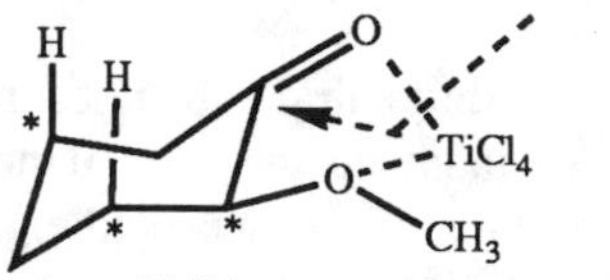

(435)

H H

O

TiCl$_4$

CH$_3$

(442)

The Lewis acid assists in the reaction by forming a chelated complex e.g. (442) which is attacked preferentially from the equatorial side by the allylsilane due to the 1,3-diaxial interaction with the hydrogens of the substrate. Table 4.21 shows some examples of the addition of allylsilanes to aldehydes.

Table 4.21. Stereoselective Additions of Allylsilanes to Carbonyl Compounds

Substrate	Reagent	Product	Selectivity erythro:threo	Reference
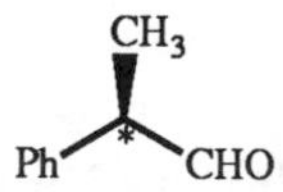 CH$_3$ Ph CHO	a	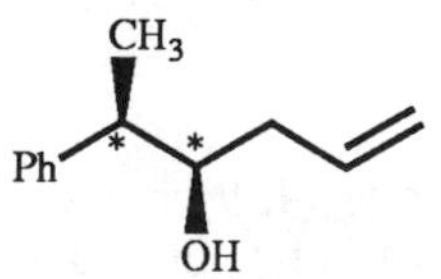 CH$_3$ Ph OH	1.6:1	[275]
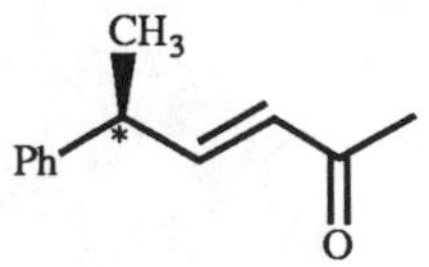 CH$_3$ Ph O	b	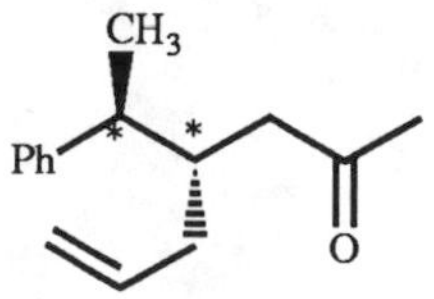 CH$_3$ Ph O	4:1	[275]

(Table 4.21. contd.)

Substrate	Reagent	Product	Selectivity erythro:threo	Reference
ᵗBuCHO	c		>99:<1 (3S, 4R) 88% e.e.	[276]
ᵗBuCHO	d		>99 : <1	[276]
	e		9:1	[277]
ⁱPrCHO	f		96% e.e.	[274]
	g		<2:>98 >98% e.e.	[278]
	h		95% (3,4 syn)	[73, 275]
	a		95% anti	[66, 206]

a: SiᵗBu, TiCl₄ b: SiᵗBu, BF₃·Et₂O,-78°C, c:

d: e: SiᵗBu f:TiCl₄, -78°C, CH₂Cl₂ g: SiᵗBu h:

4.4.7   *Palladium-Catalyzed Asymmetric Allylation*

Much attention has been focussed on devising new methodologies for asymmetric carbon-carbon bond formation reactions. Recently, palladium has been employed in asymmetric allylation of various substrates, such as enamines, chiral sulphinates etc. [279-287].

For instance, treatment of the chiral enamine (**445**), derived from the (*S*)-proline allyl ester (**444**) and 2-phenylpropionaldehyde (**443**), with tetrabis (triphenylphosphine) palladium provided *R*-(-)-2-methyl-2-phenyl-4-pentenylaldehyde (**446**) in high optical yield [279]. The reaction proceeds probably through intramolecular allylation *via* the transition state (**447**).

Table 4.22 shows some examples of Pd-catalyzed allylations.

Table 4.22. Stereoselective Pd-Catalyzed Allylations

Substrate	Reagent	Product/ Optical yield	Reference
H$_3$C—⋯—Ph, PdCl$_2$	a	H$_3$C—*⋯— Ph, CH$_2$CH=CH$_2$ 57 (S)	[284]
{H$_3$C—⋯—Ph, Pd, Ph$_2$P PPh$_2$}$^+$ BF$_4^-$	b	H$_3$C—*⋯—Ph, N(CH$_3$)$_2$ 78 (R)	[284]
{H$_3$C—⋯—Ph, Pd, Ph$_2$P PPh$_2$}$^+$ BF$_4^-$	c	H$_3$C—*⋯—Ph, CH(COOCH$_3$)$_2$ 74 (S)	[286]
p-H$_3$CC$_6$H$_4$—S⋯OCH$_2$C=C(CH$_3$)(H), O H	d	CH$_3$ (CH$_3$O$_2$C)$_2$CH—*C—CH=CH$_2$, H 83 (S)	[282]
Ph—⋯—Ph, Pd OAc	c	Ph—⋯—*Ph, H CH(COOCH$_3$)$_2$ 96 (S)	[287]

a: CH$_2$=CHCH$_2$MgCl,   b: HN(CH$_3$)$_2$,   c: NaCH(COOCH$_3$)$_2$,   d: Pd(Ph$_3$P)$_3$, CH$_2$(COOCH$_3$)$_2$,

4.4.8   Chromium(II)-Catalyzed Allylic Additions

Chromium (II)-catalysed synthesis of homoallylic alcohols has been carried out as an alternative method to the use of aldol condensations [288-291]. For instance the addition of crotylbromide in the presence of CrCl$_2$ to the aldehyde (**448**) proceeds stereoselectively to give the major *threo* diastereomer (**449**) with high diastereoselectivity [291]. A striking feature of Cr(II)-catalyzed allylation is the high stereoselectivity and chemospecificity. In general, aldehydes are more reactive towards Cr(II)-catalyzed allylations than ketones. Thus, selective attack is possible on an aldehyde carbon of a polycarbonylated compound. Further terminal ketones are allylated selectively in the presence of non-terminal ketones. Thus Cr(II)- catalyzed addition of allylbromide to 2-heptanone (450) in the presence of 4-heptanone (451) proceeds stereoselectively to the give the products (453) and (454) in a ratio of 88:12 [289]. Allylation of benzaldehyde, on the other hand, with allyl bromide in the presence of CrCl$_2$ resulted in isolation of only a single diastereomer.

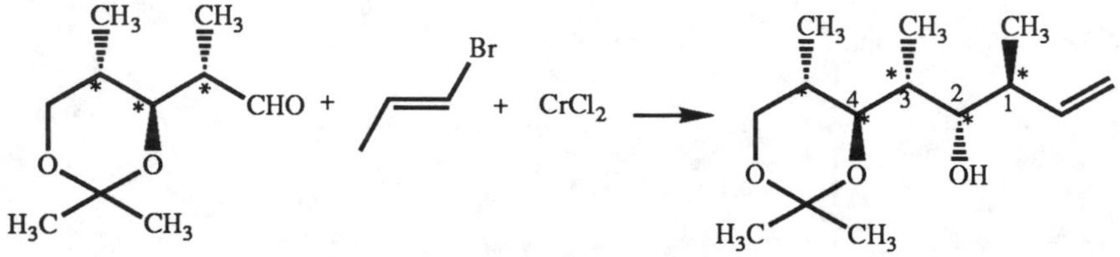

(448)

(449)   1,2 (*threo*)

(*threo* : *erythro* = 20:1)

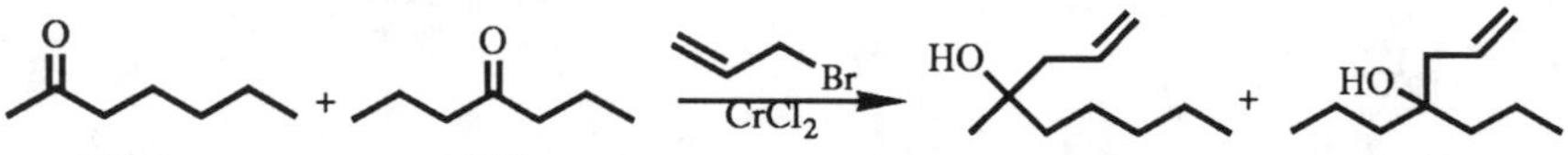

(450)   (451)   (452)   (453)

(452:453 = 88:12)

Cr-(II) catalysed allylation of α,β-unsaturated aldehydes proceeds exclusively in a 1,2-fashion as exemplified by the synthesis of artemisia alcohol (Table 4.23 entry 1).

Some examples of Cr(II)-catalyzed allylations are presented in Table 4.23.

Table 4.23. Stereoselective Cr(II)-Catalyzed Allylations

Substrate	Reagents	Product	% Selectivity	Reference
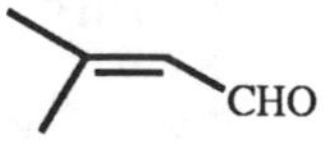	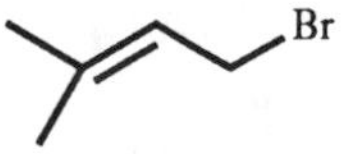 CrCl$_2$:LAH (2:1)	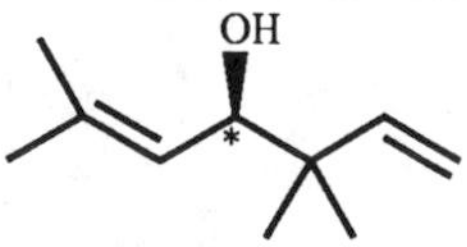	93	[289]
PhCHO	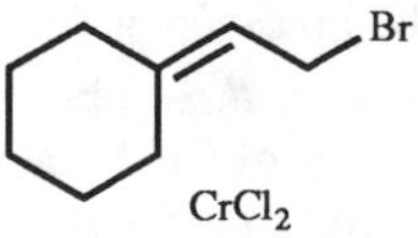 CrCl$_2$	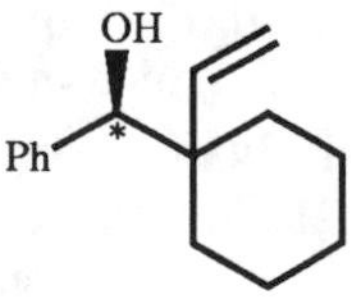	81	[289]

(Table 4.23. contd.)

Substrate	Reagents	Product	% Selec-tivity	Reference
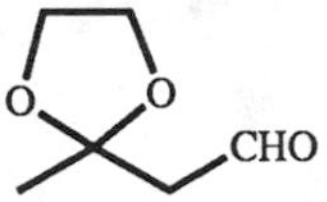	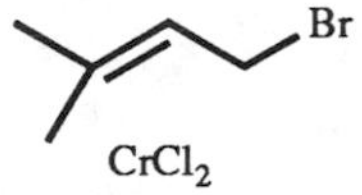	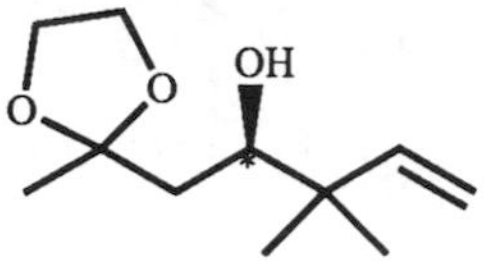	91	[289]
^{i}PrCHO	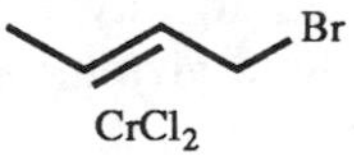	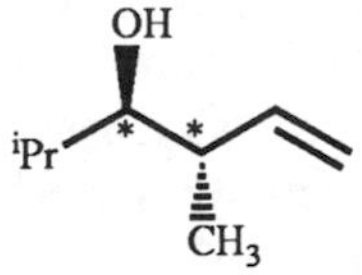	97	[288]
CH$_3$CH=CHCHO	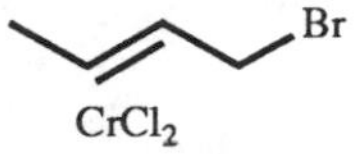	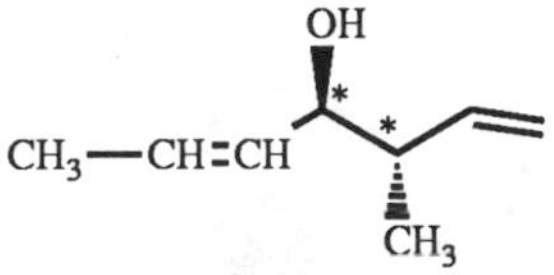	83	[288]

4.4.9   Addition of Other Allylmetals

Other metals such as Zr, Zn, Al, Li, Mg as well as Cd have been tested in asymmetric allylation reactions, albeit with modest success. Diallylzinc has been used in the reaction of the aldehyde (454) to give the major *erythro* product (455) along with the minor diastereomer (456) in a ratio of 9:1 [292].

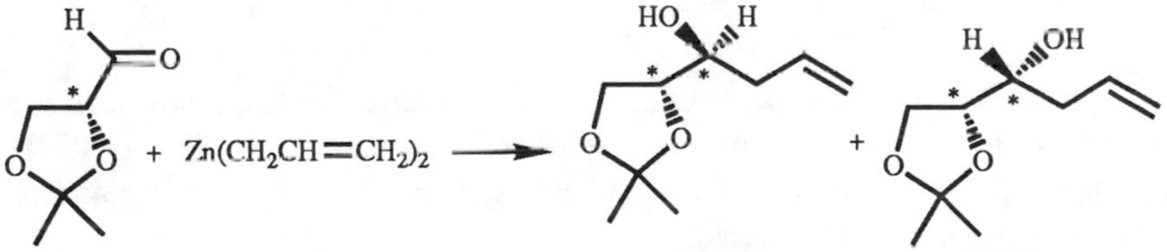

(454)   (455)   (456)

(455 : 456 = 9:1)

Allenic zinc reagents e.g. (457) have been employed in addition to various aldehydes with high *threo* selectivity [293]. Thus with propionaldehyde, the allenic zinc reagent (457) afforded the product (458) in 99% e.e. in a *threo/erythro* ratio of 96:4 [293].

$\left\{ (^nC_5H_{11})-CH=C=CH \right\} ZnCl$

(457)

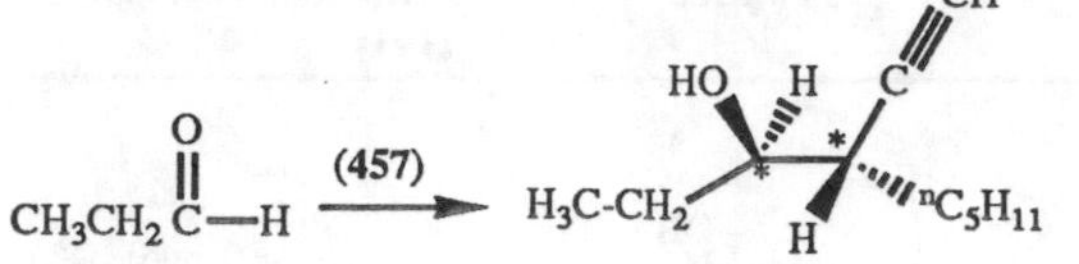

$$CH_3CH_2\overset{O}{\underset{\|}{C}}-H \quad \xrightarrow{(457)} \quad$$

(458) 99% e.e.

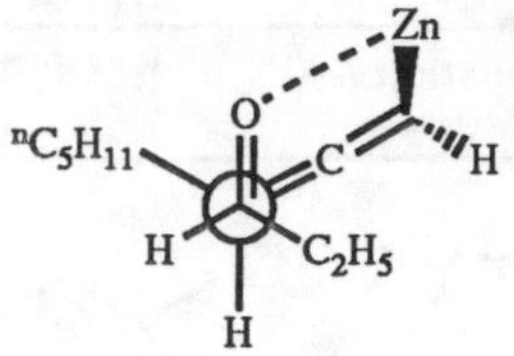

(459)

The reaction is believed to proceed through a complex (459) in which zinc is coordinated to the carbonyl group. Functionalized crotyllithium compounds e.g. (460) add to benzaldehyde stereoselectively to give the *erythro* product (461) with high diastereoselectivity. The addition is believed to proceed via a 6-membered transition complex (462).

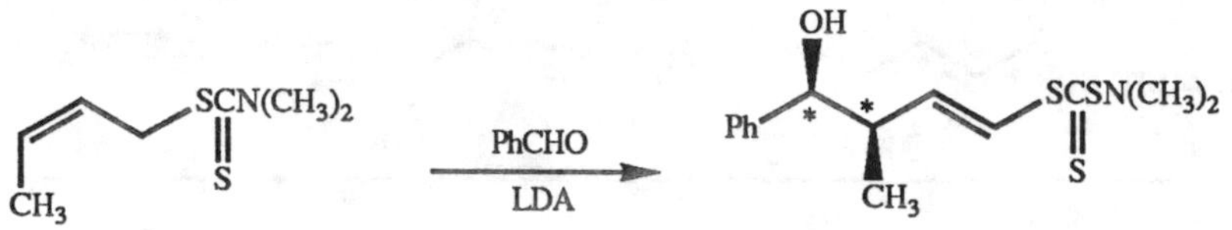

(460)

PhCHO
LDA

(461) 85% *erythro*

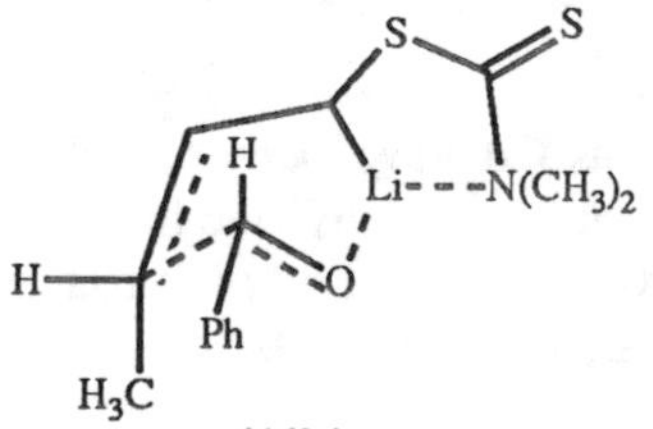

(462)

(1-Oxycrotyl)aluminium compounds such (463) have been employed in the reaction with aldehydes to give mainly *threo* products [294]. Thus with t-butyraldehyde, the compound (463) affords the (*E*)- and (*Z*)-diastereomeric products (464) and (465) in a ratio of 87:10 [294].

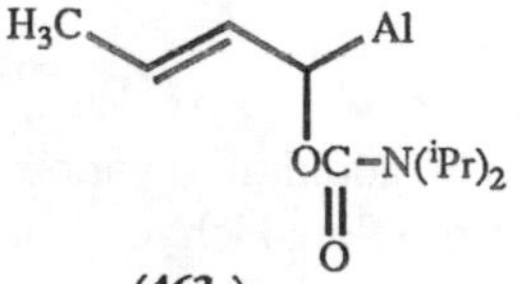

(463)

$$(CH_3)_3CC{-}H \xrightarrow{(463)} (CH_3)_3C \overset{OH}{\underset{CH_3}{\wedge}} OCON(^iPr)_2 \quad + \quad (CH_3)_3C \overset{OH}{\wedge} \underset{CH_3\ \ OCON(^iPr)_2}{}$$

$$(464)\ (E) \qquad\qquad\qquad (465)\ (Z)$$

$$(464 : 465 = 87{:}10)$$

The crotylzirconium reagent (466) is *threo* selective and adds to aldehydes to give the corresponding β-methylhomoallylic alcohols in high yields [66,295]. Thus with the aldehyde (467), the reagent (466) adds stereoselectively to give the homoallylic alcohols (468) and (469) in a ratio of 94:6 [295].

(466)

$$H_3COCCH_2CH_2CHO \xrightarrow{(466)} H_3COCCH_2CH_2 \overset{CH_3}{\underset{OH}{\wedge}} \quad + \quad CH_3OCCH_2CH_2 \overset{CH_3}{\underset{OH}{\wedge}}$$

$$(467) \qquad\qquad (468) \qquad\qquad\qquad (469)$$

$$(468 : 469 = 94{:}6)$$

Tetraallylzirconium $Zr(C_3H_5)_4$ adds to the aldehydes (470) to give the products (471) and (472) in a diastereomeric ratio of 81:19 [66]. The reaction is believed to proceed *via* an initial deprotonation to form propene and the intermediate (473). This undergoes intramolecular transfer of an allyl group *via* the bicyclic [1.3.3] transition state (474a) to give the intermediate (474b). Acid hydrolysis of (474b) affords the products.

$$H_3C \overset{}{\underset{OH}{\wedge}} \overset{H}{\underset{O}{}} \xrightarrow[\text{(ii) } H^+, H_2O]{\text{(i) } Zr(C_3H_5)_4} H_3C \overset{}{\underset{OH}{\wedge}} \overset{}{\underset{OH}{\wedge}} \quad + \quad H_3C \overset{}{\underset{OH}{\wedge}} \overset{}{\underset{OH}{\wedge}}$$

$$(470) \qquad\qquad (471) \qquad\qquad (472)$$

$$(471 : 472 = 81{:}19)$$

(473)

(474a)

(474b)

4.5   Asymmetric Alkylation Reactions

Asymmetric alkylation reactions have constituted the backbone of synthetic organic chemistry. They can be accomplished in several ways, such as by using chiral enolates, imines and enamine salts, chiral azaallyl metal reagents, chiral hydrazones, oxazolines etc.

4.5.1   Alkylation of Chiral Enolates

The alkylation of ketone enolates is among the most widely used methods for construction of carbon-carbon bonds. However due to the achiral nature of the enolate p-system, enantioselectivity cannot be accomplished with simple flexible ketones. The preparation of regioselective enolates has therefore been one of the primary goals of asymmetric alkylations [296-298]. These enolates can be obtained under conditions favouring either kinetic or thermodynamic control. For instance 2-methyl cyclohexanone (475) is converted to the corresponding enolates (476) and (477) with LDA under kinetically controlled conditions affording a ratio of 99:1 respectively [299]. By employing the base Ph_3CLi, on the other hand, the enolates (476) and (477) are obtained under thermodynamic conditions in a ratio of 10:90 respectively [300].

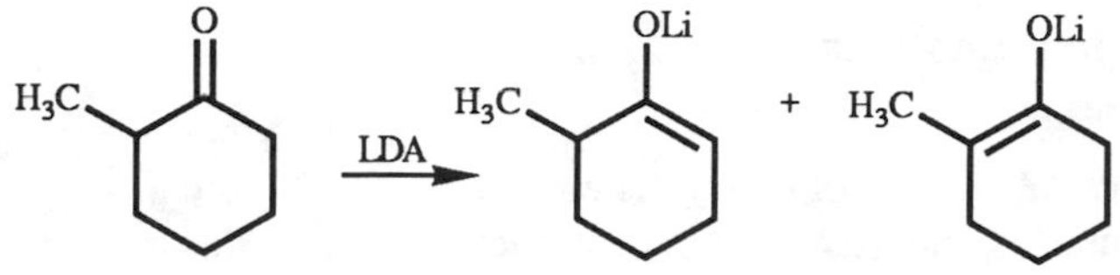

(475)   (476)   (477)

(476 : 477 = 99:1)

Deprotonation of acyclic ketones produces the regioisomeric (*E*)- and (*Z*)-enolates and by employing the appropriate reaction conditions (kinetic *versus* thermodynamic control), one may obtain the desired enolate. For instance, 3-pentanone (**478**) reacts with a sterically demanding base e.g. (**479**) to give the major (*E*)- enolate (**480**) along with the minor (*Z*)-enolate (**481**) under conditions of kinetic control in a ratio of 86:14 [301,302]

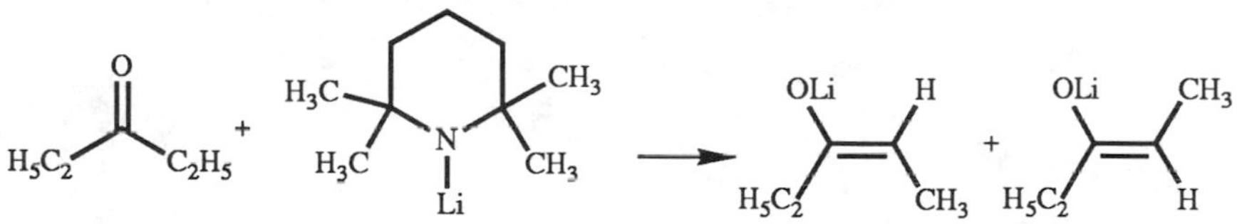

(478)   (479)   (480)   (481)

(480 : 481 = 86:14)

By employing the base (**482**) for deprotonation of the substrate (**478**) under conditions resulting in enolate equilibration, the (*Z*) isomer (**481**) is however obtained exclusively [303].

(Me$_2$PhSi)$_2$ NLi

(**482**)

Polar chelating groups in the substrates enhance the enolization selectivity.Thus, β-hydroxyesters e.g. (**483**) afford predominantly the (*Z*)-enolates (**485**) *via* the chelated lithium aldolate (**484**).

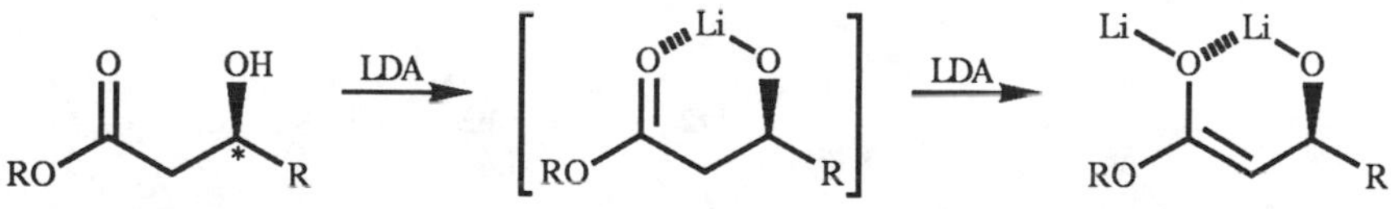

(483)   (484)   (485)

4.5.1.1   Exocyclic Enolates

In the diastereoselective alkylation of exocyclic enolates e.g. (486) the presence of the R group influences the steric course of the reaction. Thus the two transition states (487) and (488) formed from the enolate (486) offer the incoming electrophile different faces, leading to the major and minor products (489) and (490) respectively.

(486)	(487)	(488)

(489)	(490)
major	minor

For instance, alkylation of the 1,2-disubstitued cyclohexylester (491) afforded the diastereomeric products (492) and (493) in a ratio of 80:20 [304]. Similar diastereoselectivity has been achieved with 1,3- and 1,4-disubstitued cyclohexylidene enolates [304-307].

(491)	(492)	(493)

(492 : 493 = 80:20)

Table 4.24 shows some examples of alkylation of exocyclic enolates.

Table 4.24. Stereoselective Alkylation of Exocyclic Enolates

Substrate	Reagent	Product[†]	Optical Yield	Reference
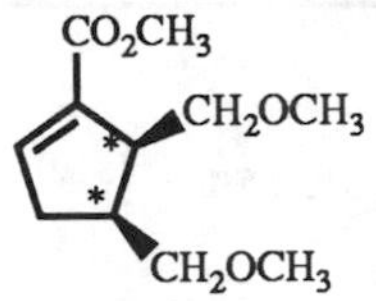	a	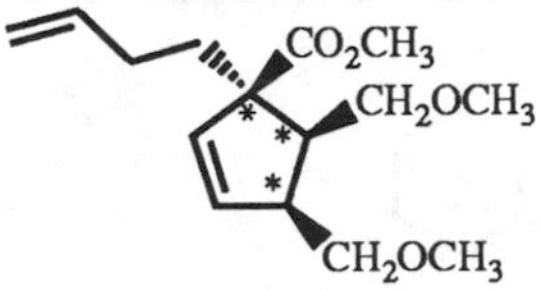	>95:<5	[308]
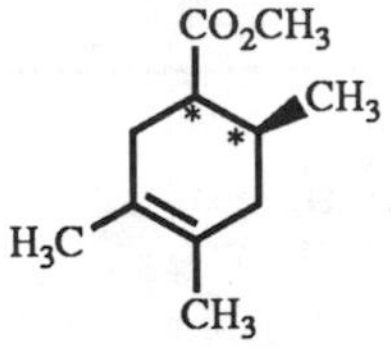	b	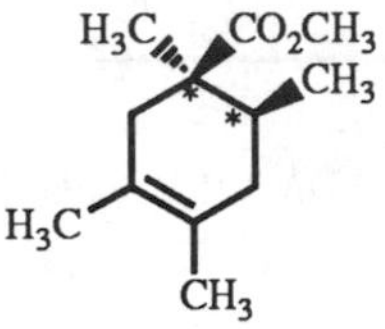	95:5	[297]
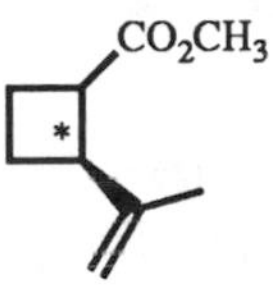	b	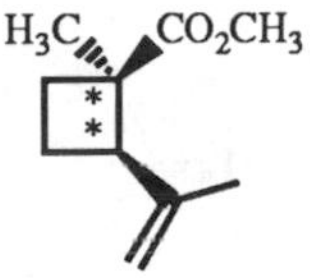	85:15	[309]
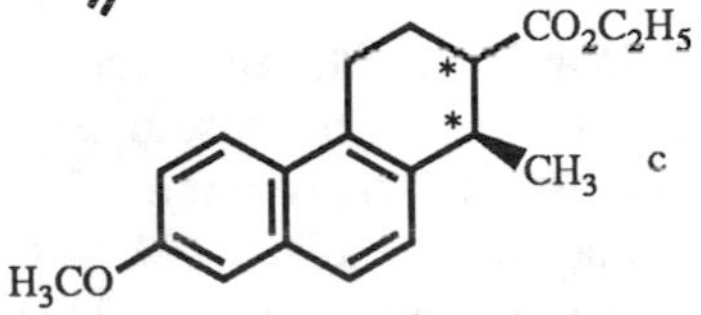	c	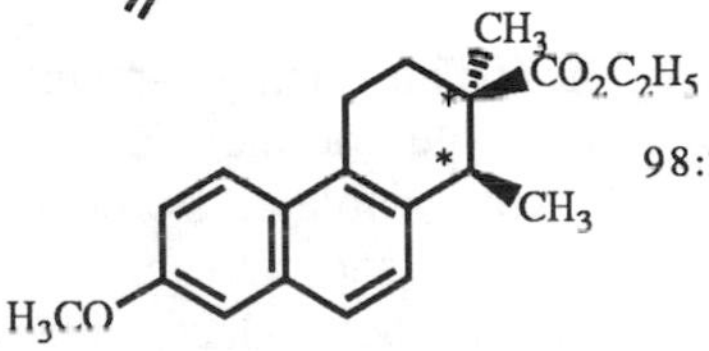	98:2	[310]
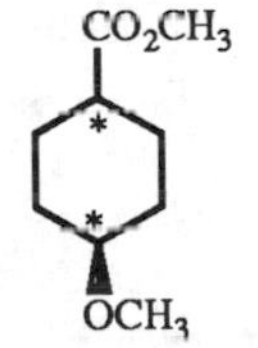	b	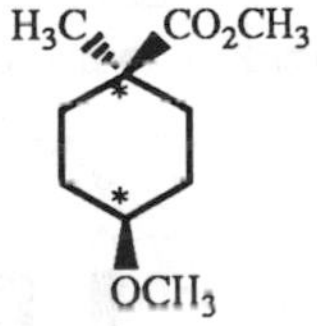	84:16	[304]
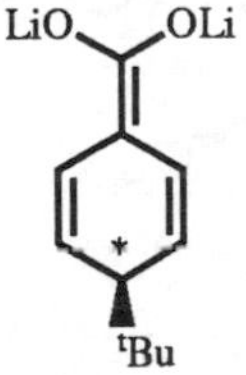	d		93:7	[307]
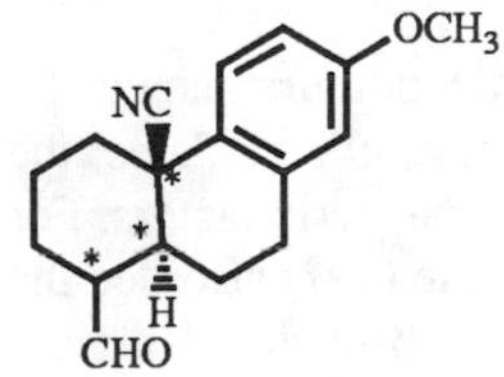	e	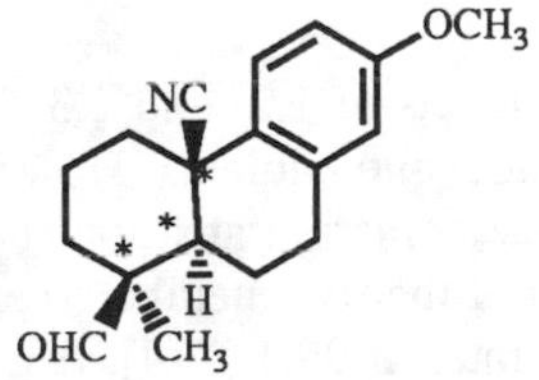	100:0	[311]

(Table 4.24. contd.)

Substrate	Reagent	Product[†]	Optical Yield	Reference
	f		50.5%	[312]

[†] only one isomer is shown

a:LDA, $_{Br}$, b:LDA, CH$_3$I, c:Ph$_3$CNa, CH$_3$I, d:^{i}PrBr, e:KO^{t}Bu, CH$_3$I,

f:Li/NH$_3$, CH$_3$I,

4.5.1.2   Endocyclic Enolates

While in exocyclic enolates and five-membered endocyclic enolates steric factors are largely responsible for the diastereoselectivity of alkylations, in six-membered endocyclic enolates stereoelectronic factors largely control the stereoselectivity. For instance alkylation of the organocopper enolate (494) with allyl iodide resulted in the products (495) and (496) in a ratio of 90:10 [313]. The high diastereoselectivity of product (495) indicates that stereoelectronic effects, although small, provide a bias for axial alkylation *via* a chair-like transition state [313b].

(494)   (495)   (496)

(495 : 496 = 90:10)

Similar good levels of 1,2- and 1,3-asymmetric induction are shown by five-membered endocyclic enolates. However as with exocyclic enolates, the stability of the diastereomeric transition states depends upon steric factors. For instance, alkylation of the five-membered endocyclic enolate (497) afforded the product (498) in a ratio of 99:1 [314]. It is evident that the β-naphthyl moiety hinders the electrophilic attack from the *syn* π face in the enolate, thereby leading to (498) preferentially.

(497)   (498) >99%

Table 4.25 shows some examples of alkylations of endocyclic enolates.

Table 4.25. Stereoselective Alkylation of Endocyclic Enolates[†]

Substrate	Reagent	Product[†]	Selectivity	Reference
	a		55:18:27[*]	[315]
	b		83:17	[316]
	c		73:27	[317]
	d		9:1	[318]

(Table 4.25. contd.)

Substrate	Reagent	Product[†]	Selectivity	Reference
	e		75:25	[319]
	f		70% trans	[320]
	g		93:7	[321]
	f		100:0	[322]
	g		>97:<3	[323]
	h		>98:<2	[324]

[†] Only the major isomer is shown.

+ Akylation at aldehydic oxygen.

a:CH₃ONa,CH₃I, b:LiNH₂, CH₃I, c:NaO ᵗAm, CH₃I, d:KOᵗBu   Cl, e:NaH,CH₃I,

f:LDA   Br, g:LDA, CH₃I, h:BrCH₂CO₂CH₃

4.5.1.3   Norbornyl Enolates

Asymmetric alkylation of norbornyl ring systems affords predominantly the *exo*-alkylated products. As shown in Fig. 4.16, both exo- and endocyclic enolates afford the *exo*- alkylated compounds as the major products [304,325-329].

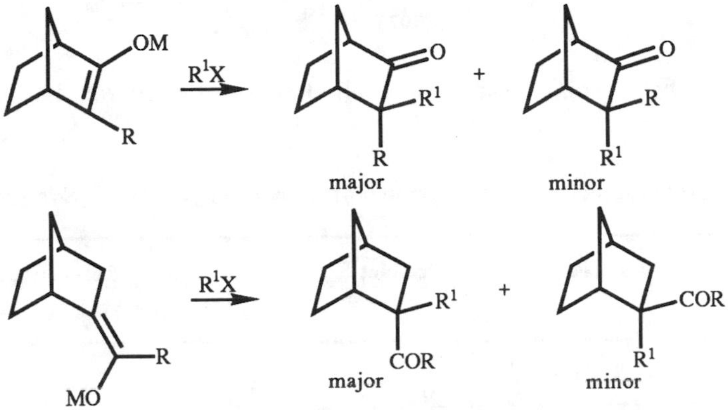

Figure 4.16. Alkylation of Exo- and Endocyclic Enolates with R^1X.

For instance methylation of the substrate (**499**) with CH_3, in the presence of LDA afforded (**500**) as the only product [325]. However substitution of the norboronyl ring at the 7 position effectively blocks the *exo* face of the enolate, thereby affording the *endo* product with high diastereoselectivity [329]. For instance the ketone (**501**) can be methylated to give the *endo* product (**502**) in an *endo-exo* ratio of 97:3 [329].

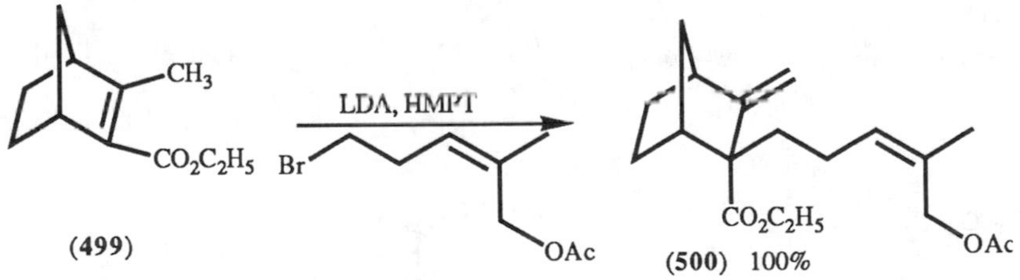

(**499**)   (**500**) 100%

(501)   (502)   >97:<3

Table 4.26 shows some examples of alkylations of norboronyl ring systems.

Table 4.26. Stereoselective Alkylation of Norbornyl Ring Systems

Substrate	Reagent	Product[†]	% Selectivity	Reference
$-CO_2CH_3$	LDA, CH_3I	$-CH_3$ CO_2CH_3	(94:6)	[304]
$-CO_2CH_3$	LDA, CH_3I	$-CH_3$ CO_2CH_3	(97:3)	[304]
$-CO_2H$	LDA, C_2H_5I	$-C_2H_5$ CO_2H	(68:32)	[304]
	LDA, CH_3I		(>3:<1)	[326]

(Table 4.26. contd.)

Substrate	Reagent	Product[†]	% Selectivity	Reference
	Ph₃CNa,CH₃I		(>97:<3)	[327]
	Ph₃CNa, CH₃I		(>97:<3)	[330]
	LDA, CH₃I		(100:0)	[328]

[†] Only the major isomer is shown

4.5.2   Alkylation of Imine and Enamine Salts

Ever since the application of imine and enamine salts in the α-alkylation of aldehydes and ketones [331], tremendous progress has been made in this field. As shown in Fig. 4.17, a wide variety of imines derived from aliphatic primary amines and enolizable aldehydes or ketones undergo enolization with RMgX. These salts react with primary and secondary alkyl halides to give monoalkylated products.

In some cases, however, Grignard reagents add to the imines rather than deprotonate them. Generally, LDA in THF is used in the deprotonation of aldimines and symmetrical ketimines, whereas methyl lithium is used for asymmetric ketimines.

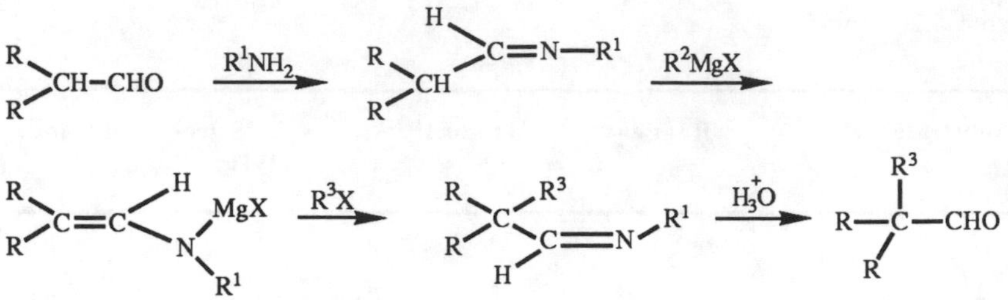

Figure 4.17. Alkylation of carbonyl compounds *via* imines.

In order to accomplish such alkylations of aldehydes and ketones in an asymmetric manner various chiral auxiliary groups such as (**503-514**) have been employed.

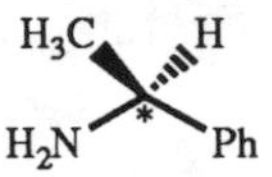

(503)

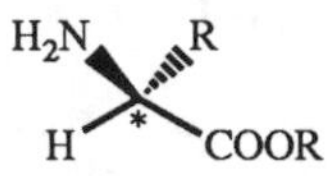

(504)

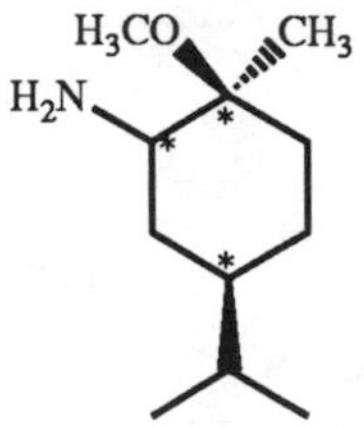

(505)

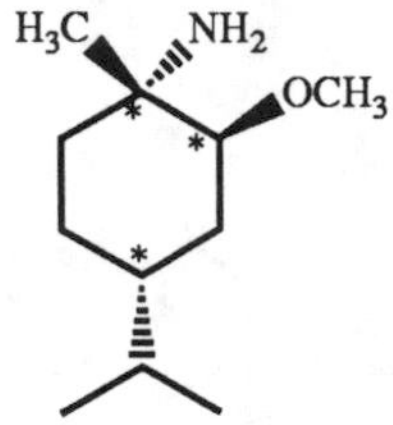

(506)

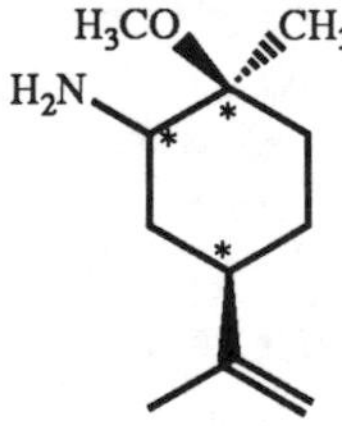

(507)

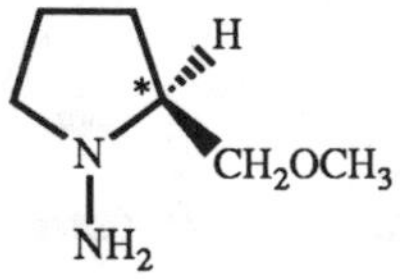

(508)

$(^nBu)_2Sn\,[N(CH_3)_2]$

(509)

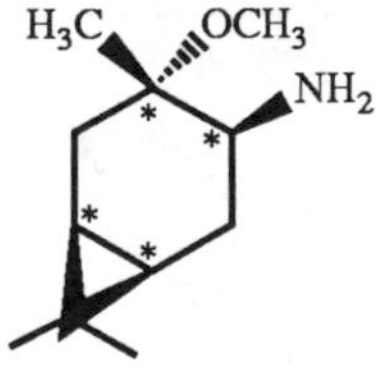

(510)

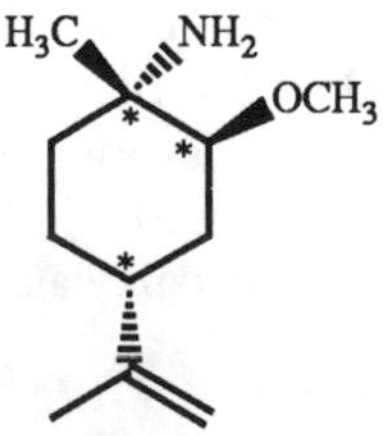

(511)

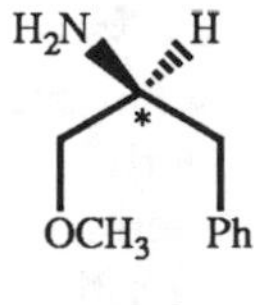

(512)

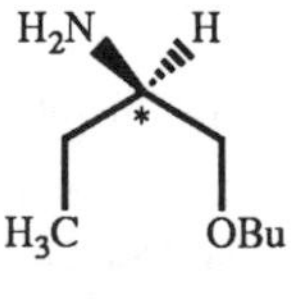

(513)

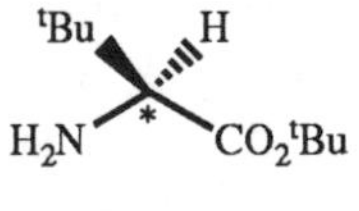

(514)

As shown in Fig. 4.18, the chiral amine (**512**) reacts with cyclohexanone to give the imine (**515**). This is converted with LDA to the lithioenamine (**516**) which forms a five-membered chelate. Alkylation of (**516**) occurs from the *si* face of the cyclohexanyl moiety affording (517) which upon hydrolysis affords the (S)-alkylated cyclohexane (518) in 87-100% e.e. [332,333].

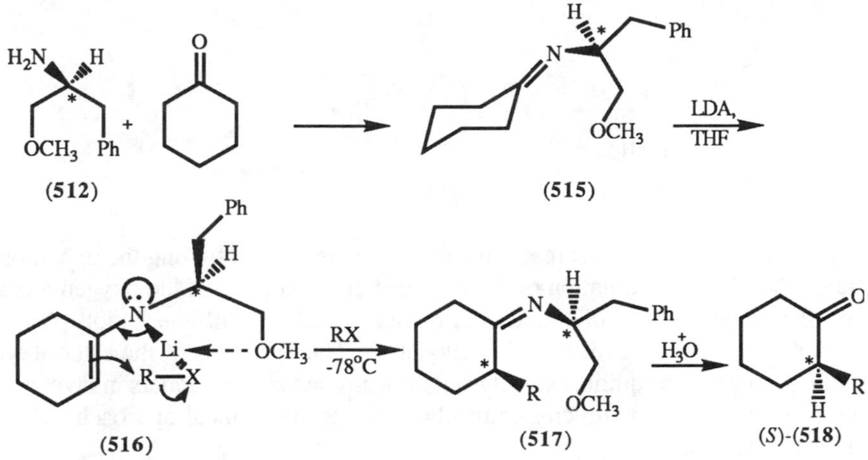

Figure 4.18. Alkylation of ketones in the presence of a chiral amine.

Similarly propionaldimine (519) is alkylated in the presence of (S) - phenethylamine (503) to give the product (520) in 88% optical purity [334]. The reaction is believed to proceed through the azaallylic anion (521) in which the alkyl halide attacks from the side opposite to the face occupied by the lithium ion. Similarly, α-alkylation of cyclohexylamine derivatives has been achieved in high optical yield by using DL-*tert*-leucine *tert* butyl ester (514) [334,335].For instance, the chiral enamine (522) afforded the alkylated product (523) in 96% optical yield [335].

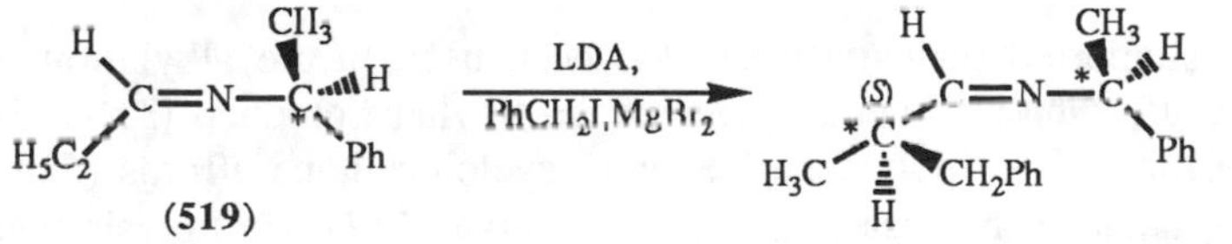

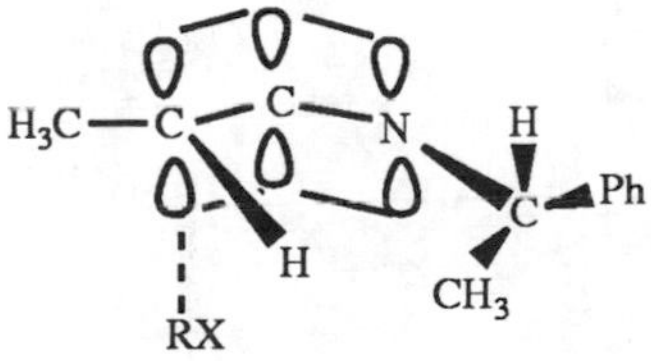

(522)   (523)   96% e.e.(S)

The selectivity in this reaction can be explained by assuming the transition state (524). The coordination of the unshared electron pairs of the oxygen atom and the leaving group of the alkylating agent with the lithium cation play a central part in this highly selective alkylation. The bulkiness of the ester alkyl group displaces the equilibrium between the *trans* and *cis* conformers in favour of the former, thereby indirectly controlling the stereochemical approach of the alkylating reagent.

(524)   (523)

Polymer supported reagents have also been used in the alkylation of cyclohexanone affording high optical yields of the alkylated products [336,337]. Thus, condensation of the polymer (525) with cyclohexanone affords (526), which is methylated in the presence of LDA to give (527). Hydrolysis then leads to α-methyl cyclohexanone in 90% e.e. [336].

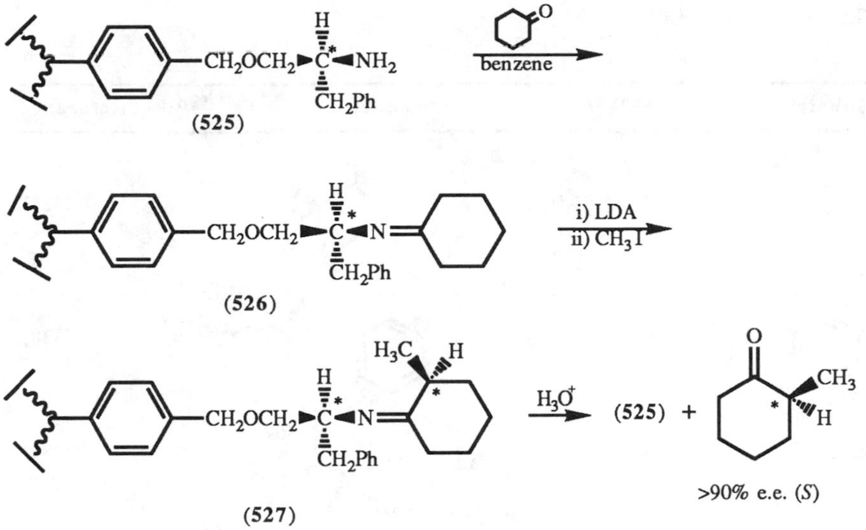

$$-CH_2OCH_2 - \overset{H}{\underset{CH_2Ph}{\overset{*}{C}}} - NH_2 \quad \xrightarrow{\text{benzene}}$$

(525)

$$-CH_2OCH_2 - \overset{H}{\underset{CH_2Ph}{\overset{*}{C}}} - N = \quad \xrightarrow[\text{ii) CH}_3\text{I}]{\text{i) LDA}}$$

(526)

$$-CH_2OCH_2 - \overset{H}{\underset{CH_2Ph}{\overset{*}{C}}} - N = \quad \xrightarrow{H_3O^+} \quad \textbf{(525)} \; +$$

(527)

>90% e.e. (S)

Table 4.27 shows some examples of asymmetric alkylations of imine and enamine substrates.

Table 4.27. Asymmetric Alkylation of Imine and Enamine Substrates

Substrate	Reagent	Product	% o.p.	Reference
 (cyclohexanone)	(512), CH$_3$I	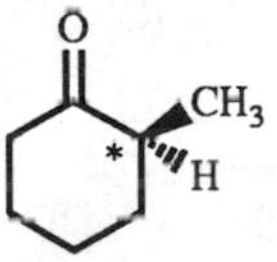 (2-methylcyclohexanone)	87 (S)	[333]
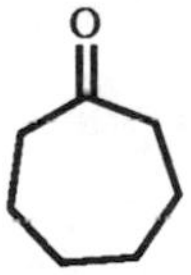 (cycloheptanone)	(512), CH$_3$I	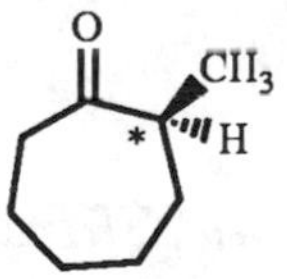 (2-methylcycloheptanone)	85 (S)	[333]
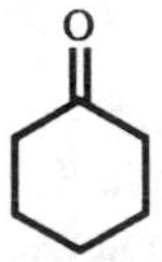 (cyclohexanone)	(512), I–CH$_2$CH$_2$–OCH$_3$ 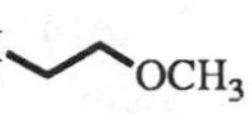	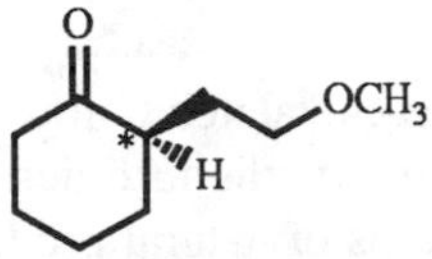 (product)	90 (R)	[333]
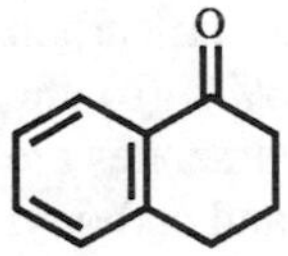 (1-tetralone)	(512), CH$_3$I	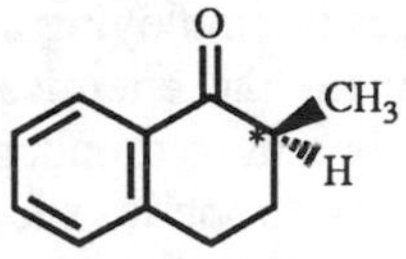 (2-methyl-1-tetralone)	79 (S)	[333]

(Table 4.27. contd.)

Substrate	Reagent	Product	%o.p.	Reference
H_5C_2–CO–C_2H_5	(512), C_2H_5 I	H_5C_2–CO–$C(CH_3)(H)(C_2H_5)$	76 (S)	[338]
(phenylacetone with Ph)	(512), CH_3 I,	(product Ph, H, CH₃)	91 (R)	[338]
ᵑHex–CH₂–CHO	(507), LiTMP, CH_3 I	ᵑHex–C(CH₃)(H)–CHO	75 (S)	[339]
CHO (cyclopentene)	(514), PhMgBr	CHO ...Ph (cyclopentane)	82 (1R,2S)	[340]
Ph–CH=CH–CHO	(514), EtMgBr	Ph–C(H)(Et)–CH₂–CHO	95 (S)	[341]
H_3C–CH=CH–CHO	(514), PhMgBr	H_3C–C(H)(Ph)–CH₂–CHO	91 (R)	[341]

4.5.3　Alkylation of Chiral Hydrazones

The introduction of α-metalated chiral hydrazones into the plethora of carbon–carbon bond forming methodologies has greatly enhanced the scope for stereoselective synthesis of natural products and biologically active compounds. Some of the advantages of employing hydrazones in asymmetric carbon-carbon bond formations are: (a) the chemical yields are high (upto 96%), (b) the hydrazone can be easily removed under mild conditions, (c) the hydrazones are more reactive towards electrophiles such as alkyl halides, oxiranes and carbonyl compounds, (d) the enolates do not take part in aldol type self-condensations and, once formed, they are stable at 25°C in the absence of water or CO_2, and (e) the formation of hydrazones is regiospecific and alkylation gives only monoalkylated products. Fig. 4.19 shows some asymmetric alkylation of carbonyl compounds *via* chiral hydrazones.

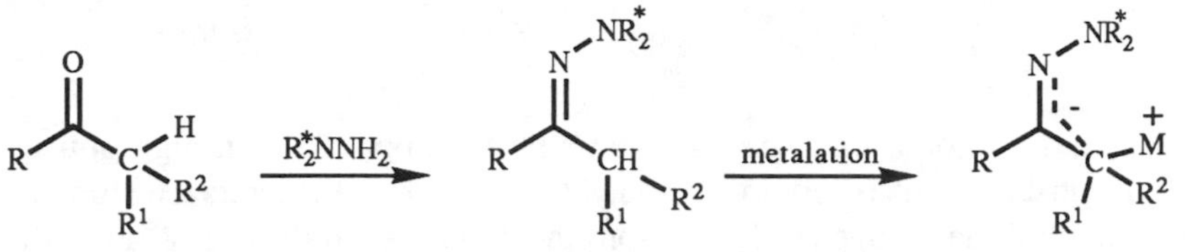

Figure 4.19. Alkylation of aldehydes and ketones *via* chiral hydrazones.

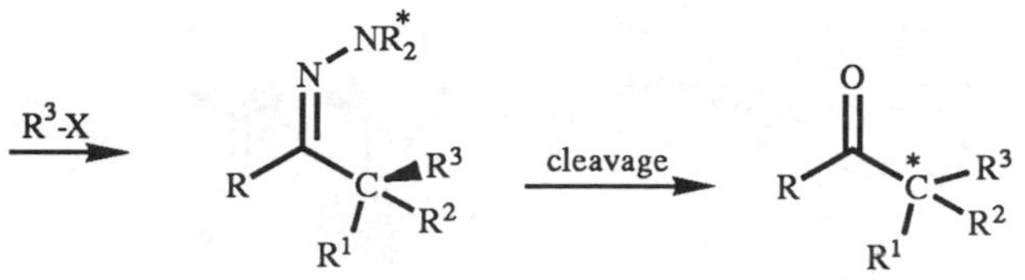

As shown in Fig. 4.19 the prochiral carbonyl compound is treated with a chiral auxiliary to afford the chiral hydrazone (**528**). This is metalated to afford the a-metalated intermediate (**529**) which is the chiral enolate equivalent. This is followed by alkylation to give the alkylated product (**530**) which upon hydrolysis affords the monoalkylated prochiral carbonyl compound (**531**). Two of the most common chiral auxiliaries used in the alkylation of chiral hydrazones are known as *S*AMP (**532**) and *R*AMP (**533**) [342-344].

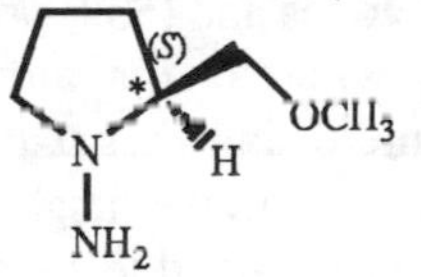

(*S*)-1-amino-2-methoxymethyl pyrrolidine
(*SAMP*) (**532**)

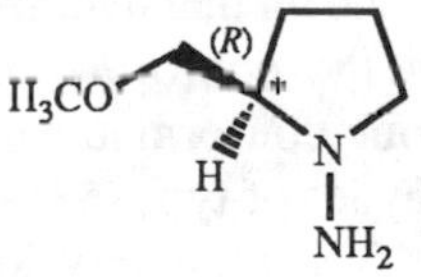

(*R*)-1-amino-2-methoxymethyl pyrrolidine
(*RAMP*) (**533**)

For instance the aldehyde (**534**) reacts with (**532**) to form the hydrazone (**535**). This is cleaved with excess CH_3I and acid hydrolysis then affords the alkylated aldehyde (**536**) in 82% diastereomeric excess (d.e.) [342].

(534) → (532) → (535) → i) Excess CH₃ I / ii) HCl → (536) 82% e.e. (S)

Similarly ketones can be alkylated with (532) or (533) in high optical yields. For instance, the α-bromo- *tert*-butyl acetate adds to the hydrazone formed from (532) and the ketone (537) to afford the alkylated product (538) in 95% e.e. [345,346].

(537) → i) (532), ii) LDA / iii) BrCH₂COO^{t}Bu / iv) O₃ → (538) > 95% e.e. (S)

(539)   (540)   (541)

a: $R^1 = CH_3$, $R^2 = H$

b: $R^1 = CH_3$, $R^2 = C_2H_5$

c: $R^1 = CH_3$, $R^2 = CH_2 = CH - CH_2$

It has been shown that hydrazones derived from ketones and (532) or (533) exist largely in the conformation shown in (539) whereas the aldehyde derivatives favour conformation (540) [347]. It is interesting that α-phenyl substituted aldehydes of type (541a - c) show selectivity only in the range of 23-31% e.e. [345]. The relatively low optical yield is probably due to partial racemization. Table 4.28 shows some examples of the alkylation of carbonyl compounds via chiral hydrazones.

Table 4.28. Alkylation of Carbonyl Substrates through Chiral Hydrazones

Substrate	Reagent	Product	% e.e.	Reference
	(532), $(CH_3)_2SO_4$		≥99 (R)	[342]
H₃C~~~CHO	(532), CH_3I	$CH_3(CH_2)_5$ *⟨CHO / CH₃⟩	86 (R)	[342]
	(532), CH_3I		86 (R)	[342]
Ph⟍⟋CHO	i) l-ephedrine ii) PhLi, iii) Pd-C/H₂	Ph, *, H, NH₂, Ph	91 (R)	[343]
H_5C_2⟋C(=O)⟋CH_3	(532), C_2H_5I	H_5C_2, C_2H_5, CH_3	94 (S)	[348]
H_3C⟍C(=O)⟋ H,H,CH_3	(532), $^nC_3H_7I$	H_3C⟍C(=O)⟋*⟍CH_3 H, CH_3	99.5 (S)	[348]
Ph⟋C(=O)⟋C_2H_5	(532), CH_3I	Ph *⟍CH_3 / CH_3	>74(R)	[348]
H_3C⟍C(=O)⟋CH_2/Ph	(532), CH_3I	H_3C⟍C(=O)⟋* H, CH_3, Ph	10 (R)	[348]
H_5C_2⟍C(=O)⟋CH_2/CH_3	(532), ⟍CH_2Br	H_5C_2⟍C(=O)⟋* H, CH_2⟍⟨cyclohexyl⟩ / CH_3	>97 (S)	[348]

(Table 4.28. contd.)

Substrate	Reagent	Product	% e.e.	Reference
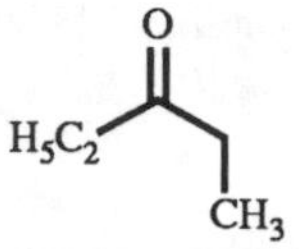	(532), $^{n}C_6H_{13}Br$	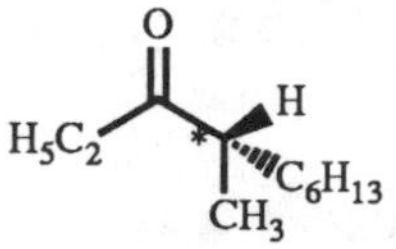	>97.5 (S)	[348]

4.5.4   Alkylation of Chiral Oxazolines

The use of chiral oxazolines in asymmetric synthesis is a relatively recent addition to the arsenal of asymmetric methodologies available to afford high optical yields [349,350]. Chiral oxazolines are now employed in the synthesis of a large variety of chiral organic compounds such as substituted acetic acids, lactones, thiiranes, chiral phthalides, binaphthyls etc.

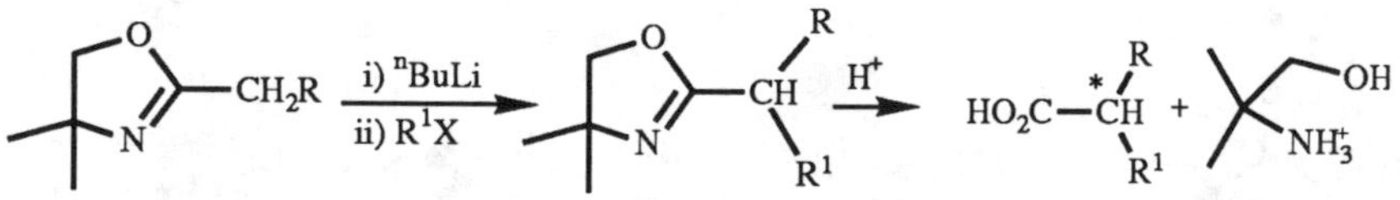

Figure 4.20. Generalized scheme for the synthesis of chiral alkanoic acids using chiral oxazolines

As illustrated in Fig. 4.20 the chiral oxazoline serves to transfer chirality from the heterocycle to the alkanoic acid in the reaction with alkyl halide. For instance, the chiral oxazoline (542) is metalated with n- butyllithium or LDA to the (E)-and (Z)-aza-enolates (543) and (544). Alkylation affords the alkylated oxazolines (545a,b) which are hydrolysed by acid to give the α,α-disubstituted carboxylic acids (546) in 72-82% e.e. [350,351].

$$(542) \qquad (E)\text{-}(543) \qquad (Z)\text{-}(544)$$

$$(545) \quad \text{a: } R^1 = \alpha \qquad (546) \ 72\text{-}82\% \ e.e. \ (S)$$
$$ \text{b: } R^1 = \beta$$

4.5.4.1   Synthesis of Alkyl Alkanoic Acids

The oxazoline (**549**), prepared by the condensation of (*1S,2S*) (+) 1 phenyl-2-amino-1,3-propanediol (**547**) with the ethyl imidate of propionitrile (**548**) , was treated with LDA to afford the lithio salts (**550**) and (**551**) by removal of the diastereotopic hydrogens. Treatment of the lithio salts with C_2H_5I afforded the oxazoline (**552**) which upon acid hydrolysis yielded 2-methylbutanoic acid (**553**) in 67% e.e [350,351]. The methoxyamino alcohol (**554**) was recovered and reused by treating it with (**548**) affording the oxazoline (**549**) for further asymmetric synthesis. It has been shown that the ratio of (*E*)-and (*Z*)-azaenolates (**550**) and (**551**) depends upon the base and the solvent used [352]. The use of LDA-THF or LDA-dimethoxyethane affords an (*E*) - (*Z*) ratio of 96:4, while n-butyllithium-THF gave a ratio of 69:31.

Acids of opposite configuration are obtained by reversing the order in which the alkyl groups are introduced into the chiral oxazoline. Thus by substituting the methyl group of the oxazoline (**549**) by an ethyl group and by using dimethyl sulfate instead of ethyl iodide for alkylation of the azaenolates, the acid (**553**) was obtained in 79% e.e. with (*R*) configuration. Table 4.29 shows some examples of the synthesis of α-alkyl alkanoic acids *via* chiral oxazolines.

Table 4.29. Enantioselective Synthesis of α,α'-Dialkylalkanoic Acid (546) via Chiral Oxazolines (542)

(542) (R)	R^1X	% e.e.	Reference
CH_3	C_2H_5	78 (S)	[351]
C_2H_5	$(CH_3)_2SO_4$	79 (R)	[351]
CH_3	$PhCH_2Cl$	72 (S)	[351]
$PhCH_2$	^nPrI	78 (R)	[351]
^nBu	$(CH_3)_2SO_4$	82 (S)	[351]
CH_3	C_2H_5I	79 (S)	[350]
CH_3	^nBu	75 (S)	[350]
Ph	^iPrI	65 (R)	[351]
$PhCH_2$	^nBuI	86 (R)	[351]
CH_3	^nPrI	78 (S)	[350]

4.5.4.2   Synthesis of α-Hydroxyacids

Chiral ketooxazolines such as **(555-560)** have been employed in the synthesis of α-hydroxy acids affording upto 87% e.e. [353].

(555)   (556)   (557)

(558)   (559)   (560)

This involves addition of Grignard and organolithium reagents to chiral α-ketooxazolines **(550-560)** resulting in α-substituted α-hydroxyoxazoline derivatives, hydrolysis of which affords α-substituted α-hydroxy acids in 30-80% e.e.

4.5.4.3   Synthesis of Butyrolactones and Valerolactones

Chiral oxazolines, such as **(561)** and **(562)** have been employed in the synthesis of 2-substituted butyrolactones and valerolactones respectively [354]. For instance, treatment of the oxazoline **(561)** with LDA and CH$_3$I afforded the alkylated derivative **(563)** which was hydrolysed to give 2-methyl butyrolactone **(564)** in 70% e.e. [354].

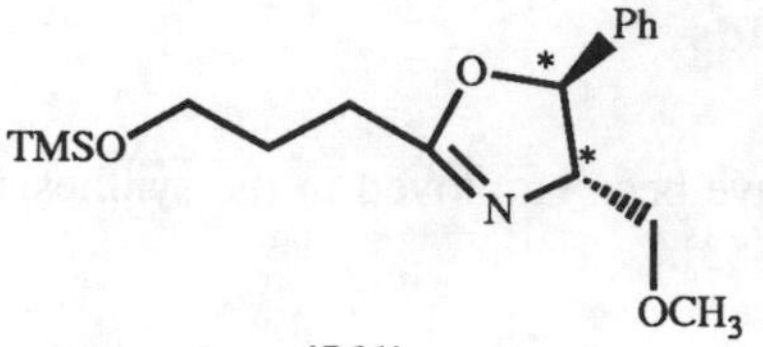

(561)

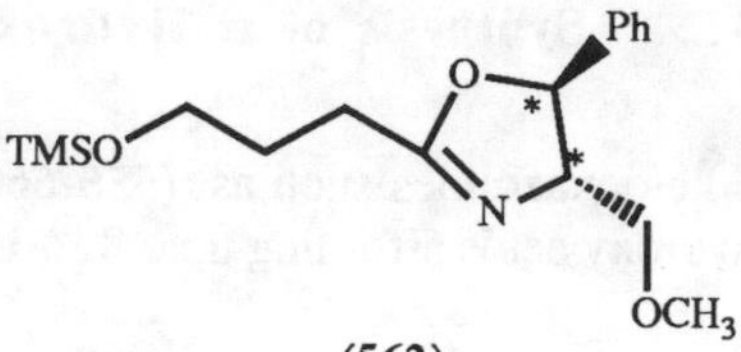

(562)

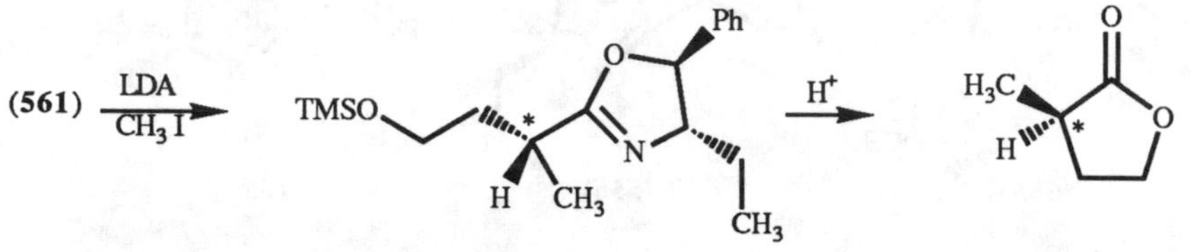

(563)

(564) 70% e.e. (R)

Table 4.30 shows some examples of the synthesis of butyrolactones and valerolactones.

Table 4.30. Synthesis of Butyrolactones and Valerolactones with Oxazolines (561) and (562)[†]

R	Oxazoline	% e.e.	Config.
C_2H_5	(561)	72	(R)
^nPr	(561)	73	(R)
allyl	(561)	72	(R)
^nBu	(561)	60	(R)
CH_3	(562)	52	(R)
C_2H_5	(562)	61	(R)
^nPr	(562)	60	(R)
allyl	(562)	72	(S)
$PhCH_2$	(562)	64	(S)

[†] from reference [354]

4.5.4.4   Synthesis of β-Alkylalkanoic Acid

Chiral β-alkylalkanoic acids are prepared from α,β-unsaturated oxazolines *via* alkylation affording optical yields of upto 99% e.e. For instance (542, R=H) is transformed via the phosphonate intermediate (654) to the α,β-unsaturated oxazoline (565). Alkylation followed by hydrolysis leads to the β-alkylalkanoic acid (566) [355,356]. The mechanism involves the metalated species (567) which is converted *via* (568), (569) and (570) to the acid (566).

Table 4.31 shows some examples of the synthesis of chiral β-substituted alkanoic acids *via* α,β-unsaturated oxazolines.

Table 4.31.   Asymmetric Synthesis of Chiral β-Substituted Alkanoic Acids[†]

R	R[1]Li	% e.e.	Config.
CH_3	Ph	98	(*S*)
C_2H_5	[n]Bu	96	(*R*)
[t]Bu	[n]Bu	98	(*R*)
Cyclohexyl	C_2H_5	99	(*R*)
$H_3COCH_2CH_2$-	[n]Pr	99	(*S*)
Ph	C_2H_5	97	(*R*)
(2-methoxyphenyl)	[n]Bu	95	(*R*)

[†] from [355]

4.5.4.5   Synthesis of Unsubstituted 1,4-Dihydropyridines

Substituted 1,4-dihydropyridines have been prepared *via* the addition of organolithium and Grignard reagents to the 4- position of 3-oxazolinyl pyridines [357,358]. The intermediates are important in the development of NADH mimics as chiral hydride transfer reagents. For example   the 3-pyridyl chiral oxazoline (**571**) is alkylated with organometallic reagents to give the 4-substituted 1,4-dihydropyridines (**574**) *via* the metalated intermediates (**572**) and (**573**) in high e.e. (see also chapter 3) [357, 358].

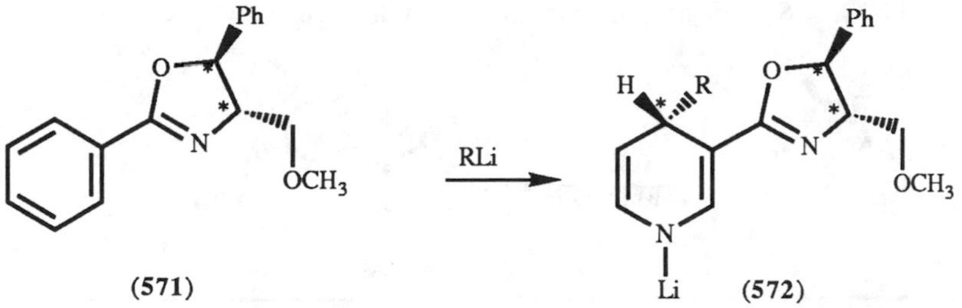

(571)   (572)

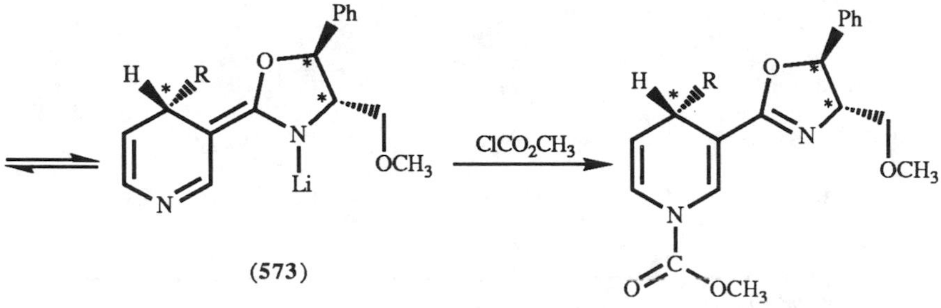

(573)   (574)

Similarly, the quinoline oxazoline (575) on treatment with CH_3Li, followed by CH_3OCOCl gave the 4-methyl dihydroquinoline (576) in a diastereomeric mixture of 98:2 [357].

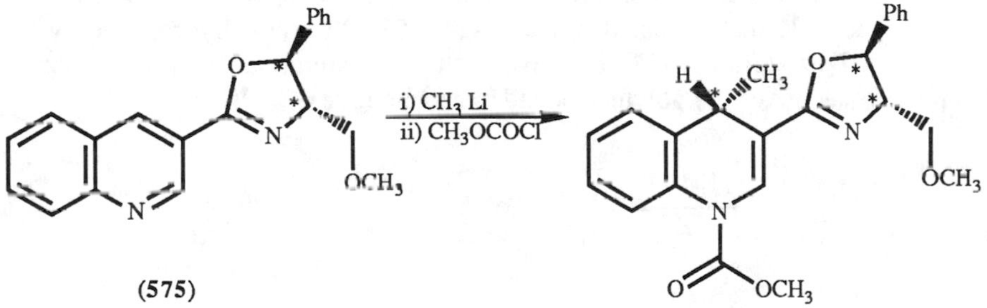

(575)   (576)

Table 4.32 shows some examples of the synthesis of 1,4- dihydropyridines.

Table 4.32.   Synthesis of Chiral 1,4-Dihydropyridines[†]

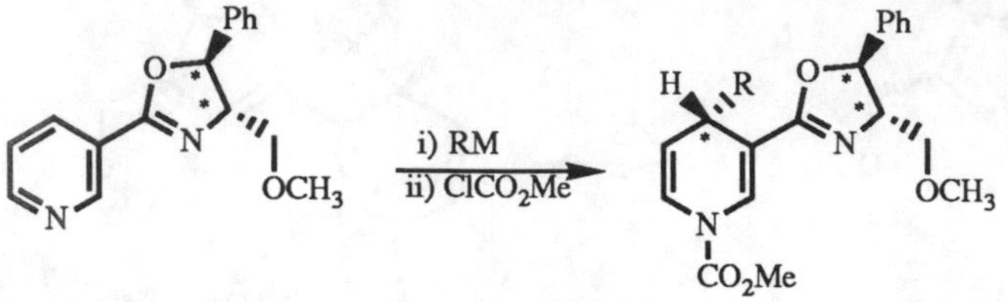

RM	Dias. ratio	Major diast.
CH$_3$Li	93: 7	(S)
CH$_3$MgCl	91:9	(S)
^{n}BuLi	97:3	(S)
C$_2$H$_5$MgBr	92:8	(S)
Ph Li	92:8	(S)

[†] from reference [357]

4.5.4.6   Synthesis of Resin-Bound Oxazolines

Resin-bound oxazolines provide another economical approach to asymmetric alkylation [359]. For instance, *trans* - (4S ,5S)-2-ethyl-4-hydroxymethyl)-5-phenyl-2-oxazoline (577) is converted to its sodium salt and attachment to polystyrene gives the polymer-bound oxazoline (578).

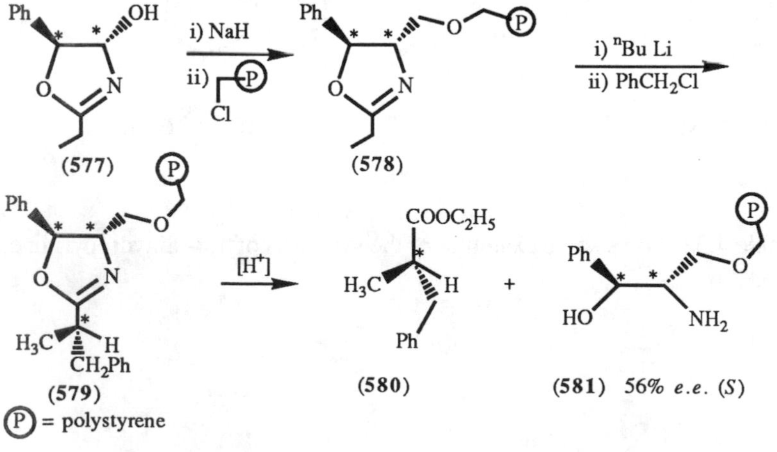

(577)

(578)

(579)

Ⓟ = polystyrene

(580)

(581) 56% *e.e.* (S)

This can be alkylated to give the oxazoline (579) which is ethanolysed under acid catalysis to give (S)-(+)-ethyl(2-methyl)-3-phenylpropanoate (580) in 56% e.e. [359]. However, this method is regarded as inferior to the homogeneous method since polymer-bound oxazolines are difficult to hydrolyse.

4.5.4.7   Alkylation via Diketopiperazines

Diketopiperazines having chiral amino acids as the chirality inducing groups may be alkylated and subsequent hydrolysis affords alkylated chiral amino acids [360-363]. Fig. 4.21 shows the various steps involved in the synthesis of amino acids (590) via diketopiperazines.

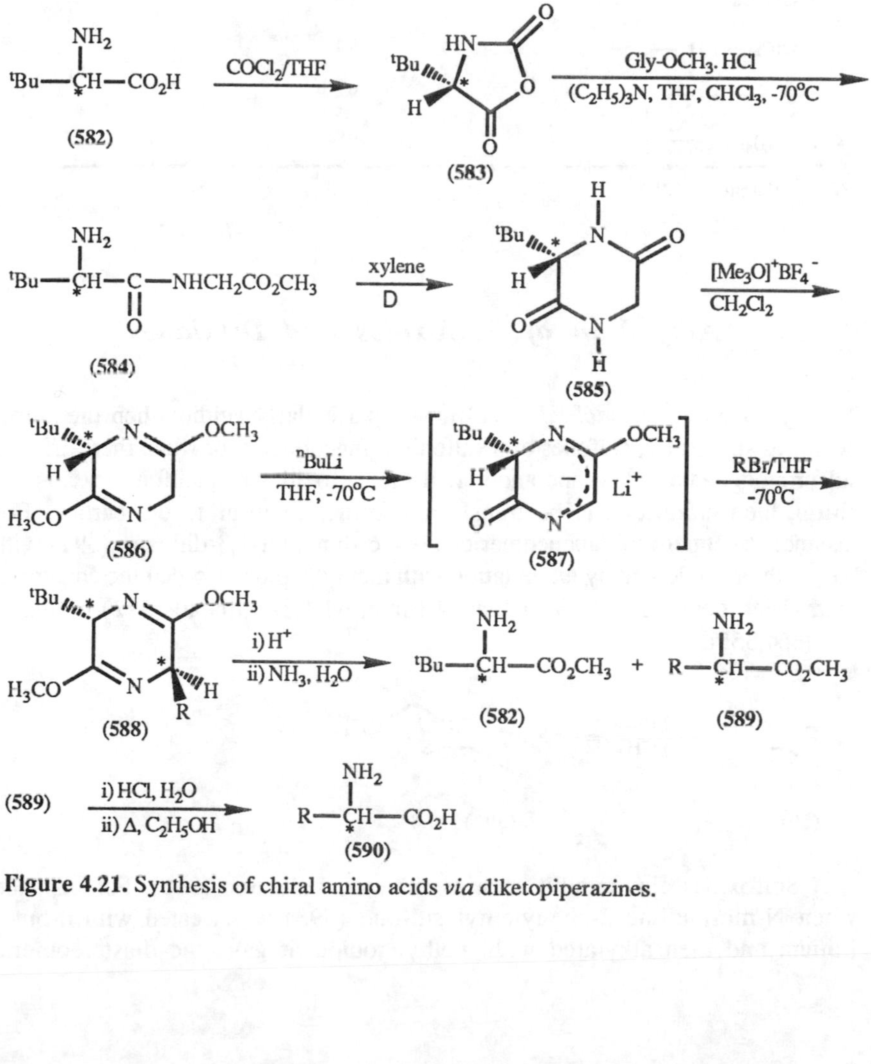

Figure 4.21. Synthesis of chiral amino acids via diketopiperazines.

Table 4.33 shows the optical yields obtained in the synthesis of amino acid methyl esters **(589)** by this method.

Table 4.33. Synthesis of Amino Acid Methyl Esters

$$\text{(588)} \quad \xrightarrow[\text{ii) NH}_3,\text{ H}_2\text{O}]{\text{i) H}^+} \quad R\text{—}\underset{*}{CH}\text{—}CO_2CH_3 \quad + \quad {}^t\text{Bu}\,\underset{*}{CH}CO_2CH_3$$

$$\text{(589)} \qquad\qquad \text{(582)}$$

R	% e.e.
PhCH$_2$	>95
- CH$_2$CO$_2$^{t}Bu	>95
CH$_2$=CH—CH$_2$—	94
CH≡C—CH$_2$—	94
CH$_3$(CH$_2$)$_5$CH$_2$—	>95

*from reference [362]

4.5.5   Alkylation of Sulfoxides and Dithianes

The hydrogen atoms *vicinal* to sulfur are particularly acidic when the sulfur occurs as sulfoxides, sulfones or a sulfonium functionality or when there are two sulfur atoms adjacent to one another. If the sulfoxides and sulfonium salts are chiral, then chirality can be transferred from the sulfur to the carbon. For instance, treatment of "anancomeric" *cis*-4,6-dimethyl-1,3-dithiane **(591)** with butyl lithium followed by methylation with methyl iodide afforded the equatorial methylated compound (γ-2,*cis*-6,*cis*-4-trimethyl-1,3-dithiane) **(592)** in 99.8% e.e. [364,365].

(591)   (592) 99.8% e.e.

Sulfoxides also exhibit such asymmetric induction [366]. For instance when N-morpholino-2-phenylethyl sulfone **(593)** was treated with n-butyl lithium and then alkylated with methyl iodide, it gave the diastereomeric products (**594**) and (**595**) in a ratio of 94:6 [366]. Similarly, methylation of *cis*-thiacyclohexane-1-oxide (**596**) afforded equatorial and axial methyl derivatives (**597**) and (**598**) in a ratio of 80:20 [367,368]. Methylation of the axial sulfoxide (**599**), on the other hand, afforded the axial methylated derivative (**600**) in quantitative yield.

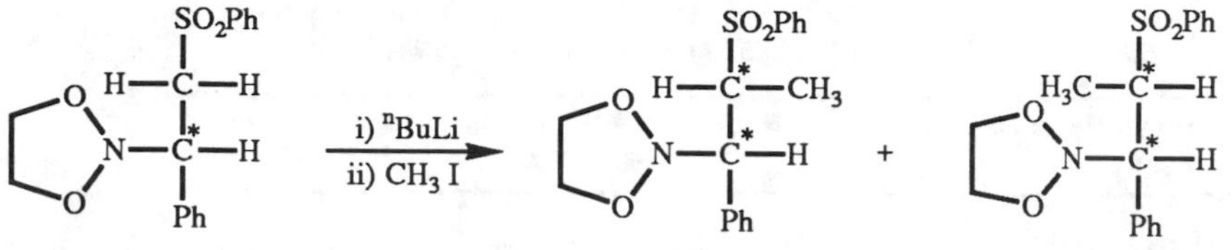

(**593**)

i) ⁿBuLi
ii) CH₃I

(**594**)

+

(**595**)

(**594** : **595** = 94:6)

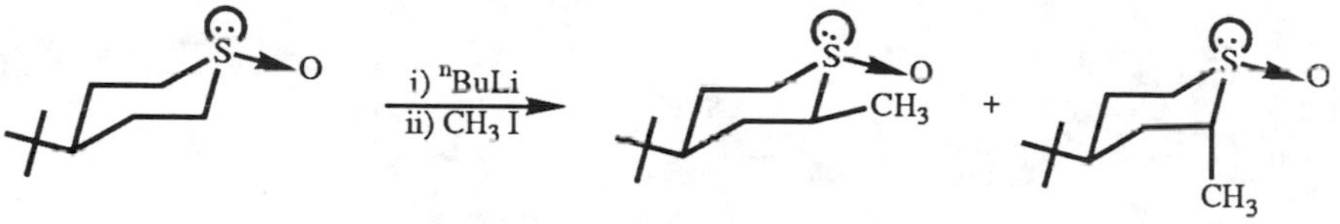

(**596**)

i) ⁿBuLi
ii) CH₃I

(**597**)

+

(**598**)

(**597** : **598** = 80 : 20)

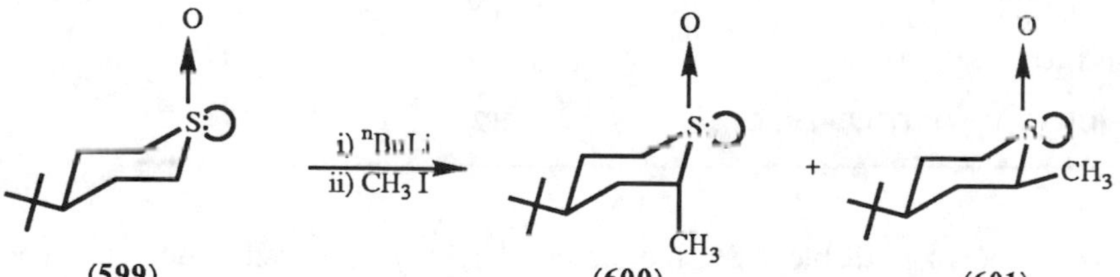

(**599**)

i) ⁿBuLi
ii) CH₃I

(**600**)

+

(**601**)

(**600** : **601** = 100:0)

Table 4.34 shows stereoselectivity of alkylation of substrates (**596**) and (**599**) with various alkylating agents.

Table 4.34.   Stereoselectivity of alkylations

(i) Alkylation of (596)

(596)   (597)   +   (598)

Alkylating/Medium Agent	Products %	
	% (597)	% (598)
$CH_3 I$ (THF)	80	20
$(CH_3O)_3 P = O$ (THF)	7	93
$(CH_3O)_3 P = O$ (THF + $LiClO_4$)	60	40

(ii) Alkylation of (599)

(599)   (600)   +   (601)

Alkylating/Medium Agent	Products %	
	% (600)	% (601)
$CH_3 I$ (THF)	100	0
$(CH_3O)_3 P = O$ (THF)	30	70
$(CH_3O)_3 P = O$ (THF + $LiClO_4$)	92	8

As shown in Table 4.26, the stereoselectivity observed under the same conditions with CH_3I is reversed with $O=P(OCH_3)_3$, i.e. 93% instead of 20% for (596) and 30% instead of 100% for (599). In contrast, in the presence of excess of $LiClO_4$ the stereoselectivity with trimethylphosphate is again similar to that of CH_3I for (599) and somewhat less for (596). These results are rationalized in terms of steric conformational effects as depicted in transition states (602) and (603) for the substrates (596) and (599) respectively. The approach of CH_3I towards the substrates (596) and (599) is governed by steric hindrance, and therefore occurs *trans* to the chelated face (path A), since in the *quasi* equatiorial approach (along path E) the electrophile is hindered by the axial hydrogens at the β' carbon and by the ^{t}Bu group.

(602)   (603)

In the case of trimethylphosphate, which requires coordination to Li⁺, the reagent attacks along the less hindered path A (*cis* to the S → O bond) *via* the transition state (602). Similar stereoselectivity of carbanions α to sulfoxides has been studied in acyclic substrates [369,370]. For instance benzyl *tert.*-butyl sulfoxide (604) could be methylated to give (605) in 98% e.e. [369,370].

(604)   (605) 98% e.e.

The stereoselectivity in this reaction is related to the chelation ability of the proton donor. It has been proposed that an sp^2 carbanion engaged in an ion pair tends to react with *retention with chelating reagents* while it reacts with *inversion with non-chelating reagents*. The sp^3 carbanion, on the other hand, reacts with retention with all reagents. The transition state (606) has been proposed to be responsible for the high stereoselectivity.

(606)

4.5.6   Michael Addition Reactions

The Michael reaction has been extensively studied for producing chiral adducts. There are three ways in which the Michael reaction has been utilised in asymmetric synthesis: (i) addition of chiral anions, (ii) addition of achiral anions complexed to chiral ligands, and (iii) addition of achiral anions to chiral Michael acceptors.

4.5.6.1   Addition of Chiral Anions

In this method, chiral α-sulfinyl esters as well as α,β-unsaturated esters have been employed as chiral anions in the asymmetric Michael addition. For instance, addition of the chiral α-sulfinyl ester (607) to the ethylenic ester (608) yielded the product (609) after desulfurization in 24% e.e. [372]. However this method generally affords low optical yields (< 25 % e.e.).

(607)   +   (608)   i) NaH / ii) Raney - Ni

(609) 24% e.e. (S)

4.5.6.2   Addition of Achiral Anions Complexed with Chiral Ligands to Prochiral Michael Acceptors

Addition of achiral anions complexed with chiral ligands to prochiral Michael acceptors is one of the most widely used methods for asymmetric carbon-carbon bond formations. However the optical yields were initially low since the influence of the chiral environment decreases on going from the ground state to the transition state [373-375]. The shortcomings involved in using chiral leaving groups for asymmetric induction have been surmounted by using (i) bulky chiral leaving groups [376-379], (ii) designing reactions involving cyclic transition states [380, 381] and (iii) inserting another step before the leaving group is eliminated [382,383].

For instance, using chiral ligands such as (610) alanine is obtained by methylation of the anion (611) in 40% e.e. [379].

(610)

(611)   i) (610), HMPT   ii) H⁺   alanine 40% e.e. (S)

Recently high optical yields have been obtained in the asymmetric nitroolefination of lactone enolates [384]. The reaction of the nitroenamine (612) with the lactone enolate (613) in the presence of Zn^{2+} afforded the product (614) in 96% e.e. [384].

(612)   +   (613)   (614) 96% e.e. (S)

4.5.6.3 Addition of Achiral Anions to Michael Acceptors Having One or More Chiral Centres

This approach involves addition of organometallic reagents to chiral Michael substrates to give high optical yields of the alkylated products [385,386]. For instance, (S)-$(+)$-2,2(p-tolylsufonyl)-2-cyclopentenone (**615**) reacts with CH_3MgI and derivatization with dinitrophenylhydrazine affords (S)-(**616**) in 72% e.e. [385].

$$\begin{array}{ccc} \text{(615)} & \begin{array}{l} \text{i) } CH_3MgI \\ \text{ii) } H^+ \\ \text{iii) Al-Hg} \\ \text{iv) } 2,4(NO_2)_2\text{-}C_6H_3NNH_2 \end{array} & \text{(616) 72\% e.e. (S)} \end{array}$$

The (R)-enantiomer of product (**621**) can be obtained by using divalent metal halides e.g. $ZnBr_2$ which form chelation complexes, thereby blocking one side from attack prior to the alkylation. Similarly, α,β-unsaturated carboxylic acid amides (R = CH_3, C_2H_5, n-Bu,Ph) (**617**) derived from (l)-ephedrine react with Grignard reagents e.g. ^nBuMgBr to afford the β-substituted alkanoic acids (**621**) via the intermediates (**618-620**) in 79-99% optical yields [386]. The optical yields in these reactions depend upon the solvent used. Diethyl ether gave high e.e. whereas THF lowered the optical yields by coordinating to the metal salts, thereby weakening the chelate complex (**618**).

4.5.6.4   Additions Using Optically Active Transition Metal-Ligand Catalysts

In this method, transition metal catalysts such as Ni(acac)$_2$, Co(acac)$_2$ etc. are used with chiral ligands to effect addition of cyclic as well as open-chain 1,3 dicarbonyl compounds to α,β-unsaturated compounds [387-391]. For instance, addition of (624) to methyl vinyl ketone (623) in the presence of (S,S)-1,2-diphenyl ethanediamine (622) and Co(acac)$_2$ proceeds *via* the complex (625) to give the product (626) in 66% e.e. [390]. Similarly, chiral crown ethers e.g. (627), (628) complexed to potassium bases such as KNH$_2$ or KOC(CH$_3$)$_3$ catalyse the Michael addition of β-ketoesters to methyl vinyl ketone and of phenyl acetic esters to methyl acrylate affording the corresponding products in 60-96% e.e. [392].

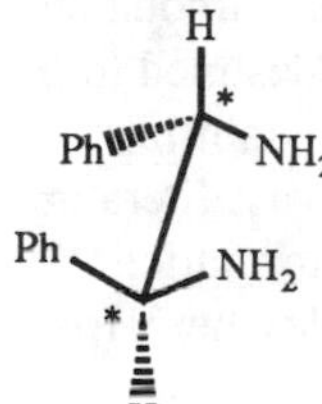

(622)

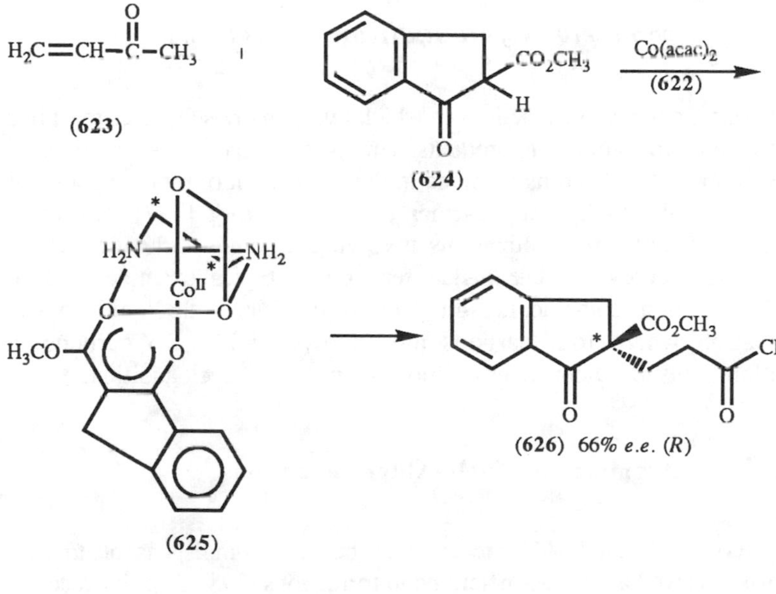

(623)

(624)

(625)

(626) 66% e.e. (R)

(R)

(627)   (628)

4.6 Pericyclic Reactions

No other reaction type can be better qualified for asymmetric carbon-carbon bond formations than that of pericyclic reactions which are inherently destined to be stereoselective by proceeding through a cyclic transition state, stabilized by optimum overlap of all involved orbitals [393-396]. From the vast literature which continues to accrue in the area of carbon-carbon bond formations by pericyclic reactions, it is pertinent to present selected examples here from asymmetric cycloaddition and sigmatropic rearrangements.

4.6.1   Asymmetric Cycloaddition Reactions

Cycloadditions are pericyclic reactions in which two or more substrates react in a concerted manner to form cyclic products. This type of reaction is classified by either the number of p electrons involved in the cyclization or by the number of ring atoms contributed by each partner to the new ring [397]. However, electrocyclic reactions, i.e cyclizations involving a single molecule are not included in this category. The Diels-Alder reaction is a typical example in which the two classifications coincide, i.e. 4π electrons combine with 2π electrons in ring formation where four carbons are involved with two carbons in cycloaddition. Thus the Diels-Alder reaction is a $\pi s^4 + \pi s^2$ cycloaddition.

4.6.1.1   Asymmetric Diels-Alder Reactions

Since its discovery the Diels-Alder reaction has become an indispensable tool for carbon-carbon as well as carbon- hetero bond formations [398,399]. The reaction is highly stereospecific and regioselective.

Diels-Alder reactions are generally accelerated by Lewis acids, while high pressure has also been used to facilitate the reaction. There are a number of ways to effect asymmetric Diels-Alder reactions: (i) addition of a prochiral diene to a chiral dienophile, (ii) addition of a chiral diene to a dienophile, and (iii) the use of chiral catalysts.

4.6.1.1.1 Addition to Chiral Dienophiles

Chiral dienophiles are simply made by appending a chiral prosthetic group to a dienophile. The prosthetic group must in turn fulfill the following criteria: (i) it should provide a high yield with quantitative and predictable π-face differentiation, (ii) be capable of attachment and removal with *retention* of the induced configuration, (iii) permit facile purification of the major cycloadduct, (iv) impose crystallinity on intermediates, and (v) be readily available in an enantiomerically pure form.

Most of the work on asymmetric Diels-Alder reactions has been done on addition of 1,3-dienes to chiral conjugated carboxylic esters. Table 4.35 shows some common chiral auxiliaries used in the asymmetric Diels-Alder reaction.

Table 4.35. **Chiral Auxiliaries Used in Asymmetric Diels-Alder Reactions**

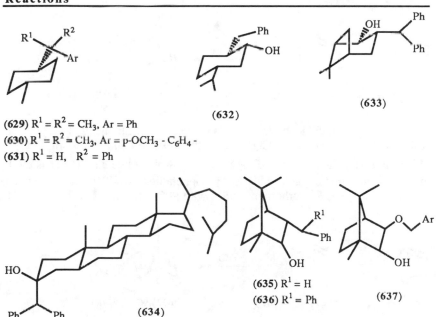

(629) $R^1 = R^2 = CH_3$, Ar = Ph
(630) $R^1 = R^2 = CH_3$, Ar = p-OCH$_3$ - C$_6$H$_4$ -
(631) $R^1 = H$, $R^2 = Ph$

(632)

(633)

(634)

(635) $R^1 = H$
(636) $R^1 = Ph$

(637)

(Table 4.35. contd.)

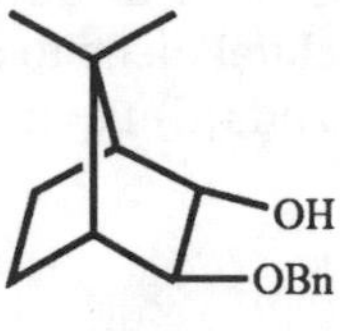

(638)

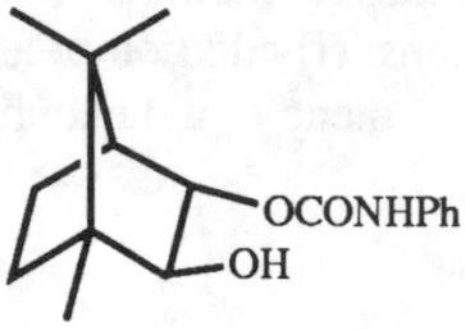

(639)

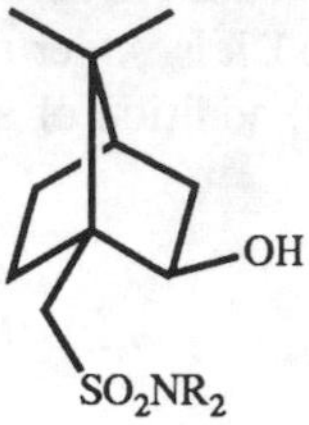

(640) R = ^{i}Pr
(641) R = cyc-hex

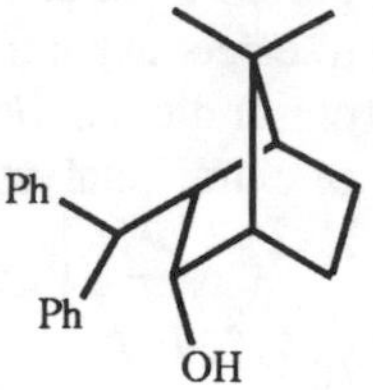

(642)

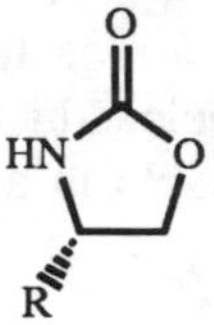

(643) R = ^{i}Pr
(644) R = Bu

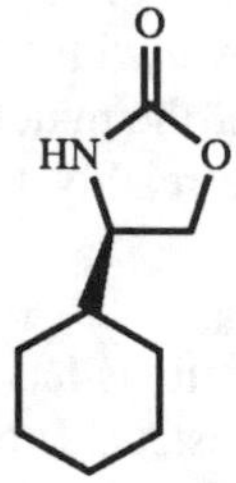

(645)

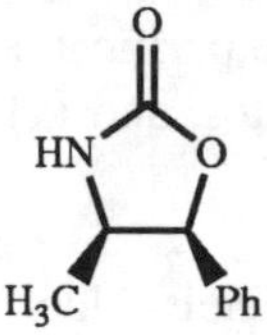

(646)

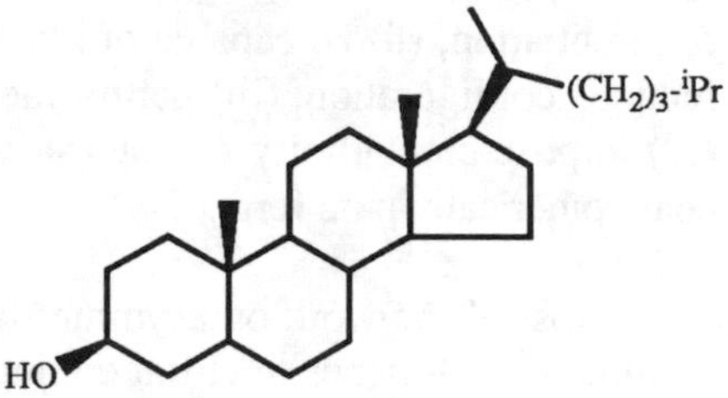

(647)

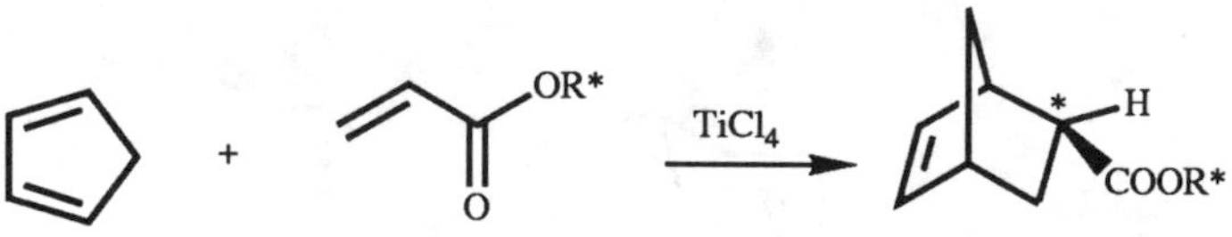

(648)

(649, R* = 629)

TiCl$_4$

(650) 89% e.e. (R)

For instance, cyclopentadiene (648) reacts with the menthyl acrylic ester (649) in the presence of TiCl$_4$ to give the *endo* product (650) in 89% e.e. [400]. Similarly the acrylate (651), having (-)-8-phenylmenthol (632) as the R* group on the ester oxygen, adds to cyclopentadiene (648) in the presence of TiCl$_4$ to give the (2R)-norbornene (652) with 90% asymmetric induction [401].

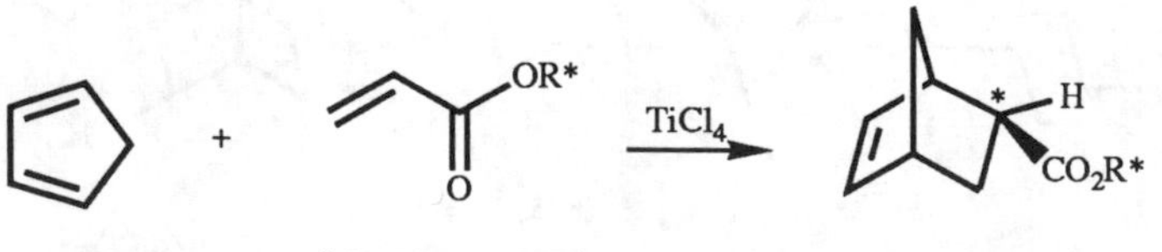

(648)

(651, R* = 632)

TiCl$_4$

(652) 90% e.e. (R)

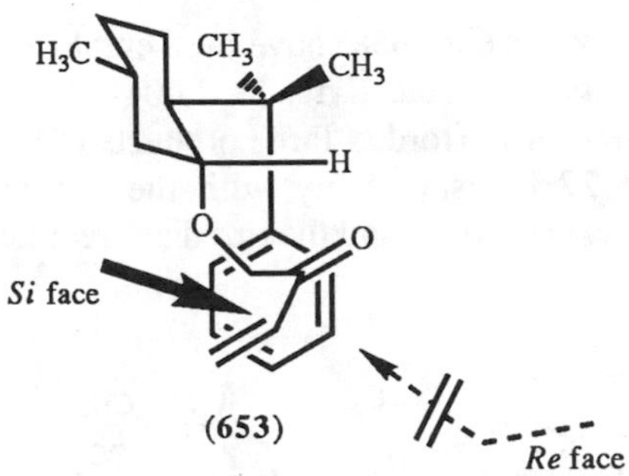

(653)

Si face

Re face

The high asymmetric induction exhibited by the chiral acrylate esters has been explained by assuming the preferred acrylate conformation (653) in the transition state. Accordingly, the phenyl ring shields one face of the dienophile, i.e., C(α)-*re* face by π,π-orbital overlap (more effectively in the presence of a Lewis acid) thereby directing the diene addition to the dienophile-*si*-face. Similarly, chiral dienophiles have been developed in which the chiral auxiliary is directly attached to the three-carbon enone unit [402]. For instance, the chiral enone (654) reacted with butadiene to give the products (655) and (656) in a ratio of > 100 :<1 [402].

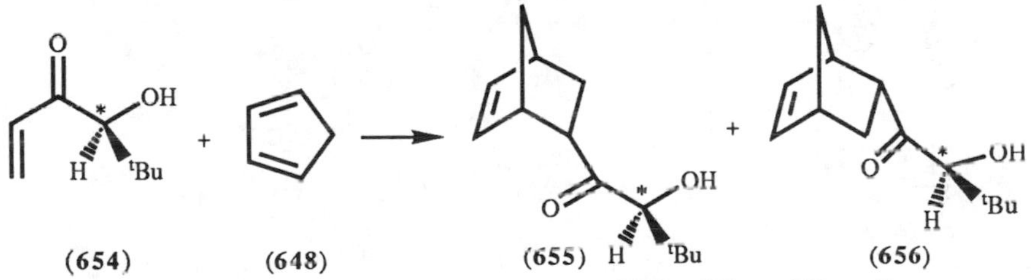

(654)   (648)   (655)   (656)

(655 : 656 = >100 : <1)

A remarkable feature of this reaction is that it proceeds in high stereoselectivity without the aid of an external catalyst. This is attributed to the formation of a five-membered chelate (657) through hydrogen bonding which effectively blocks the free rotation along the C-(C=O)-C* axis, thus highly differentiating the two diastereotopic faces of the enone system.

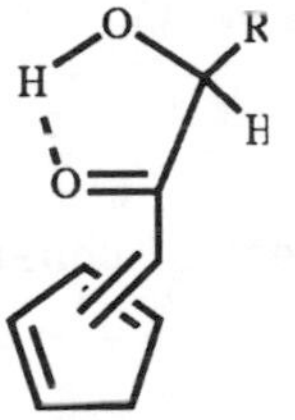

(657)

Similarly, chiral α,β-unsaturated sulfoxides e.g. (658) have been employed in the Diels-Alder reaction showing high diastereoselectivity [403,404]. Thus, the cyclopentadiene and the (E) sulfoxide (658) afforded three products (659-661) in chemical yields of 63,15 and 22% respectively with the *endo* diastereomer (659) in 62% diastereomeric excess (d.e.) and the *exo* diastereomer (661) in 100% d.e. [404].

The reaction is believed to proceed *via* the transition states (662) and (663) where the α,β- unsaturated sulfoxide takes the (S)-*trans* conformation with respect to the S-O and the C=C bonds.

Table 4.36 shows some examples of Diels-Alder reactions of chiral dienophiles with simple dienes.

Table 4.36. Diels-Alder Reactions of Chiral Dienophiles with Dienes

Chiral dienophile	R*	Diene	Optical yield (config.)	Reference
(fumarate diester, OR*)	(635)	(butadiene)	12.8 (1S,2S)	[405]
(glyoxylate, COOR*)	(635)	(2,3-dimethylbutadiene)	20.9 (R)	[406]
(acrylate, OR*)	(643)	(cyclopentadiene)	88 (S)	[400]
(crotonate, OR*)	(648)	(cyclopentadiene)	82.7 (S)	[400]
(Bu^t camphor-derived ester)		(cyclopentadiene)	99% d.e. (endo:exo = 98:2)	[407]
(Bu^t camphor-derived ester)		(cyclopentadiene)	99.3% d.e. (endo:exo = 96:4)	[407]
(hydroxy ketone, Bu^t, OH, H)		(cyclopentadiene)	>99 (S)	[408]
(pyrrolidinone, CH$_2$OCH$_3$, CH$_3$)		(cyclopentadiene)	95% d.e. (endo:exo = 97:3)	[409]

4.6.1.1.2   Addition to Chiral Dienes

Asymmetric Diels-Alder reactions of chiral dienes with prochiral dienophiles have been less extensively studied. High pressure (upto 15 Kbar) has been employed in asymmetric Diels-Alder reactions of chiral dienes with *p*-benzoquinone, but the optical yields are 2-50% d.e. [410]. However, in some cases, diastereomeric excess of >99 % has been achieved. For instance, addition of the chiral diene (**664**) to acrolein (**665**) in the presence of Lewis acid affords the product (**666**) in 64% d.e. [411]. Similarly, with juglone (**667**) the chiral diene (**665**) reacts in the presence of boron triacetate to give the product (**668**) in 98% d.e. [412].

The results are explained on the basis of a π-stacking model, in which conformation (**669**) of the chiral diene is preferred over (**670**). In the latter conformation, the large group L projects towards the diene encountering a severe non-bonded interaction. Such a non-bonded interaction is significantly less in (**669**) since it is the small group S which projects towards the diene. This model is supported by the complete inertness of the diene (**671**) towards Diels-Alder reaction, since the diene is sandwiched between two aromatic rings, and hence both the faces of the diene are shielded. However, the π-stacking model has been questioned lately on the ground that substitution of cyclohexyl ring in place of the phenyl group does not influence the asymmetric induction. It has been suggested that in the transition state, conformation (**672**) is responsible for the observed asymmetric induction [412]. In this case, the bulky phenyl group lies nearly perpendicular to the approximate plane defined by the diene moiety in the transition state. Attack on the diene face opposite the phenyl group is therefore strongly favoured. This so-called "perpendicular" model has been further supported by the X-ray structures of the products which show that the phenyl group faces towards the CO_2 group.

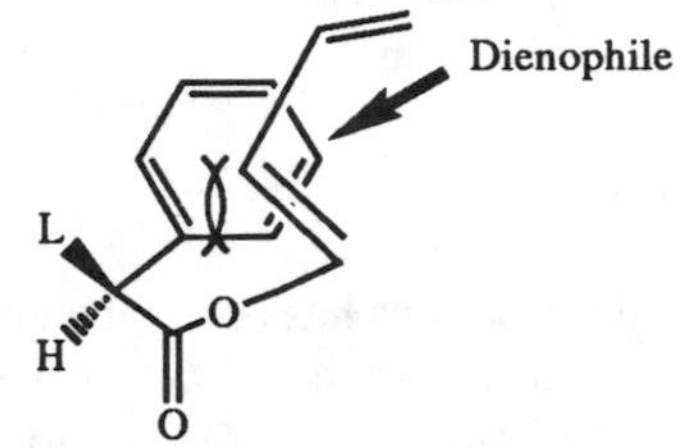

(670) 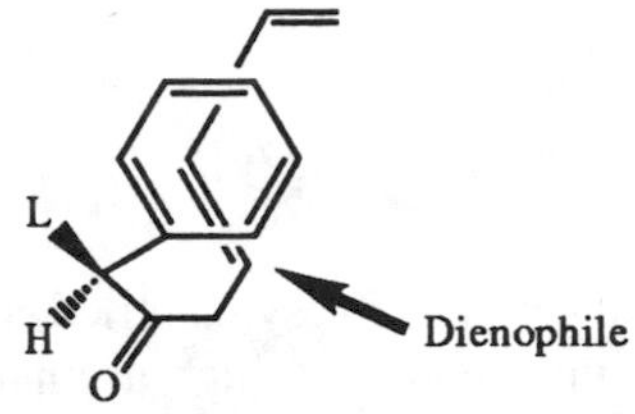   (669)

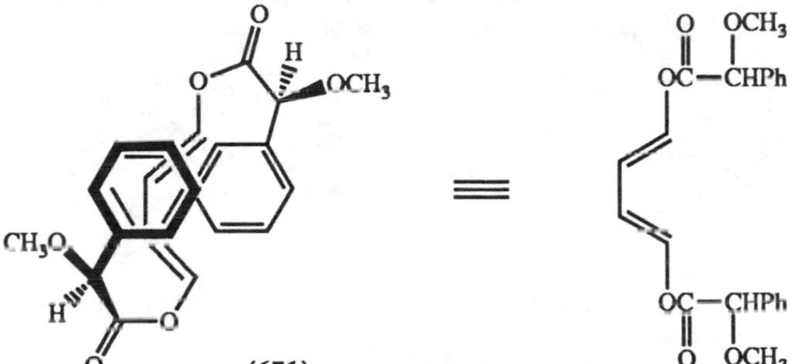

(671) 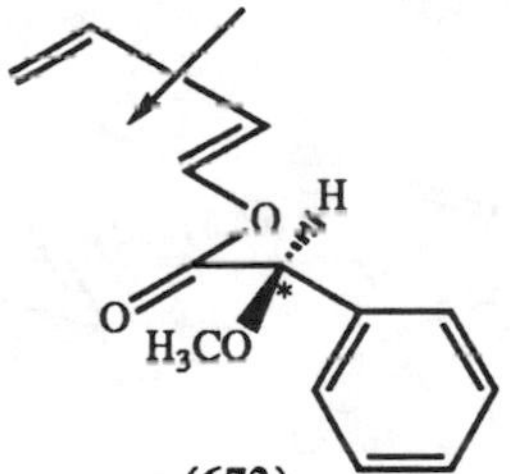   (672)

The use of chiral dienes with chiral dienophiles has also been studied and excellent results have been achieved with double asymmetric induction operating in the Diels-Alder reaction [413]. Thus the reaction of the chiral diene (**664**) with the chiral dienophile (**673**) afforded the product (**674**) in 99.2% d.e. [413]. This is an example of the asymmetric induction occurring due to a matched pair (see section 4.3.). The mismatched pair using (R)-(**664**) and (S)-(**673**) afforded low diastereoselectivities [413].

(664)   +   (673)   →   (674) >99.2% d.e.

4.6.1.1.3   Chiral Catalysts

The use of chiral catalysts in asymmetric Diels-Alder reactions has been recently demonstrated with modest to high asymmetric induction [414-418]. Various chiral titanium reagents (675-677) were employed in asymmetric Diels-Alder reactions affording optical yields of up to 92% e.e. [414]. Thus the reaction of substrate (678) with cyclopentadiene in the presence of catalyst (675) afforded the product (679) in 92% e.e. with *endo/exo* ratio of 90:10 [414].

(675)   (676)   (677)

(678)   (648)   (675)   (679) 92% d.p.

Chiral borane reagents (680-684) have also been employed in asymmetric Diels-Alder reactions [415,416]. For example, the reaction of methacrolein (685) with cyclopentadiene in the presence of (684) afforded the *exo* aldehyde (686) in 90% e.e. with an *exo* selectivity of 97.4 % [416]. With reagents (680-683) the optical yield was upto 28.5 % [415].

(680) R = Cl
(681) R = Br [S(CH₃)₂]

(682) R = Cl
(683) R = Br

(684)

(685)   **(648)**   **(686)** 90% e.e.

Recently, synthetic receptors e.g. **(687)**, **(688)** have been developed with suitably placed binding groups which modulate the rate of the cycloaddition reaction by selectively binding to the different structures on the reaction pathway [419]. Thus the intramolecular Diels-Alder reaction of the disubstituted N-furfurylfumaramide derivative **(689)** proceeds *via* the transition state **(690)** in the presence of **(687)** to yield the product **(691)** with the *S-cis/S-trans* ratio of >97:3 [419].

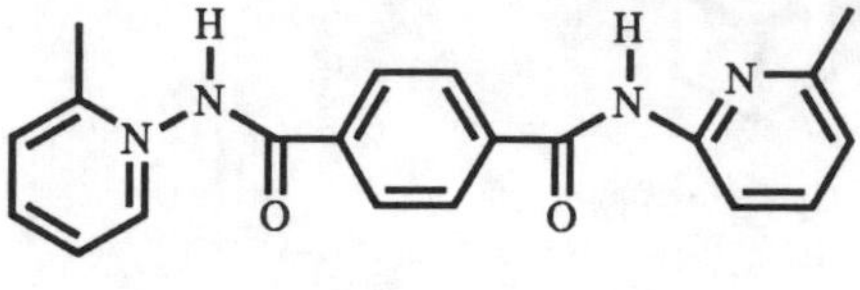

(687)

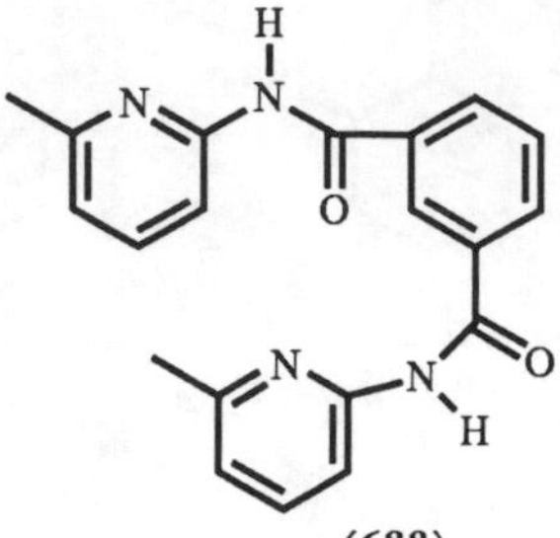

(688)

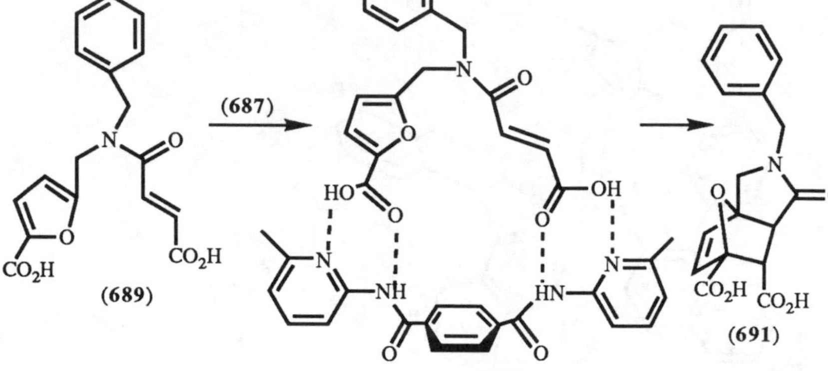

(689) → **(687)** → **(690)** → **(691)**

Table 4.37 shows some examples of asymmetric Diels-Alder reactions.

Table 4.37.   Asymmetric Diels-Alder Reactions

Substrate	Reaction condition	Product[†]	Selectivity	Reference
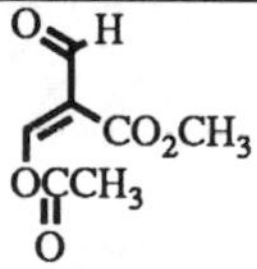	a	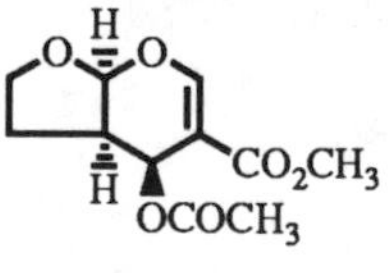	(3:1)	[419]
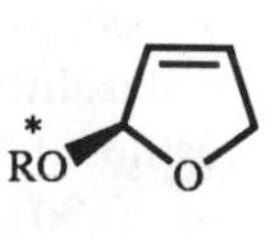	b	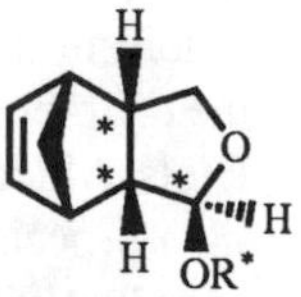	>96% d.e.	[420]
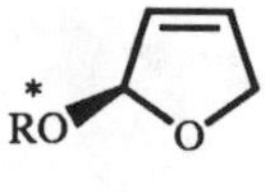	c	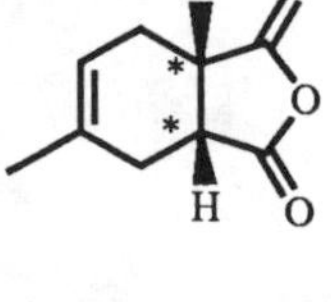	>96% d.e.	[420]

(Table 4.37. contd.)

Substrate	Reaction condition	Product[†]	Selectivity	Reference
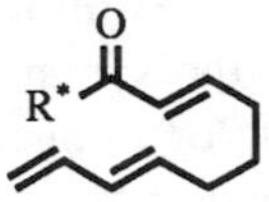	d	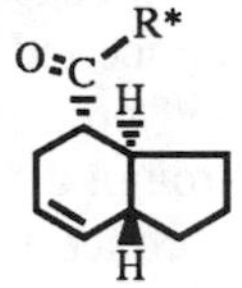	(95:5)	[421]
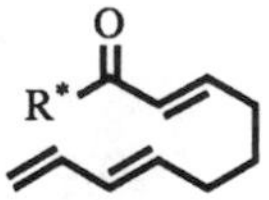	e	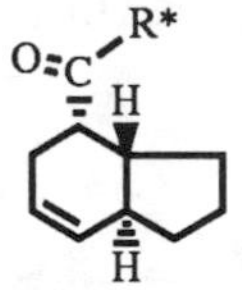	(97:3)	[421]
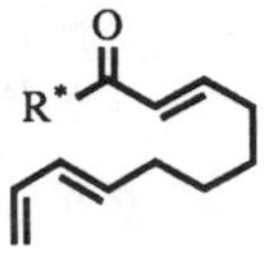	d	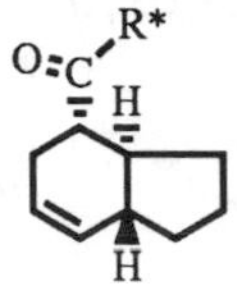	(97:3)	[421]
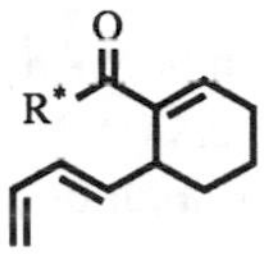	e	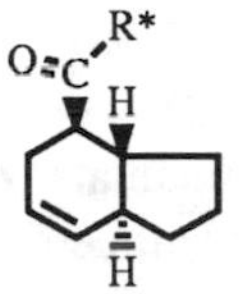	(94:6)	[421]
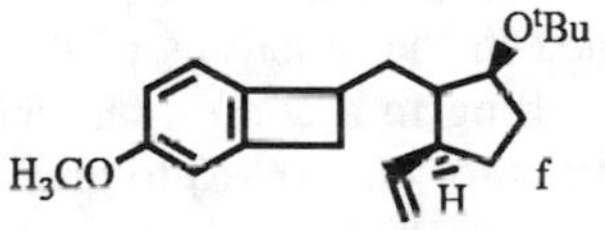	f	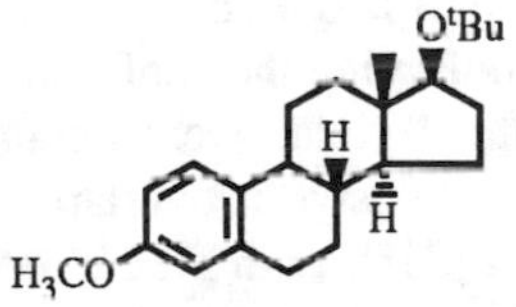	(98.5:1.5)	[422, 423]
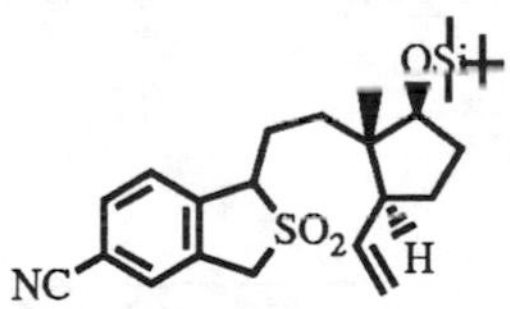	g	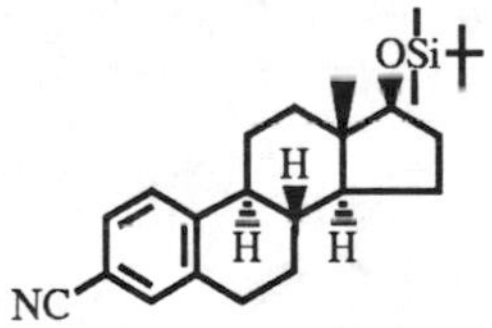	80%	[424]

† only the major isomer is shown

a:  , b: R* = 1-menthyl,  , toluene, 110°C, c: R* = 1-menthyl, 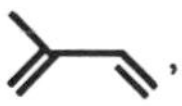 , toluene, 110°C,

d: R* = , (CH₃)₂AlCl, CH₂Cl₂, -30°C, e: R* = , (CH₃)₂AlCl, CH₂Cl₂, -30°C,

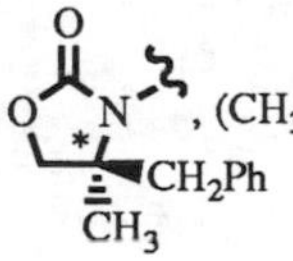

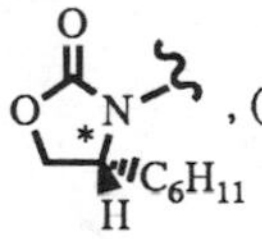

f:180°C,   g:213°C,

4.6.2   Asymmetric [2+2] Cycloadditions

[2+2] Cycloaddition reactions can be carried out either photochemically in a concerted manner or thermally in a non-concerted way, generally in the presence of Lewis acids as catalysts. The concerted process is a $[\pi s^2 + \pi a^2]$ allowed reaction, whereas the non-concerted process proceeds through the symmetry forbidden $[\pi s^2 + \pi s^2]$ pathway through a diradical intermediate or through a zwitter ion. However, both the concerted and the non-concerted [2+2] cycloaddition reactions can be subjected to asymmetric induction [425].

(692)

i) hv
ii) MeOH, H⁺

(693) 86% e.e.   (694)

(693 : 694 = 2 : 1)

One of the first asymmetric [2+2] cycloadditions was the photochemical irradiation of L-erythritol-1,4-dicinnamate (692) leading to products (693) and (694) in a ratio of 2:1 with 86% asymmetric induction [426]. The high optical yield achieved in this reaction is due to the intervening chiral chain connecting the cycloaddends thereby enforcing the conformation leading to chiral products. In similar cases, high optical yields were obtained in the reactions of *trans*-stilbene (695), which has extensive π-overlap resulting in a conformationally rigid exciplex (excited complex), with methyl bornyl formate (696) to give the μ- and δ-truxinates (697) and (698) in 90% e.e. and 10% e.e. respectively [427].

(695)   (696)

i) hv
ii) CH₃OH, H⁺

(697) 90% e.e.   (698)

Employing dibornyl fumarate, on the other hand, led to the δ-truxinate derivative in only 16% e.e. Since no μ-truxinate derivative is formed, its absence is explained by steric hindrance. The lack of cooperativity in the asymmetric induction for the doubly chiral ester was shown by the fact that there was no increase in optical yield of the truxinate derivative on replacing a methyl group with a bornyl group, indicating the existence of asymmetry in the transition state leading to the products. This implies that the reaction proceeds through a nonconcerted 1,4-diradical intermediate, since the concerted pathway has a C_2 symmetry as shown in Fig. 4.22.

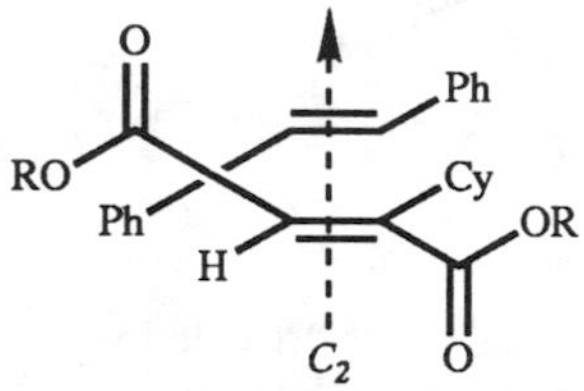

Figure 4.22. Symmetry (C_2) of concerted reaction.

Similarly, high asymmetric induction has been reported in the photochemical addition of the phenylglyoxylic acid ester [428]. Thus, the ester (699) reacts with tetramethylethylene (700) to give the oxetane (701) in 96% d.e. [428].

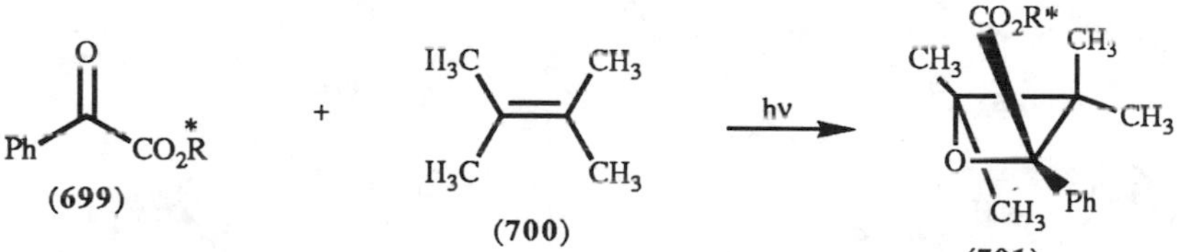

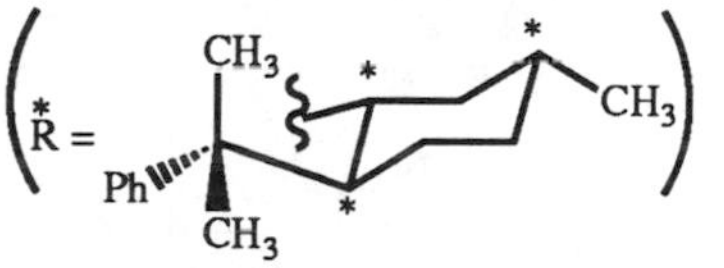

Lewis acid catalysed [2+2] cycloaddition of ketones with Schiff bases affords β-lactams in high optical yields [429-431]. Thus the ketone derivative (702) reacts with the Schiff base (703) in the presence of $TiCl_4$ to afford the β-lactams (704) and (705) with high (98%) asymmetric induction [430].

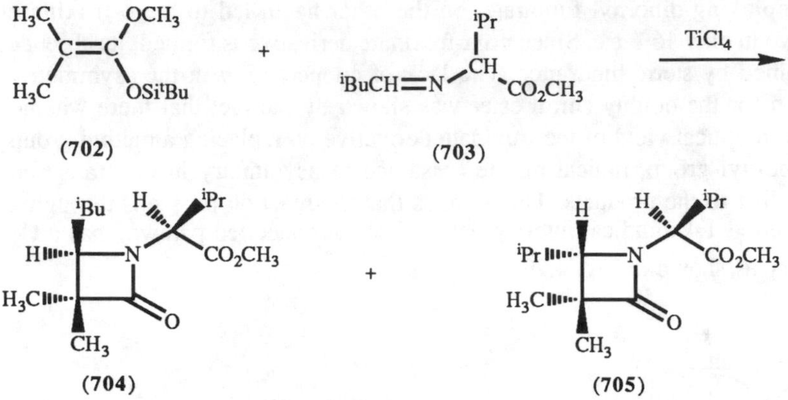

(702)   (703)

(704)   +   (705)

(704 : 705 = >99 : <1)

The reaction is believed to proceed through a sterically less crowded rigid bicyclic transition state (706) with the Lewis acid acting as an effective "template" to give the product (704). Steric hindrance, on the other hand, destabilizes the transition state (707) leading to the minor product (705).

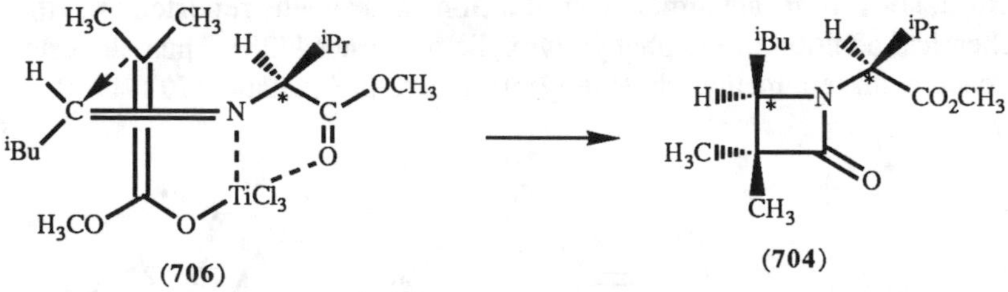

(706)   (704)

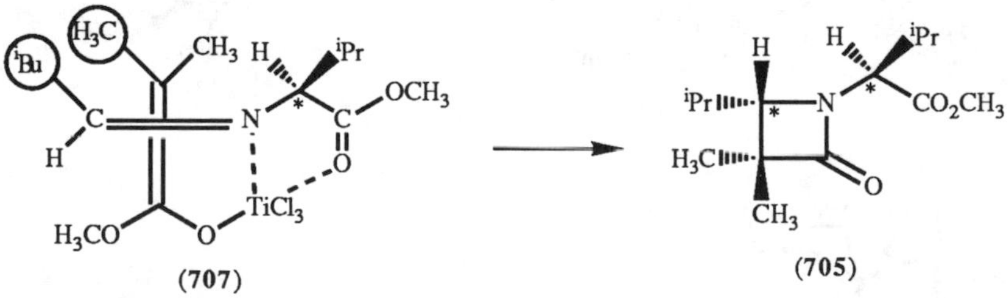

(707)   (705)

Similarly, significant asymmetric induction has been observed in the cycloaddition of dichloroketene (708) with chiral enol ethers [432]. Thus the enol ether (709) (where $R^* = 640$) reacted with (708) in the presence of Zn - Cu at 20°C to afford the products (710) and (711) in a diastereomeric ratio of 90:10. The diastereoselectivity can be reversed by using other chiral auxiliaries. Thus with $R^* = (712)$, the corresponding cyclobutanones were obtained in a ratio of 10 :90.

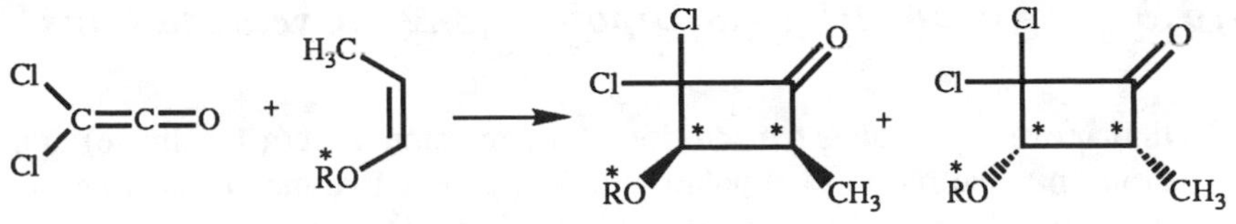

(708)   (709) (R* = 640)   (710)   (711)

(710 : 711 = 90 : 10)

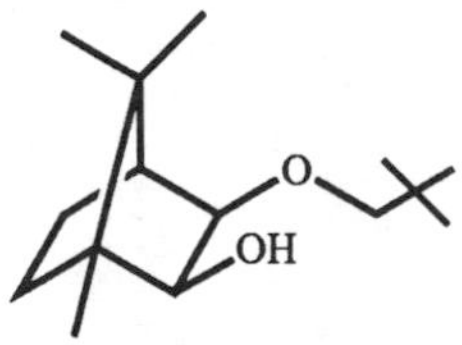

(712)

Optically active cinchona alkaloids have been used in the [2+2] cycloaddition of ketenes to chloral [433]. Thus ketene (713) adds to chloral (714) in the presence of (+)-quinidine (715) to give the (S)-(-)-(716) in 98% e.e. [433]. This is an example of a $\pi s^2 + \pi s^2$ thermally forbidden cycloaddition reaction. It has been suggested that the chiral centre adjacent to the tertiary nitrogen determines the chirality of the products, but the exact mechanism of this reaction is not known [433].

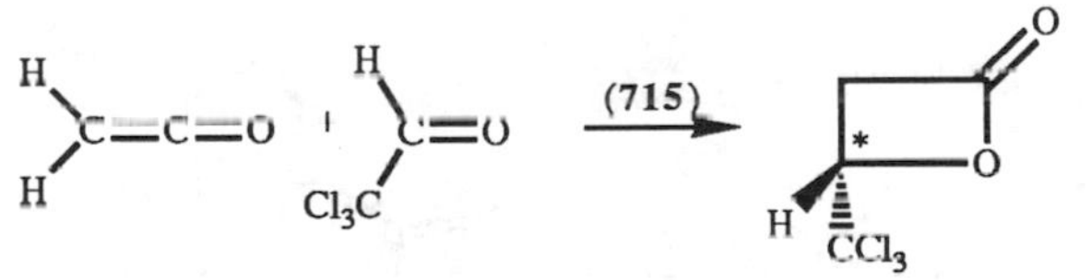

(713)   (714)   (716) 98% e.e. (S)

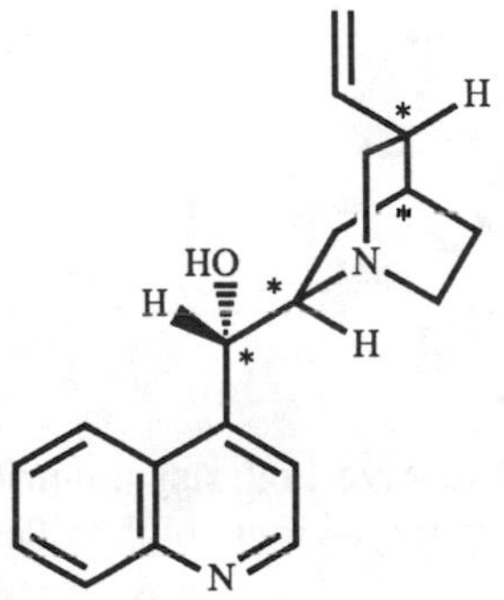

(+)-(715)

4.6.3   *Asymmetric 1,3-Dipolar [3+2] Cycloadditions*

1,3-Dipolar cycloaddition is the addition of a three-atom unit (a 1,3-dipole) to a two-atom unit (the olefin or dipolarophile) to give a five-membered ring, as shown in Fig. 4.13. It is also referred to as a [3+2]-cycloaddition [434].

Figure 4.23. 1,3-Dipolar cycloaddition.

The 1,3-dipole is a species represented by zwitterionic resonance structures. There are four π-electrons in the 1,3-dipole combining with two π-electrons of the dipolarophile in a *supra-supra* fashion and by analogy with the Diels-Alder reaction it is regarded as a $[\pi s^4 + \pi s^2]$ cycloaddition.

The 1,3-dipolar cycloaddition of nitrile oxide with unsymmetric olefins affords isoxazolines in high regioselectivity [435,436]. Thus the 1,3-dipolar cycloaddition of the nitrile oxide (**717**) with the olefin (**718**) afforded the isoxazoline (**719**), which can be converted to the 3-hydroxy ketone (**720**) with a high degree (95%) of stereoselectivity [436].

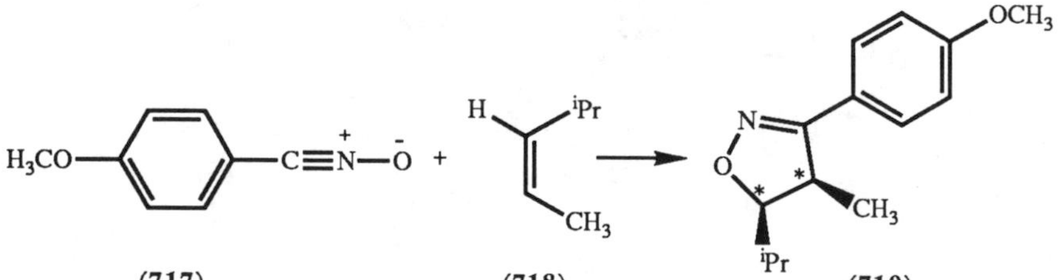

(**717**)   (**718**)   (**719**)

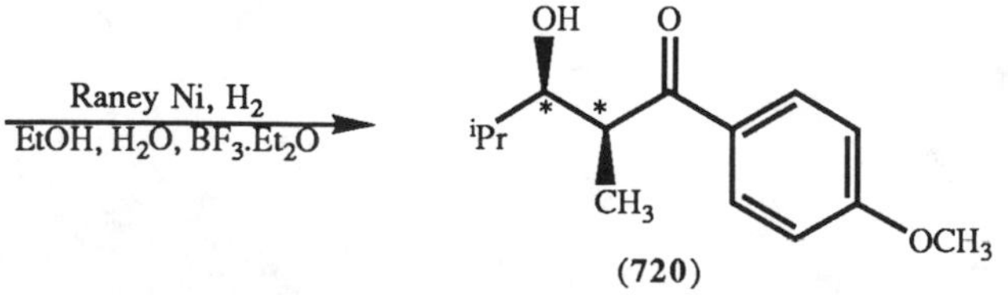

(**720**)

Similarly acyclic nitrones have been reported to give high asymmetric induction in the 1,3-dipolar cycloaddition of (*R*)-(+)-p-tolyl vinyl sulphoxides

[437]. The sulfoxide (721) reacts with the C,N-diphenyl nitrone (722) to give the product (723) in 99% e.e. [437]. The sulfoxide (721) has been employed in the synthesis of (1S)-(-)-2α-tropanol (724) which has the opposite configuration to that of natural cocaine [438].

(721)   (722)   (723) >99% e.e.

(724)

The sulfoxide (721) reacts with 1-methyl-3-oxido pyridinium (725) to give 36% of the *exo* product (726) and 29% of the *endo* product (727) [438].

(721)   (725)   (726) *exo*   (727) *endo*

Chiral nitrones have also been employed in the 1,3-dipolar cycloaddition with olefins [439]. For instance, the chiral nitrone (728) reacts with styrene in a *cis/trans* ratio of 87:13. Fig. 4.24 shows the mode of attack of the substrates. Thus *exo- si* attack affords the major product (3R,5S) (731).

1,3-Dipolar cycloaddition has also been employed in the stereoselective synthesis of various compounds in conjunction with the tandem Diels-Alder-Michael or [2,3]-sigmatropic rearrangements [440,441].

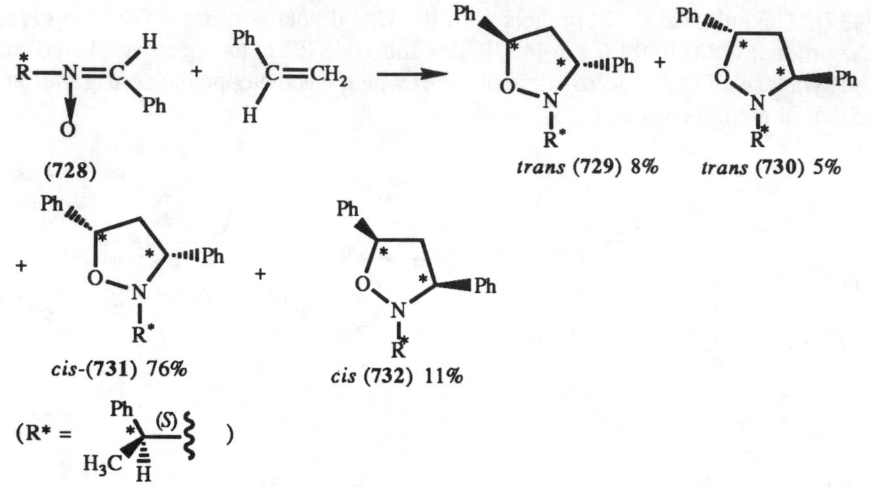

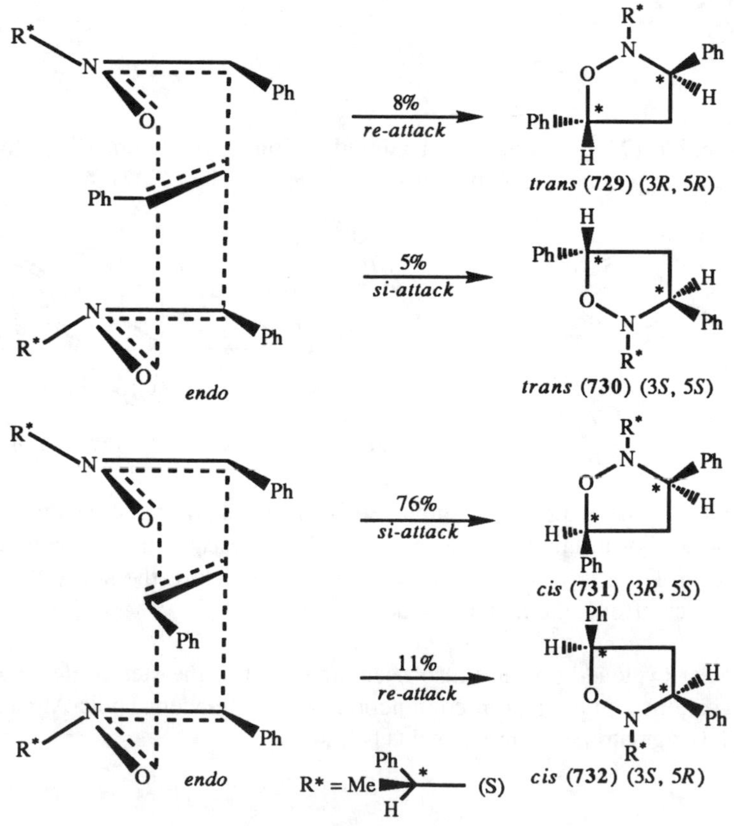

Figure 4.24. Mode of attack in 1,3-dipolar cycloaddition.

Recently, excellent diastereofacial control has been achieved in the 1,3-dipolar cycloaddition of photochemically generated azomethine ylides e.g (**734**) to chiral acryloyl sultam (**735**) to give the *exo, re* product (**736**) with (-)-(**735**) and *exo- si* product (**737**) with (+)-(**735**) [442].

4.6.4   *Sigmatropic Rearrangements*

A sigmatropic rearrangement is defined as a concerted reaction involving migration of a sigma-bond that is flanked by one or more conjugated systems to a new position within the system. The system is numbered by starting at the atoms attached to the migrating sigma-bond. The reaction is termed [1,j] sigmatropic shift when the bond migrates from position [1,1] to position [1,j] [443-445]. The migrating group can be hydrogen, carbon or a heteroatom. Fig. 4.25 (A) shows a generalized [1,5] sigmatropic (*suprafacial*) shift of a hydrogen. However an *antarafacial* shift may also take place.

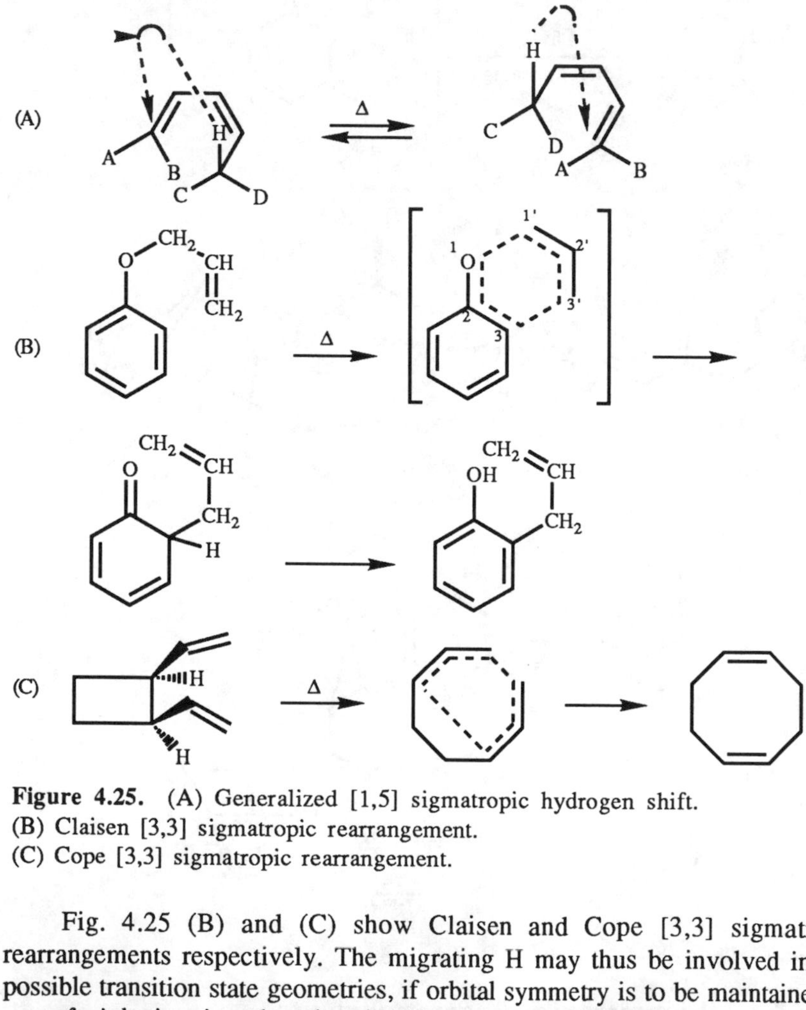

Figure 4.25. (A) Generalized [1,5] sigmatropic hydrogen shift.
(B) Claisen [3,3] sigmatropic rearrangement.
(C) Cope [3,3] sigmatropic rearrangement.

Fig. 4.25 (B) and (C) show Claisen and Cope [3,3] sigmatropic rearrangements respectively. The migrating H may thus be involved in two possible transition state geometries, if orbital symmetry is to be maintained. In *suprafacial* migration, the migrating H atom maintains contact at all times with the same face of the π system, while in an *antarafacial* migration the hydrogen must maintain contact with both faces of the π system simultaneously. If the migrating atom is capable of undergoing stereochemical inversion, then the selection rules allow two distinct *suprafacial* or *antarafacial* pathways with attendant retention or inversion at the migrating centre. Table 4.38 shows selection rules for [1, j] thermal sigmatropic migrations.

Table 4.38. HOMO Selection Rules for [1,j] Sigmatropic Migrations

j	Allowed	Forbidden
3	*Antara* - retention	*Antara* - inversion
	Supra - inversion	*Supra* - retention
5	*Antara* - inversion	*Antra* - retention
	Supra - retention	*Supra* - inversion
7	*Antara* - retention	*Antara* - inversion
	Supra - inversion	*Supra* - retention

An alternative explanation considers that thermal sigmatropic migration occurs via aromatic transition states [446]. Cyclic transition state, may be classified as either aromatic or antiaromatic depending on the number of π electrons involved and whether the substrate has Hückel (zero or even number of twists in the cyclic array) or Möbius (odd number of twists) topology. Möbius transition states will be aromatic for $[4n]\pi$ electrons with one sign reversal in the cycle, while Hückel transition states will be aromatic for $[4n + 2]\pi$ electrons with no sign reversals. Table 4.39. shows selection rules derived from the Hückel - Möbius concept, which are identical with those derived from the HOMO approach given in Table 4.38 but the former are preferred because the necessity of knowing the individual symmetries of HOMOs are avoided.

Table 4.39. Selection Rules for Thermal [1,j] Sigmatropic Migrations

No. of electrons	System type	j	Allowed Migration
4	Möbius	3	*antara* - retention
			supra - inversion
6	Hückel	5	*Antara* - inversion
			Supra - retention
4n	Möbius	(4n - 1)	*Antara* - retention
			Supra - inversion
4n + 2	Hückel	(4n + 1)	*Antara* - inversion
			Supra - retention

4.6.4.1   [3,3] Sigmatropic Rearrangements

Asymmetric induction in the Cope rearrangement has been studied in the thermal conversion of optically active dienes [447,448]. Thus *trans*-3-methyl-3-phenylhepta-1,5-diene (**738**) rearranged at 250°C to a mixture of *cis*- and *trans*-3-methyl-6-phenylhepta-1,5-dienes (**739**) and (**740**) in a ratio of 87:13 and optical yields of 96% and 94% e.e. respectively [447]. There are four possible transition state conformations resembling puckered cyclohexane rings of which the most stable chair conformations (**741**) and (**742**) are shown.

The preference for the formation of (**739**) may correspond to the free energy difference between the transition states (**741**) in which the phenyl group group is equatorial and (**742**) in which the phenyl group is axially oriented.

Allyl vinyl ethers undergo Claisen rearrangement and in cyclic systems suprafacial transposition of the π-electron orbitals occurs with complete transfer of chirality. Thus the vinyl ether (**743**) rearranges via the transition state (**744**) to (-)-cyclopentene-3- acetaldehyde (**745**) in 81% yield [449].

(+)-**743**   **744**   (-)-**745** 81%

The complete transfer of chirality demonstrates that the new carbon-carbon bond formed is *cis* to the carbon oxygen bond being broken, which is consistent with the concerted mechanism. Similarly, the Claisen rearrangement of vinyl ethers (**746**) afforded the rearranged product (**747**) in 95% e.e. [450]. The reaction is believed to proceed through a chair-like transition state (**748**) having *pseudo* axial substituents, thus minimizing non-bonded interactions.

(**746**)   (**747**) 95%

(**748**)   (**747**)

Another variant of the Claisen rearrangement is the acid - catalyzed thermal reaction of allylic alcohols e.g., (*S*)-(-)-trans-3-penten-2-ol (**749**) with triethyl orthoacetate to afford the vinyl ether (**750**) which rearranges *in situ* on heating to the (*R*)-(+)-ethyl ester (**751**) in 90% e.e. [451].

(**749**)   (**750**)   (**751**) 90% e.e. (*R*)

Similarly [3,3] sigmatropic rearrangement of allylic esters as the enolate anions or the corresponding silylketene acetals produces γ,δ-unsaturated acids in high yield [452-454]. Thus the (S)-methoxyacetate (752) afforded the product (754) *via* the metal-chelated (Z) enolate (753) in 68% e.e.

Aza-Claisen rearrangements have been studied with TiCl$_4$ as catalyst [455] constituting the so-called "charge induced pericyclic reactions" [456] and "auxiliary reagent mediated reactions" [457]. Thus N-allylketene -N,O-acetal (755) afforded the oxazolines (756) and (757) in a ratio of 87:13 and 74% d.e. [456].

Table 4.40 shows some examples of [3,3] sigmatropic rearrangements.

Table 4.40.   [3,3] Sigmatropic Rearrangements

Substrate	Reaction Conditions	Product[†]	% Selectivity	Reference
	a		95% syn	[458]

(Table 4.40. contd.)

Substrate	Reaction Conditions	Product[†]	% Selectivity	Reference
	b		98% syn, 91% e.e.	[451]
	c		(96 : 4)	[459]
	d		(98 : 2)	[459]
	e		(86 : 14)	[460]
	e		(91 : 9)	[460]
	f		(92 : 8)	[461]

(Table 4.40. contd.)

Substrate	Reaction Conditions	Product[†]	% Selectivity	Reference
	g		(70 : 30)	[461]
	h		90% e.e. (S)	[454]
	i		95.4% e.e.(S)	[462]

[†] Only the major isomer is shown

a:Δ, b:i) [(CH₃)₃Si]₂NLi, THF, ii) (CH₃)₃ SiCl, 25°C, c:(CH₃)₂SO, 100°C, d:NaH, (CH₃)₂SO, e:LiTMP, Δ, f:LDA, g:LDA, (CH₃)₃SiCl, h:HMPA-THF, ᵗBu(CH₃)₂SiCl, i:Zn/Cu, CCl₃COCl, reflux, 4h,

4.6.4.2   [2,3] Sigmatropic (Wittig) Rearrangements

The [2,3] sigmatropic reaction is defined as a reaction involving six electrons which proceeds through a cyclic transition state leading to thermal isomerization [463,464]. Fig. 4.26 shows a generalized scheme for the [2,3] sigmatropic rearrangement, where X is sulfur or selenium and Y is carbanion site, oxygen, a non-bonding electron pair or ylide. There is a great variety of reactions of type I and II which generally occur at lower temperatures than the [3,3] sigmatropic rearrangements.

A special class of [2,3] sigmatropic rearrangements which involves oxycarbanions, i.e. X = oxygen, Y = carbanion is generally known as the [2,3] - Wittig sigmatropic rearrangement. As shown in Fig. 4.27 A, this is formally a [2,3]- sigmatropic version of the classic Wittig rearrangement which is a 1,2-alkyl shift of oxycarbanions (Fig. 4.27B) [465].

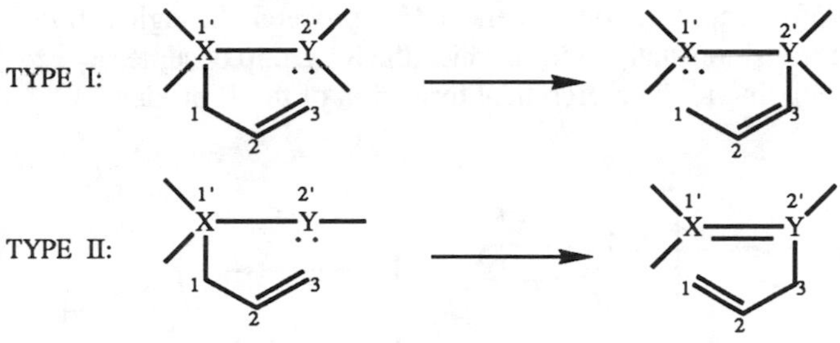

Figure 4.26. [2,3] Sigmatropic rearrangement

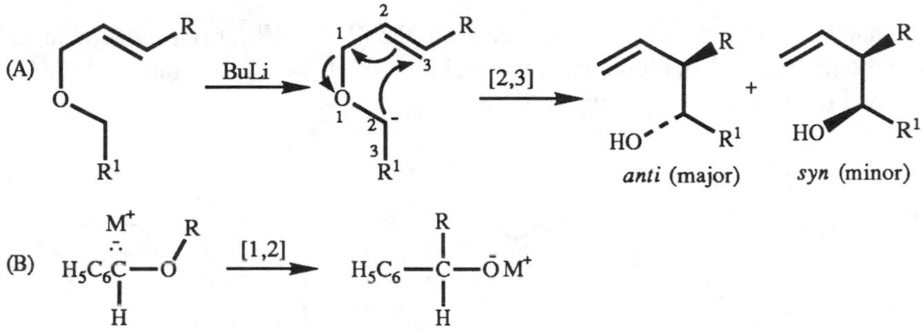

Figure 4.27. Wittig rearrangements.

While the [2,3] sigmatropic Wittig rearrangement is a concerted thermally allowed reaction, the [1,2]-Wittig rearrangement is non-concerted and proceeds *via* a diradical mechanism [465]. It is generally accompanied by the [2,3]-Wittig sigmatropic rearrangement especially at higher temperatures. The latter has also to compete with the [3,3] Claisen rearrangement.

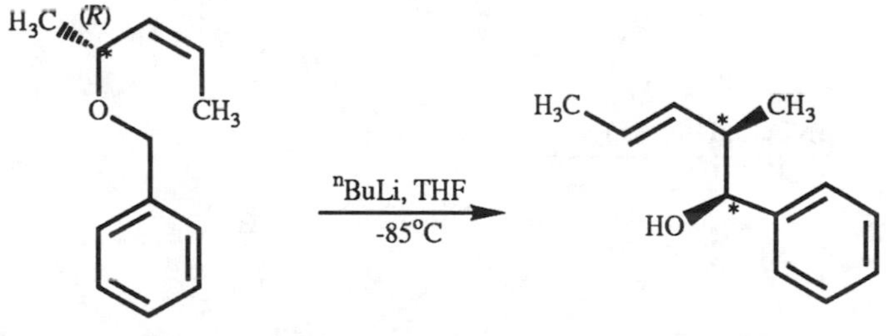

(R) (Z) - **(758)**   *(E)*-(*1S, 2R*)-**(759)**

The [2,3] Wittig rearrangement of propargylic and allylic ethers affords mainly the corresponding *E*-olefins with high stereoselectivity [466-469]. Thus the Z-benzyl ether (**758**) undergoes a [2,3] - Wittig rearrangement when a large excess of butyllithium is used to afford (*E*)-(*1S*), (*2R*)-(**759**) in high geometric purity (99% E) [469]. The reaction is believed to proceed through a five-membered envelope transition state (760) in which the R^1 group occupies an *exo* orientation, thereby leading to the preferential formation of the E-product.

(760)

However tin-substituted ethers undergo the [2,3] - Wittig rearrangement to give the (Z)-homoallyl alcohols as the major products [470]. For instance (761) afforded (Z)/(E) (762) in a ratio of 97:3.

(761)   (762) $Z/E = 97:3$ (763)

The high preference for the formation of (Z) products has been explained by assuming a transition state e.g. (763) having the butyl group in a *pseudo*-axial orientation in which the steric interaction of the butyl substituent is much less severe than the alternative equatorially oriented butyl group which would have a severe steric interaction with the vinyl methyl group.

(764)   (765) 78% e.e. (*erythro*)

The asymmetric [2,3] - Wittig rearrangement involving a chiral enolate as the migrating terminus has been reported to show a high *erythro* selectivity [471]. Thus the [2,3]- Wittig rearrangement of the chiral oxazoline (764) afforded the chiral α-hydroxy ester (765) in high optical yield [471].

(766)

The high degree of diastereoface selection and *erythro* selectivity is explanied by implicating a metal chelated (*E*)-enolate (766) as the transition state undergoing the [2,3] shift from the bottom side (*re*-face), while the oxazoline ring occupies a *pseudo*-axial position and the methoxy methyl group is at the *re*-face. However, employing KH in the absence or presence of 18-crown-6 afforded an entirely different course of diastereoface selection, with somewhat reduced diastereoselectivity [472]. This has been rationalized in terms of top-side (*si*- face) rearrangement as shown in (767) in which the enolate-K$^+$ ion is favourably located on the top-side (*si*-face) of the chelated (*R*)-enolate to avoid the dicationic repulsion.

(767)   (768)

In the case of KH/crown-induced process, the [2,3] shift might also proceed preferentially from the top-side (*re*-face) of the non- chelated Z-enolate as shown in (768) where the steric hinderance between the allyloxy moiety and the "octopus arms" of 18-crown-6 is minimized.

Similarly Zr-enolates of the (*E*)-alkenyl-oxyacetyl compounds rearrange with high diastereoface and *erythro* selection [473a]. Thus substrate (769) rearranges in the presence of dicyclopentadienylzirconium dichloride (Cp$_2$ZrCl$_2$) to afford (770) in high geometrical yield.

(769)   (770)

The high selectivity of zirconium enolates is explained by assuming a transition state such as (771) which is sterically less crowded and leads to complete diastereoface (*erythro*) selection.

(771)

Recently it has been shown that N-substituted β-methyldimethyl-ammonium ylides undergo facile [2,3] sigmatropic rearrangement to give trisubstituted olefins in upto 100% stereoselectivity [473b]. Ylides with strong electron-withdrawing substituents (-COCH$_3$, or -CO$_2$C$_2$H$_5$) in the α position afford exclusively the *E*-olefins, while those with a vinyl group carrying an ester moiety at the β-position give the corresponding Z-olefins. For instance, the ammonium salt (772) reacts with potassium *tert*-butoxide in DMF to form presumably the ammonium ylide intermediate (773) which undergoes spontaneous [2,3] sigmatropic rearrangement (2h, -50°C) to give the E-ester (774) in 84% yield with 100% stereoselectivity [473 b].

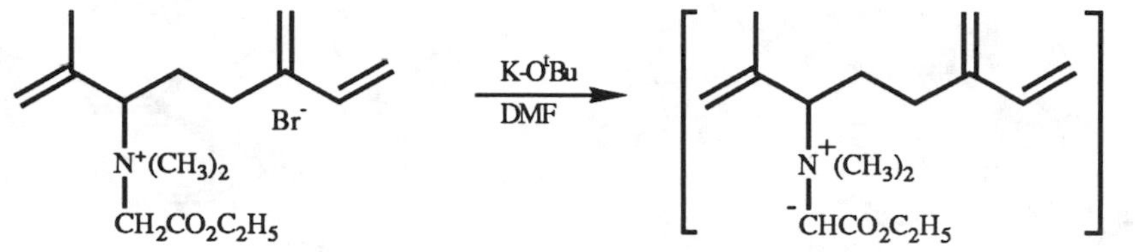

K-O⁻Bu
DMF

Br⁻

N⁺(CH₃)₂

CH₂CO₂C₂H₅

(772)

N⁺(CH₃)₂

⁻CHCO₂C₂H₅

(773)

2h,-50°C

CO₂C₂H₅

(H₃C)₂N

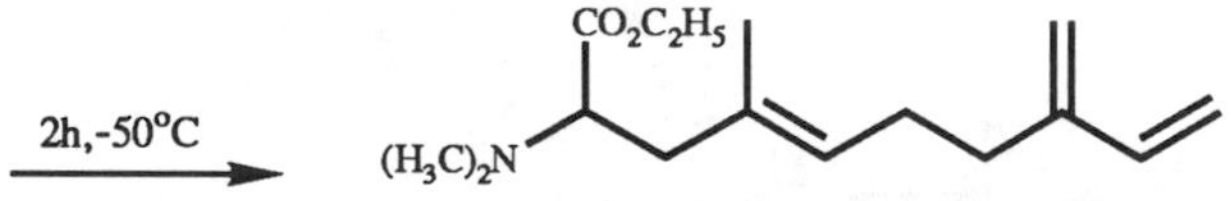

(774)   84% yield (Z:E = 0:100)

4.6.4.2.1   Allylsulfenate Rearrangements

Allyl arylsulfoxides e. g. (775) undergo rapid racemisation through concerted [2,3] sigmatropic shifts *via* the sulfinate-type intermediate (776) to form (777).

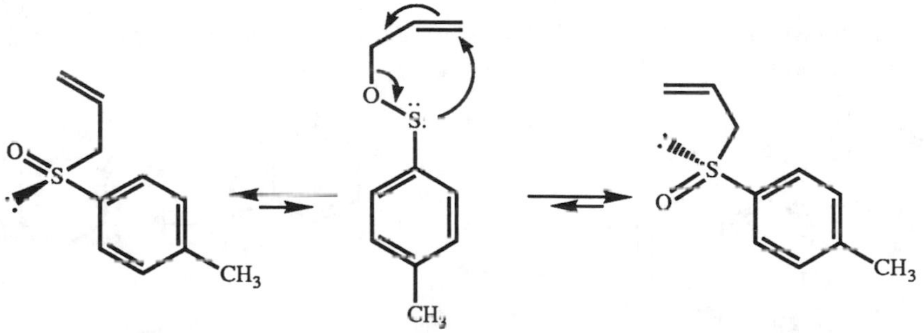

O=S

CH₃

(775) (R)

O
S

CH₃

(776)

S
O

CH₃

(777) (S)

Suitable substituents at the β-position of the allyl group allow the rearrangement to proceed stereoselectively, thereby affording allyl alcohols in high optical yields [474-477]. Thus (R)-(778) afforded the (S)-alcohol (781) *via* the *exo*-allylsulfoxide (779) and the sulfenate intermediate (780) in 60% e.e. [475].

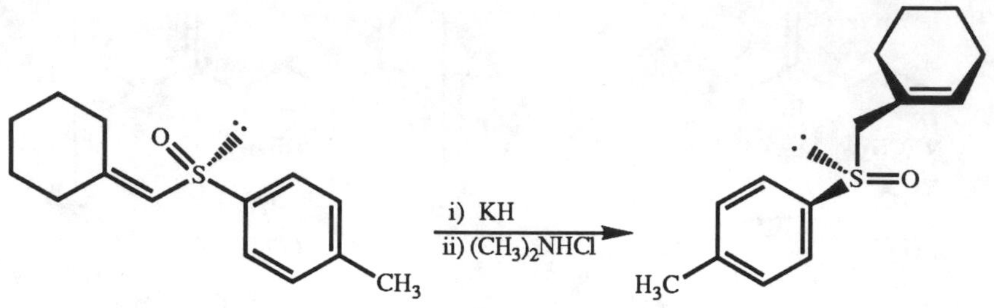

(778) (R)   (779) exo (R)

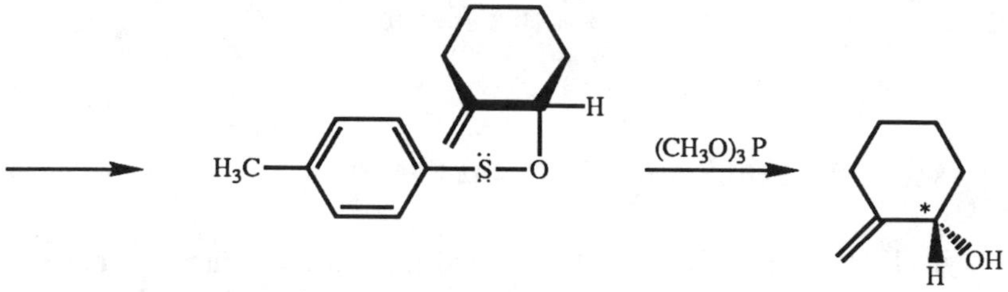

i) KH
ii) (CH₃)₂NHCl

(780) (S)   (781) 66% e.e. (S)

(CH₃O)₃ P

Similarly the sulfoxide (782) undergoes rearrangement to the alcohol (784) *via* the sulfenate (783) in 90% e.e. [477].

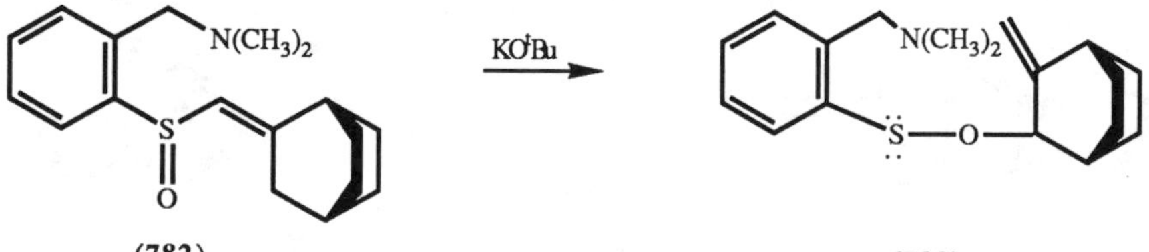

(782)   (783)

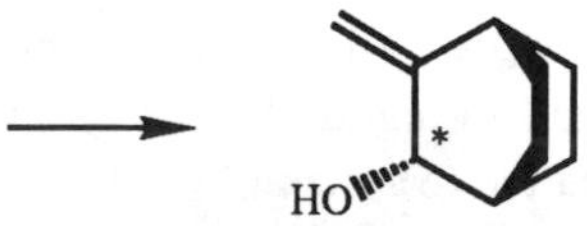

KO'Bu

(784) >90% e.e. (S)

Table 4.41 shows some examples of [2,3] sigmatropic rearrangements.

Table 4.41.   [2,3]-Sigmatropic Rearrangements

Substrate	Reaction Conditions	Product[†]	Selectivity Config.	Reference
	a		87(S)[+]	[478]
	a		24(R)[+]	[478]
	b		erythro : threo 95:5 52(R)[+]	[479]
	c		98:2 60(S)[+]	[479]
	d		erythro 97(2S)[+]	[464]
	e		92(S)[+]	[480]
	f		65 (E)[+]	[481]
	g		erythro 100%	[482]
	h		threo:erythro (91:9)	[483]

(Table 4.41. contd.)

Substrate	Reaction Conditions	Product[†]	Selectivity Config.	Reference
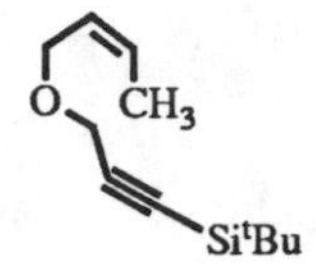	i	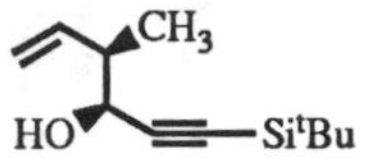	erythro 100%	[484]

† Only the major isomer is shown
+ % e.e. of the major isomer
a:DMF, 90-120°C,   b:i) LDA, -85°C, ii) H₃O⁺, iii) CH₂N₂,   c:LDA, Cp₂ZrCl₂,   d:-70°C, HMPA, THF,

e:(OCH₃)₂CH N(CH₃)₂, 130°C f:BuLi, HMPA, -80°C g:ⁿBuLi, -80°C, h:ⁿBuLi, -78°C, i:ⁿBuLi, -85°C

4.6.5   Ene Reactions

The ene reaction involves the reaction of an electrophilic alkene (the enophile) with a substrate having an allylic hydrogen (the ene) resulting in carbon-carbon bond formation, double bond migration and hydrogen transfer, as shown in Fig. 4.28 [485].

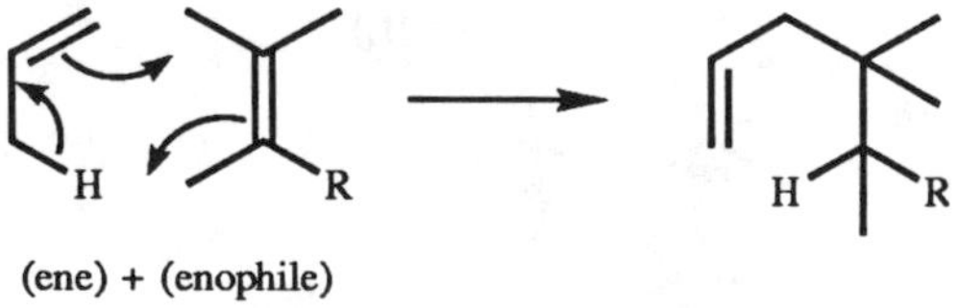

(ene) + (enophile)

Figure 4.28. Generalized ene reaction.

Like the Diels-Alder reaction, the ene reaction is a 6-electron electrocyclic reaction but the two p-electrons of the diene in the Diels-Alder reaction are replaced here by the two electrons of the allylic C-H bond. Thus the activation energy of the ene reaction is generally higher and Diels-Alder reaction competes with the ene reaction at lower temperatures. The ene reaction is also catalysed by Lewis acids and the reaction may be either concerted or non-concerted *via* free radicals. The ene reaction may be inter- or intramolecular.

4.6.5.1   Intermolecular Ene Reactions

One of the first studies on asymmetric induction in the intermolecular ene reaction involved the reaction of pent-1-ene (785) with (-)- menthyl glyoxylate (786) in the presence of a Lewis acid catalyst to afford the product (787) in upto 31% optical yield [486].

| (785) | (786) | (787) 31% e.e. (S) |

The product (787) having (S) configuration at the newly formed asymmetric centre was obtained with the Lewis acids BF_3 and $TiCl_4$ whereas with $AlCl_3$ the product with the (R) configuration was obtained. The same reaction using cyclohexane and trans-but-2-ene as substrates in the presence of 8-phenyl menthyl glyoxylate afforded the corresponding products with asymmetric induction of above 93% [487].

| (788) | (789) | (790) |

(789:790 = 97:3)

Similarly, the thermal ene addition of choral (CCl_3CHO) to (-)-β-pinene (788) in the presence of ferric chloride as catalyst afforded the products (789) and (790) in a ratio of 97:3 [488]. The same reaction catalysed by $TiCl_4$ afforded the adduct (789) exclusively. The results have been explained by assuming a transition state such as (791) in which the enophile is a chloral- Lewis acid complex, where the Lewis acid is anti to the bulky CCl_3 group.

(791)

The diethylaluminium chloride catalyzed ene reaction between the (20*E*)-3β-acetoxy-5,17(20)-pregnadiene (792) and methylpropiolate (793) yielded stereospecifically the product (794) in 94% yield [489,490]. The (Z)-substrate afforded (794) having the (*R*) configuration.

(792)   (793)   (794) 20(*S*) 100% d.e.

4.6.5.2   Intramolecular Ene Reactions

The intramolecular ene reaction allows the absolute sense of induction to be controlled by varying enoate geometry, the chirality directing group being recycled. For instance, thermal cyclization of the substrate (795) afforded the product (796) exclusively [491]. The reaction is believed to proceed through the transition state (797) which is in the *endo* orientation, thus minimizing the nonbonded interaction exhibited by the *exo* orientation of the side chain.

High asymmetric induction has also been reported in the Lewis acid catalyzed intramolecular ene reaction [492]. Thus substrate (798) afforded the products (799) and (800) in a ratio of 95:5 [492].

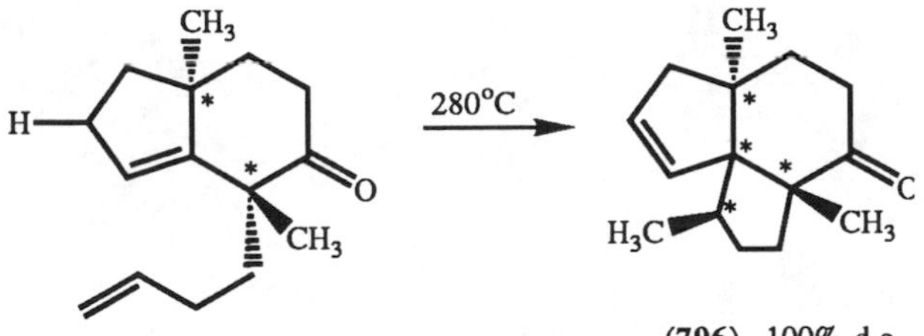

(795)

280°C

(796)   100% d.e.

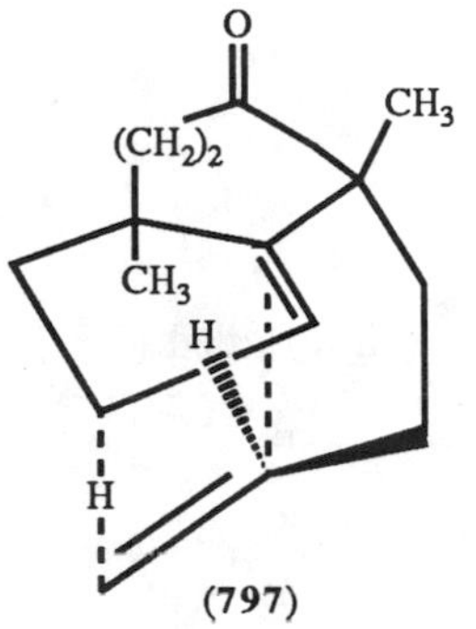

(797)

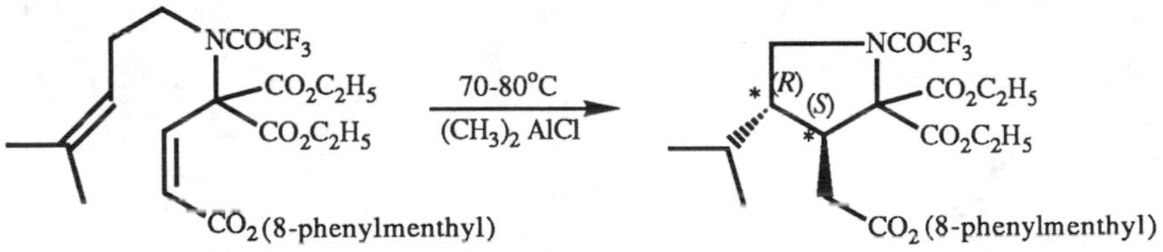

(798)

70-80°C
(CH₃)₂ AlCl

(799)

+

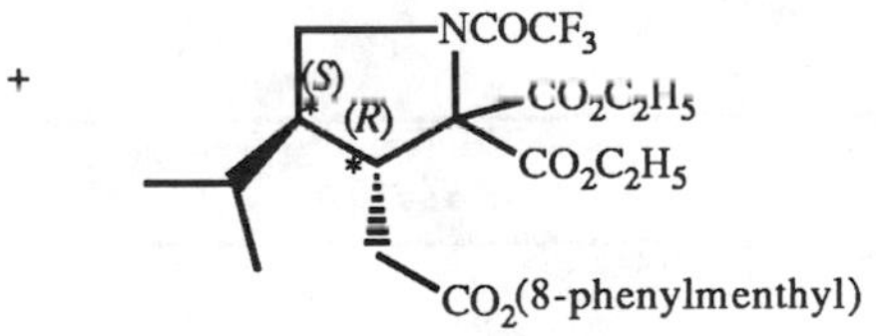

(800)

The high asymmetric induction in this reaction has been rationalized by a transition state (801) in which the enoate CO group is *anti*planar with the $C_\alpha = C_\beta$ and *syn*planar with the alkoxy-C,H-bond, while the phenyl ring shields the C_β - *si* face of the enophile [493].

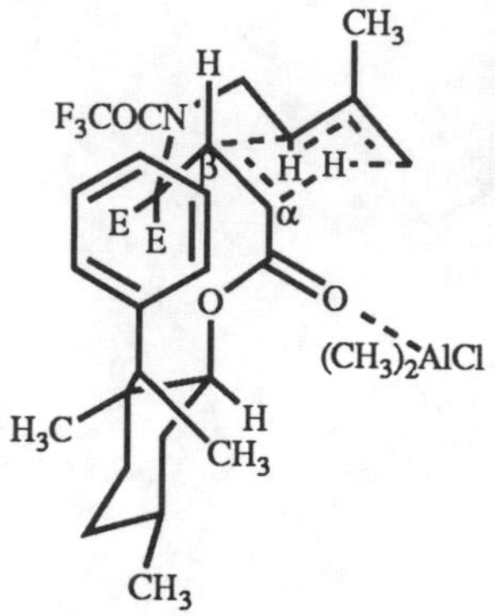

CH$_3$

F$_3$COCN

H

E

E

O

O

(CH$_3$)$_2$AlCl

H$_3$C

H

CH$_3$

CH$_3$

β

α

H H

(801)

The ene reaction of (**802**) in toluene at 250°C for 16h afforded [3.3.3]-propellane (**803**), which is an intermediate in the synthesis of the naturally occurring sesquiterpene modephene (**804**), in 76% yield [494].

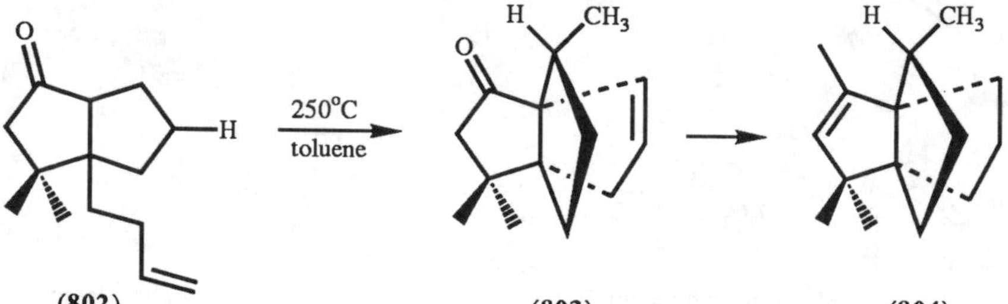

O

H

250°C
toluene

O

H CH$_3$

H CH$_3$

(802)   **(803)**   **(804)**

Table 4.42 shows some examples of the ene reaction.

Table 4.42. Ene Reactions

Substrate	Reaction Conditions	Product	% e.e. Config.	Reference
 (cyclohexene)	a	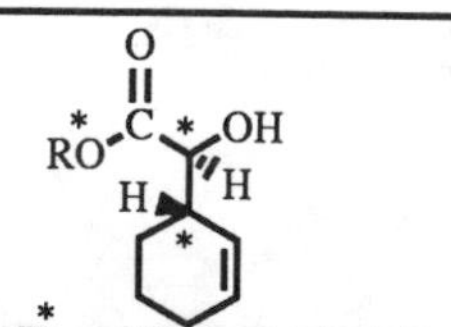 O ‖ RO–C*–*OH, H, H, (R* = 8-Phenylmenthyl)	97.2	[487]
 (pentene)	b	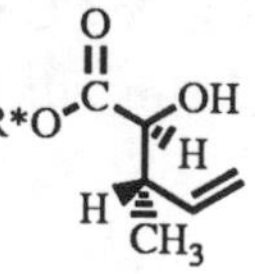 O ‖ R*O–C–OH, H, CH$_3$	93	[487]

(Table 4.42. contd.)

Substrate	Reaction Conditions	Product	% e.e. Config.	Reference
	c		95(*syn*)	[495]
	d		100(*cis*)	[496]
	d		89(*cis*)	[496]
	e		100%	[497]

a: $H-\overset{O}{\overset{\|}{C}}-\overset{O}{\overset{\|}{C}}-O-$(8-phen.menth), b: $H-\overset{O}{\overset{\|}{C}}-\overset{O}{\overset{\|}{C}}-OR*$, c: , d:280°C, e:130°C,
SnCl$_4$ SnCl$_4$ C$_2$H$_5$AlCl$_2$

4.7 References

1.   E. Fischer, *Ber. Deutsch. Chem. Ges.*, **27**, 3189 (1894).

2.   G. Bredig and P.S. Fiske, *Biochem. Z.*, **46**, 7 (1912).

3.   P.S. Fiske, Thesis, ETH Zürich (1911).

4.   L. Rosenthaler, *Biochem. Z.*, **14**, 238 (1908); **17**, 257 (1909); **19**, 186 (1909).

5.   V. Prelog and M. Wilhelm, *Helv., Chim. Acta,* **37**, 1634 (1954).

6.   J. Oku and S. Inoue, *J.Chem.Soc.Chem.Commun,* 229 (1981).

7.   Y. Kobayashi, S. Asada, I. Watanabe, H. Hayashi, Y. Motoo and S. Inoue, *Bull. Chem. Soc. Jpn.,* **59**, 893 (1986).

8.   K. Narasaka, T. Yamada, and H. Minamikawa, *Chemistry Lett.,* 2073 (1987).

9.   H. Minamikawa, S. Hayakawa, T. Yamada, N. Iwasawa and K. Narasaka, *Bull. Chem. Soc. Jpn.,* 4379 (1988).

10.   A.G. Olivero, B. Weidmann and D. Seebach, *Helv. Chim. Acta,* 2485 (1981).

11.   M. Pfau, G. Revial, A. Guingant and J. d'Angelo, *J. Amer. Chem. Soc.,* **107**, 273 (1985).

12.   D. Enders, *Chemtech,* 504 (1981).

13.   A.I. Meyers, D.R. Williams, G.W. Erickson, S. White and M. Druelinger, *J. Amer. Chem. Soc.,* **103**, 3081 (1981).

14.   K. Tamioka, K. Ando, Y. Tamemasa and K. Koga, *J. Amer. Chem. Soc.,* **106**, 2718 (1984).

15.   S. Hashimoto and K. Koga, *Tetrahedron Lett.,* 573 (1978).

16.   D. Enders and H. Eichenauer, *Chem. Ber.,* **112**, 2933 (1979).

17.   P.G. Duggan and W.S. Murphy, *J. Chem. Soc. Perkin Trans. I,* 634 (1976).

18.   P. Duhamel, J.Y. Valnot and J. J. Eddine, *Tetrahedron Lett.,* 2863 (1983).

19.   P. Duhamel, J. J. Eddine and J.Y. Valnot, *Tetrahedron Lett.,* 2355 (1984).

20.   J.C. Fiaud, *Tetrahedron Lett.,* 3495 (1975).

21.   K. Saigo, H. Koda and H. Nohita, *Bull. Chem. Soc. Jpn.,* **52**, 3119 (1979).

22. S. Julia, A. Ginebreda, J. Guixer and A. Thomas, *Tetrahedron Lett.*, 3709 (1980).

23. T. Mukaiyama, K. Soai, T. Sato, H. Shimizu, and K. Suzuki, *J. Amer. Chem. Soc.*, **101**, 1455 (1979).

24. T.D. Inch, G.J. Lewis, D.L. Sunsbury and D.J. Sellers, *Tetrahedron Lett.*, **41**, 3657 (1969).

25. T. Mukaiyama, K. Suzuki, K. Soai and T. Sato, *Chemistry Lett.*, 447 (1979).

26. W.S. Johnson, B. Frei and A.S. Gopalan, *J.Org.Chem.*, **46**, 1512 (1981).

27. M. Asami and T. Mukaiyama, *Chemistry Lett.*, 17 (1980).

28. B. Weidmann, L. Wilder, A.G. Olivero, C.D. Maycock and D. Seebach, *Helv. Chim. Acta,* **64**, 357 (1981).

29. E.J. Corey and N.W. Boaz, *Tetrahedron Lett.*, **26**(49), 6015, 6019 (1985).

30. G. Hallnemo and C. Ullenius, *Tetrahedron Lett.*, **27**(3), 395 (1986).

31. E.J. Corey and N.W. Boaz, *Tetrahedron Lett.*, **25**(29), 3063 (1984)

32. E.J. Corey, R. Naef and F.Z. Hannon, *J. Amer. Chem. Soc.*, **108**, 7114 (1986).

33. E.L. Eliel, J.K. Koksimies and B.Lohri, *J. Amer. Chem. Soc.*, **100**, 1614 (1978).

34. J.E. Lynch and E.L. Eliel, *J. Amer. Chem. Soc.*, **106**, 2943 (1984).

35. E.L. Eliel and S. Morris-Natschke, *J. Amer. Chem. Soc.*, **106**, 2937 (1984).

36. K.Ko, W.J. Frazee and E.L. Eliel, *Tetrahedron,* **40**, 1333 (1984).

37. J. Yamada, H. Sato and Y. Yamamoto, *Tetrahedron Lett.*, **30**(41), 5611 (1989).

38. N.Kokuni, Y. Matsuda and T. Kaneko, Jpn. Kokai, Tokkyo Koho J.P., 313, 445 [89,313,445 (Cl. Co.7 C33/20), 18 Dec. 1989, Appl. 88/143, 781, 13 June (1988).

39.   K. Soai, S. Yokohama, K. Ebihara and T. Hayasaka, *J. Chem. Soc., Chem. Commun.* 1690 (1987).

40.   K. Soai, A. Ookawa, T. Kaba and K. Ogawa, *J. Amer. Chem. Soc.,* **109**, 7111 (1987).

41.   K. Soai, A. Ookawa, K. Ogawa and T. Kaba, *J. Chem. Soc., Chem. Commun.,* 467 (1987).

42.   K. Soai and S. Niwa, *Chemistry Lett.,* 481 (1989).

43.   K. Soai, S. Yokohama, T. Hayasaka and K. Ebihara, *Chemistry Lett.,* 843 (1988).

44.   D. Seebach, G. Crass, E. Wilka, D. Hilvert and E. Brunner, *Helv. Chim. Acta,* **62**, 2695 (1979).

45.   J.P. Mazaleyrat and D.J. Cram, *J. Amer. Chem. Soc.,* **103**, 4585 (1981).

46.   J.D. Morrison, H.S. Mosher, : "Asymmetric Organic Reactions", Prentice-Hall, Englewood Cliffs, (1971).

47.   Y. Izumi, A. Tai., "Stereodifferentiating Reactions", Academic Press, New York, (1977).

48.   J. Apsimon, "The Total Synthesis of Natural Products", Vol. 1-5, Wiley, New York, (1973-1983).

49.   D.J. Cram and F.A. Abd Elhafez, *J. Amer. Chem. Soc.,* **74**, 5828 (1952).

50.   D.A. Evans, J.V. Nelson and T.R. Taber, *Top. Stereochem.,* **31**, 1 (1982).

51.   R.W. Hoffmann, *Angew. Chem.,* **21**, 555 (1982).

52.   W.C. Still and J.H. McDonald, *Tetrahedron Lett.,* 1031 (1980).

53.   A. Gaset, M.T. Maurette and A. Lattes, *C. R. Acad. Sci. Sec.C.,* **270**, 72; **270**, 2002 (1970).

54.   A. Gaset, P. Andoye and A. Lattes, *J. Appl. Chem. Biotechnol.,* **25**, 13 (1975).

55.   P. Duhamel, L. Duhamel and J. Gralak, *Tetrahedron Lett.*, 2285 (1977).

56.   E.L. Eliel, in "Asymmetric Synthesis", Vol.2., (J.D. Morrison, ed.), p.125, Acadmic Press, New York, (1983).

57.   W.C. Still and J.A. Schneider, *Tetrahedron Lett.*, **21**, 1035 (1980).

58.   M.T. Reetz, R. Steinbach, J. Westermann, R. Urz, B. Wenderoth and R. Peter, *Angew. Chem.*, **21**, 135 (1982).

59.   M.T. Reetz, R. Steinbach, J. Westermann and R. Peter, *Angew. Chem.*, **19**, 1011 (1980).

60.   B. Weidmann and D. Seebach, *Helv. Chim. Acta*, **63**, 2451 (1980).

61.   M.T. Reetz, J. Westermann and R. Steinbach, *Angew. Chem.*, **92**, 933 (1980).

62.   P.A. Bartlett, *Tetrahedron*, **36**, 3 (1980).

63.   T.J. Leitereg and D.J. Cram, *J. Amer. Chem. Soc.*, **90**, 4011, 4019 (1968).

64.   M.T. Reetz, *Top. Curr. Chem.*, **106**, 1 (1982).

65.   B. Weidmann, and D. Seebach, *Angew. Chem.*, **22**, 31 (1983).

66.   M.T. Reetz and A. Jung, *J. Amer. Chem. Soc.*, **105**, 4833 (1983).

67.   M.T. Reetz, R. Steinbach, J. Westermann, R. Urz, B. Wenderoth and R. Peter, *Angew. Chem. Suppl.*, 257 (1982).

68.   A. McKenzie, *J. Chem. Soc.*, **85**, 1249 (1904).

69.   V. Prelog, *Helv. Chim. Acta*, **36**, 308 (1953).

70.   J.K. Whitesell, D. Deyo and A. Bhattacharya, *J. Chem. Soc., Chem. Commun.*, 802 (1983).

71.   J.K. Whitesell, A. Bhattacharya and K. Henke, *J. Chem. Soc., Chem. Commun.*, 988 (1982).

72.   A.I. Meyers and J. Slade, *J. Org. Chem.*, **45**, 2785 (1980).

73.   M.T. Reetz, K. Kesseler, S. Schmidtberger, B. Wenderoth and R. Steinbach, *Angew. Chem.* **22**, 989 (1983).

74.   G. Sterk and J.M. Stryker, *Tetrahedron Lett.*, **24**(45), 4887 (1983).

75.   E.C.Ashby and J.T. Laemmle, *Chem. Rev.*, **75**, 521 (1975).

76.   A. Dormond, A. Aaliti and C. Moise, *J. Org. Chem.*, **53**, 1034 (1988).

77.   G.A. Molander, E.R. Burkhardt and P. Weinig, *J. Org. Chem.*, **55**, 4990 (1990).

78.   Mioskowski and G. Solladie, *J. Chem. Soc., Chem. Commun.*, 162 (1977).

79.   A.I. Meyers, M.A. Hanagan, L.M. Trefonas and R.J. Baker, *Tetrahedron*, **39**(12), 1991 (1983).

80.   G. Solladie, F.M. Moghadam, C.Luttmann and C. Mioskowski, *Helv. Chim. Acta*, **65**, 1602 (1982).

81.   A. Gaset, P. Audoye and A. Lattes, *J. Appl. Chem. Biotechnol.* **25**, 13 (1975).

82.   M. Tramontini, *Synthesis*, 605 (1982) (and references therein).

83.   D.L. Temple and J. Sam, *J. Heterocyclic Chem.*, **7**, 847 (1970).

84.   J.W. Cornforth, R.H. Cornforth and K.K. Mathew, *J. Chem. Soc.*,112 (1959).

85.   D.J. Cram and K.R. Kopecky, *J. Amer. Chem. Soc.*, **81**, 2748 (1959).

86.   M. Cherest, H. Felkin and N. Prudent, *Tetrahedron Lett.*, 2199 (1968).

87.   N.T. Anh, *Top. Curr. Chem.*, **88**, 145 (1980).

88.   M.T. Reetz, K. Kesseler, S. Schmidtberger, B. Wenderoth and R. Steinbach, *Angew. Chem. Suppl.*, 1511 (1983).

89.   M.L. Wolfrom and S. Hanessian, *J. Org. Chem.*, **27**, 1800 (1962).

90.   T.D. Inch, *Carbohydrate Res.*, **5**, 45 (1967).

91.   G. Parrinello and J.K. Stille, *J. Amer. Chem. Soc.*, **109**, 7122 (1987).

92.   T. Hayashi, M. Tanaka and I. Ogata, *J. Mol. Cat.*, **26**, 17 (1984).

93.   C.U. Pittman Jr., Y. Kawabata and L.I. Flowers, *J. Chem. Soc., Chem. Commun.*, 473 (1982).

94.   G. Gonsiglio, P. Pino, L.I. Flowers and C.U. Pittman, Jr., *J. Chem. Soc., Chem. Commun.*, 612 (1983).

95.   P. Pino, G. Consiglio, C. Botteghi and C. Salomon, *Adv. Chem. Ser.*, No. **132**, 295 (1974).

96.   C. Botteghi, G. Consiglio and P. Pino, *Chimia*, **26**, 141 (1972).

97.   M. Tanaka, Y. Ikeda and I. Ogata, *Chemistry Lett.*, 1115 (1975).

98.   C.F. Hobbs and W.S. Knowles, *J. Org. Chem.*, **46**, 4422 (1981).

99.   A. Stefani and D. Tatone, *Helv. Chim. Acta*, **60**, 518 (1977).

100.   Y. Becker, A. Eisenstadt and J.K. Stille, *J. Org. Chem.*, **45**, 2145 (1980).

101.   C. Botteghi and C. Salone, *Tetrahedron Lett.*, 4285 (1974).

102.   C. Botteghi, M. Branca, and A. Saba, *J. Organomet. Chem.*, **184**, C17 (1980).

103.   D. Evans, J.A. Osborn, and G. Wilkinson, *J. Chem. Soc.*, A, 3133 (1968).

104.   G. Strukul, M. Bonivento, M. Graziani, E. Cernia and N. Palladino, *Inorg. Chim. Acta*, **12**, 15 (1975).

105.   G. Parrinello, R. Deschenaux, and J.K. Stille, *J. Org. Chem.*, **51**, 4189 (1986).

106.   J.K. Stille, in "Catalysis of Organic Reactions", (R.L. Augustine ed.), Marcel Dekker, New York, (1985).

107.   J.K. Stille, *Pure Appl. Chem.,* **54**, 99 (1982).

108.   J.K. Stille, S.J. Fritschel, N. Takaishi, T. Masuda, H. Imai, and C.A. Bertelo, *Ann. N.Y. Adad. Sci.,* **333**, 35 (1980).

109.   S.J. Fritschel, J.J.H. Ackerman, T. Keyser and J.K. Stille, *J. Org. Chem.,* **44**, 3152 (1979).

110.   E. Bayer and V. Schurig, *Chem.Tech*, 212 (1976).

111.   G. Consiglio, C. Botteghi, C. Salomon and P. Pino, *Angew. Chem.*, **12**, 669 (1973).

112.   C. Botteghi, G. Consiglio and P. Pino, *Chimia*, **27**, 477 (1973).

113.   G. Consiglio, *Helv.Chim.Acta*, **59**, 124 (1976).

114.   G. Consiglio, and P. Pino, *Chimia*, **30**, 193 (1976).

115.   G. Consiglio, *J. Organomet. Chem.*, **132**, C26 (1977).

116.   T. Hayashi, M. Tanaka and I. Ogata, *Tetrahedron Lett.*, 3925 (1978).

117.   I. Ogata, *Shokubai*, **24**, 205 (1982).

118.   (a) C.H. Heathcock, : in "Asymmetric Synthesis", (J.D. Morrison, ed.,) Vol.3, p.111. Academic Press, New York, (1984); (b) C.H. Heathcock, : in "Asymmetric Synthesis", (J.D. Morrison, ed.,) Vol.3, p.1. Academic Press, New York, (1984); (c) C.H. Heathcock: in "Comprehensive Carbanion Chemistry," (E. Buncel, T. Durst, eds.), Elsevier, Amsterdam, Part B, Chap.4, p.177, (1984).

119.   C.H. Heathcock, M.C. Pirrung, J. Lampe, C.T. Buse and S.D. Young, *J. Org. Chem.,,* **46**, 2290 (1981).

120.   P. Fellmann and J.E. Dubois, *Tetrahedron,* **34**, 1349 (1978).

121.   C.H. Heathcock, C.T. Buse, W.A. Kleschick, M.C. Pirrung, J.E. Sohn and J.Lampe, *J. Org. Chem.,* **45**, 1066 (1980).

122.   (a) For a recent review on enolates, see: "Comparenshive Organic Synthesis", (B.M. Trost and I. Fleming, eds.), Vol.2, (C.H. Heathcock, ed.),

p.99 and 181, (1983); (b) P.L. Stotter, M.D. Friedman and D.E. Minter, *J. Org. Chem.*, **50**, 29 (1985); (c). C.T. Buse and C.H. Heathcock, *J. Amer. Chem. Soc.*, **101**, 7723 (1979); (d) R.E. Ireland, R.H. Mueller and A.K. Willard, *J. Amer. Chem. Soc.*, **98**, 2868 (1976); (e). A.I. Meyers and P. Reider, *J. Amer. Chem. Soc.*, **101**, 2501 (1979); (f) C.H. Heathcock, M.C. Pirrung, S.H. Montgomery and J. Lampe, *Tetrahedron*, **37**, 4087 (1981).

123.   (a) I. Fleming and J.K. Kilburn, *J. Chem. Soc., Chem. Commun.*, 305 (1986); (b) D.A. Widdowson, G.H. Wiebecke and D.J. Williams, *Tetrahedron Lett.*, **23**, 4285 (1982); (c) H.M. Shieh and G.D. Prestwich, *J. Org. Chem.*, **46**, 4319 (1981); (d) F. Babudri, L.DiNunno and S. Florio, *Tetrahedron Lett.*, **24**(36), 3883 (1983); (e) T. Welch and S. Eswarakrishnan, *J. Org. Chem.* **56**, 5403 (1985); (f) H.H. Otto, R. Mayrhofer and H.J. Bergmann, *Liebigs Ann.Chem.*, 1152 (1983); (g) F. DiNinno, T.R. Beattie and B.G. Christensen, *J. Org. Chem.*, **42**, 2960 (1977); (h) P. Beslin, P.Metzer, Y. Vallee and J.Vialle, *Tetrahedron Lett.*, **24**, 3617 (1983); (i) P. Beslin and Y. Vallee, *Tetrahedron*, **41**, 2691 (1985).

124.   J. Mulzer, M. Zippel, G. Bruntrup, J. Segner and J. Finke, *Liebigs Ann. Chem.*, 1108 (1980).

125.   (a) Y. Tamaru, Y. Amino, Y. Furukawa, M. Kagotani and Z. Yoshida, *J. Amer. Chem. Soc.*, **104**, 4018 (1982); (b) Y. Tamaru, T. Harada, S. Nishi, M. Mizutani, T. Hioki and Z.I. Yoshida, *J. Amer. Chem. Soc.*, **102**, 7806 (1980).

126.   Y. Tamaru, T. Hioki and E.I. Yoshida, *Tetrahedron Lett.*, **25**, 5793 (1984).

127.   (a) E.J. Corey, C.-M. Yu and S.S. Kim, *J.Amer.Chem.Soc.*, **111**, 5495 (1989). (b) H.C. Brown, R.K. Dhar, R.K. Bakshi, P.K. Pandiaraja and B. Singaram, *J. Amer. Chem. Soc.*, **111**, 3441 (1989).

128.   E.J. Corey and S.S.Kim, *J. Amer. Chem. Soc.*, **112**, 4976 (1990).

129.   S. Masamune, W. Choy, F.A.J. Kerdesky and B. Imperiali, *J. Amer. Chem. Soc.*, **103**, 1566 (1981).

130.   D.A. Evans and J. Bartroli and T.L. Shih, *J. Amer. Chem. Soc.*, **103**, 2127 (1981).

131.   M.T. Reetz, E. Rivadeneira and C. Niemeyer, *Tetrahedron Lett.*, **31**, 3863 (1990).

132.   I. Peterson, J.M. Goodman, M.A. Lister, R.C. Schumann, C.K. McClure and R.D. Norcross, *Tetrahedron* **46**, 4663 (1990).

133.   (a) Y. Tamura, T. Hioki, S. Kawamura, H. Satomi and Z. Yoshida, *J. Amer. Chem. Soc.,* **106**, 3876 (1984);   (b) J.E. Lynch, R.P. Volante, R. W. Wattley and I. Shinkai, *Tetrahedron Lett.,* **28**(13); 1385 (1987).

134.   (a) C. Siegel and E.R. Thornton, *J. Amer. Chem. Soc.,* **111**, 5722 (1989); (b) R.O. Duthaler, A. Hafner and M. Riediker, *Pure Appl. Chem.,* **62**, 631 (1990);   (c) M.T. Reetz,"Organotitanium Reagents in Organic Synthesis", Springer-Verlag, Berlin, (1986).

135.   P.J. Murphy, G. Proeter and A.T. Russell, *Tetrahedron Lett.,* **28**(8), 2037 (1987).

136.   (a) M. Nerz-Stormis and E.R Thornton, *Tetrahedron Lett.,* **27**(8), 897 (1986); (b) R.O. Duthala, P. Herold, W. Lottenbach, K. Oertle, and M. Riediker, *Angew.Chem.,* **101**(4), 495 (1989).

137.   C.Siegel and E.R. Thornton, *Tetrahedron Lett.,* **27**(4), 457 (1986).

138.   J.S. Panek and O.A. Bula, *Tetrahedron Lett.,* **29**(14), 1661 (1988).

139.   C. Palazzi, L. Colombo and C. Gennari, *Tetrahedron Lett.,* **27**(15), 1735 (1986).

140.   A.I. Meyers and Y. Yamamoto, *Tetrahedron* **40**, 2309 (1984).

141.   J.V. Jephcote, A.J. Pratt and E.J. Thomas, *J. Chem. Soc., Chem. Commun.,* 800 (1984).

142.   S. Masamune, T. Kaiho and D.S. Garvey, *J. Amer. Chem. Soc.,* **104**, 5521 (1982).

143.   D. Boschelli, J.W. Ellingboe and S. Masamune, *Tetrahedron Lett.,* 3395 (1984).

144.   J.M. Brown and I. Cutting, *J. Chem. Soc., Chem. Commun.,* 578 (1985).

145.   G. Helmchen, U. Leikanf and I. Tanfer-Knöpfel, *Angew.Chem.,* **24**, 874 (1985).

146.   Y. Ito, T. Katasuki and M. Yamaguchi, *Tetrahedron Lett.*, 4643 (1985).

147.   G.Gennari, A. Bernardi, L. Colombo and C. Scolastico, *J. Amer. Chem. Soc.*, **107**, 5812 (1985).

148.   T.A. Spencer, R.W. Britton and D.S. Watt, *J. Amer. Chem. Soc.*, **89**, 5727 (1967).

149.   (a) K.K. Heng and R.A.J. Smith, *Tetrahedron* **35**, 425 (1979);   (b) A. Balsamo, P.L. Barili, P. Crotti, M. Feretti, B. Macchia and F. Macchia, *Tetrahedron Lett.*, 1005 (1974).

150.   (a) W.R. Vaughan and H.P. Knoess, *J.Org.Chem.*, **35**, 2394 (1970);   (b) F.H. van der Steen, J.T.B.H. Jastrzebski and G. van Koten, *Tetrahedron Lett.*, **29**, 2467 (1988);   (c) J.K. Gawronsky, *Tetrahedron Lett.*, **25**, 2605 (1984).

151.   H.O. House, "Modern Synthetic Reactions", 2nd ed., pp. 629 W.A. Benjamin Reading Mass., (1972).

152.   M. Guette, J. Capillon and J.P. Guette, *Tetrahedron* **29**, 3659 (1973).

153.   H.O. House, D.S. Crumrine, A.Y. Teranishi and H.D. Olmstead, *J. Amer. Chem. Soc.*, **95**, 3310 (1973).

154.   W.A.Kleschick, C.T. Buse and C.H. Heathcook, *J. Amer. Chem. Soc.*, **99**, 247 (1977).

155.   S. Shenvi and J.K. Stille, *Tetrahedron Lett.*, **23**(6), 627 (1982).

156.   S.S. Labadie and J.K. Stille, *Tetrahedron,* **40**(12), 2329 (1984).

157.   Y.Yamamoto, H., Yatagai and K. Maruyama, *J. Chem. Soc. Chem. Commun.*, 162 (1981).

158.   T. Mukaiyama, K. Banno and K. Narasaka, *J. Amer. Chem. Soc.*, **96**, 7503 (1974).

159.   For a review on silylenol ethers see: P. Brownbridge, *Synthesis*, **1**, 85 (1983).

160.   T.H. Chan, T. Aida, P.W.K. Lau, V. Gorys and D.N. Harpp, *Tetrahedron Lett.*, **42**, 4029 (1979).

161.   E. Nakamura, M. Shimizu, I. Kuwajima, J. Sakata, K. Yokoyama and R. Noyori, *J. Org. Chem.*, **48**, 932 (1983).

162.   S. Murata, M. Suzuki and R. Noyori, *J. Amer. Chem. Soc.*, **102**, 3248 (1980).

163.   R. Noyori, I. Nishida and J. Sakata, *J. Amer. Chem. Soc.*, **103**, 2106 (1981).

164.   D.A. Evans and L.R. McGee, *Tetrahedron Lett.*, **21**, 3975 (1980).

165.   (a) D.A. Evans and L.R. McGee, *J. Amer. Chem. Soc.*, **103**, 2876 (1981); (b) Y.Yamamoto and K. Maruyama, *Tetrahedron Lett.*, **21**, 4607 (1980).

166.   (a) M. Ertas and D. Seebach, *Helv. Chim. Acta*, **68**, 961 (1985);   (b) Y. Naruse, J. Ukai, N. Ikeda and H. Yamam , *Chemistry Lett.*, 1451 (1985);   (c) A. Itoh, S. Ozawa, K. Oshima, K. Takai and S. Ozawa, *Tetrahedron Lett.*, **21**, 361 (1980);   (d) J. Tsuji, T. Yamada, M. Kaito and T. Mandai, *Tetrahedron Lett.*, 2257 (1979);   (e) G. Iwasaki and M. Shibasaki, *Tetrahedron Lett.*, **28**, 3257 (1987);   (f) G.A. Slough, R.G. Bergman and C.H. Heathcock, *J.Am.Chem.Soc.*, **111**, 938 (1989);   (g) M.T. Reetz and A.E. Vougioukas, *Tetrahedron Lett.*, **28**, 793 (1987);   (h) T. Imamoto, T. Kusumoto and M. Yokoyama, *Tetrahedron Lett.*, **24**, 5233 (1983);   (i) J.J. Doney, R.G. Bergman and C.H. Heathcock, *J.Am.Chem.Soc.*, **107**, 3724 (1985);   (j) C.H. Heathcock, J.J.Doney and R.G. Bergman, *Pure Appl. Chem.*, **57**, 1789 (1985);   (k) Y. Yamamoto, and K. Maruryama, *J. Am. Chem. Soc.*, **104**, 2323 (1982);   (l) K.H. Theopold, P.N. Becker and R.G. Bergman, *J. Am. Chem. Soc.*, **104**, 5250 (1982).

167.   D.A. Evans, J. Bartroli and T.L. Shih, *J. Amer. Chem. Soc.*, **103**, 2127 (1981).

168.   G. Solladie, in "Asymmetric Synthesis" (J.D. Morrison, ed.) Vol.2 Part A, Chapter 6, Academic Press, New York, (1983).

169.   G. Solladie, C. Frech and G. Demailly, *Nouv. J. Chim.*, **9**, 21 (1985).

170.   C. Moiskowski and G. Solladie, *Tetrahedron*, **36**, 227 (1980).

171.   G. Solladie, *Chimia*, 38 (1984).

172.   S. Kobayashi and T. Mukaiyama, *Chemistry Lett.*, 297 (1989).

173.   L.S. Liebeskind and M.E. Welker, *Tetrahedron Lett.*, 25(39), 4341 (1984).

174.   (a) S.G. Davies, I.M. Dordor and P. Warner, *J. Chem. Soc., Chem. Commun.*, 956 (1984); (b) S.G. Davies, I.M. Dordor-Hedgecock, P. Warner, R.H. Jones and K. Prout, *J. Organomet. Chem.*, 285, 213 (1985).

175.   N. Aktogu, H. Felkin and S.G. Davies, *J. Chem. Soc., Chem. Commun.*, 1303 (1982).

176.   W. Ando and H.Tsumaki, *Chemistry Lett.*, 1409 (1983).

177.   C.H. Heathcock, M.C. Pirrung, C.T. Buse, J.P. Hagen, S.D. Yong and J.E. Sohn, *J. Amer. Chem. Soc.*, 101, 7077 (1979).

178.   C.H. Heathcock and C.T. White, *J. Amer. Chem. Soc.*, 101, 7076 (1979).

179.   D.A. Evans and T.R. Taber, *Tetrahedron Lett.*, 21, 4675 (1980).

180.   D. Seebach, V. Ehrig and M. Teschner, *Liebigs Ann.Chem.*, 1357 (1976).

181.   D.A. Evans, E. Vogel and J.V. Nelson, *J. Amer. Chem. Soc.*, 101, 6120 (1979).

182.   H. Eichenauer, E. Friedrich, W. Lutz, and D. Enders, *Angew.Chem.*, 17, 206 (1978).

183.   D.E. van Horn and S. Masamune, *Tetrahedron Lett.*, 2229 (1979).

184.   A.I. Meyers and G. Knaus, *Tetrahedron Lett.*, 1333 (1974).

185.   A.I. Meyers and Y. Yamamoto, *J. Amer. Chem. Soc.*, 103, 4278 (1981).

186.   A. Togni, S.D. Pastor and G. Ribs, *Helv.Chim.Acta*, 72, 1471 (1989).

187.   Y. Ito, M. Sawamura, E. Shirakawa, K. Hayashizaki and T. Hayashi, *Tetrahedron Lett.*, 29(2), 235 (1988).

188.   Y. Ito, M. Sawamura, and T.Hayashi, *Tetrahedron Lett.*, 28, 6215 (1987).

189.   Y. Ito, M. Sawamura and T. Hayashi, *Tetrahedron Lett.*, **29**, 239 (1988).

190.   T. Hayashi, *Pure Appl. Chem.*, **60**, 7 (1988)

191.   (a) Y. Ito, M. Sawamura and T. Hayashi, *J. Amer. Chem. Soc.*, **108**, 6405 (1986); (b) M.Sawamura, H. Hamashima and Y. Ito, *J. Org. Chem.*, **55**, 5935 (1990).

192.   C.H. Heathcock and L.A. Flippin, *J. Amer. Chem. Soc.*, **105**, 1667 (1983).

193.   S. Masamune, *Aldrichimica Acta*, **11**, 23 (1978).

194.   E. Nakamura, Y. Horiguchi, J.J. Shimada and I. Kuwajima, *J. Chem. Soc., Chem. Commun.*, 796 (1983).

195.   (a) C. Mukai, W.J. Cho and M. Hanaoka, *Tetrahedron Lett.*, **30**, 7435 (1989);   (b) C. Mukai, W.J. Cho, I.J. Kim, M. Kido and M. Hamaoka, *Tetrahedron*, **47**, 3007 (1991).

196.   For a review on double stereo-differentiation; see: S. Masamune, W. Choy, J.S. Petersen and L.R. Sita, *Angew.Chem.*, **24**, 1 (1985).

197.   S. Masamune, S.A. Ali, D.L. Snitman and D.S. Garvey, *Angew.Chem.*, **19**, 557 (1980).

198.   Y. Yamamoto and K. Maruyama, *Tetrahedron Lett.*, **21**, 4607 (1981).

199.   E. Nakamura and I. Kuwajima, *Tetrahedron Lett.*, 3343 (1983).

200.   D.A. Evans, J.V. Nelson, E. Vogel and T.R. Taber, *J. Amer. Chem. Soc.*, **103**, 3099 (1981).

201.   C. Goasdoue, N. Goasdoue and M. Gaudemar, *Tetrahedron Lett.*, 4001 (1983).

202.   C. Gennari, S. Cardani, L. Colombo and C. Scolastico, *Tetrahedron Lett.*, **25**, 2283 (1984).

203.   A.I. Meyers and R.P. Reider, *J. Amer. Chem. Soc.*, **101**, 2501 (1979).

204.   T.Mukaiyama, H. Uchiro and S. Kobayashi, *Chemistry Lett.*, 1147 (1990).

205.   T. Mukaiyama, H. Uchiro and S.Kobayashi, *Chemistry Lett.*, 1001 (1989).

206.   M.T. Reetz, K. Kesseler and A. Jung, *Tetrahedron Lett.*, **25**(7), 729 (1984).

207.   D.A. Evans and J. Bertroli, *Tetrahedron Lett.*, **23**, 807 (1982).

208.   B.M. Mikhailov, *Usp.Khim.*, **45**, 1102 (1976); *Engl.Transl.* p.557; *Organomet. Chem. Rev.*, A8, 1 (1972).

209.   G.W. Kramer and H.C. Brown, *J. Organomet. Chem.*, **132**, 9 (1977).

210.   W. Neugebauer and P.von R. Schleyer, *J. Organomet. Chem.*, **198**, C1 (1980).

211.   M. Schlosser and M. Stähle, *J. Organomet. Chem.*, **220**, 277 (1981).

212.   M. Schlosser and J. Hartmann, *J. Amer. Chem. Soc.*, **98**, 4674 (1976).

213.   T.B. Thompson and W.T. Ford, *J. Amer. Chem. Soc.*, **101**, 5459 (1979).

214.   J. Slutsky and H. Kwart, *J. Amer. Chem. Soc.*, **95**, 8678 (1973).

215.   J.A Verdone, J.A. Maugravite, N.M. Scarpa, and H.G. Kuivila, *J. Amer. Chem. Soc.*, **97**, 843 (1975).

216.   E. Matarasso-Tchiroukhine and P. Cadiot, *J. Organomet. Chem.*, **121**, 155 (1976).

217.   M. Yamaguchi and T. Mukaiyama, *Chemistry Lett.*, 993 (1980).

218.   R.W. Hoffmann and H.J. Zeiß, *J. Org. Chem.*, **46**, 1309 (1981).

219.   K.G. Hancock and J.D. Kramer, *J. Organomet. Chem.*, **64**, C29 (1974).

220.   (a) L.S. Hegedus, *J.Orgamomet Chem.Libr.*, **1**, 329 (1976); (b) P. Heimback, P.W. Jolly and G. Wilke, *Adv. Organomet. Chem.*, **8**, 29 (1970).

221.   F. Sato, S. Iijima and M. Sato, *Tetrahedron Lett.*, **22**, 243 (1981).

222.   H. Lehmkuhl and S.Fustero, *Liebigs Ann. Chem.*, 1371 (1980).

223.   M. Stahle, J. Hartmann and M. Schlosser, *Helv. Chim. Acta.*, **60**, 1730 (1977).

224. A. Cutler, D. Ehntholt, W.P.Giering, P. Lennon, S. Raghu, A. Rosan, M. Rosenblum, J. Tancrede and D. Wells, *J. Amer. Chem. Soc.*, **98**, 3495 (1976).

225. G. Wittig, *Angew. Chem.*, **70**, 65 (1958).

226. W. Tochtermann, *Angew. Chem.*, **5**, 351 (1966).

227. D.A. Evans, J.V. Nelson and T.R. Taber, *Top. Stereochem.*, 13 (1982).

228. P. Vittorelli, H.-J. Hensen and H. Schmid, *Helv. Chim. Acta*, **58**, 1293 (1975).

229. K. König, and W.P. Neumann, *Tetrahedron Lett.*, 495 (1967).

230. C. Cervens and M. Pereyre, *J. Organomet. Chem.*, **26**, C4 (1971).

231. Y. Naruta, S. Ushida and K. Maruyama, *Chemistry Lett.*, 919 (1979).

232. A. Hosomi, H. Iguchi, M. Endo and H. Sakurai, *Chemistry Lett.*, 977 (1979).

233. G. Deleris, J. Dunogues and R. Calas, *J. Organomet. Chem.*, **93**, 43 (1975).

234. A. Hosomi and H. Sakurai, *Tetrahedron Lett.*, 1295 (1976); 2589 (1978).

235. I. Ojima, M. Kamugai and Y. Miyazawa, *Tetrahedron Lett.*, 1385 (1977).

236. Y. Yamamoto, H. Yatagai, Y. Naruta and K. Maruyama, *J. Amer. Chem. Soc.*, **102**, 7107 (1980).

237. H.C. Brown and P.K. Jadhav, *J. Amer. Chem. Soc.*, **105**, 2092 (1983).

238. (a) Y.Yamamoto, T. Komatsu and K.Maruyama, *J. Chem. Soc., Chem. Commun.*, 191 (1983); (b) M.M. Midland and S.P. Preston, *J. Amer. Chem. Soc.*, **104**, 2330 (1982).

239. Y. Yamamoto, H. Yatagai and K. Maruyama, *J.Amer.Chem.Soc.*, **103**, 3229 (1981); (b) H.C. Brown and P.K. Jadhav, *J. Org. Chem.*, **49**, 4089 (1984).

240. Y.Yamamoto, T. Komatsu and K. Maruyama, *J. Amer. Chem. Soc.*, **106**, 5031 (1984).

241. H.C. Brown, K.S. Bhat and R.S. Randad, *J. Org. Chem.*, **54**, 1570 (1989).

242. W.R. Ronsh, A.D. Palkowitz and M.A.J. Palmer, *J. Org. Chem.*, **52**, 316 (1987).

243. H.C. Brown and P.K. Jadhav, *J. Amer. Chem. Soc.*, **108**, 293 (1986).

244. J. Garcia, B.M. Kim and S. Masamune, *J. Org. Chem.*, **52**, 4831 (1987).

245. T. Herold and R.W. Hoffmann, *Angew.Chem.*,**17**, 768 (1979).

246. R.W. Hoffmann and W.Ladner, *Tetrahedron Lett.*, **48**, 4653 (1979).

247. J.C. Stowell, "Carbanions in Organic Synthesis", Wiley, New York, (1979).

248. M.T. Reetz, *Top. Curr. Chem.*, **106**, 1 (1982).

249. M.T. Reetz, "Organotitanium Reagents in Organic Synthesis", Springer, Berlin, (1986).

250. M.T. Reetz and B. Wenderoth, *Tetrahedron Lett.*, **23**, 5259 (1982).

251. M.T. Reetz, J.Westermann, R.Steinbach, B. Wenderoth, R. Peter, R. Ostarek and S. Maus, *Chem.Ber.*, **118**, 1421 (1985).

252. M.T. Reetz, R. Steinbach, J. Westermann, R. Peter and B. Wenderoth, *Chem.Ber.*, **118**, 1441 (1985).

253. F. Sato, K. Iida, S. Iijima, H. Moriya and M. Sato, *J. Chem. Soc., Chem. Commun.*, 1140 (1981).

254. L. Widler and D. Seebach, *Helv. Chim. Acta*, **65**, 1085 (1982).

255. Y. Yamamoto, W. Ito and K. Maruyama, *J. Chem. Soc. Chem. Commun.*, 1004 (1984).

256. K. Furuta, M. Ishiguro, R. Haruta, N. Ikeda and H. Yamamoto, *Bull. Chem. Soc. Jpn.*, **57**, 2768 (1984).

257. F. Sato, H. Uchiyama, K. Lida, Y. Kobayashi and M. Sato, *J. Chem. Soc., Chem. Commun.*, 921 (1983).

258.   M. Riediker and R.O. Duthaler, *Angew.Chem.*, **28**, 494 (1989).

259.   M.T. Reetz, R. Steinbach, B. Wenderoth and J. Westermann, *Chem. and Ind.*, (London) 541(1981).

260.   C. Servens and M. Pereyre, *J. Organomet. Chem.*, **35**, C20 (1972).

261.   G.E. Keck and E. P.Boden, *Tetrahedron Lett.*, **25**, 265 (1984).

262.   G.E. Keck and E.P. Boden, *Tetrahedron Lett.*, **25**, 1879 (1984).

263.   G.E. Keck and D.E. Abbott, *Tetrahedron Lett.*, **25**, 1883 (1984).

264.   M. Koreeda and Y. Tamaka, *Abstr.ORGN* 300, 180. Meeting, *Amer. Chem. Soc.*, San Francisco/Las Vegas, (1980).

265.   K. Maruyama, Y. Ishihara and Y. Yamamoto, *Tetrahedron Lett.*, **22**, 4235 (1981).

266.   Y.Yamamoto and K. Maruyama, *Heterocycles,*, **18**, 357 (1982).

267.   Y. Yamamoto, N. Maeda and K. Maruyama, *J. Chem. Soc., Chem. Commun.*, 774 (1983).

268.   Y. Yamamoto, H. Yatagai, Y. Ishihara, M. Maeda and K. Maruyama, *Tetrahedron* , **40**, 2239 (1984).

269.   J. Otera, Y. Kawasaki, H. Mizuno and Y. Shimizu, *Chemistry Lett.*, 1529 (1983).

270.   H. Yatagai, Y. Yamamoto, and K. Maruyama, *J. Amer. Chem. Soc.*, **102**, 4548 (1980).

271.   Y. Yamamoto, H. Yatagai and K. Maruyama, *J. Amer. Chem. Soc.*, **103**, 3229 (1981).

272.   A.J. Pratt and E.J. Thomas, *J. Chem. Soc., Chem. Commun.*, 1115 (1982).

273.   T. Hayashi, K. Kabeta, I. Hamachi and M. Kumada, *Tetrahedron Lett.*, **28**, 2865 (1983).

274.   T. Hayashi, M. Konishi and M. Kumada, *J. Org. Chem.*, **48**, 281 (1983).

275.   C.H.Heathcock, S. Kiyooka and T.A. Blumenkopf, *J. Org. Chem.*, **49**, 4214 (1984).

276.   T. Hayashi, M. Konishi and M. Kumada, *J. Amer. Chem. Soc.*, **104**, 4963 (1982).

277.   S. Kiyooka and C.H. Heathcock, *Tetrahedron Lett.*, **24**, 4765 (1983).

278.   K. Mikami, T. Maeda, N. Kishi and T. Nakai, *Tetrahedron Lett.*, **25**, 5151 (1984).

279.   K. Hiroi, J. Abe, K. Suya and S. Sato, *Tetrahedron Lett.*, **30**, 1543 (1989).

280.   K. Hiroi, *Yakigosei Kagaku Kyokai Shin*, **44**, 907 (1986).

281.   K. Hiroi, R. Kitayama and S. Sato, *J. Chem. Soc., Chem. Commun.*, 303 (1984).

282.   K. Hiroi, R. Kitayama and S. Sato, *Chemistry Lett.*, 929 (1984).

283.   K. Hiroi and K. Makino, *Chemistry Lett.*, 617 (1986).

284.   T. Hayashi, M. Konishi and M. Kumada, *J. Chem. Soc., Chem. Commun.*, 107 (1984).

285.   B.M. Trost and N.R. Schmuff, *Tetrahedron Lett.*, 2999 (1981).

286.   T. Hayashi, T. Hagihara, M. Konishi and M. Kumada, *J. Amer. Chem. Soc.*, **105**, 7767 (1983).

287.   G. Consiglio and R.M. Waymouth, *Chem. Rev.*, **81**, 257 (1989).

288.   T. Hiyama, K. Kimura and H. Nozaki, *Tetrahedron Lett.*, 1037 (1981).

289.   Y. Okude, S. Hirano, T. Hiyama and H. Nozaki, *J. Amer. Chem. Soc.*, **99**, 3179 (1977).

290.   C.T. Buse and C.H. Heathcock, *Tetrahedron Lett.*, 1685 (1978).

291.   M.D. Lewis and Y. Kishi, *Tetrahedron Lett.*, **23**, 2343 (1982).

292. J. Mulzer, M. Kappert, G. Huttner, and I. Jibril, *Angew.Chem.*, **23**, 704 (1984).

293. G.Zweifel and G. Hahn, *J. Org. Chem.,* **49**, 4567 (1984).

294. D. Hoppe and F. Lichtenberg, *Angew.Chem.*, **21**, 372 (1982).

295. Y. Yamamoto and K. Maruyama, *Tetrahedron Lett.*, **22**, 2895 (1981).

296. J.D.'Angelo, *Tetrahedron* , **32**, 2979 (1976).

297. D.A. Evans, "Asymmetric Synthesis" (J.D. Morrison, Ed), Vol. 31 p. 1-110, (1984).

298. D. Caine, in "Carbon-Carbon Bond Formation", (R.L. Augustine, ed.), Vol.1, p.85, Dekker New York, (1979).

299. H.O. House, L.J. Czuba, M. Gall and H.D. Olmstead, *J. Org. Chem.,* **34**, 2324 (1969).

300. B.J. L. Huff, Ph.D. Dissertation, Georgia Institute of Technology, Atlanta (1968).

301. Z.A. Fataftah, I.E. Kopka and M.W. Rathke, *J. Amer. Chem. Soc.,* **102**, 3959 (1980).

302. E. Nakamura, K. Hashimoto and I. Kuwajima, *Tetrahedron Lett.*, 2079 (1978).

303. S. Masamune, J.E. Ellingboe and W. Choy, *J. Amer. Chem. Soc.,* **104**, 5526 (1982).

304. A.P. Krapcho and E.A. Dundulis, *J. Org. Chem.,* **45**, 3236 (1980).

305. H.O. House and T.M. Bare, *J. Org. Chem.,* **33**, 943 (1968).

306. F.E. Ziegler and P.A. Wender, *J. Amer. Chem. Soc.,* **93**, 4318 (1971).

307. H.von Bekkum, C.B. van den Bosch, G.van Minnen Pathuis, J.C. Demos and A.M.van Wijk, *Recl.Trav.Chim. Pays-Bas.,* **90**, 137 (1971).

308. F. Plavac and C.H. Heathcock, *Tetrahedron Lett.*, 2115 (1979).

309.   R.D. Clark, *Syn. Commun.*, **9**, 325 (1979).

310.   J.A. Hogg, *J. Amer. Chem. Soc.*, **70**, 161 (1948).

311.   W. Nagata, T. Sugasawa, M. Narisada and T. Wakabayashi, *J. Amer. Chem. Soc.*, **89**, 1483 (1967).

312.   H.O. House, R.C. Strickland and E.J. Zaiko, *J. Org. Chem.*, **41**, 2401 (1976).

313.   (a) R.K. Boeckman, Jr, *J.Org.Chem.*, **38**, 4450 (1973).   (b) E.J. Corey and R.A. Sneen, *J. Amer. Chem. Soc.*, **78**, 6269 (1956).

314.   G.H. Posner, M.J. Chapdelaine and C.M. Lentz, *J. Org. Chem.*, **44**, 3661 (1979).

315.   M. Pesaro, G. Bozzato and P. Schudel, *J. Chem. Soc., Chem. Commun.*, 1152 (1968).

316.   M.E. Kuehne, *J. Org. Chem.*, **35**, 171 (1970).

317.   J.M. Conia and P. Briet, *Bull. Soc. Chim. Fr.*, 3881, 3886 (1966).

318.   R.E. Ireland, P.S. Grand, R.E. Dickerson, J. Bordner and D.R. Rydjeski, *J. Org. Chem.*, **35**, 570 (1970).

319.   E.J. Corey, R.Z. Mitra and H. Uda, *J. Amer. Chem. Soc.*, **86**, 485 (1964).

320.   S. Takano, N. Tamura and K. Ogasawawa, *J. Chem. Soc., Chem. Commun.*, 1155 (1981).

321.   V. Jager and W. Schwab, *Tetrahedron Lett.*, 3129 (1978).

322.   D. Seebach, M. Boes, R. Naef and W.B. Schweizer, *J. Amer. Chem. Soc.*, **105**, 5390 (1983).

323.   P.A. Bartlett and C.F. Pizzo, *J. Org. Chem.*, **46**, 3896 (1981).

324.   H.O. House and C.J. Blankley, *J. Org. Chem.*, **32**, 1741 (1967).

325.   K. Sato, *Chemistry Lett.*, 1183 (1981).

326. P.A. Grieco, C.S. Pogonowski and M. Miyashita, *J. Chem. Soc., Chem. Commun.*, 592 (1975).

327. E.J. Corey, R. Hartmann and P.A. Vatakencherry, *J. Amer. Chem. Soc.*, **84**, 2611 (1962).

328. P.A. Grieco, Y. Ohfune, Y. Yokoyama and W. Owens, *J. Amer. Chem. Soc.*, **101**, 4380 (1979).

329. B.M. Trost, P.R. Bernstein and P.C. Funfschilling, *J. Amer. Chem. Soc.*, **101**, 4378 (1979).

330. R.L. Snowdown, P. Sonnay and G. Ohloff, *Helv. Chim. Acta,* **64**, 25 (1981).

331. G. Stork and S.R. Dowd, *J. Amer. Chem. Soc.*, **85**, 2178 (1963).

332. A.I. Meyers, D.R. Williams and M.D. Druelinger, *J. Amer. Chem. Soc.*, **98**, 3072 (1976).

333. A.I. Meyers, D.R. Williams, G.W. Erickson, S. White and M. Druelinger, *J. Amer. Chem. Soc.*, **103**, 3081 (1981).

334. R. Fraser, F. Akiyama and J. Banville, *Tetrahedron Lett.*, **41**, 3929 (1979).

335. S. Hashimoto and K. Koga, *Chem. Pharm. Bull.*, **27**, 2760 (1979).

336. C.R. McArthur, P.M. Worster, J.-L. Jiang and C.C.Leznoff, *Can.J.Chem.*, **60**, 1836 (1982).

337. P.M. Worster, C.R. McArthur and C.C. Leznoff, *Angew.Chem.*,**18**, 221 (1979).

338. A.I. Meyers and D.R. William, *J. Org. Chem.*, **43**, 3245 (1978).

339. A.I. Meyer, Z. Brick and G.W. Erickson, *J. Chem. Soc., Chem. Commun.*, 566 (1979).

340. S. Hashimoto, H. Kogan, K. Tomioka and K.Koga, *Tetrahedron Lett.*, 3009 (1979).

341. S. Hashimoto, S. Yamada and K. Koga, *Chem. Pharm. Bull.*, **27**, 771 (1979).

342. D. Enders and H. Eichenauer, *Chem. Ber.*, **112**, 2933 (1979).

343. H. Takahashi and H. Inagaki, *Chem. Pharm. Bull.*, **30**, 922 (1982).

344. S.I. Pennanen, *Acta Chem. Scand.*, **B35**, 555 (1981).

345. D. Ender, in "Asymmetric Synthesis", (J.D. Morrison, ed.), p.275 Academic Press, Inc, (1985).

346. H.M. Fales, M.S. Blum, R.M. Crewe and J.M. Brand, *J. Insect Physiol.*, 1077 (1972).

347. G.J. Karabatsos and R.A. Taller, *Tetrahedron*, **24**, 3923 (1968).

348. D. Enders, H. Eichenauer, U. Bans, H. Schubert and K.A.M. Kremer, *Tetrahedron*, **40**(8), 1345 (1984).

349. For a review see: K.A. Lutomski, in "Asymmetric Synthesis", (J.D. Morrison, ed.), vol.3, Academic Press, Inc., p.213, (1984).

350. A.I. Meyers, G. Knaus and K. Kamata, *J. Amer. Chem. Soc.*, 268 (1974).

351. A.I. Meyers, G. Knaus, K. Kamata and M.E. Ford, *J. Amer. Chem. Soc.*, **98**, 567 (1976).

352. M.A. Hoobler, D.E. Bergbreiter and M. Newcomb, *J. Amer. Chem. Soc.*, **100**, 8182 (1978).

353. A.I. Meyers and J. Slade, *J. Org. Chem.*, **45**, 2785 (1980).

354. A.I. Meyers, Y. Yamamoto, E.D. Mihelich and R.A. Bell, *J. Org. Chem.*, **45**, 2792 (1980) p.73.

355. A. I. Meyers, R.K. Smith and C.E. Whitten, *J. Org. Chem.*, **44**, 2250 (1979).

356. A.I. Meyers and C.E. Whitten, *J. Amer. Chem. Soc.*, **97**, 6266 (1975).

357. A.I. Meyers, N.R. Natale, D.G. Wettlaufer, S. Rafii and J. Clardy, *Tetrahedron Lett.*, 5123 (1981).

358. A.I. Meyers and N.R. Natale, *Heterocycles*, **18**, 13 (1982).

359.   A.R. Colwell, L.R. Duckwall, R. Brooks and S.P. McManus, *J. Org. Chem.*, **46**(15), 3097 (1981).

360.   (a) U.Schöllkopf, U. Groth and C. Deng, *Angew.Chem.*, **20**, 798 (1981); (b) U. Schöllkopf, *Pure Appl.Chem.*, **55**, 1799 (1983); (c) *ibid, Top. Curr. Chem.*, **109**, 65 (1983); (d) *ibid, Tetrahedron*, **39** 2085 (1983); (e) U. Schöllkopf, W. Hartwig, and U. Groth, *Angew. Chem.*, **18**, 863 (1979).

361.   U. Schöllkopf, U. Groth, K.O. Westphalen and C. Deng, *Synthesis*, 969 (1981).

362.   J. Nozlak and U. Schöllkopf, *Synthesis*, 861 (1982).

363.   J. Nozulak and U. Schöllkopf, *Synthesis*, 866 (1982).

364.   E.L. Eliel, A.A. Hartmann and A.G. Abatjoglou, *J. Amer. Chem. Soc.*, **96**, 1807 (1974).

365.   E.L. Eliel, *Tetrahedron* **30**, 1503 (1974).

366.   J.J. Eisch and J.E. Galle, *J.Org.Chem.*, **45**, 4534 (1980).

367.   C. Chassaing, R. Lett and A. Marquet, *Tetrahedron Lett.*, **5**, 471 (1978).

368.   R. Lett and A. Marquet, *Tetrahedron Lett.*, **19**, 1579 (1975).

369.   J.F. Biellmann and J.J. Vicens, *Tetrahedron Lett.*, **5**, 467 (1978).

370.   T. Durst, R. Viau and M.R. McClory, *J. Amer. Chem. Soc.*, **93**, 3077 (1971).

371.   K. Nishihata and M. Nishio, *Tetrahedron Lett.*, 1695 (1976).

372.   F.Matloubi and G. Solladie, *Tetrahedron Lett.*, **23**, 2141 (1979).

373.   P.G. Duggan and W.S. Murphy, *J. Chem. Soc., Perkin Trans.I.*, 634 (1976).

374.   P.G. Duggan and W.S. Murphy, *J.Chem.Soc.Chem. Commun.*, 263 (1976).

375.   S.I. Bhole and V.N. Goyte, *Ind.J.Chem.*, **20B**, 218, 222 (1981).

376.   P. Duhamel, J.J. Eddine and J.Y. Valnot, *Tetrahedron Lett.*, **28**, 3801 (1987).

377.   G. Boche and W. Schrott, *Tetrahedron Lett.*, **23**, 5403 (1982).

378.   P. Duhamel, J.J. Eddine and J.Y. Valnot, *Tetrahedron Lett.*, **25**, 2355 (1984).

379.   P. Duhamel, J.Y. Volnot and J.J. Eddine, *Tetrahedron Lett.*, **23**, 2863 (1982).

380.   S.Sakane, J.Fujiwara, K. Maruoka and H. Yamamoto, *Tetrahedron*, **42**, 2193 (1986).

381.   S. Sakane, J. Fujiwara, K. Muruoka and H. Yamamoto, *J. Amer. Chem. Soc.*, **105**, 6154 (1983).

382.   J.M. Wilson and D.J. Cram, *J. Org. Chem.*, **49**, 4930 (1984).

383.   J.M. Wilson and D.J. Cram, *J. Amer. Chem. Soc.*, **104**, 881 (1982).

384.   K. Fuji, M. Node, H. Nagasawa, Y. Naniwa, T. Taga, K. Machida and G. Snatzke, *J. Amer. Chem. Soc.*, **111**, 7921 (1989).

385.   G.H. Posner, J.P. Mallamo, M. Hulce and L.L. Frye, *J. Amer. Chem. Soc.*, **104**, 4180 (1982).

386.   T. Mukaiyama and N. Iwasawa, *Chemistry Lett.*, 913 (1981).

387.   C.P. Fei and T.H. Chan, *Synthesis*, 467 (1982).

388.   T. Saegusa, Y. Ito, S. Tomita and H. Kinoshita, *Bull. Chem. Soc. Jpn.*, **45**, 496 (1972).

389.   J.H. Nelson, P.N. Howells, G.C. Delullo, G.L. Landen and R.A. Henry, *J. Org. Chem.*, **45**, 1246 (1980).

390.   H. Brunner and B. Hammer, *Angew.Chem.*, **23**, 312 (1984).

391.   K. Watanabe, K. Miyazu and K. Irie, *Bull. Chem. Soc., Jpn.*, **55**, 3212 (1982).

392.   D.J. Cram and G.D.Y. Sogal, *J. Chem. Soc., Chem. Commun.*, 625 (1981).

393.   W.Oppolzer, *Angew.Chem.*, **23**, 876 (1984).

394.   H. Wurziger, *Kontakte*, 3 (1984).

395.   For an earlier review, see: J.B. Hendrickson, *Angew.Chem.*, **13**, 47 (1974).

396.   R.B. Woodward and R. Hoffmann, *Angew.Chem.*, **8**, 781 (1969).

397.   R. Huisgen, *Angew.Chem.*,, **7**, 321 (1968).

398.   O. Diels and K. Alder, *Justus Lieb. Ann. Chem.*, **460**, 98 (1928)

399.   For a review see: J. Sauer and R. Sustmann, *Angew.Chem.*, **19**, 779 (1980).

400.   W. Oppolzer, M. Kurth, D. Reichlin, C. Chapuis, M. Mohnhaupt and F. Moffat, *Helv.Chim.Acta*, **64**(8), 2802 (1981).

401.   W. Oppolzer, M. Kurth, D. Reichlin and F. Moffatt, *Tetrahedron Lett.*, 2545 (1981).

402.   W. Choy, L.A. Reed and S. Masamune, *J. Org. Chem.*, **48**, 1137 (1983).

403.   S. Danishefsky, T. Harayama and R.K. Singh, *J. Amer. Chem. Soc.*, **101**, 7008 (1979).

404.   T. Koizumi, I. Hakamada and E. Yoshii, *Tetrahedron Lett.*, **25**(1), 87 (1984).

405.   J. Jurczak, *Bull. Chem. Soc. Jpn.*, **52**, 3438 (1979).

406.   J. Jurczak and M. Thocz, *J.Org.Chem.*, **44**, 3347 (1979).

407.   W. Oppolzer and C. Chapuis, *Tetrahedron Lett.*, **24**(43), 4665 (1983).

408.   R. Gleiter and M.C. Böhm, *Pure Appl. Chem.*, **55**, 237 (1983).

409.   N. Ikota, *Chem. Pharm. Bull.*, **37**(8), 2219 (1989).

410.   W.G. Dauben and R.A. Bunce, *Tetrahedron Lett.*, **23**(47), 4875 (1982).

411.   B.M. Trost, D.O' Krongly and J. Belletire, *J. Amer. Chem. Soc.*, **102**, 7595 (1980).

412.   C. Siegel and E.R Thornton,*Tetrahedron Lett.*, **29**(41), 5225 (1988).

413. S. Masamune, L.A. Reed, J.T. Davis and W. Choy, *J. Org. Chem.*, **48** (23), 4441 (1983).

414. K. Narasaka, M. Inoe and N. Okada, *Chemistry Lett.*, 1109 (1986).

415. G. Bir and D. Kaufmann, *Tetrahedron Lett.*, **28**(7), 777 (1987).

416. D. Kaufmann and R. Boese, *Angew.Chem.*, **29**(5), 545 (1990).

417. M. Bednarski and S. Danishefsky, *J. Amer. Chem. Soc.*, **105**, 3716 (1983).

418. T. Bauer, C. Chapuis, J. Kozak and J. Jurczak, *Helv. Chim. Acta.*, **72**, 482 (1989).

419. S.C. Hirst and A.D. Hamilton, *J.Amer.Chem.Soc.*, **113**, 382 (1991); (b) L.F. Tietze and K.H. Glüsenkamp, *Angew.Chem.*, **22**, 887 (1983).

420. J.C. de Jong, F. van Bolhuis and B.L. Feringa, *Tetrahedron Asymmetry*, **2**(12), 1247 (1991).

421. D.A. Evans, K.T. Chapman and J. Bisala, *Tetrahedron Lett.*, **25**(37), 4071 (1984).

422. W. Oppolzer, *Synthesis*, 793 (1978).

423. W. Oppolzer, *Pure and Appl. Chem.*, **53**, 1181 (1981).

424. W. Oppolzer and D.A. Roberts, *Helv. Chim. Acta,* **63**(6), 1703 (1980).

425. J. March: "Advanced Organic Chemistry" p.760, Wiley-Interscience Publication, New York, (1985) (and references cited therein).

426. B.S. Green, A.T. Hagler, Y. Rabinson and M. Rejto, *Isr. J. Chem.*, **15**(1-2), 124 (1977), *C.A.* **88**, 23288f (1978).

427. L.M. Tolbert and M.B. Ali, *J. Amer. Chem. Soc.*, **104**, 1742 (1982).

428. H. Kock, J. Runsink and H.D. Scharf, *Tetrahedron Lett.*, 3217 (1983).

429. I. Ojima and S. Inaba, *Tetrahedron Lett.*, **21**, 2077 (1980).

430.   I. Ojima and S. Inaba, *Tetrahedron Lett.*, 2081 (1980).

431.   C. Gluchowski, L. Cooper, D.E. Bergbreiter and M. Newcomb, *J. Org. Chem.*, **45**, 3413 (1980).

432.   A.E. Greene and F. Charbonnier, *Tetrahedron Lett.*, **26**(45), 5525 (1985).

433.   H. Wynberg and E.G.J.Staring, *J. Amer. Chem. Soc.*, **104**, 166 (1982).

434.   For a review, see: R. Huisgen, *J. Org. Chem.*, **41**(3), 403 (1976).

435.   K. Bast, M. Christl, R. Huisgen, W. Mack and R. Sustmann, *Chem. Ber.*, **106**, 3258, 3275 (1973).

436.   S.F. Martin and B. Dupre, *Tetrahedron Lett.*, **24**(13), 1337 (1983).

437.   T. Koizumi, H. Hirai and E. Yoshii, *J. Org. Chem.*, **47**, 4004 (1982).

438.   T. Takahashi, K. Kitano, T. Hagi, H. Nihonmatsu and T. Koizumi, *Chemistry Lett.*, 597 (1989).

439.   C. Belzecki and I. Panfil, *J. Org. Chem.*, **44**(8), 1212 (1979).

440.   G. Donegan, R. Grigg and F. Heaney, *Tetrahedron Lett.*, **30**(5), 609 (1989).

441.   A. Padwa, B.H. Norman and J. Perumattam, *Tetrahedron Lett.*, **30**(6), 663 (1989).

442.   P. Garner and W.B. Ho, *J.Org.Chem.*, **55**, 3973 (1990).

443.   R.B. Woodward and R. Hoffmann, "The Conservation of Orbital Symmetry", p.114, Verlag Chemie, Academic Press Inc., (1970).

444.   T.L. Gilchrist and R.C. Storr, "Organic Reactions and Orbital Symmetry", p.242, Cambridge University Press, (1979).

445.   C.W. Spangler, *Chem.Rev.*, **76**(2), 187 (1976).

446.   For review on Claisen rearrangement see: F.E. Ziegler, *Chem. Rev.*, **88**(8), 1423 (1988).

447.   R.K. Hill and N.W. Gilman, *J. Chem. Soc., Chem. Commun.*, 619 (1967).

448.   H.J. Hansen and H. Schmidt, *Tetrahedron*, **30**, 1959 (1974).

449.   R.K. Hill and A.G. Edwards, *Tetrahedron Lett.*, **44**, 3239 (1964).

450.   K.K. Chan, N. Cohen, J.P. De Nobel, A.C. Specian, Jr. and G. Saucy *J. Org. Chem.*, **41**(22), 3497 (1976).

451.   R.K. Hill, R. Soman and S.Sawada, *J. Org. Chem.*, **37**(23), 3737 (1972).

452.   R.E. Ireland and R.H. Mueller and A.K. Willard, *J. Amer. Chem. Soc.*, **98**, 2868 (1976).

453.   T. Fujisawa, K. Tajima and T. Sato, *Chemistry Lett.*, 1669 (1984).

454.   M. Nagatsuma, F. Shirai, N. Sayo and T. Nakai, *Chemistry Lett.*, 1393 (1984).

455.   R.K. Hill and H.N. Khatri, *Tetrahedron Lett.*, **45**, 4337 (1978).

456.   K. Widmer, J. Zsindely, H.-J. Hansen and H. Schmid, *Helv. Chim. Acta*, **56**(1), 75 (1973).

457.   M.J. Kurth and O.H. W. Decker *Tetrahedron Lett.*, **24**(42), 4535 (1983).

458.   P.A. Bartlett and W.F. Hahne, *J. Org. Chem.*, **44**, 882 (1979).

459.   S.E. Denmark and M.A. Harmata, *J. Org. Chem.*, **48**, 3370 (1983).

460.   S.R. Wilson and R.S. Myers, *J. Org. Chem.*, **40**(22), 3309 (1975).

461.   T. Sato, K. Tajima and T. Fujisawa, *Tetrahedron Lett.*, **24**(7), 729 (1983).

462.   R. Ochrlein, R. Jeschke, B. Ernest and D. Bellus, *Tetrahedron Lett.*, **30**(27), 3517 (1989).

463.   R.W. Hoffmann, *Angew.Chem.*, **18**(8), 563 (1979).

464.   T. Nakai and K. Mikami *Chem. Rev.,* **86**, 885 (1986).

465.   (a) For review see: U. Schöllkopf, *Angew.Chem.,* **9**, 763 (1970);   (b) G. Wittig and L. Löhmann, *Liebigs Ann. Chem.,* **550**, 260 (1942).

466.   N. Sayo, F. Shirai and T Nakai, *Chemistry Lett.,* 255 (1984).

467.   N. Sayo, E-I., Kitahara and T. Nakai, *Chemistry Lett.,* 259 (1984).

468.   T. Nakai, K. Mikami and S. Taya, *J. Amer. Chem. Soc.,* **103**, 6492 (1981).

469.   T. Nakai, K. Mikami, S. Taya, Y. Kimura and T. Mimura, *Tetrahedron Lett.,* **22**, 69 (1981).

470.   W.C. Still and A.Mitra, *J. Amer. Chem. Soc.,* **100**, 1927 (1978).

471.   K. Mikami, K. Fujimoto, T. Kasuga and T. Nakai, *Tetrahedron Lett.,* **25**, 6011 (1984).

472.   K. Mikami, T. Kasuga, K. Fujimoto and T. Nakai, *Tetrahedron Lett.,* **27**(35), 4185 (1986).

473.   (a) N. Uchikawa, T. Hanamoto, T. Katsuki, and M. Yamaguchi, *Tetrahedron Lett.,* **27**(38), 4577 (1986);   (b) K. Honda, S. Inoue and K. Sato, *J. Org. Chem.,* **57**, 428 (1992).

474.   P. Bickart, F.W. Carson, J. Jacobus, E.G. Müller and K. Mislow, *J. Amer. Chem. Soc.,* **90**, 4869 (1968).

475.   R.W. Hoffmann, R. Gerlach and S.Goldmann, *Tetrahedron Lett.,* 2599 (1978).

476.   S.Goldman, R.W. Hoffmann, N. Maat and K. J.Geueke, *Chem. Ber.,* **113**, 831 (1980).

477.   R.W. Hoffmann, S.Goldman, R. Gerlach and N. Maak, *Chem. Ber.,* **113**, 845 (1980).

478.   K. Hiroi, P.Kitayama and S. Sato, *J. Chem. Soc., Chem. Commun.,* 1470 (1983).

479. K.Mikami, O. Takahashi, T. Kasuga and T. Nakai, *Chemistry Lett.*, 1729 (1985).

480. K.K. Chan and G. Saucy, *J. Org. Chem.*, **42**, 3828 (1977).

481. J.A. Marshall and T.M. Jenson, *J. Org. Chem.*, **49**, 1707 (1984).

482. V.J. Rautenstrach, *J. Chem. Soc., Chem. Commun.*, 4 (1970).

483. K.Mikami, Y. Kimura, N. Kishi and T. Nakai, *J. Org. Chem.*, **48**, 279 (1983).

484. (a) K. Mikami, K. Azuma and T. Nakai, *Chemistry Lett.*, 1379 (1983); (b) *ibid Tetrahedron*, **40**, 2303 (1984).

485. W. Oppolzer and V. Snieckus, *Angew.Chem.*, **17**, 476 (1979).

486. O. Achmatowicz, Jr., and B. Szechner, *J. Org. Chem.*, 37(7), 964 (1972).

487. J.K. Whitesell , A. Bhattacharya, D.A. Aguilar and K. Henke, *J. Chem. Soc., Chem. Commun.*, 989 (1982).

488. G. Bryon and B. Wallace, *J. Chem. Soc., Chem. Commun.*, 382 (1977).

489. W.G. Dauben and T. Brookhart, *J. Org. Chem.*, **47**, 3921 (1982).

490. W.G. Dauben and T. Brookhart, *J. Amer. Chem. Soc.*, **103**, 237 (1981).

491. W. Oppolzer, K. Bättig and T. Hudlicky, *Helv.Chim.Acta*, **62**(5), 1493 (1979).

492. W. Oppolzer, C. Robbiani and K. Bättig, *Helv.Chim.Acta*, **63**(7), 2015 (1980).

493. W. Oppolzer, C.Robbiani and K. Bättig, *Tetrahedron* 40(8), 1391 (1984).

494. W. Oppolzer and K. Bättig, *Helv. Chim. Acta*, 64(7), 2489 (1981).

495.   J.V. Duncia, P.T. Lansbury, Jr., T. Miller and B.B. Snider, *J. Amer. Chem. Soc.*, **104**, 1930 (1982).

496.   W. Oppolzer, *Pure Appl. Chem.*, **53**, 1181 (1981).

497.   W. Oppolzer and K. Thirring, *J. Amer. Chem. Soc.*, **104**, 4978 (1982).

5 Asymmetric Oxidations

Asymmetric oxidation is currently the target of extensive investigations in organic synthesis and efforts have been devoted to the development of efficient oxidations which are highly enantioselective [1,2]. Asymmetric oxidations have been carried out with several substrates of diverse nature involving synthesis of 1,2-diols, oxidation of sulfides, epoxidation etc.

5.1 Asymmetric Epoxidation

Asymmetric epoxidation is of immense synthetic value due to the potential for conversion of the resulting epoxides into many different functional groups, and the formation of two contiguous chiral centres. The reaction is mild and can be carried out at low temperatures. Another aspect of asymmetric epoxidation is that the stereochemical outcome can be predicted with ease and accuracy.

5.1.1 Asymmetric Epoxidation of Allylic Alcohols

5.1.1.1 Katsuki-Sharpless Epoxidation

The asymmetric epoxidation of allylic alcohols has taken a new turn after the discovery by Katsuki and Sharpless that in the presence of titanium tetra-isopropoxide, L-(+)- or D-(-)-diethyl tartrate (DET) and tert.butyl hydroperoxide

$Ti(O^iPr)_4,(+)DET$
TBHP

(1) (2) (2S, 3S) 95% e.e.

(TBHP), allylic alcohols can be epoxidized with very high enantioselectivities [3,4]. For instance, epoxidation of the allyl alcohol (1) with $Ti(O^iPr)_4$, (+)-DET and TBHP afforded (2) in 95% e.e. [3]. As shown in Figure 5.1, when the allyl alcohol is drawn with the hydroxymethyl group at the lower right, the natural (L)-(+)-tartrate directs epoxidation to the bottom face of the olefin, whereas the synthetic analog (D)-(-)-tartrate does the opposite.

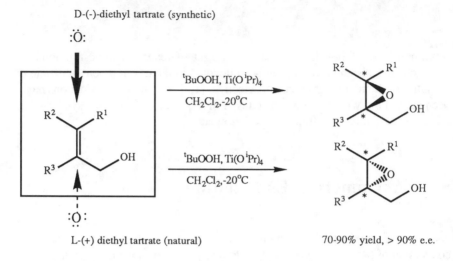

Figure 5.1. Asymmetric epoxidation of prochiral allylic alcohols

However the use of stoichiometric amounts of titanium-tartrate complex rendered the reaction less suitable for developing on a commercial scale. This difficulty was surmounted by the addition of molecular sieves (3Å or 4Å) to the reaction system which enabled the standard DEPT/Ti(O^{i}Pr)$_4$ complex to work as a catalyst (5-10 mol%) [5]. This catalytic version of the Katsuki -Sharpless epoxidation has been the basis for the commercial production of glycidol by the Arco Chemical Co.

The use of tartramide ligands, such as tartaric acid -N,N- dibenzylamide (3) with Ti(O^{i}Pr)$_2$ in a ratio of 1:2 afforded inverse enantiofacial selection [6]. Thus (E)-α-phenylcinnamyl alcohol (4), when treated with Ti(O^{i}Pr)$_4$ and (3) in a ratio of 2:1 afforded the product (5) in 82% e.e.

(3)

(4)   (5) 82% e.e.

Under standard reaction conditions, however, i.e. taking Ti:tartramide (2:2.4), the asymmetric epoxidation of (4) afforded (6) in 96% e.e.[*] The opposite enantioface selection exhibited by the two systems by simply changing the ratio of Ti: tartramide indicates the existence of two dinuclear complexes as the active species.

(4)   (6) 96% e.e.

Based on X-ray crystal structural analysis of Ti-tartrate complexes, Sharpless recently proposed the active species to have structure (7) [7, 8].

(7)

The essential features of the mechanism of asymmetric epoxidation are: (a) There is a complicated mixture of equilibrating complexes with the Ti-tartrate catalyst. The concentration of each complex is determined by thermodynamic stability, (b) Ti-tartrate affords consistently high enantioselectivities for widely different substrate structures by virtue of a combination of stereoelectronic and steric factors.

---

[*] The standard conditions in asymmetric epoxidation include the use of slightly greater than 1 equiv. of tartramide per equiv. of Ti. It is denoted as 2:2, because it is the complex with 2:2 Ti:tartramide stoichiometry that is active.

(8)   (9)   (10)   (11)

Some assumptions made in this regard are: (i) the alkyl peroxide is activated by bidentate coordination to the Ti(IV) centre, (ii) the olefinic moiety is constrained so that it attacks the coordinated peroxide along the O-O bond axis, and (iii) the epoxide C-O bonds are formed simultaneously. However, the ultimate asymmetric determinant(s) have not yet been conclusively identified and Corey has recently proposed an ion-pair transition state [9] .

A double Katsuki-Sharpless epoxidation has been applied affording diepoxides in high optical yields [10]. For instance the trienol (8), upon asymmetric epoxidation afforded the enantiomers (9) and (10) as well as the *meso*-isomer (11). Calculations showed that at 90% e.e., a ratio of (9):(10):(11) was 361:38:1. Thus the enantiomeric purity of (9) turned out to be 99.45%. Table 5.1. shows some examples of the Katsuki-Sharpless epoxidation.

Table 5.1.   Katsuki-Sharpless Epoxidations*

Substrate	Product % e.e.	Reference
	 >95	[3]
	94	[3]

(Table 5.1. contd.)

Substrate	Product % e.e.	Reference
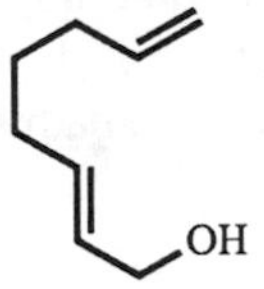	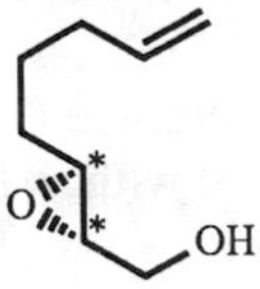 >95	[4]
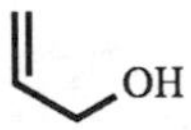 OH	 OH 95	[11]
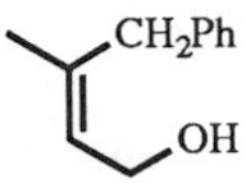 CH₂Ph OH	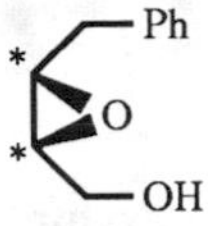 Ph O OH 91	[12]
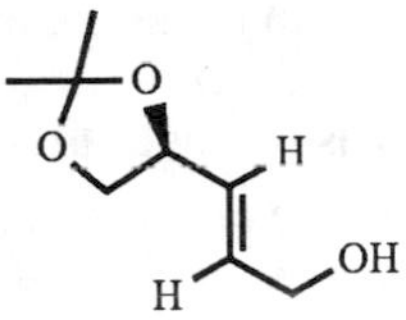	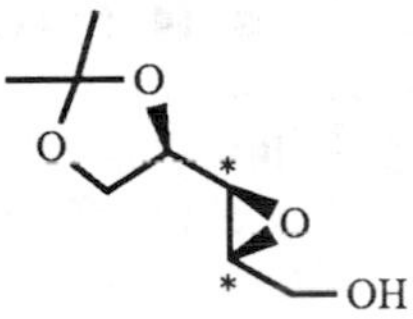 >95	[13]
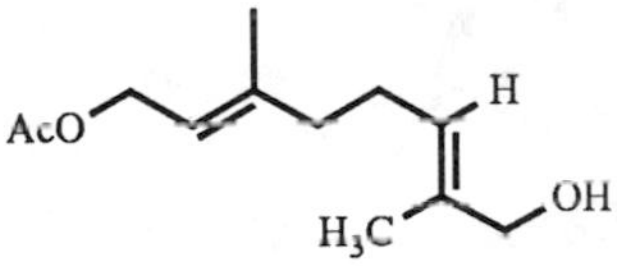	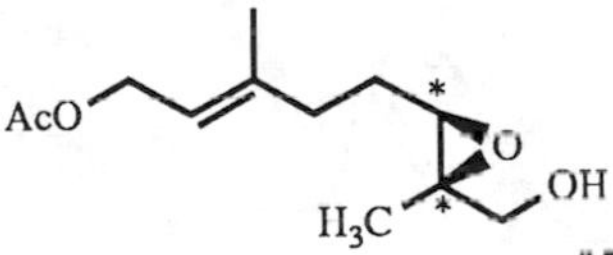 >95	[3]
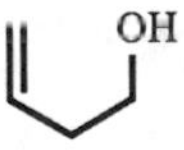 OH	 OH 55	[14]
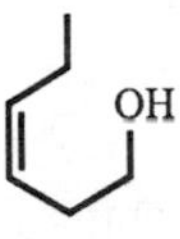 OH	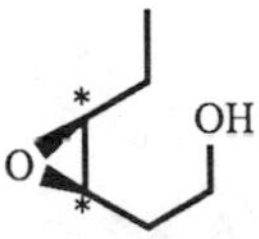 OH 50	[14]†
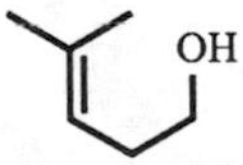 OH	 OH 27	[14]

† Homoallylic alcohols exhibit opposite stereochemistry towards L(+)- and D(-)-DET than allylic alcohols.

* Reagent = (+)-DET

5.1.2 *Asymmetric Epoxidation of other Substrates*

Several other methods have been developed for asymmetric epoxidations. For instance, vanadium (V) catalyzes the asymmetric epoxidation of *cis*-homo-allylic alcohols in high *anti* selectivity [15]. Thus, the *cis* homoallylic alcohol (12) is epoxidized with *tert*.butyl hydroperoxide (TBHP) in the presence of vanadium (V) to afford the epoxy alcohol (13) with an *anti:syn* ratio of 30:1 [15].

(12)   (13) (*anti* : *syn* = >30:1)

Another method of effecting asymmetric epoxidation involves the use of chiral iron porphyrins affording optical yields of upto 51% e.e. [16]. Thus iron-tetraphenylporphyrin complexes (14) and (15) were employed in the asymmetric epoxidation of styrenes. The complex (14) gave styrene oxide with 31% e.e. The reaction using (15) and PhIO or 2,4,6-(CH3)3 C6H2IO as oxidants afforded styrene oxide with 48% e.e., *p*-chlorostyrene oxide with 51% e.e. and 1-octene oxide with 20% e.e. [16].

Another approach involved the use of chiral oxodiperoxo-Mo complex (16) in the epoxidation of the olefin (17) which gave the epoxy product (18) in 53% e.e. [17].

(16)

(17)   (18) 53% e.e.

Terminal olefins have also been oxidized biochemically using microorganisms [18,19]. For instance, hexadec-1-ene (19) is oxidized to 1,2-epoxy hexadecene (20) by *Corynebacterium equi* in 41% yield and 100% e.e. [18].

(19)   (20) 100% e.e. (*R*)

Another approach toward stereoselective epoxidation involves Pd(II)-catalyzed oxidation of 1,5-dienes in the presence of chiral acids which affords asymmetric induction of upto 22% [20]. Thus, *cis*-1,2-divinyl cyclohexane (21) afforded (22) in the presence of Pd(OAc)$_2$ and (23) as the chiral acid in 11% e.e.

(21)   (22) 11% e.e.

Another method of catalytic asymmetric epoxidation involves the use of polyamino acids such as poly-(S)- alanine (24) in a triphase system affording high optical yields of the epoxy products [21]. Thus chalcone (25) is epoxidized to (26) with the aid of (24) in 96% e.e. The enantioselectivity decreases with lower degrees of polymerization of (24). The catalyst can be recovered and recycled, but a decrease in chemical and optical yields is observed.

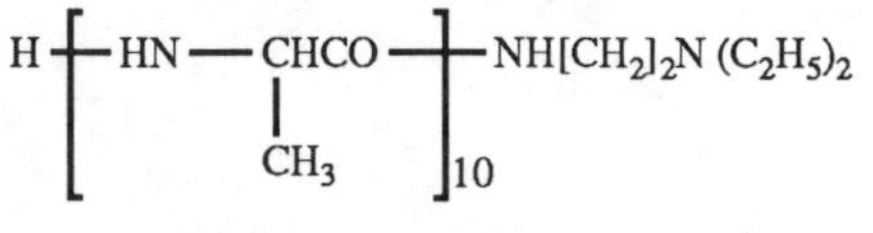

$$\text{H}\left[\text{HN}-\underset{\underset{\text{CH}_3}{|}}{\text{CHCO}}\right]_{10}\text{NH[CH}_2]_2\text{N (C}_2\text{H}_5)_2$$

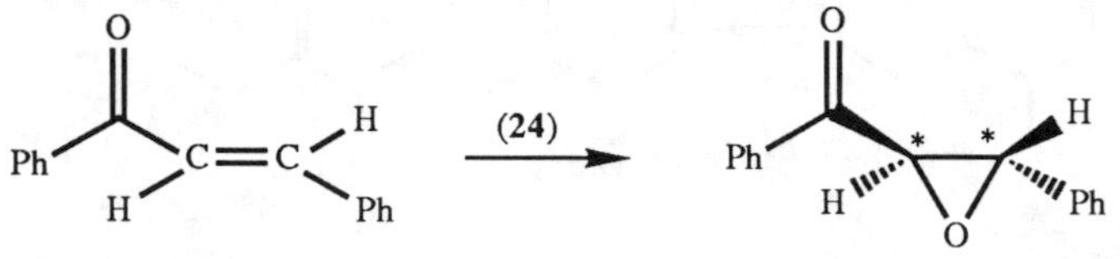

(24)

(25)   (24) →   (26) 96% e.e.

Unfunctionalized alkenes undergo asymmetric epoxidation in the presence of chiral 2-sulfonyloxaziridines to afford the corresponding epoxides in upto 40% e.e [22]. For instance chiral 2-sulfonyloxaziridine (27) employed in the oxidation of (28) afforded the product (29) in 35% e.e.

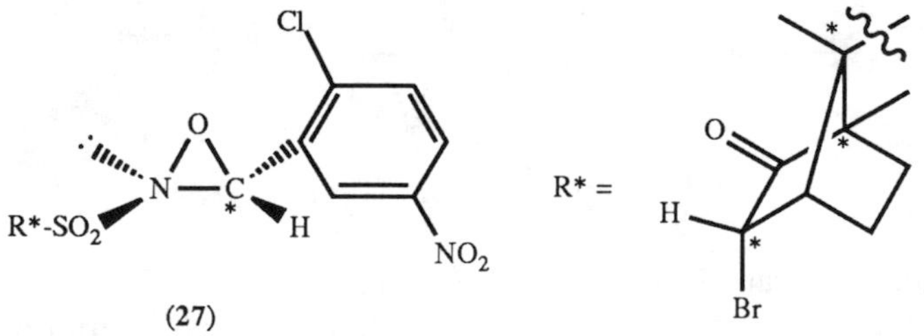

(27)   R* =

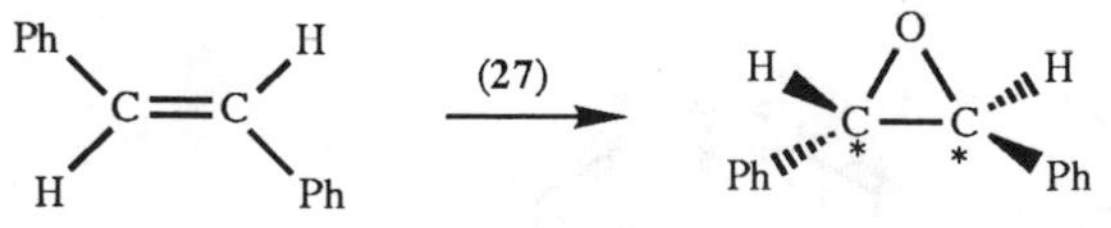

(28)   (27) →   (29) 35% e.e. (R, R)

5.2 Asymmetric Oxidation of Sulfides

Prochiral sulfides can be oxidized to chiral sulfoxides, important intermediates in asymmetric synthesis, by DEPT-Ti complexes in optical yields of upto 90% [23-27]. The use of a homogeneous reagent, which is the Sharpless reagent [Ti(O^{i}Pr)$_4$ + (mol. eql. DETP + 2 (^{t}BuOOH)] modified by 1 mol. eq. H$_2$O led to oxidation of arylalkyl sulfides to the corresponding sulfoxides in upto 95% e.e. [23,24]. Thus, methyl p-tolyl sulfide (30) was oxidized to sulfoxide (31) with the modifed Sharpless reagent in 93% e.e. The reaction was sensitive to steric hindrance and the enantioselectivity decreased with bulky groups.

(30)   (31) 93% e.e. (R)

For instance, isopropyl p-tolyl sulfide afforded the corresponding sulfoxide in 63% e.e. whereas n-butyl- and benzyl p- tolyl sulfoxides were afforded in 20% and 7% e.e. respectively [24]. The reaction involves electrophilic attack of the oxidant on the sulfur atom of the sulfide. Another approach involves direct oxidation of the sulfides by *tert*. butyl hydroperoxide in the presence of Ti(O^{i}Pr)$_4$ and (+)-DET affording the corresponding sulfoxides in upto 88% e.e. [26]. Other metal alkoxides, e.g. V and Mo afforded much lower optical yields.

Similarly, asymmetric oxidation of sulfides with organic hydroperoxides in the presence of catalytic amounts of optically active Schiff base oxovanadium (IV) complexes afforded chiral sulfoxides in 20-40% e.e. [28].

5.3 Asymmetric Oxidation of Selenides

The preparation of optically active selenoxides had been attempted in the seventies [29] while optically active sulfoxides were developed in the early twenties [30]. The difficulty lies in the facile racemization of the selenoxides through achiral hydrates [31]. Bulky substituents have therefore been added to the selenium moiety to prevent racemization and optically active selenoxides have been obtained with 1% to 33.7 % e.e. [32,33]. For instance, the selenide (32) is oxidized with Sharpless reagent to the optically active selenoxide (33) in 32.7% e.e. [33].

Optically active selenoxides have also been obtained by kinetic resolution, affording 5-11% e.e. [34] or by complexation with optically active diols [35].

(32)   (33) 32.7% e.e. (R)

5.4   Asymmetric Hydroxylations

The asymmetric synthesis of α-hydroxycarbonyl compounds and 1,2-diols is important in organic synthesis and notable successes have been achieved in this field relatively recently.

The synthetic utility of 2-(phenyl sulfonyl)-3-phenyl oxaziridine (34) and analogs, a new class of aprotic and neutral oxidizing reagents, has been established in asymmetric hydroxylations [36-38]. The high stereoselectivity exhibited by (34) is illustrated by oxidation of the enolate of lactone (35), generated by treating the lactone with potassium hexamethyldisilazane (KHMDS), to afford the hydroxylactone (36) in 100% e.e. and in 91% yield [36]. Similarly, (34) has been employed in the oxidation of chiral azaenolate (39) prepared from the ketone (37) and SAMP (38) affording (40) in 96% e.e. This undergoes oxidative cleavage with ozone to yield α-hydroxy ketones (41) in high enantiomeric purity [37].

(35)   (34)   (36) 100% e.e.

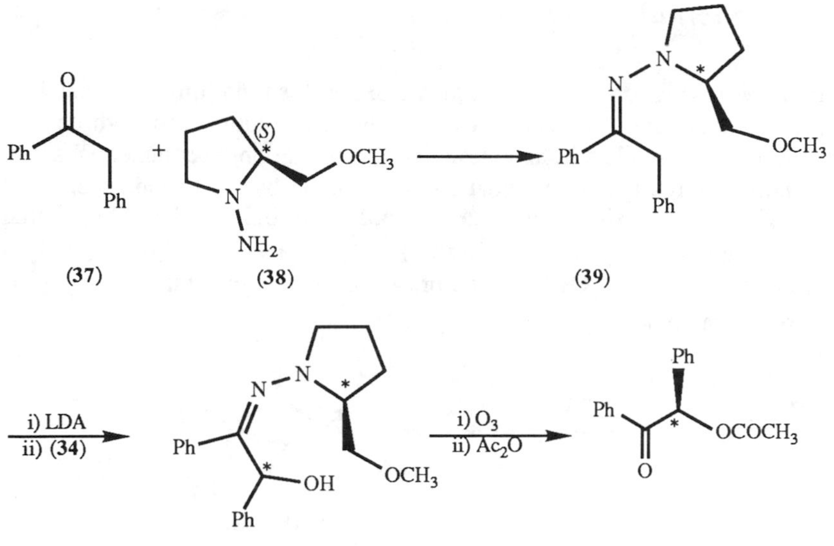

(37)   (38)   (39)

(40) >96% e.e.   (41) >96% e.e. (R)

Similarly, α-silylated ketones are oxidised to α-hydroxyketones in high optical yields [39]. Thus (42) on treatment with a base followed by oxidation with (34) gives (43) which is hydrolysed with acid followed by flash chromatography to give (44). Subsequent desilylation with aq. HBF$_4$ afforded (45) in 98% e.e.

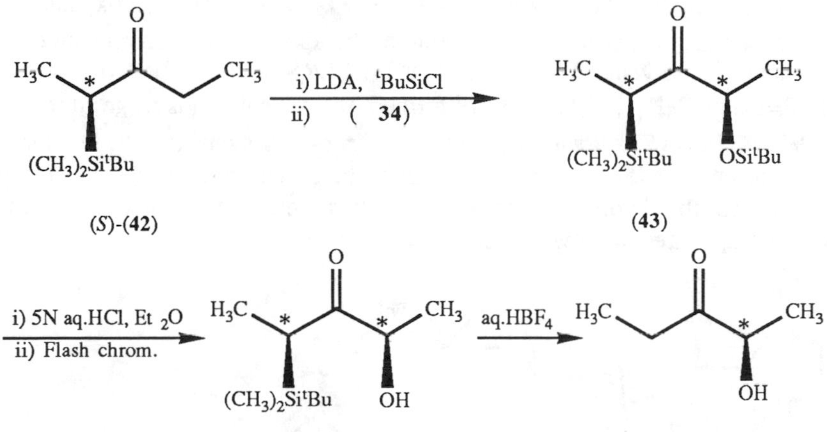

(S)-(42)   (43)

(44)   (45) >98% c.c. (R)

5.4.1 *Vicinal Hydroxylations*

Vicinal hydroxylation with osmium tetroxide (osmylation) is a standard procedure in organic synthesis. However the toxic nature and high cost of osmium tetroxide renders it less attractive for utilization on commercial scale. An important inroad into this goal has been made by " a ligand accelerated catalysis". Thus by using cinchona alkaloids with 0.2- 0.4% OsO_4, vicinal diols were obtained in 20-88% e.e. [40,41]. For instance, (E)-stilbene (46) is oxidized with catalytic OsO_4 in the presence of (47) to afford the (R), (R)-dihydroxy compound (48) in 88% e.e. [40].

(46)

(48) 88% e.e. (R, R)

(47)

$R^1 = p$-chlorobenzoyl

$R^2 =$

Recently, new chiral amines (49-54) have been employed in the asymmetric dihydroxylation of alkenes with OsO_4 [42,43]. The hydroxylation in this reaction is dependent upon the nature of the alkyl group in chiral amines as well as on the solvents. With R=CH_3 or Pr, (R,R)-diols are formed, whereas with R=Bu or C_5H_{11}, (S,S)-diols are formed, the optical yields range from 7 to 100 % e.e. Thus (E)-stilbene is oxidised to *threo*-hydrobenzoin (48) in toluene in the presence of (54) in 100% e.e. [43]. The structure of the active species involved in the hydroxylation reaction has been determined by X-ray crystallography and is shown in (55).

(49) R = CH_3, (52) R = Bu,
(50) R = C_2H_5, (53) R = pentyl,
(51) R = Pr,   (54) R = neohexyl

(55)

Other chiral ligands have also been employed in the asymmetric osmylation of chiral α, β-unsaturated esters affording high *anti:syn* diastereo-selection [44]. Thus, the ester (57) is oxidised to give (58) and (59) in the presence of OsO$_4$ and (56) in a diastereomeric ratio of 45:1 [43].

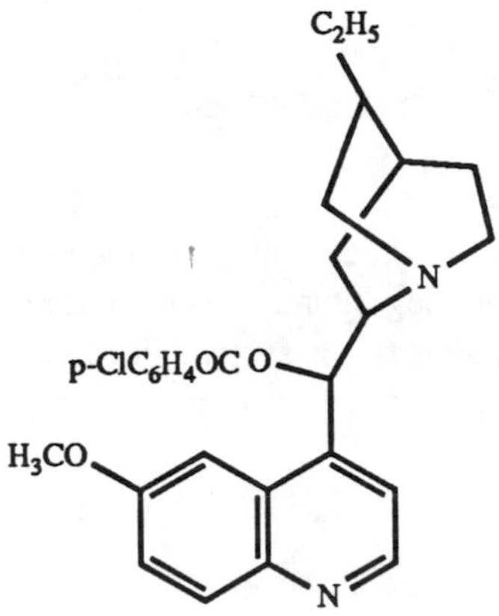

(56)

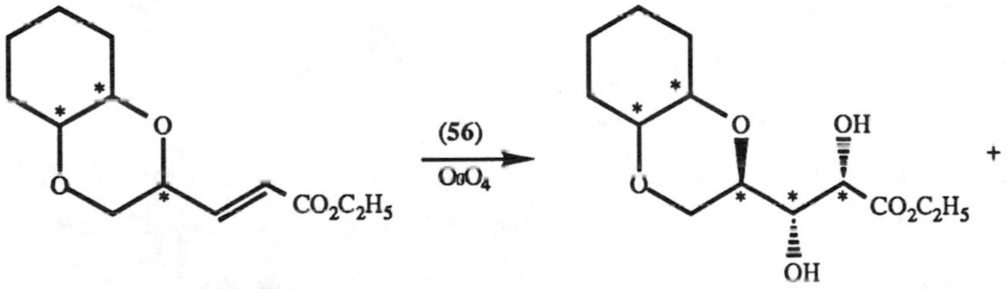

(57)   (58)

(58 : 59 = 45:1)

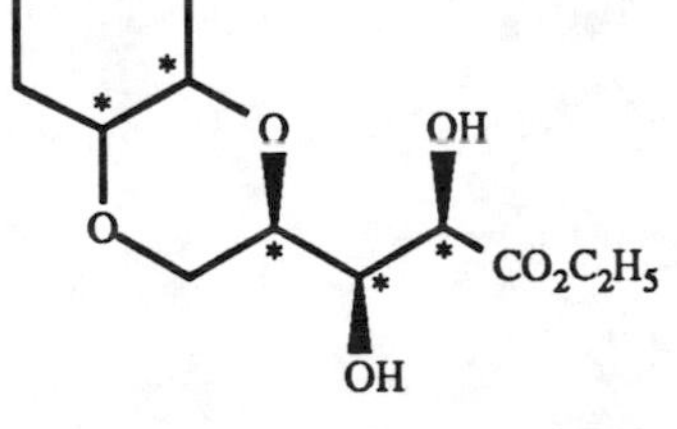

(59)

Similarly oxygen-substituted allylic silanes undergo vicinal hydroxylations with very high levels of diastereofacial selectivities in the presence of catalytic amount of OsO$_4$ to give predominantly 1,2-*anti*-1,2,3-triols [45]. For instance, the substrate (60) is oxidized to the *anti* and *syn* products (61) and (62) in a ratio of 97:3 [45].

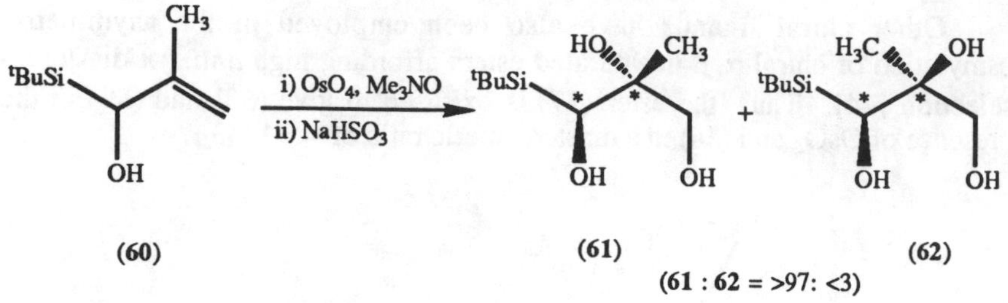

(60)   (61)   (62)

$(61 : 62 = >97: <3)$

However (Z)-crotylsilanes showed lower selectivities. The poor selectivity of the Z-isomers may be due to the steric destabilizing interactions between the vinyl methyl and the C-OR' substituents eclipsing the π-system in (63), as shown in Fig. 5.2.

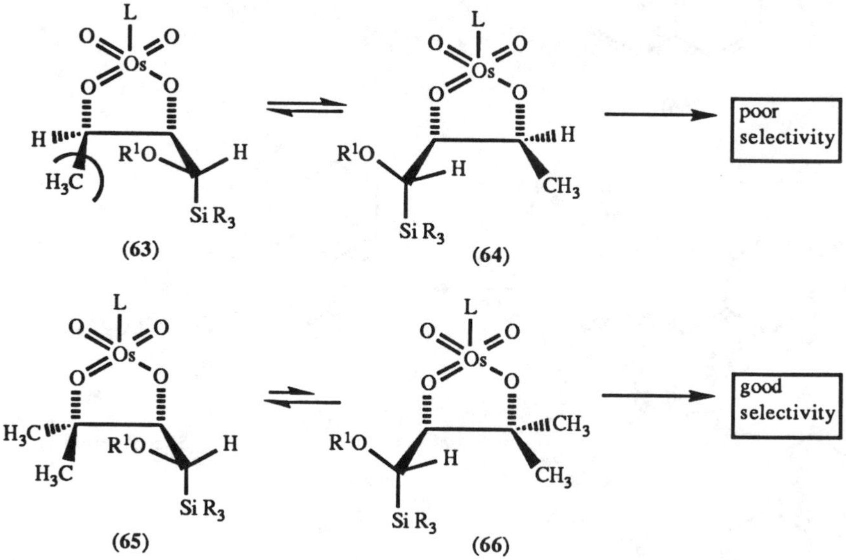

Figure 5.2.   Facial bias in osmylation.

Steric interaction in (Z)-alkenes (63) and (64) results in diminished face selectivity whereas the destabilizing interaction is less severe for the E stereoisomers, (65) and (66) giving a higher selectivity.

5.5 Asymmetric Oxidation of Aromatic Substrates via Donor-Acceptor Interaction

Donor-acceptor interaction plays an important role in enzymatic systems and based on enzymatic models, certain reagents have been developed which are capable of recognizing different positions on aromatic substrates [46-49]. The use of 2,3-dichloro-5,6-dicyanobenzoquinone (DDQ) (67) in the asymmetric oxidation of benzylic carbon has afforded diastereoselectivity of upto 66% [49]. The reagent (67) is known to form charge-transfer complexes with aromatic substrates [49]. Hence using chiral substrates, the difference in the stability of the intermediate complexes (68) and (69) would lead to asymmetric oxidation at the benzylic position of the substrate, as shown in Fig. 5.3.

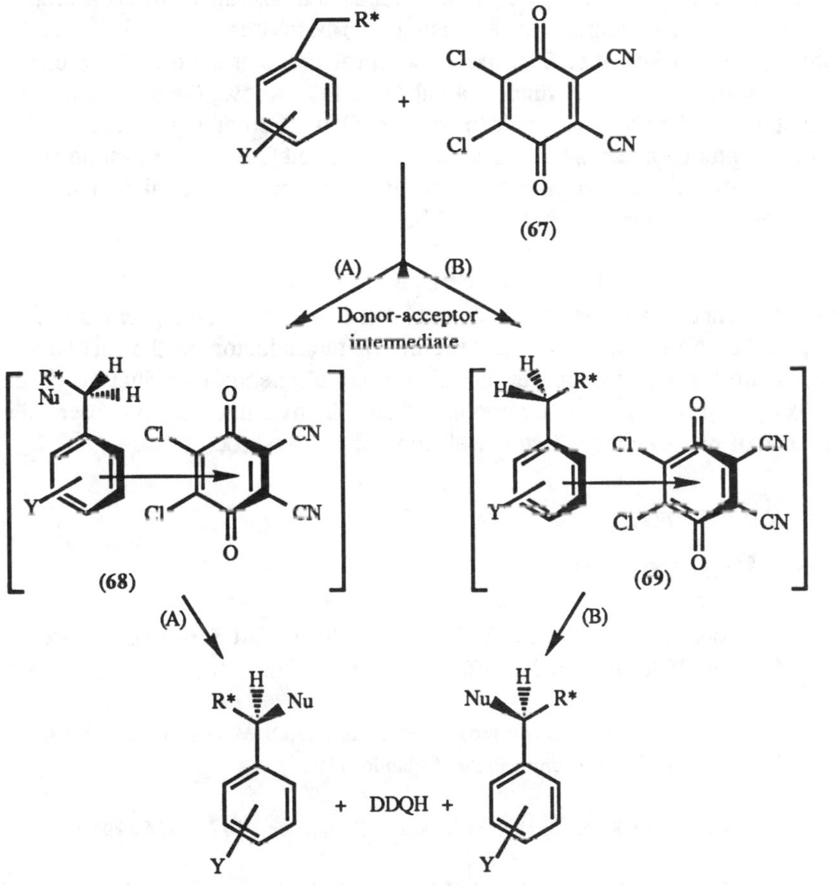

Figure 5.3. Mechanism of oxidation by DDQ (Y = electron releasing group).

For instance, substituted phenylacetic acid (70) is oxidized to the mandelic acid derivative (71) using DDQ and acetic acid in 67% d.e. [49].

(70)   (71) 67% d.e.

(R* = (-)-8-phenylmenthyl)

The rate and diastereoselectivity of oxidation by DDQ are dependent mainly on the diastereofacial bulking of the inductor as demonstrated by the differences observed for diastereomeric excess using (+)-bornyl ester (5% d.e.) and (-)-isobornyl ester (54% d.e.).The use of a chiral inductor such as (-)-8-phenyl menthol which specifically hinders a substrate face, allows the approach of the reagent during the donor-acceptor to be controlled affording high induction. The donor-acceptor complex is formed at the less hindered face of the substrate during the course of the reaction, therefore permitting selective removal of one of the diastereotopic hydrogens from the prochiral methylene.

Thus the asymmetric centre is created in two distinct steps. The first step involves a donor-acceptor interaction between two reagents, the geometry of the complex being dependent on the structure of the inductor. In the second step DDQ removes a hydrogen atom with the simultaneous introduction of the acetoxy group from the bulkier opposite face. The overall induction observed is the result of the combined diastereoselection of the two processes.

5.6   References

1.   B. Bosnich, ed., "Asymmetric Catalysis", NATO ASI Series 1E: Applied Sciences 103, Martinus Nijhoff, Dordrecht, (1986).

2.   B.E. Rossiter: in "Asymmetric Synthesis", (J.D. Morrison, ed.), p.193, Vol. 5, Orlando: Academic Press, Orlando, (1985).

3.   T. Katsuki and K. B. Sharpless, *J.Amer.Chem.Soc.*, **102**, 5974 (1980).

4.   B.E. Rossiter, T. Katsuki and K.B. Sharpless, *J.Amer.Chem.Soc.*, **103**, 464 (1981).

5.   R.M. Hanson and K.B. Sharpless, *J.Org.Chem.*, **51**, 1922 (1986).

6.   D.L. Lu, R.A. Johnson, M.G. Finn and K.B. Sharpless, *J.Org.Chem.*, **49**, 731 (1984).

7.   M.G. Finn and K.B. Sharpless, *J.Amer.Chem.Soc.*, **113**, 113 (1991).

8.   S.S. Woodward, M.G. Finn and K.B. Sharpless, *J.Amer.Chem.Soc.*, **113**, 106 (1991).

9.   E.J. Corey, *J.Org.Chem.*, **55**, 1693 (1990).

10. T.R. Hoye and J.C. Suhadolnik, *J.Amer.Chem.Soc.*, **107**, 5312 (1985).

11. K.B. Sharpless, S.S. Woodward and M.G. Finn, *Pure Appl. Chem.*, **55**, 1823 (1983).

12. K.B. Sharpless, C.H. Behrens, T. Katsuki, A.W. M. Lee, V.S. Martin, S.M. Viti, F.J. Walker and S.S. Woodward, *Pure Appl.Chem.*, **55**, 589 (1983).

13. T. Katsuki, A.W.M. Lee, P.Ma, V.S. Martin, S. Masamune, K.B. Sharpless, D. Tuddenham and F.J. Walker, *J.Org.Chem.*, **7**, 1373 (1982).

14. B.E. Rossiter and K.B. Sharpless, *J.Org. Chem.*, **49**, 3707 (1984).

15. T. Hanamoto, T. Katsuki and M. Yamaguchi, *Tetrahedron Lett.*, **28**(49), 6191 (1987).

16. J.T. Groves and R.S. Myers, *J. Amer.Chem.Soc.*, **105**, 5791 (1983).

17. E. Broser, K. Krohn, K. Hintzer and V. Schurig, *Tetrahedron Lett.*, **25**, 2463 (1984).

18. H. Ohta and H. Tetskawa, *J.Chem.Soc.Chem.Commun.*, 849 (1978).

19. S.W. May and R.D. Schwartz, *J.Amer.Chem.Soc.*, **96**, 4031 (1974).

20. A. Heumann and C. Moberg, *J.Chem.Soc.Chem.Commun.*, 1516 (1988).

21. S. Julia, J. Guixer, J. Masana, J. Rocas, S. Colonna, R. Annuziata and H. Molinari, *J.Chem.Soc. Perkin Trans I*, 1317 (1982).

22. F.A. Davis, M. E. Harakal and S.B. Awad, *J.Amer.Chem.Soc.*, **105**, 3123 (1983).

23. P. Pichen, E. Dunach, M.N. Deshmukh and H.B. Kagan, *J.Amer.Chem.Soc.*, **106**, 8188 (1984).

24. P. Pittchen and H.B. Kagan, *Tetrahedron Lett.*, **25**(10), 1049 (1984).

25. H.B. Kagan, E. Dunach, C. Nemecek, P.Pitchen, O. Samuel and S.H. Zhao, *Pure Appl. Chem.*, **57**, 1911 (1985).

26. F. Di Furia, G. Modena and R. Seraglia, *Synthesis*, 325 (1984)

27. E. Dunach and H.B. Kagan, *Nouv. J.Chem.* **9**, 1 (1985).

28. K. Nakajima, M. Kojima and J. Fujita, *Chemistry Lett.*, 1483 (1986).

29. D.N. Jones, D. Mundy and R. D. Whitehouse, *J.Chem.Soc.Chem. Commun.*, 86 (1970).

30. P.W.B. Harrison, J. Kenyon and H. Phillips, *J.Chem.Soc.*, 2079 (1926).

31. M. Oki and H. Iwamura, *Tetrahedron Lett.*, 2917 (1966).

32. M. Kobayashi, H. Ohkubo, and T. Shimizu, *Bull.Chem.Soc.Jpn.*, **59**, 503 (1986).

33. T. Shimizu, M. Kobayashi and N. Kamigata, *Bull.Chem.Soc.Jpn.*, **62**, 2099 (1989).

34. F.A. Davis, O.D. Stringer and J.P. McCauley Jr., *Tetrahedron*, **41**, 4747 (1985).

35. F. Toda and K. Mori, *J.Chem.Soc.Chem.Commun.*, 1357 (1986).

36. F.A. Davis, L.C. Vishwakarma and J.M. Billmers, *J.Org.Chem.*, **49**, 3241 (1984).

37. D. Enders and V. Bhushan, *Tetrahedron Lett.*, **29**(20), 2437 (1988).

38. F.A. Davis and M.C. Weismiller, *J.Org.Chem.*, **55**, 3715 (1990)

39. B.B. Lohray and D. Enders, *Helv.Chim. Acta*, **72**, 980 (1989).

40. E.N. Jacobsen, I. Marko, W.S. Mungall, G. Schröder and K.B. Sharpless, *J.Amer.Chem.Soc.*, **110**, 1968 (1988).

41. B.B. Lohray, T.H. Kalantar, B.M. Kim, C.Y. Park, T. Shibata, J.S.M. Wai and K.B. Sharpless, *Tetrahedron Lett.*, **30**(16), 2041 (1989).

42. M. Hirama, T. Oishi and S. Ito, *J.Chem.Soc.,Chem.Commun.*, 665 (1989).

43. T. Oishi and M. Hirama, *J.Org.Chem.*, **54**, 5834 (1989).

44. R. Annunziata, M. Cinquini, F. Cozzi and L. Raimondi, *Tetrahedron,* **44**, 6897 (1988).

45. J.S. Panek and P.F. Cirillo, *J.Amer.Chem.Soc.*, **112**, 4873 (1990).

46. A. Guy, M. Lemaire and J.P. Guette, *J.Chem.Soc.,Chem.Commun.*, 8 (1980)

47. M. Lemaire, A. Guy and J.P.Guette, *Bull.Soc.Chim.Fr.*, 477 (1985).

48. M. Lemaire, A. Guy, D. Imbert and J.P. Guette, *J.Chem.Soc.Chem. Commun.,* 741 (1986).

49. A. Guy, A. Lemor, D. Imbert and M. Lemaire, *Tetrahedron Lett.*, **30**, 327 (1989).

6   Asymmetric Carbon-Heteroatom Bond Formations

Among the many methods available for the stereoselective carbon- heteroatom bond formations, only those involving formation of C-O, C-N, C-S, C-P and C-H bonds will be discussed.

6.1   Carbon-Oxygen Bond Formation

A number of methods are available for the stereoselective formation of carbon-oxygen bonds. These include halolactonization and hydroboration.

6.1.1   Asymmetric Halolactonization

Asymmetric halolactonization, i.e. the addition of a halogen atom and a carboxy group across a double bond in an *anti* manner resulting in the formation of an α-halolactone has been extensively studied to prepare optically active substrates such as α-hydroxy acids, epoxides etc. [1-5]. The method has been particularly useful in the total synthesis of prostaglandins [6,7]. For instance, the halolactonization of proline amides proceeds stereospecifically affording α-hydroxy acids in 89-98% e.e upon dehalogenation and hydrolysis [1]. (S)-(-)-tigloylproline (3) (formed by the reaction of tigloyl chloride (1) with (S)-proline (2)) undergoes bromolactonization to afford (4) and (5) in a ratio of 94.5:5.5 [3].

(1)	(2)	(3)

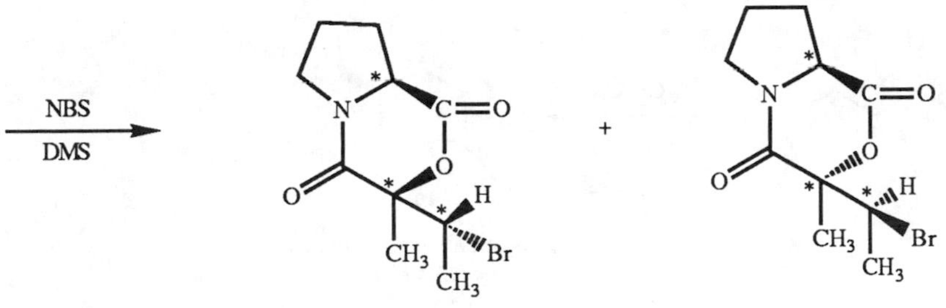

(4)   (5)

(4:5 = 94.5: 5.5)

The reaction leading to the predominant diastereomer is believed to proceed via the intermediate (6) [3]. Debromination of (4) followed by acid hydrolysis afforded the R- α-hydroxy acid (7) in 89% e.e.

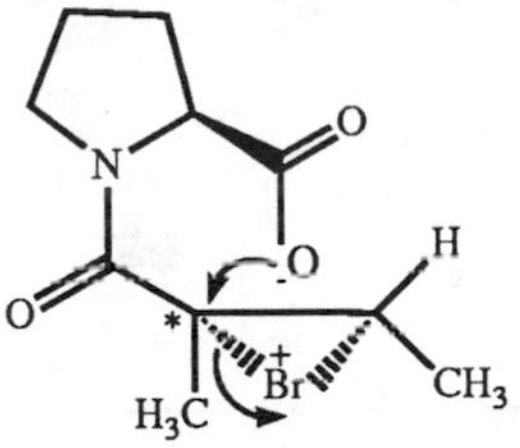

(6)

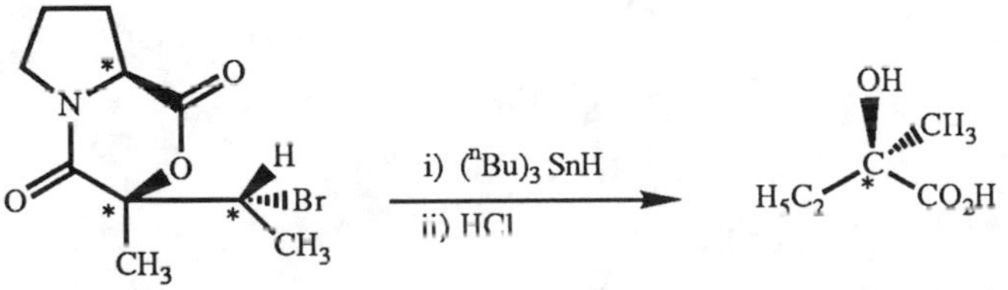

(4)   (7) 89% e.e. (R)

Similarly, iodolactonization of acyclic olefinic and hydroxy olefinic acids affords iodolactones which can be converted to epoxides upon methanolysis [4,5]. By varying the reaction conditions, either of the two products (i.e. kinetic or thermodynamic) are obtained. For instance, the substrates (8) and (10) afford the thermodynamic product (9) and the kinetic product (11) which are disposed at C-3 and C-4 as *trans* and *cis* respectively.

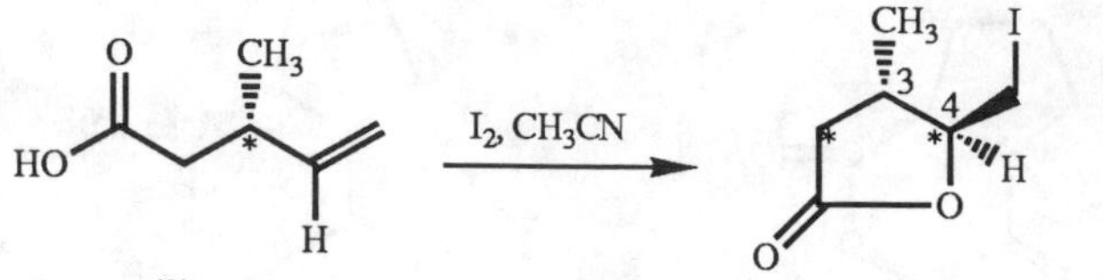

(8)   (9)

thermodynamic product
(3,4 *trans*)

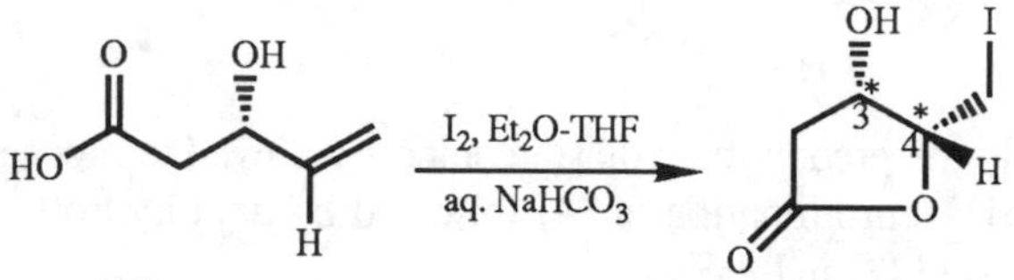

(10)   (11)

kinetic product
(3,4-*cis*)

Table 6.1. gives some examples of asymmetric halolactonizations.

Table 6.1. Asymmetric Halolactonizations

Substrate	React. Cond.	Product[†]	Optical Yield	Reference
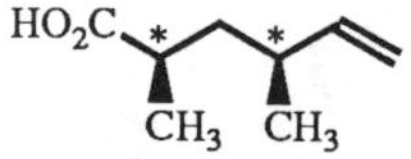	a	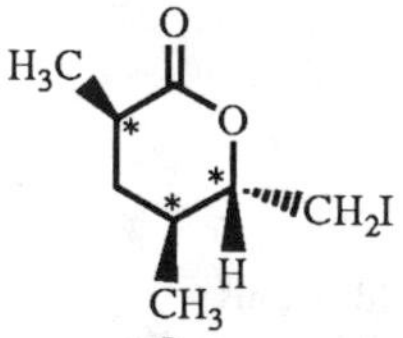	(20 : 1)	[5]
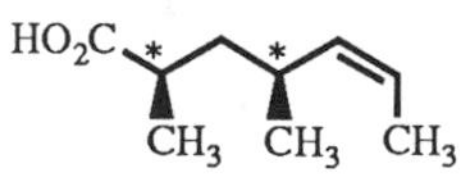	a	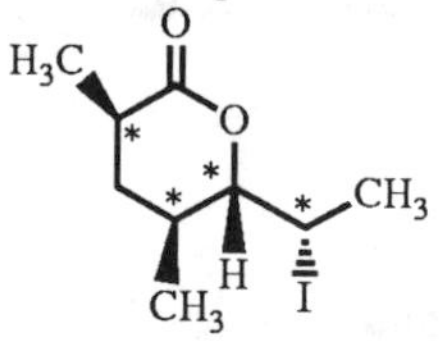	(20 : 1)	[8]
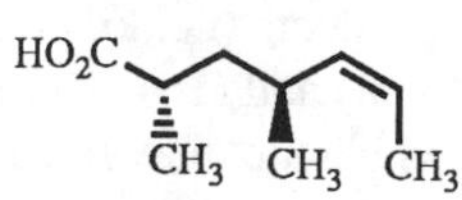	b	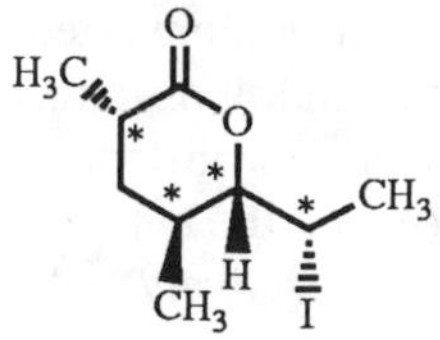	(20 : 1)	[8]

(Table 6.1. contd.)

Substrate	React. Cond.	Product[†]	Optical Yield	Reference
(pyrrolidine N-acyl, CO$_2$H, CH$_3$, H, Ph)	c	(bicyclic lactone; CH$_3$, H, Br, Ph)	(99 : 1)	[3]
(H$_3$C, H$_3$C, N(CH$_3$)$_2$, vinyl)	d	(tetrahydrofuranone; H$_3$C, H$_3$C, 2, 4, I)	(2,4 trans) (97 : 3)	[9]
(PhCH$_2$, N(CH$_3$)$_2$)	e	(Ph, lactone, I)	> 99 : <1	[9]
(OH, CO$_2$H, CH$_3$)	f	(OH, I, CH$_3$, lactone)	100% cis	[9]
(OH, ^{n}Bu, CH$_3$)	g	(OH, OH, CH$_3$, ^{n}Bu, H, I)	99%	[10]
(OSi, HO$_2$C, CH$_3$, CH$_3$)	h	(SiO, I, H, CH$_3$, CH$_3$, lactone)	96:4	[4]
(OH, HO$_2$C, CH$_3$)	h	(OH, I, H, CH$_3$, H, lactone)	100 : 0	[4]

([†]only major isomer is shown)

a = I$_2$, CH$_3$, CN, 0°C,   b = I$_2$, CH$_3$CN, NaHCO$_3$, 0°C   c = NBS, ^{t}BuOK, -20°C → rt

d = i) I$_2$, ii) Bu$_3$SnH,   e = I$_2$ (CH$_3$O)$_2$ C$_2$H$_4$: H$_2$O (1 : 1),   f = I$_2$, Et$_2$O, THF, H$_2$O+NaHCO$_3$

g = I$_2$, H$_2$O,   h = I$_2$, NaHCO$_3$

6.1.2 Asymmetric Hydroboration

Another method of carbon-oxygen bond formation is through hydroboration followed by oxidation. An optically active alkylborane may be used for the asymmetric induction [13]. For instance diisopinocampheylborane (13) formed by the reaction of α-pinene (12) with diborane proved to be effective in inducing asymmetry, particularly in unhindered (Z)-olefins [12].

Thus hydroboration of *cis*-2-butene (14) with (13) afforded the organoborane intermediate (15) which was oxidized with alkaline H_2O_2 to give (R) (-)-2-butanol (16) in 87% e.e. [13]. The optical purity of the α-pinene used was 93%. Increasing its optical purity to 99% increased the optical yield of the alcohol to 98.1% e.e. [11].

The high asymmetric induction exhibited by (13) is attributed to the participation of a four-centre transition state (18) [14] between diisopinocampheylborane, which is depicted as (17) in its most stable conformation having boron as diequatorial, and the substrate (14).

The formation of such a rigid transition state is highly influenced by steric factors of both the substrate and the reagent and in the preferred transition state such as **(18)** the methyl group of *cis*-2-butene is positioned away from the more bulky methylene group at C_4 and towards the smaller hydrogen atom at C_3'.

Similarly, monoisopinocampheylborane **(19)** is the least hindered and most reactive of all the chiral hydroborating agents, particularly for *trans* alkenes for which the diisopinocampheylborane **(13)** is ineffective [11]. Thus *trans*-2-butene **(20)** reacts with **(19)**, and after oxidation affords 2-butanol **(21)** in 73% e.e. [15].

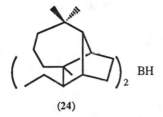

      i) **(19)**
      ii) H_2O_2

(19)   **(20)**   **(21)**   73% e.e. (*S*)

Similarly *trans*-2,2,5,5-tetramethyl-3-hexene **(22)** on reaction with **(19)** afforded the alcohol **(23)** in 92% e.e.

      i) **(19)**
      ii) H_2O_2

(22)   **(23)**   92% e.e. (*R*)

Another promising hydroborating agent is dilongifolylborane **(24)** which is very effective in hydroborating *cis*-alkenes and trisubstituted alkenes [16].

BH

(24)

Table 6.2. gives some examples of the asymmetric formation of carbon-oxygen bonds through hydroboration followed by oxidation.

Table 6.2.   Carbon - Oxygen Bond Formation Through Hydroboration - Oxidation

Substrate	Borane	React. Cond.	Product*	% Select.	Reference
(structure)	(24)	30°C, THF, H_2O_2	H_3C, OH, H, H_3C, CH_3	70 (R)	[16]
(structure)	(24)	H_2O_2	H_3C, OH, H, H_3C, CH_3	78 (R)	[16]
Ph, CH_3	(13)	-25°C, THF, H_2O_2	OH, H, Ph, CH_3	63 (S)	[17]
(structure)	(13)	-25°C, THF, H_2O_2	HO, H	93.2 (R)	[17]
Ph (structure)	(19)	-25°C, H_2O_2	H, OH, Ph	76 (R)	[18]
CH_3 (structure)	(19)	-25°C, H_2O_2	CH_3, H, OH	72 (1S, 2R)	[18]
Ph (structure)	(19)	-25°C, H_2O_2	Ph, HO	(100) (1S, 2R)	[18]
(structure)	(24)	30°C, THF, H_2O_2	H, OH	75 (R)	[11]
(structure)	(13)	H_2O_2	HO	96 (1S, 2S)	[19]
$CH_2CO_2CH_3$ (structure)	(13)	H_2O_2	OH, $CH_2CO_2CH_3$	(R, R)	[20]

6.2   Carbon-Nitrogen Bond Formation

The stereoselective formation of carbon-nitrogen bonds has been generally studied in cases which result in cyclization to give heterocyclic compounds. These are often important intermediates in the synthesis of biologically active natural products [21]. Various methods have been accordingly developed for the formation of nitrogen heterocycles. These include halocyclization and mercuricyclization.

6.2.1   Halocyclization

N-Chloroamines undergo intramolecular cyclization with olefinic bonds in the presence of Lewis acids affording very high stereospecificities (upto 100% d.p.) [22]. For instance, the N-chloroamine (24) when reacted with CuCl/CuCl$_2$ in a molar ratio of 1 : 0.1 : 1 underwent homolytic fission of the N-Cl bond affording the metal-complexed aminyl radical (25) which cyclised by intramolecular addition to double bond followed by *anti*-attack of the chlorine at the radical (26) to give the product (27) in 100% diastereomeric purity.

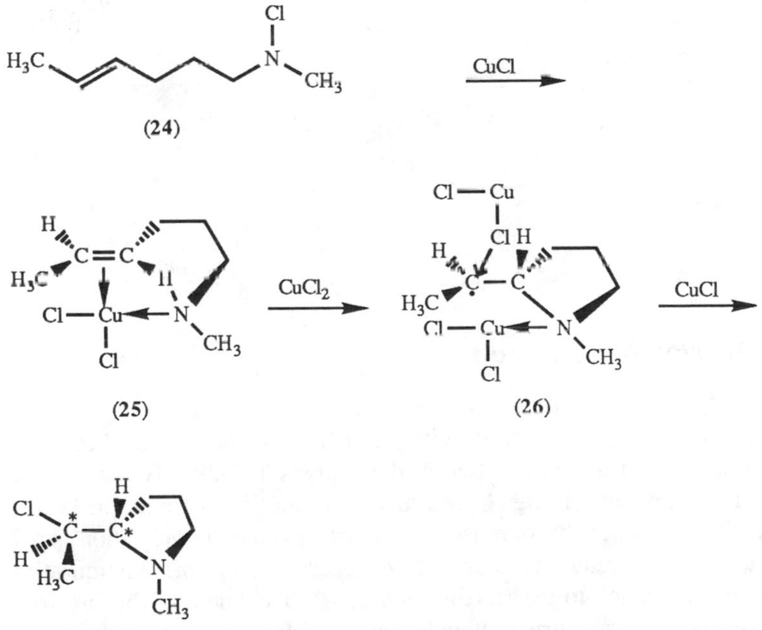

(24)

(25)   (26)

(27) *anti* (100% d.p.)

6.2.1.1   Iodolactamization

Recently, iodolactamization has been employed in the stereoselective construction of carbon-nitrogen bonds affording high diastereomeric yields of the products, since amides are less susceptible to oxidation than amines [23-25]. For instance the thioimidate (28) undergoes iodolactamization to form the products (29) and (30) in a diastereomeric ratio of 12:1 [24].

(28)   (29)   (30)

(29 : 30 = 13:1)

The high 1,3-*trans* selectivity has been rationalized by assuming a cyclic transition state (31) in which the iodomethyl group bears a *quasi*-equatorial orientation, while the C-3 methyl group is axial to avoid 1,2-interaction with the thiomethyl group [26].

(31)

6.2.2   Mercuricyclization

Mecuricyclization for forming carbon-nitrogen bonds has been studied in detail with   unsaturated amines and other N-derivatives [27-29]. However, the demercuration step involving borohydride reduction often spoils the regioselectivity and hence the stereoselectivity of the mercuricyclization, since the product undergoes reductive elimination under a variety of demercuration conditions with reversion to the starting materials. Treatment of the substrate (32), for instance, with mercuric acetate followed by demercuration with sodium borohydride afforded the piperidine acetonides (33) and (34) in a ratio of 21:1 [28].

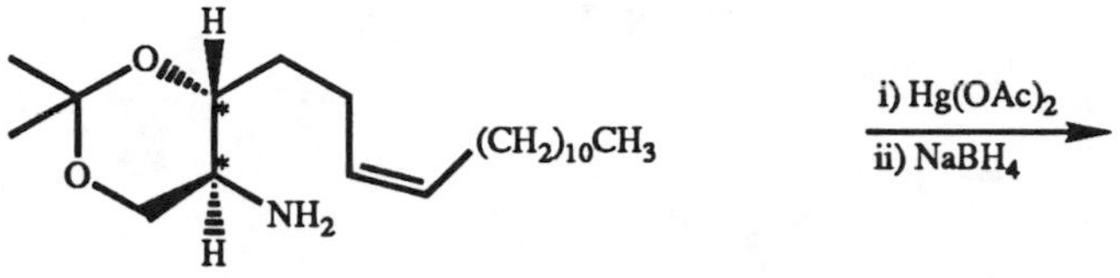

(32)

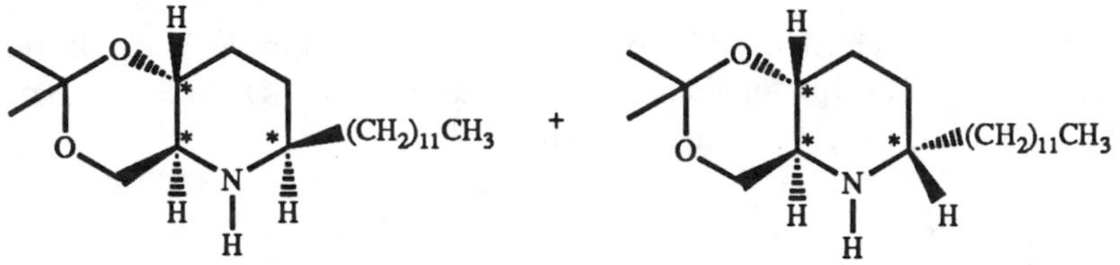

i) Hg(OAc)$_2$
ii) NaBH$_4$

(33)

(34)

(33 : 34 = 21:1)

Similarly, the δ-alkenylimine derivative (**35**) underwent cyclization with Hg(OAc)$_2$ followed by NaBH$_4$ reduction to give the *trans* product (**36**) in 98% yield [27].

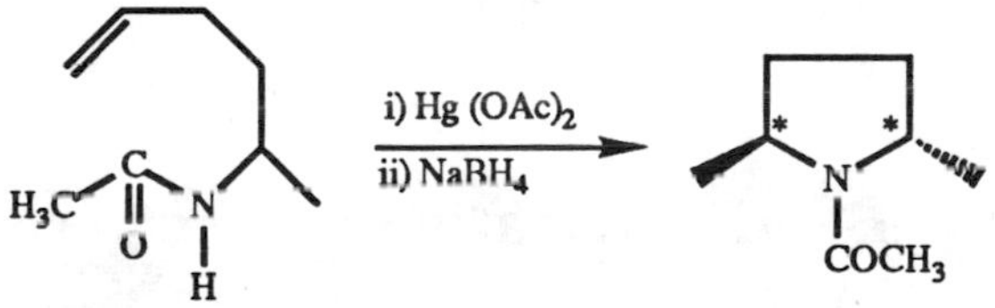

i) Hg (OAc)$_2$
ii) NaBH$_4$

(35)

(36)   98% *trans*

The stereochemistry of the cyclization was rationalized by assuming a preferred conformation of a chair-like transition state (**37**) taking part in the reaction.

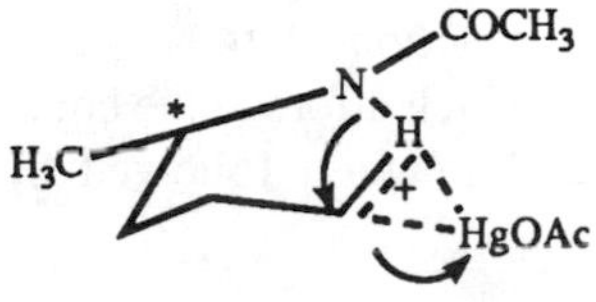

(37)

Other methods of asymmetric carbon-nitrogen bond formation include the reaction of allylic acetates with amines in the presence of palladium supported on cross-linked polystyrene or by silica gel catalysis [30]. For example the compound (**38**) reacts with diethylamine in the presence of polystyrene - supported Pd to give exclusively the *cis* product (**39**) [30].

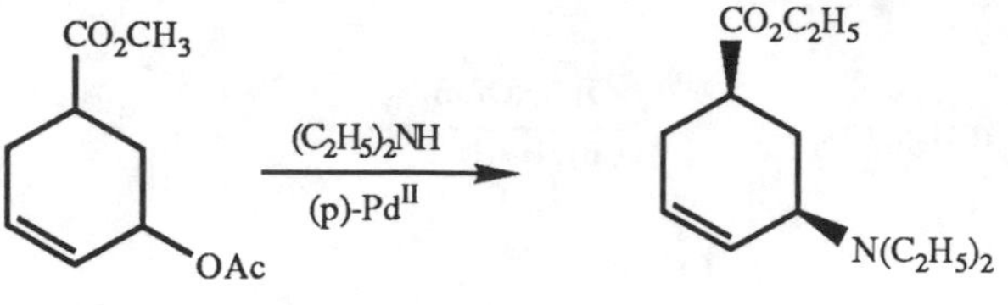

(38)

(39)   100% *cis*

Similarly, phenyl selenylchloride has been employed in the N-cyclization of urethanes [31]. Thus the urethane (40) cyclizes to form (41) in the presence of PhSeCl affording the product in 86% yield [31]

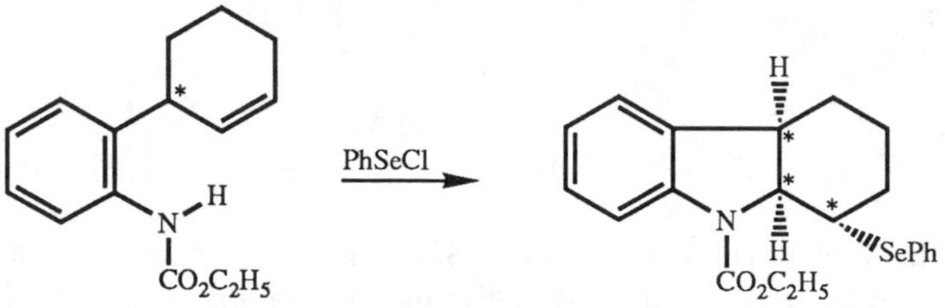

(40)

(41)   86%

6.3   Carbon-Sulfur Bond Formation

Various methods have been developed to accomplish stereoselective carbon-sulfur bond formations, including the enantioselective Michael addition of thiols [32], the use of sulfonyl chloride in the asymmetric synthesis of chiral sulfoxides [33,34], as well as the protiothiolactonization of thioamides [9]. For instance, the chiral ethanolamine (42) reacts with $SOCl_2$ to give a diastereomeric mixture of (43) and (44) in a ratio of 72:28 respectively. Reaction of (43) with a Grignard reagent gave the open-chain sulfinamide (45) which afforded the chiral sulfoxide (45) after treatment with excess Grignard reagent, followed by hydrolysis.

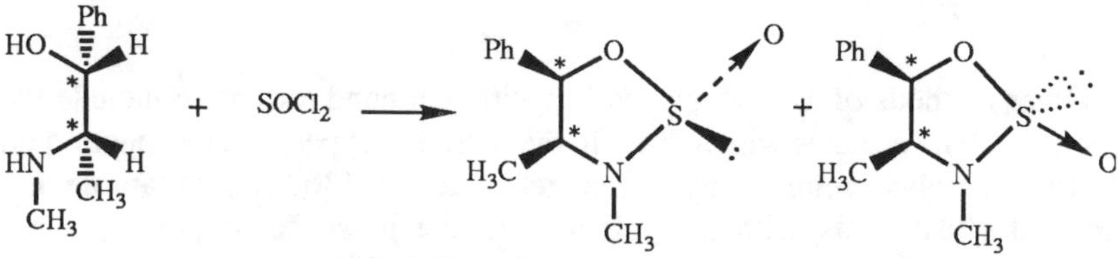

(42)

(43)

(44)

(43)   (45)

(46)   100% o.p.

Another method for asymmetric carbon-sulfur bond formation is the enantioselective Michael addition of thiols to cycloalkenones in the presence of a chiral catalyst. The thiol (47), for instance, reacts with 2-cyclohexene-1-one (48) in the presence of the 4-hydroxypyrrolidine derivative, (49) to give the product (50) in 88% e.e. [32].

(47)   (48)   (50)   88% e.e. (R)

The stereoselectivity of the reaction has been explained by assuming the formation of an ammonium thiolate complex (51) by the thiol (47) and the catalyst (49), which is stabilized by hydrogen bonding.

(51)

Subsequently, cyclohexenone approaches the complex (51) through formation of a hydrogen bond between its carbonyl group and the hydroxyl group of the catalyst. Finally attack of the thiolate anion takes place at the *re*-face which is sterically favoured as shown in (52) to give the product (50) having *R* configuration of the C-S bond.

(52)   (50)

6.4 Carbon-Phosphorus Bond Formation

Very few examples of asymmetric carbon-phosphorus bond formation are known [35-38]. For instance, (*S*)-(+)-phosphinothricin (55), which has strong antifungal and herbicidal properties, has been prepared by asymmetric Michael addition in high optical yield [35]. Thus addition of the chiral Schiff base (54) to the vinyl phosphorus compound (53) afforded the product (55) in 79% optical purity and in 66% yield [35].

(53)   (54)   (55)   79% o.p.

Similarly, optically active α-aminophosphonic acid has been prepared by asymmetric induction [36,38] as well as by resolution of the racemate [39]. Thus, the Schiff base (56) reacted with diethylphosphite to afford after, hydrolysis and hydrogenolysis, the optically active aminophosphonic acid (57) in 65-70% yield [36].

$$H_5C_6-CH=N-CH \overset{CH_3}{\underset{Ph}{|}} \quad \begin{array}{l} \text{i) } (C_2H_5O)_2POH \\ \text{ii) } H^+, H_2O \\ \text{iii) } H_2 \, (Pd) \end{array} \longrightarrow$$

$$\underset{H_5C_6}{\overset{H}{\diagdown}} \overset{PO_3H_2}{\underset{*}{\underset{NH_2}{\diagup}}}$$

(56)

(57) 65-70% (*R*)

6.5 Stereoselective C-H Bond Formation and Proton Migration

Although hydrogen is generally not considered as a "heteroatom", it is appropriate to discuss in this section some asymmetric C-H bond formations except the ene reaction which has already been discussed in section 4.6.5.

Allylsilanes undergo intramolecular rearrangement upon protonation leading to products with 1,3-asymmetric induction [40,41]. Thus, substrate (**58**) undergoes protodesilylation with rearrangement upon protonation using $BF_3/AcOH$ in CH_2Cl_2 to afford (**59**) in 85% yield and with a selectivity of (78:1).

(58)

(59) (2*S*, 4*R*)

Similarly, the alcohol (**60**) upon protonation with BF_3, AcOH in CH_2Cl_2 afforded the diastereomeric products (**61**) and (**62**) in a ratio of 9:1 [41].

(60)

(61)

(62)

(**61** : **62** = 90 : 10)

It is interesting that the corresponding acetates undergo protodesilylation with reversed stereoselectivity. The observed stereoselectivity has been rationalized in terms of a six-membered transition state for the alcohol (60), such as (63), and an eight-membered transition state such as (64) for its acetate derivative. However, the reason for the reversal of stereoselectivity is not fully understood.

(63)   (64)

A similar proton migration occurs in the thermal conversion of hydrandane (65) to the less stable *trans* diastereomer (66) in 90% stereoselectivity and 80% yield [42].

(65)   (66)

Inter- and intramolecular hydride shifts also occur in the addition of alkenes to α,β- unsaturated ketones and aldehydes in the presence of Lewis acids [43]. The intramolecular 1,2-hydride shift proceeds with high regio- and stereo-specificity. Thus (67) undergoes a double 1,2-hydride shift via (68) to give exclusively the product (69). Its isomer with a β-oriented isopropyl chain, which would have resulted from a 1,3-hydride shift, is not observed.

(67)   (68)   (69)

Allylic hydrogen migration of tertiary and secondary allylamines occurs in the presence of chiral rhodium catalysts affording the corresponding (*E*)-enamines and imines respectively [44]. Thus nerylamine (**70**) undergoes isomerization to give the optically active (*E*)-enamine (**71**) in the presence of [Rh -(+) BINAP]$^+$ catalyst in 93% e.e. [44].

(**70**)   (**71**)   93% e.e. (*R*)

Geranylamine (**72**) on the other hand afforded the (*E*)- enamine (**73**) with the opposite configuration.

(**72**)   (**73**)   96% e.e. (*S*)

6.6   References

1.   S. Terashima, S.S. Jew, *Tetrahedron Lett.*, **11**, 1005 (1977).

2.   S.S. Jew, S. Terashima and K. Koga, *Tetrahedron* , **35**, 2337 (1979).

3.   S.S. Jew, S. Terashima and K. Koga, *Tetrahedron*, **35**, 2345 (1979).

4.   A.R. Chamberlin, M. Dezube and P. Dussault, *Tetrahedron Lett.*, **22**(46), 4611 (1981).

5.   P.A. Bartlett and J. Myerson, *J.Amer.Chem.Soc.*, **100**, 3950 (1978).

6.   A. Mitra, : "The Synthesis of Prostaglandins" p.98-104, Wiley, New York, London, Sydney, Toronto, (1977).

7.   E.J. Corey, N.M. Weinshenker, T.K. Schaaf, and W. Huber, *J.Amer.Chem. Soc.*, **91**, 5675 (1969).

8.   P.A. Bartlett, D.P. Richardson and J. Myerson, *Tetrahedron*, **40**, 2317 (1984).

9.   Y. Tamura, M. Mizutani, Y. Furukawa, S. Kuwamura, Z. Yoshida, K. Yanagi, and M. Minobe, *J.Amer.Chem.Soc.*, **106**, 1079 (1984).

10.   A.R. Chamberlin and R.L. Mulholland Jr., *Tetrahedron*, **40**(12), 2297 (1984).

11.   H.C. Brown and P.K. Jadhav, : in "Asymmetric Synthesis", (J.D. Morrison, ed.), Vol.2, p.1, Academic Press, London, (1983).

12.   H.C. Brown and G. Zweifel, *J.Amer.Chem. Soc.*, **83**, 486 (1961).

13.   G. Zweifel, N.R. Ayyangar, T. Munekata and H.C. Brown, *J.Amer.Chem.Soc.*, **86**, 397 (1964).

14.   H.C. Brown and G. Zweifel, *J.Amer.Chem.Soc.*, **83**, 2544 (1961).

15.   H.C. Brown, P.K. Jadhav and A.K. Mandal, *Tetrahedron*, **37**, 3547 (1981).

16.   P.K. Jadhav and H.C. Brown, *J.Org.Chem.*, **46**(14), 2988 (1981).

17.   H.C. Brown, M.C. Dessai and P.K. Jadhav, *J.Org.Chem.*, **47**, 5065 (1982).

18.   H.C. Brown, P.K. Jadhav and A.K. Mandal, *J.Org.Chem.*, **47**, 5074 (1982).

19.   J.J. Patridge, N.K. Chadha and M.R. Uskokovic, *J.Amer.Chem.Soc.*, **95**, 532 (1973).

20.   J.J. Patridge, N.K. Chadha and M.R. Uskokovic, *J.Amer.Chem.Soc.*, **95**, 7171 (1973).

21.   P.A. Bartlett : in "Asymmetric Synthesis" (J.D. Morrison, ed.), Vol.3. Part B, pp.411, Academic Press, London, (1984).

22.   J.-L. Bougeois, L. Stella and J.M. Surzur, *Tetrahedron Lett.*, **22**, 61 (1981).

23.   H. Takahata, K. Yamazaki, T. Takamatsu, T. Yamazaki and T. Momose, *J.Org. Chem.*, **55**, 3947 (1990).

24.   H. Takahata, T. Takamatsu, Y.S. Chen, N. Ohkubo, T. Yamazaki and T. Momose, *J.Org.Chem.*, **55**, 3792 (1990).

25.   H. Takahata, T. Takamatsu and T. Yamazaki, *J.Org.Chem.*, **54**, 4812 (1989).

26.   H. Takahata, T. Takamatsu, M. Mozumi, Y.S. Chen, T. Yamazaki and K. Aoe, *J.Chem.Soc.Chem.Commun.*, 1627 (1987).

27.   K.E. Harding and S.R. Burks, *J.Org.Chem.*, **46**, 3920 (1981).

28. Y. Saitoh, Y. Moriyama, T. Takahashi and Q. Khuong-Huu, *Tetrahedron Lett.*, **21**, 75 (1980).

29. J. Ambuhl, P.S. Pregosin, L.M. Venanzi, G. Consigli, F. Bachechi and L. Zambonelli, *J.Organomet.Chem.*, **181**, 255 (1979).

30. B.M. Trost and E. Keinan, *J.Amer.Chem.Soc.*, **100**, 7779 (1978).

31. D.L.J. Clive, V. Farina, A. Singh, C.K. Wong, W.A. Kiel and S.M. Menchen, *J.Org.Chem.*, **45**, 2120 (1980).

32. K. Suzuki, A. Ikegawa and T. Mukaiyama, *Bull.Chem.Soc. Jpn.*, **55**, 3277 (1982).

33. F. Wudle and T.B.K. Lee, *J.Amer.Chem.Soc.*, **95**, 6349 (1973).

34. K. Hiroi, S. Sato and R. Kitayama, *Chemistry Lett.*, 1595 (1980).

35. N. Minowa, M. Hirayama and S. Fukatsu, *Tetrahedron Lett.*, **25**(11), 1147 (1984).

36. T. Glowiak, W.S. Dobrowolska, J. Kowalik, P. Mastalerz, M. Soroka and J. Zon', *Tetrahedron Lett.*, **45**, 3965 (1977).

37. I.A. Natchev, *Bull.Chem Soc Jpn.*, **61**, 3699 (1988).

38. W.F. Gilmore and H.A. McBride, *J.Amer.Chem.Soc.*, **94**, 4361 (1972).

39. S.V. Roghozin, V.A. Darankov, and Yu.P. Belov, *Izv.Akad.Nauk, USSR, Ser.Khim.*, 955 (1973).

40. S.R. Wilson and M.F. Price, *J.Amer.Chem.Soc.*, **104**, 1124 (1982).

41. S.R. Wilson and M.F. Price, *Tetrahedron Lett.*, **24**(6), 569 (1983).

42. E.J. Corey and T.A. Engler, *Tetrahedron Lett.*, **25**(2), 149 (1984).

43. (a) B.B. Snider, D.J. Rodini and J. van Straten, *J.Amer.Chem.Soc.*, **102**, 5872 (1980); (b) D. Potin, K. Williams and J. Rebek, Jr., *Angew.Chem.*, 1420 (1990).

44. K. Tani, T. Yamagata, S. Akutagawa, H. Kumobayashi, T. Taketomi, H. Takaya, A. Miyashita, R. Noyori and S. Otsuka, *J.Amer. Chem. Soc.*, **106**, 5208 (1984).

7   Enzyme-Catalyzed Reactions

Enzymes exert complete control over the synthesis of biomolecules which are usually formed in enantiomerically pure states. This is due to the fact that enzymes are themselves made up of single enantiomers of amino acids and therefore serve as chiral templates for the synthesis of only one enantiomer of the product. This ability of enzymes has attracted chemists to devise analogous methods which mimic the action of enzymes for the asymmetric synthesis of natural products [1-4].

There is an ever-increasing demand for enantiomerically pure compounds many of which are being used in chemotherapy, molecular electronics and optical data storage and enzymic reactions can be employed for their syntheses. The enormous potential of enzymes in exerting control over several stereochemical aspects in single-step reactions is illustrated by the several classes of enzymes now used for achieving asymmetric synthesis [5].

1. *Oxidoreductases.* They catalyze oxidation-reduction reactions, e.g.

$$R_3C-H \longrightarrow R_3C-OH$$

$$R_3C-OH \longrightarrow R_2C=O$$

$$RCH_2-CH_2R \longrightarrow RCH=CHR$$

2. *Hydrolases.* Enzymes of this group hydrolyse esters, amides, peptides, glycosides, anhydrides, etc.

3. *Transferases.* They are involved in the transfer of groups, such as acyl, phosphoryl, aldehyde, keto, etc. from one molecule to another.

4. *Isomerases.* These enzymes induce isomerizations, including double bond migrations, *cis-trans* isomerizations, etc.

5. *Ligases.* Enzymes of this group catalyse formations of C-O, C-S, C-N, C-C and O-P bonds.

6. *Lyases.* These enzymes catalyse addition reactions to C=C, C=N and C=O groups, as well as the reverse processes.

Generally, the most useful groups of enzymes used in asymmetric synthesis are the *oxidoreductases, hydrolases* and *lyases.*

7.1 Enzyme Specificity

The specificity of an enzyme is generally considered to be its selectivity not only towards a particular reaction, but also towards the structure and stereochemistry of the reactants and products. However, the asymmetric synthetic potential of enzymes depends largely on their stereospecificities. These may be considered in terms of either *enantiomeric specificity* or *prochiral specificity.*

7.1.1 Enantiomeric Specificity of Enzymes

When asymmetric synthesis is carried out in the presence of an enantiomerically specific enzyme, the reaction involving transformation of a racemate stops at the point when all the reactive enantiomer has been converted, i.e. when half of the racemate has reacted. In such reactions, the maximum attainable yield is 50%, that is, the other enantiomer is discarded. In order to increase the optical yield, the unreactive enantiomer may be recycled *via* racemization.

For instance, the racemic (±)-N-acylchloro derivative (1) of D- isodehydro-valine (2), which is a penicillin synthon, is converted by reaction with the enzyme hog kidney acylase followed by acid hydrolysis to enantiomerically pure (2) in 60% yield [6].

(±)-(1)   (D)-(2) 60%

Similarly, the *N*-acetyl derivative of α- amino acid (3), on treatment with *Aspergillus* acylase afforded (2S, 3R)-(4) in 55% yield [7].

(±) - (3)   (2S, 3R) - (4)

Enantiomerically pure amino acids have also been obtained through enzymatic hydrolysis of hydantoins, e.g. D-(5) present in the racemic mixture is hydrolysed with hydantoinase to afford the carbamate D-(6) which on alkaline hydrolysis afforded the D-amino acid D-(7) [8-10]. The unreactive L-enantiomer is recycled by *in situ* racemization either non-enzymically or by employing racemase.

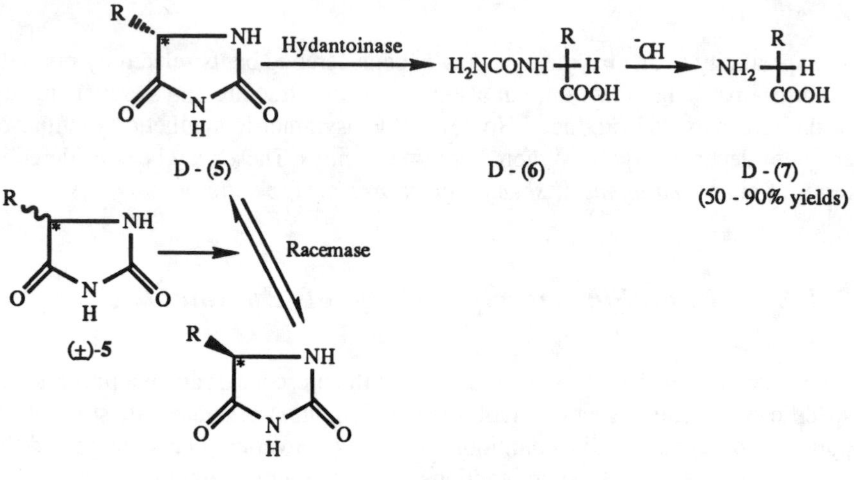

D - (5)

D - (6)

D - (7)
(50 - 90% yields)

(±)-5

L - (5)
(R=CH₃, CH₂OH, CH₂Ph, etc.)

In another approach, racemic acetyl-phenylalanine (8) is reacted with *p*-toluidine (9) in the presence of papaine to give acetyl-L-phenylalanine-*p*-toluidide (10), which is insoluble in the reaction medium and is therefore precipitated. The D-(8) isomer is recycled via racemisation and the yield of the L-*p*-toluidide obtained is 82% [11].

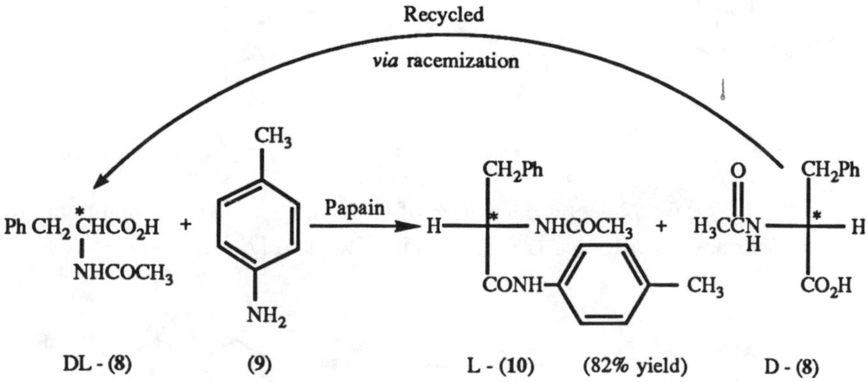

Recycled

via racemization

DL - (8)   (9)   L - (10)   (82% yield)   D - (8)

The enzyme *Aspergillus* aminoacylase has been employed in the resolution of (±)-2-chloroacetaminoadipic acid (11) to give the product (S)-12 in 96% yield [12].

(±)-(11)   (S) - (12) (96%)

The microbe *Gliocladium roseum* specifically hydrolyses the ester group attached to the centre of a substrate having R stereochemistry. Thus *cis*-2,4-dimethylglutarate (13) is hydrolysed to give (2R,4R)-(14) in 50% yield, which is hydrogenated to give the product (15) in 98% e.e. [13].

(±) - (13)   (2R, 4R) - (14)

(15) >98% e.e.

Porcine pancreatic lipase (PPL) catalyzes the hydrolysis of racemic diesters to give the products in high optical yield [14]. Thus (R,S)-dimethylbenzyl succinate (16) on treatment with PPL afforded the (S)-product (17) in 98% e.e. and in 90% yield [14].

(R, S) - (16)   (S) - (17) > 98% e.e.

Other enzymes have been used to enantioselectively hydrolyse various substrates in high optical yields [15-17]. For instance, *Pseudomonas fluorescens* lipase (PFL) catalyzes the hydrolysis of (±)-acetoxycyclopentanes, e.g. (18) to give (19) in 99% e.e. [15].

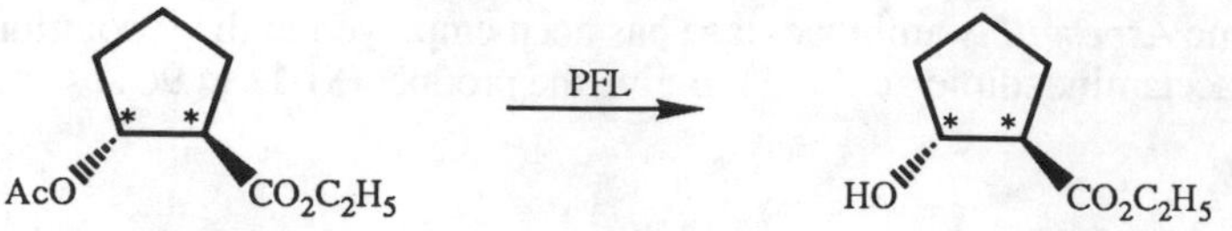

PFL

AcO   CO$_2$C$_2$H$_5$

(±) - (18)

HO   CO$_2$C$_2$H$_5$

(19)   > 99% e.e.

Table 7.1 shows some examples of the enantioselective formation of optically active products through various enzymatic systems.

Table 7.1. Enantioselective Reactions of Enzyme Catalyzed Substrates

Substrate (±)	Enzyme system	Product	% e.e.	Reference
CH$_3$CO—O CO$_2$C$_2$H$_5$	a	HO CO$_2$C$_2$H$_5$	> 99 (1S, 2R)	[15]
H H CO$_2$CH$_3$ OCOCH$_3$	a	H H CO$_2$CH$_3$ OH	77 (2R, 3R)	[15]
H H CO$_2$CH$_3$ OCOCH$_3$	a	H H CO$_2$CH$_3$ OH	45 (2R, 3R)	[15]
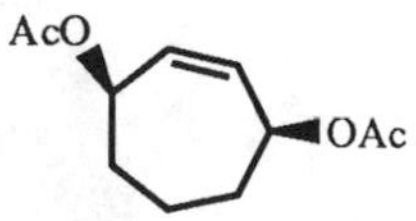 AcO OAc	b	 HO OAc	39 X	[18]
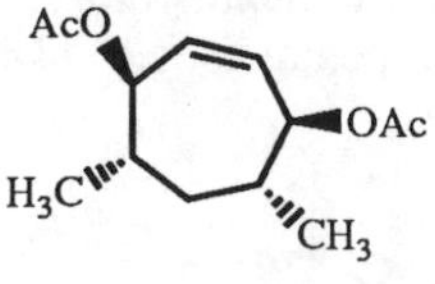 AcO OAc H$_3$C CH$_3$	c	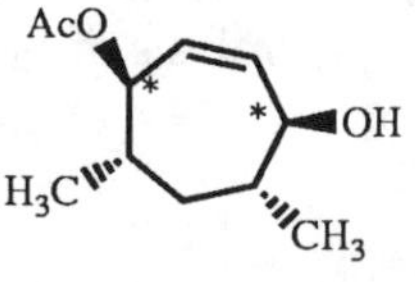 AcO OH H$_3$C CH$_3$	60 X	[18]

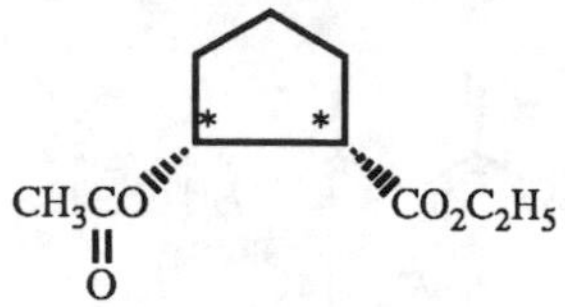

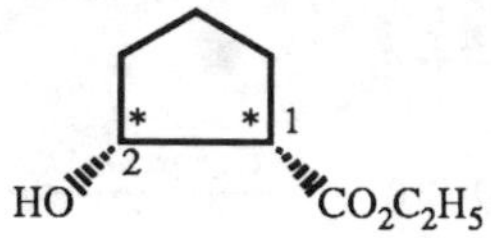

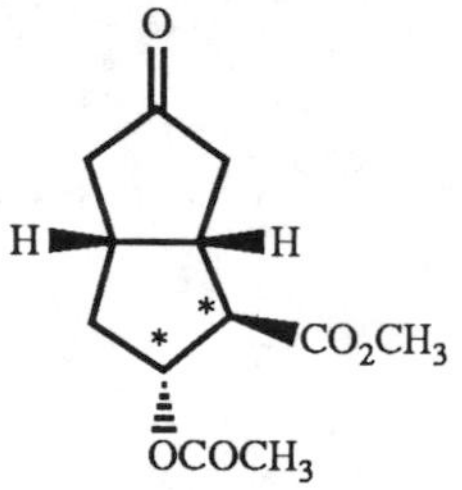

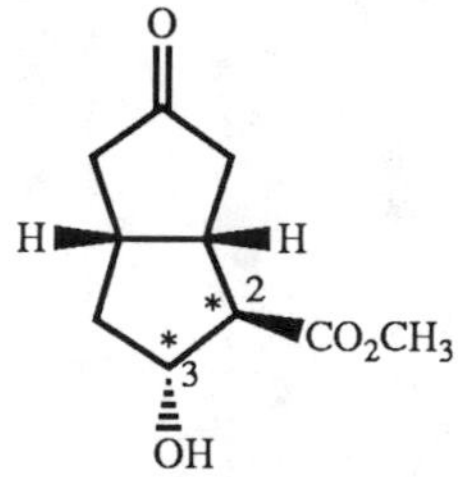

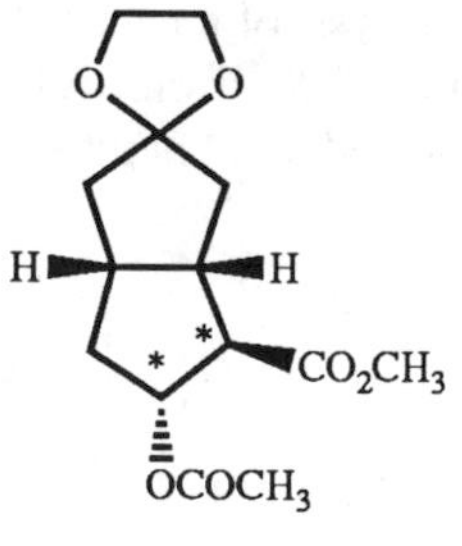

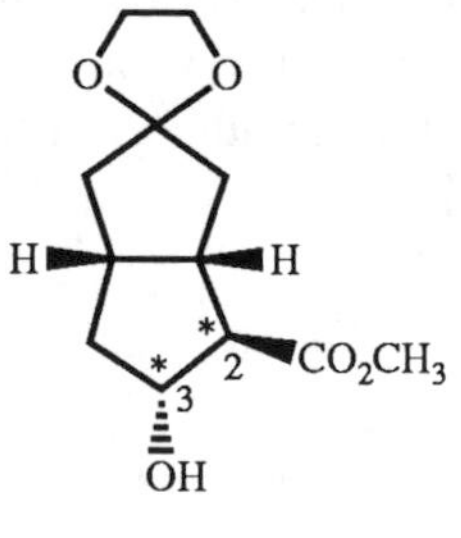

(Table 7.1. contd.)

Substrate (±)	Enzyme system	Product	% e.e.	Reference
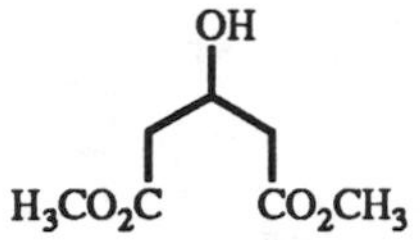 (structure: OH; H₃CO₂C, CO₂CH₃)	d	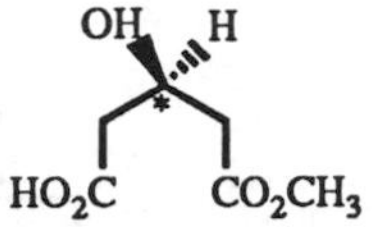 (structure: OH, H; HO₂C, CO₂CH₃)	60 (R)	[16]
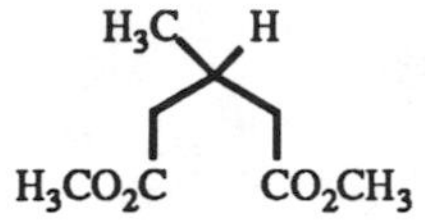 (structure: H₃C, H; H₃CO₂C, CO₂CH₃)	e	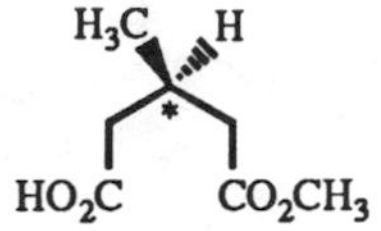 (structure: H₃C, H; HO₂C, CO₂CH₃)	79 (R)	[19]
(structure: H, OAc; H₅C₂, C≡CH)	f	(structure: H, OAc; H₅C₂–C–C≡CH) + (structure: AcO, H; H₅C₂–C–C≡CH)	72 (R) / 54 (S)	[20]
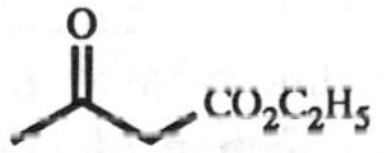 (structure: O ketone, CO₂C₂H₅)	g	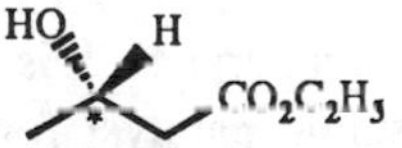 (structure: HO, H; CO₂C₂H₅)	100 (S)	[21]

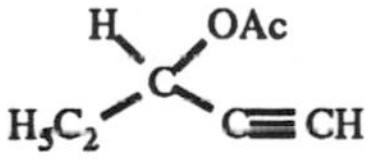

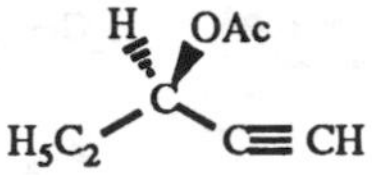

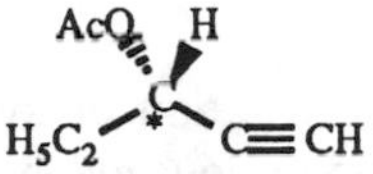

a = PFL,   b = (Electric eel acetylcholineesterase),   c = *Candida cylindracea* lipase,
ACE

d = α-Chymotrypsin,   e = PLE,   f = *B.subtilis*   g = *Saccharomyces bailii*
var. *niger*,

X = % yield reported

7.1.2   Prochiral Stereospecificity

7.1.2.1   Additions to Stereoheterotopic Faces

Enzymes possessing prochiral stereospecificities are valuable assets in the synthesis of optically active products, since the problem of enantiomeric separation can be avoided. Moreover enzymes with opposite enantiotopic stereospecificities provide the opportunity to synthesize the other mirror image, thereby allowing both enantiomers to be obtained in high optical purity by using the appropriate enzymes. For instance, chloropyruvic acid (20) is reduced with D- or L- lactate dehydrogenase affording D- and L- chlorolactic acid (21) in 53 and 52% yield respectively [22].

$$
\begin{array}{c}
\underset{\text{dehydrogenase}}{\overset{\text{L-Lactate}}{\longrightarrow}} \quad \text{HO}-\overset{\text{COOH}}{\underset{\text{H}}{\overset{|*}{\text{C}}}}-\text{CH}_2\text{Cl} \quad 52\% \text{ yield} \\
\text{L-(21)}
\end{array}
$$

ClCH$_2$ C COOH
‖
O

(20)

$$
\begin{array}{c}
\underset{\text{dehydrogenase}}{\overset{\text{D-Lactate}}{\longrightarrow}} \quad \text{ClH}_2\text{C}-\overset{\text{COOH}}{\underset{\text{H}}{\overset{|*}{\text{C}}}}-\text{OH} \quad 53\% \text{ yield} \\
\text{D-(21)}
\end{array}
$$

Another stereospecific reaction of synthetic importance is the enzyme catalyzed addition of HCN to various aldehydes affording cyanohydrins in high optical yields [23]. The enzyme D- oxynitrilase, readily obtained from almonds, has been employed for this purpose. It is immobilised with cellulose-based ion-exchanger to give a stable catalyst for the continuous production of D-α-hydroxynitriles, e.g. (22) to (23), which can be derivatized to other useful chiral synthons.

(22)   (23) 97% e.e. (R)

Similarly the yeast-mediated condensation of fermentatively generated acetaldehyde with aldehydes, e.g. (24) affords initially the acyloin (25) which is further reduced with enantiotopic specificity to give the α-tocopherol chromanyl moiety precursor (26) in 20-25% yield [24-26].

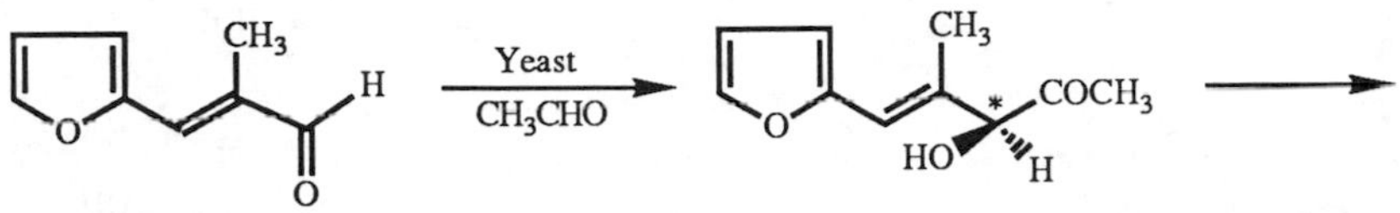

(24)   (25)

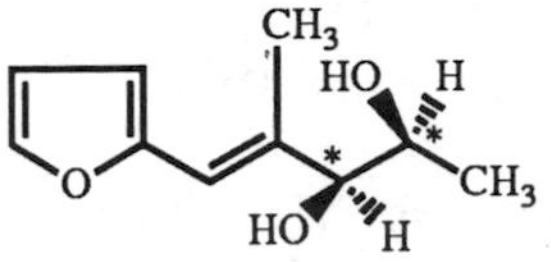

(26)

An interesting example of the stereospecificity of enzymes is illustrated by the pig liver-farnesyl pyrophosphate synthetase which catalyzes the coupling of geranyl pyrophosphate (27) with 3-methylpent-3-enyl pyrophosphate (28) in a highly stereospecific manner, E-(28) affording S-(29) and Z-(28) affording R-(29) exclusively [27, 28]. The enzyme attacks C-4 at the re-face of (E)- and (Z)-(28) affording the (S) and (R) products respectively.

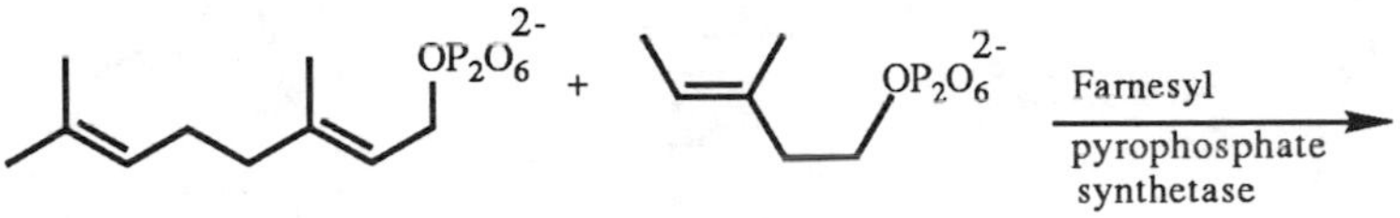

(27)   (E) - (28)

(S) - (29)

Many L-amino acids have been synthesized using pyridoxal phosphate (30)-dependent enzymes [9, 29-32]. Thus, the intermediate (32) is formed by the condensation of pyruvic acid (31), ammonia and (30) in the presence of β-tyrosinase or tryptophanase.

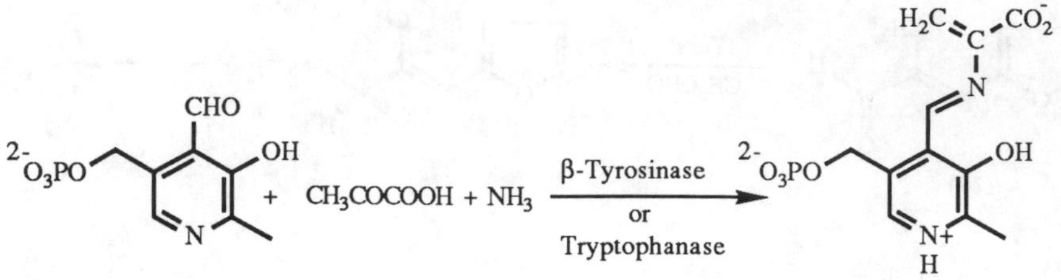

(30)   (31)   (32)

The amino acrylic intermediate (32) then combines with RH, e.g. phenol R=PhOH (33) to give the imine (34) which is hydrolysed enzymatically to give the amino acid tyrosine (35) with the recovery of pyridoxal phosphate (30). By this method, the L-amino acids (36-38) have been prepared in 30-90% yields [31,32].

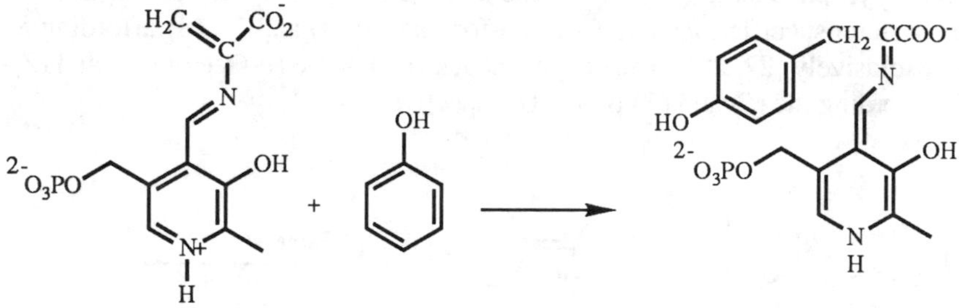

(32)   (33)   (34)

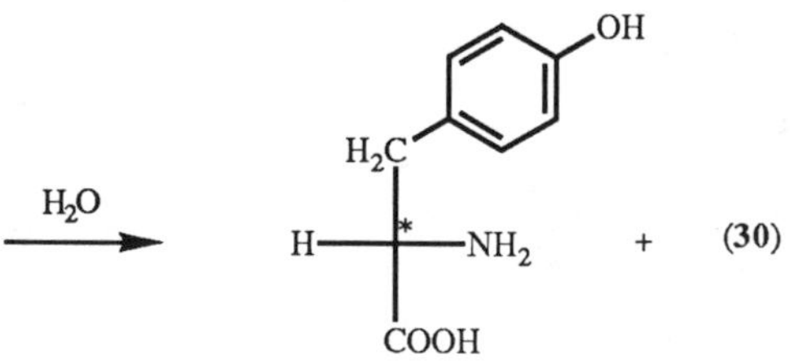

(35)

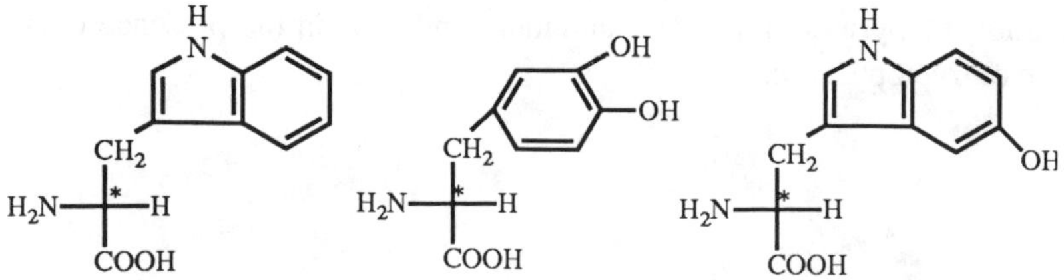

(36)   (37)   (38)

Commercial preparation of L-amino acids has also been achieved by the highly stereospecific enzyme-catalyzed addition to carbon-carbon double bonds [33-37]. Thus fumaric acid (39) undergoes addition of NH_3 in the presence of aspartase to give L-aspartic acid (40) in 95% yield [34].

(39)   (40) 95% yield

Similarly, additions to unactivated double bonds e.g. in oleic acid (41) are catalyzed by enzymes to give the R-10-hydroxy stearic acid (42) in 22% yield [37]. In all these reactions, addition to the double bond proceeds in an *anti* fashion.

(41)   (42) 22% (R)

The enzyme-catalyzed epoxidation constitutes another key transformation carried out with unactivated substrates affording the products in high optical yields [38]. This is in contrast to the stereospecific chemical epoxidations which require allylic alcohol activation [39]. For instance, *Penicillium spinulosum* catalyzes the epoxidation of (43) affording the epoxide (44) in 90% e.e.

(43)   (44) 90% e.e.

Similarly, double epoxidation of 1,7-nonadiene (45) with *Pseudomonas oleovorans* afforded (46) in 84% e.e.

(45)   (46) 84% e.e.

Another useful enzyme-catalyzed stereospecific reaction is the asymmetric reduction of carbon-carbon double bonds [40-42]. These are generally effected either by using NADH-dependent enoate reductases or by employing hydrogenases together with molecular hydrogen. Thus employing hydrogenase from *Clostridium* species (**47**) is reduced by molecular hydrogen to (**48**) in high yield [40].

(**47**)   (**48**)

Similarly by employing enoate reductase (**49**) is reduced to (**50**) in 95% yield.

(**49**)   (**50**) 95% yield

Chiral sulfoxides have been prepared by enantiotopically specific oxidations of sulfides [43,44]. Employing the appropriate organism, either the (*R*) or (*S*) enantiomer can be prepared. This is illustrated by the oxidation of sulfide (**51**) affording the (*R*) and (*S*) sulfoxides (**52**) in reasonable yields [44].

(**51**)   (*S*) - (**52**)   50% yield

(*R*) - (**52**)   60% yield

7.1.2.2   Stereoheterotopic Groups and Atoms

Another useful aspect of enzyme-catalyzed reactions is their ability to discriminate between enantiotopic or diastereotopic groups or atoms affording complete asymmetric induction on symmetrical substrates. This has been aptly demonstrated by the oxidation of symmetric diols such as (53) with horse liver alcohol dehydrogenase (HLADH) [46,47]. The reaction proceeds with pro-S selectivity affording the intermediate (54) which undergoes intramolecular aldol reaction to ultimately give the product (55) in high yield.

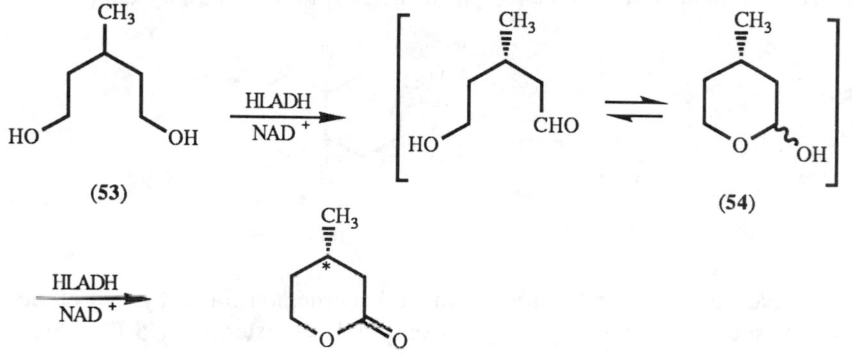

(53)

(54)

(55)   70% yield, 87% e.e.

Similarly, the organism *Gluconobacter scleroideus* has been employed in the asymmetric oxidation of (56) affording (S)- mevalonolactone (57) in 79% e.e. [48].

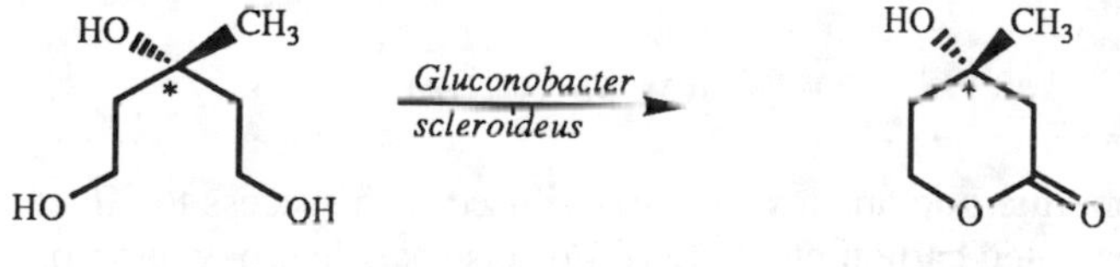

(56)

(57)

Another enzyme, NAD/H-independent galactose oxidase (besides HLADH) catalyzes the stereospecific oxidation of glycerol (58) to L-glyceraldehyde (59) in high yields [49].

(58)   (59)

The asymmetric reduction of symmetrical compounds can be accomplished by alcohol dehydrogenases with high enantiotopic selectivity [50,51]. For instance, decalindione (60) is reduced with HLADH to (61) in 89% yield [50].

(60)   (61)   89% yield

Similar enantiotopic specificities have been demonstrated by hydrolytic enzymes, such as chymotrypsin (CT) or pig liver esterases (PLE) which hydrolyse diesters with pro-S specificity affording half esters in good optical yields [52,53]. Thus, hydrolysis of the glutarate diester (62) with PLE afforded the 3(R)-half ester (63) in 94% yield [52].

(62)   (63) (R) 94% yield

Hydrolyases from microorganisms can also be used with success to effect hydroxylation at unactivated carbon atoms [54]. For instance, hydroxylation of the pinane derivative (64) with *Sporotrichum sulfurescens* afforded the monohydroxylated product (65) in 62% yield [55].

(64)   (65)

It has been observed that hydroxylases are enantiotopically specific, replacing a methylene hydrogen by a hydroxyl group with *retention* of configuration [54]. Similarly diastereotopic specificity has also been observed in hydroxylases [54,56]. For instance, camphor (66) is hydroxylated with *Pseudomonas putida* to (67) in 95% yield [57].

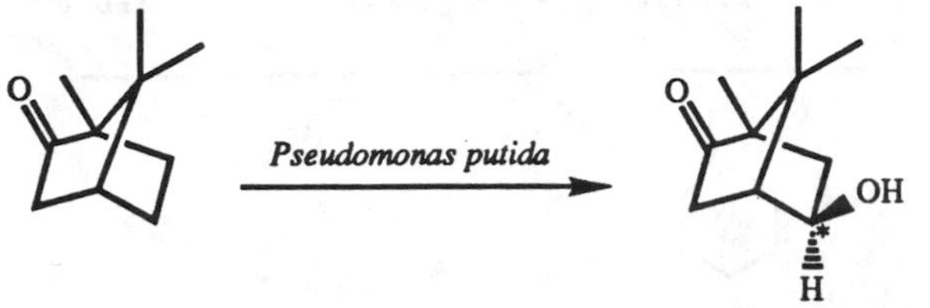

(66)   (67) 95% yield

It is often possible to predict the position and stereochemistry of hydroxylation for many organisms in certain compounds [54]. Lipoxygenase from potatoes introduces a peroxy group with enantiotopic specificity [58]. Thus arachidonic acid (68) is converted to the corresponding (S)-5-peroxy product (69).

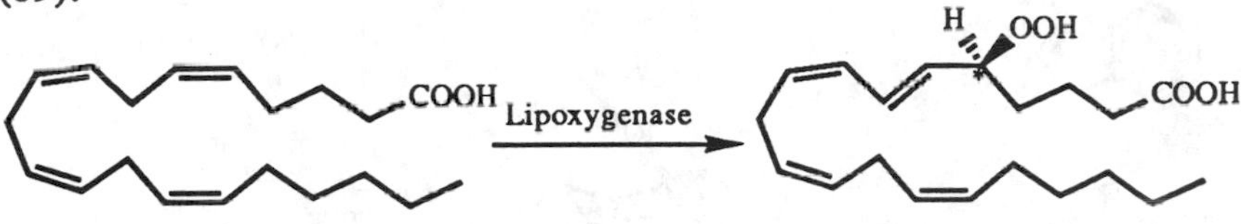

(68)   (69)

Similarly the enzyme cyclase from *Cephalosporium acremonium* catalyzes the diastereotopic cyclization of certain peptides which can be employed in the synthesis of penicillins [59]. For instance, the peptide (70) is cyclized to isopenicillins with the stereospecific loss of the pro-S hydrogen to give the product (71) in quantitative yield.

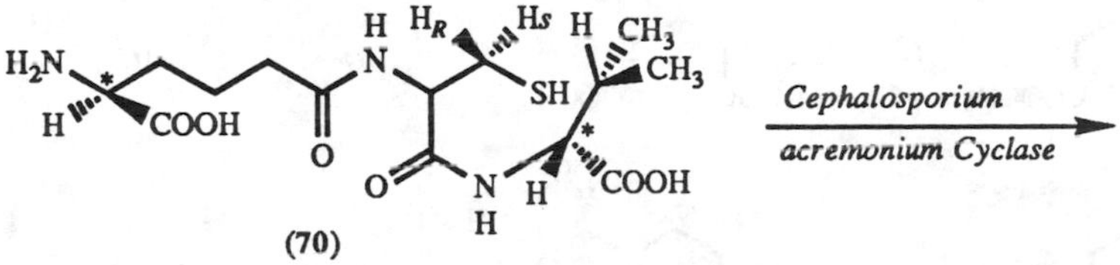

(70)

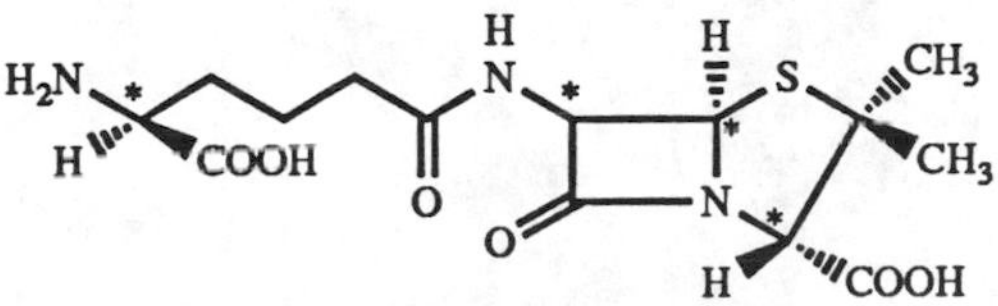

(71) 100% yield

Table 7.2 gives some examples of the enantio- and diastereotopic specificities of enzymes.

Table 7.2. Prochiral Stereospecificity of Enzymes

Substrate	Enzyme/ Organism	Product	% Yield	Reference
	a		> 99 (R)	[60]
	b		> 99	[61]
	c		≥98% (R)	[62]
	d		90%	[54]
	e		70%	[66]
	f		78%	[67]
	c		≥98 (1S, 2R)	[62]

(Table 7.2. contd.)

Substrate	Enzyme/ Organism	Product	% Yield	Reference
Ph CH₂—⌈—OH / ⌊—OH (PhCH₂ with OH, OH)	g	CH₂OH / Ph CH₂▸*C◂H / CH₂OAc	97 (R)	[63]
N₃⌣C(=O)⌣OCOCH₃	h	OH / N₃⌣*⌣OCOCH₃	> 96 (S)	[64]
OCOCH₃ / CH₃ (cyclohexene)	i	O / *⌣CH₃ / H	83 (S)	[64]

a = Baker's yeast,   b = *Hansenula anomala*,   c = *Pseudomonas putida*,   d = *Sporotrichum sulfurescens*

e = Yeast,   f = α-Chymotrypsin,   g = CH₂ = CH— OAc,   h = Baker's yeast,   i = *Pichia miso*
Lipase from
Pseudomonas
fluorescens

7.2   Meso Compound Transformations

The remarkable ability of enzymes to discriminate between enantiotopic groups of symmetrical substrates, e.g. meso compounds is being exploited with ever increasing frequency in asymmetric synthesis. *Meso* diesters, e.g. (72) have been hydrolysed stereospecifically with esterases from several microorganisms, such as *Gliocladium roseum* affording the product (73) in high yield [68].

$$H_3CO_2C \overset{*}{\diagup}\diagdown\overset{*}{\diagup} CO_2CH_3 \xrightarrow[\text{esterase}]{\textit{Gliocladium roseum}} H_3CO_2C \overset{*}{\diagup}\diagdown\overset{*}{\diagup} CO_2H$$

(72)   (73)

The absolute configuration of (73) was established by its reduction with LiBH$_4$ affording the lactone (2R,4S)-(74). Since LiBH$_4$ selectively reduces carboxylic acid esters and leaves the carboxylic acid groups intact, it was concluded that esterase from *G.roseum* had hydrolysed the the pro-*R* ester group of (72). Pig liver esterase (PLE), on the other hand, catalyzed the stereospecific hydrolysis of the pro-*S* ester group of (72) thereby allowing the enantiomer of (73) to be prepared.

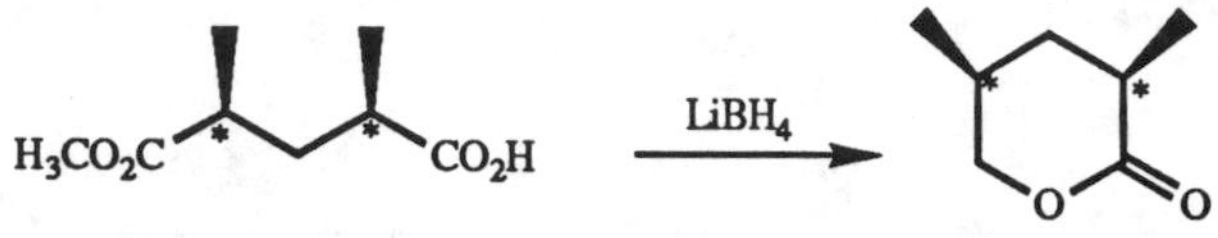

(73)

(74) > 98% e.e. (2R, 4S)

Similarly, pig liver esterase (PLE) has been employed in the stereospecific hydrolysis of (75) affording the product (76) in 96% yield [69].

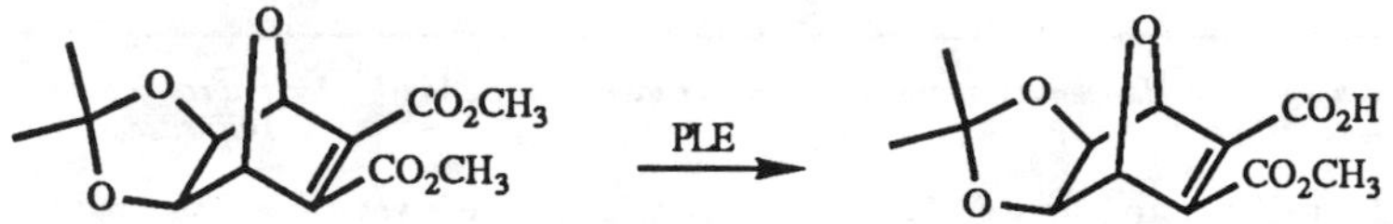

(75)

(76) 96% (77% e.e.)

Horse liver alcohol dehydrogenase (HLADH) is a very versatile enzyme in effecting transformations of *meso* compounds [70-74]. Thus, *meso* oxabicyclic diol (77) is oxidized with HLADH to afford the product (80) *via* the intermediates (78) and (79) in 83% yield and 98% e.e. [70, 72].

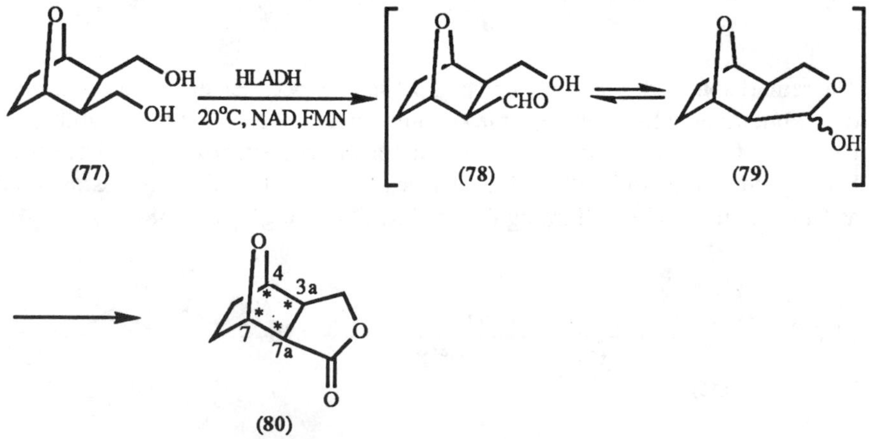

(77)   (78)   (79)

(80)

(3$_a$R, 4R, 7S, 7$_a$R)
(83% yield, >98% e.e.)

Similarly, the diol (81) was oxidized with HLADH *via* the intermediates (82) and (83) affording the product (84) in 100% e.e. [71].

(81)   (2S, 3R) - (82)   (3S, 4R) - (83)

(3S, 4R) - (84) 100% e.e.

Other enzyme systems have also been tested with success in the transformation of *meso* compounds. Thus, the epoxide (85) is hydrolysed with epoxide hydrolase to give the product (86) in 100% yield and 70% e.e. [75].

(85)   (86)   70% e.e.

7.3   Multienzyme Systems

One of the distinguishing features of enzymatic systems is their mutual compatibility, because most enzymes operate satisfactorily in aqueous solution between pH 6.5 and 8.5 and at room temperature. Thus it is possible to assemble complex systems containing multiple enzymes providing complementary catalytic activities.

Several multienzymatic systems have been tested involving generally between 3 and 10 cooperating enzymes which is the maximum limit in complexity that can be tolerated in most synthetic applications [76-80]. For instance N-acetyllactosamine (91), a disaccharide of glycoproteins has been synthesized by multienzyme synthesis involving the use of six cooperating enzymes [77]. Fig. 7.1 shows the synthesis starting from glucose -6-phosphate

(87) which is converted to glucose-1-phosphate (88) with the enzyme phosphoglucomutase (PGM). The compound (88) when treated with uridine triphosphate (UTP) in the presence of uridine-5'- phosphate glucose pyrophosphorylase (UDPGP) afforded (89). This reaction proceeds with the hydrolysis of pyrophosphate using pyrophosphatase (PPase) in order to drive the reaction forward. The compound (89) undergoes epimerisation to (90) with UDP- galactose-4-epimerase (UDPGE) and coupling of (90) with N-acetyl glucosamine (91) by galactosyl transferase (gal transferase) afforded the disaccharide N-acetyllactosamine (92) in 75% yield. The UTP was regenerated from UDP using pyruvate kinase (PK) with phosphenol pyruvate (PEP) $\longrightarrow$ pyruvic acid (PA) transformation.

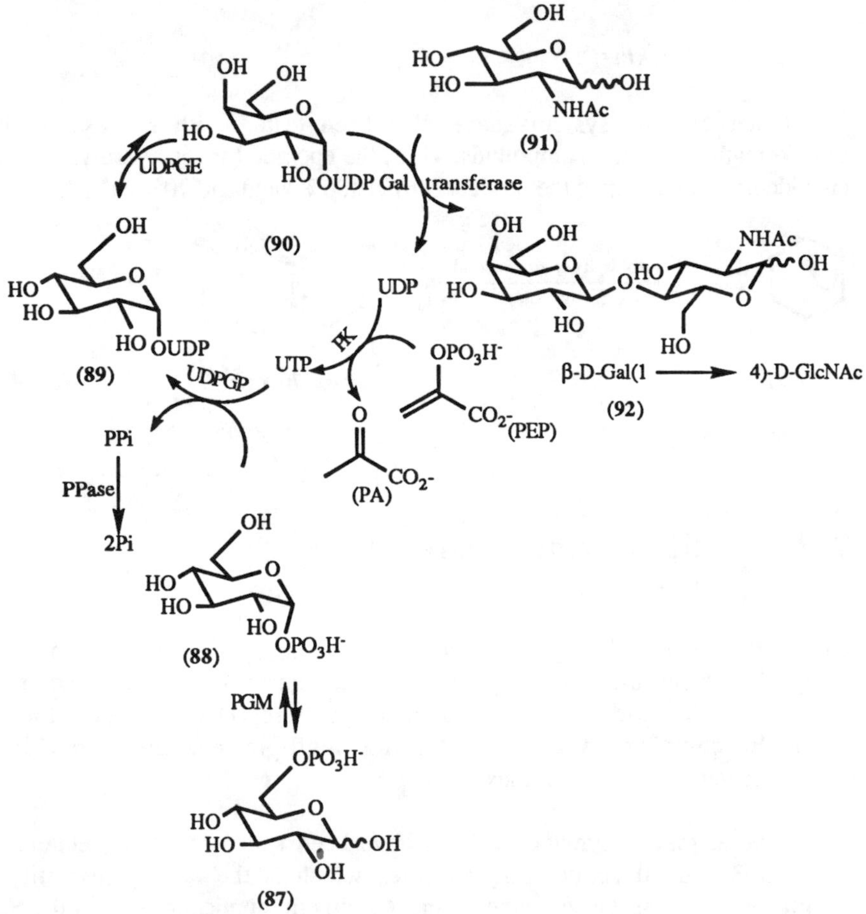

Figure 7.1. Synthesis of N-acetyllactosamine (92) (for abbreviations, see text).

Such a synthesis based on multienzyme cooperating systems has distinct advantages over the corresponding "chemical" synthesis which can be very complicated on account of multiple protection/deprotection steps and by the difficulty in controlling the regioselectivity. However commercial exploitation of procedures for the preparations of optically active products involving multienzyme systems remains to be fully realised and such systems may be increasingly applied in the future.

7.4   References

1.   For a review, see J.B. Jones, in "Asymmetric Synthesis", (J.D. Morrison, ed.), Vol.5, p.309, Academic Press, Orlando, (1985).

2.   G.M. Whitesides and C.-H. Wong, *Angew.Chem.*, **24**, 617 (1985).

3.   C.-H. Wong, *Science*, **244**, 1145 (1989).

4.   (a) A. Pluckthun, *Kontakte*, **2**, 40 (1990).   (b) G.Guanti, L. Baufi and E. Narisano, *J.Org.Chem.*, **57**, 1540 (1992) and references therein.

5.   "Enzyme Nomenclature", New York: Academic Press, 1978.

6.   J.E. Baldwin, M.A. Christie, S.B. Haber and L.I. Kruse, *J.Amer.Chem.Soc.*, **98**, 3045 (1976).

7.   K. Mori and H. Iwasawa, *Tetrahedron*, **36**, 2209 (1980).

8.   M. Guivarch, C. Gillonier and J.C. Brunie, *Bull.Soc.Chim.Fr.*, 91 (1980).

9.   H. Yamada, : in "Enzyme Engineering", (I. Chibata, S. Fukui and L.B. Wingard, eds.), Vol.6, p.97, Plenum Press, New York, (1982).

10.   F. Cecere, G. Galli, and F. Morisi, *FEBS Lett.*, **57**, 192 (1975).

11.   H.T. Huang and C. Niemann, *J.Amer.Chem.Soc.*, **73**, 475 (1951).

12.   K. Mori, : in "Studies in Natural Products Chemistry", (Atta-ur-Rahman, ed.), Vol.1, Stereoselective Synthesis (Part A), p 677, Amsterdam: Elsevier, (1988).

13.   C.-H. Chen, Y. Fujimoto and C.J. Sih, *J.Amer.Chem.Soc.*, **103**, 3580 (1981).

14.   E. Guibe-Jampel, G. Rousseau and J. Salaun, *J.Chem.Soc.Chem.Commun.*, 1080 (1987).

15.   Z.F. Xie, H. Suemune and K. Sakai, *J.Chem.Soc.Chem.Commun.*, 838 (1987).

16.   P. Mohr, L. Rösslein and C. Tamm, *Helv.Chim.Acta*, **70**, 142 (1987).

17.   V. Kerscher and W. Kreiser, *Tetrahedron Lett.*, **28**(5), 531 (1987).

18.   A.J. Pearson, H.S. Bansal and Y.-S. Lai, *J.Chem.Soc.Chem.Commun.*, 519 (1987).

19.   L.K.P. Lam, R.A.H.F. Hui and J.B. Jones, *J.Org.Chem.*, **51**, 2047 (1986).

20.   K. Mori and H. Akao, *Tetrahedron*, **36**, 91 (1980).

21.   T. Sugai, M. Fujita and K. Mori, *Nippon Kagaku Kaishi*, 1315 (1983).

22.   B.L. Hirschbein and G.M. Whitesides, *J.Amer.Chem.Soc.*, **104**, 4458 (1982).

23.   W. Becker and E. Pfeil, *J.Amer.Chem.Soc.*, **88**, 4299 (1966).

24.   (a). For an excellent recent review on Baker's yeast mediated transformations, see: R. Csuk and B.I. Glänzer, *Chem.Rev.*, **91**, 49 (1991).   (b). C. Fuganti, P. Grassili and G. Marioni, *J.Chem.Soc.Chem.Commun.*, 205 (1982).

25.   C. Fuganti, P. Grassili and G. Marioni, *Chem.Ind.*, (London), 983 (1977).

26.   R. Bernardi, C. Fuganti, P. Grassili and G. Marioni, *Synthesis*, 50 (1980).

27.   M. Kobayashi, T. Koyama, K. Ogura, S. Seto, F.J. Ritter and I.E.M. Bruggermann-Rotgans, *J.Amer.Chem.Soc.*, **102**, 6602 (1980).

28.   T. Koyama, A. Saito, K. Ogura and S. Seto, *J.Amer.Chem.Soc.*, **102**, 3614 (1980).

29.   H. Yamada, S. Takahashi, K. Yoshiaki and H. Kumagai, *J.Ferment.*, **56**, 484 (1978).

30.   H. Yamada and H. Kumagai, *Pure Appl.Chem.*, **50**, 1117 (1978).

31.   S. Fukui, S. Ikeda, M. Fujimura, H. Yamada and H. Kumagai, *Eur.J.Biochem.*, **51**, 155 (1975).

32.   S. Fukui, *et.al, Eur.J.Appl.Microbiol.*, **1**, 25 (1975).

33.   Hill, R.L. and Teipel, J.W. in "The Enzymes", (P.D. Boyer, ed.), Vol.5 (3rd Ed), p.539, Academic Press, New York, (1971).

34.   I. Chibata, T. Tosa and T. Sato, *Methods Enzymol.*, **44**, 739 (1976).

35.   I.A. Rosa and K.R. Hanson, *Tech.Chem,.* **10**, 507 (1976).

36.   K. Yamamoto, T. Tosa, K. Yamashita and I. Chibata, *Eur.J.Appl. Microbiol.*, **3**, 169 (1976).

37.   W.G. Niehaus, A. Kisic, A. Torkelson, D.J. Bednarzyk and G.J. Schroepfer, *J.Biol.Chem.*, **245**, 3790 (1970).

38.   M.J.de Smet, B. Witholt and H. Wynberg, *J.Org.Chem.*, **46**, 3128 (1981).

39.   T. Katsuki and K.B. Sharpless, *J.Amer.Chem.Soc.*, **102**, 5974 (1980).

40.   H. Simon, H. Günther, J. Bader and W. Tischer, *Angew.Chem.*, **20**, 861 (1981).

41.   H.G. W. Leuenberger, W. Boguth, E. Widmer and R. Zell, *Helv.Chim.Acta*, **59**, 1832 (1976).

42.   H.G.W. Leuenberger, W. Boguth, R. Barner, M. Schmid and R. Zell, *Helv.Chim.Acta* **62**, 455 (1979).

43.   E. Abushanab, D. Reed, F. Suzuki and C.J. Sih, *Tetrahedron Lett.*, **37**, 3415 (1978).

44.   R.S. Phillips and S.W. May, *Enzyme.Microb.Technol.*, **3**, 9 (1981).

45.   J.B.Jones, and K.P. Lok, *Can.J.Chem.*, **57**, 1025 (1979).

46.   A.J. Irwin and J.B. Jones, *J.Amer.Chem.Soc.*, **99**, 556 (1977).

47.   A.J. Irwin, K.P. Lok, K.W.C. Huang and J.B. Jones, *J.Chem.Soc.Perkin Trans.I.*, 1636 (1978).

48.   H. Ohta, H. Tetsukawa and N. Noto, *J.Org.Chem.*, **47**, 2400 (1982).

49.   A.M. Klibanov, B.N. Alberti and M.A. Marletta, *Biochem.Biophys.Res. Commun.*, **108**, 804 (1982).

50. M. Nakazaki, H. Chikamatsu and M. Taniguchi, *Chemistry Lett.*, 1761 (1982).

51. D.R. Dodds and J.B. Jones, *J.Chem.Soc.Chem.Commun.*, 1080 (1982).

52. M. Ohno, S. Kobayashi, T. Iimori, W.-F. Wang, and T. Izawa, *J.Amer.Chem.Soc.*, **103**, 2405 (1981).

53. F.-C. Huang, L.F.H. Lee, R.S.D. Mittal, P.R. Ravikumar, J.A. Chan, C.J. Sih, E.Caspi and C.R. Eck, *J.Amer.Chem.Soc.*, **97**, 4144 (1975).

54. R.A. Johnson, in "Oxidation in Organic Chemistry", (W.S. Trehanovsky, ed.), Part C, p.131-210, Academic Press, New York, (1978).

55. R.A. Johnson, M.E. Herr, H.C. Murray and G.S. Fonken, *J.Org.Chem.*, **35**, 622 (1970).

56. A. Ciegler, "Microbial Transformation of Terpenes", CRC Handbook of Microbiology, **4**, 449, (1974).

57. C.-A. Yu and I.C. Gunsalus, *Biochem.Biophys.Res.Commun.*, **40**, 1431 (1970).

58. E.J. Corey, J.O. Albright, A.E. Barton and S. Hashimoto, *J.Amer.Chem.Soc.*, **102**, 1435 (1980).

59. G.A. Bahadur, J.E. Baldwin, J.J. Usher, E.P. Abraham, G.S. Jayatilake and R.L. White, *J.Amer.Chem.Soc.*, **103**, 7650 (1981).

60. M. Bucciarelli, A. Forni, I. Moretti and G. Torre, *J.Chem.Soc.Chem. Commun.*, 456 (1978).

61. N. Shimizu, T. Ohkura, H. Akita, T. Oishi, Y. Litaka and S. Inayama, *Chem.Pharm.Bull.*, **37** (3), 712 (1989).

62. D.R. Boyd, R. Austin, S. Mcmordie, N.D. Sharma, H. Dalton, P. Williams and R.O. Jenkins, *J.Chem.Soc.,Chem.Commun.*, 339 (1989).

63. K. Tsuji, Y. Terao and K. Achiwa, *Tetrahedron Lett.*, **30**(45), 6189 (1989).

64. T. Sato, T. Mizutani, Y. Okumura and T. Fujisawa, *Tetrahedron Lett.*, **30**, 3701 (1989).

65. H. Ohta, K. Matsumoto, S. Tsutsumi and T. Ihori, *J.Chem.Soc.Chem.Commun.*, 485 (1989).

66.   D.W. Brooks, P.G. Grothaus and W.L. Irwin, *J.Org.Chem.*, **47**, 2820 (1982).

67.   E. Santaniello, M. Chiari, P. Ferraboschi and S. Trave, *J.Org.Chem.*, **53**, 1567 (1988).

68.   C.-S. Chen, Y. Fujimoto and C.J. Sih, *J.Amer.Chem.Soc.*, **103**, 3580 (1981).

69.   Y. Ito, T. Shibata, M. Arita, H. Sawai and M. Ohno, *J.Amer.Chem.Soc.*, **103**, 6739 (1981).

70.   J.B. Jones and C.J. Francis, *Can.J.Chem.*, 2578 (1984).

71.   G.S.Y. Ng, L.-C. Yuan, I.J. Jakovac and J.B. Jones, *Tetrahedron* 40(8), 1235 (1984).

72.   K.P. Lok, I.J. Jakovac and J.B. Jones, *J.Amer.Chem.Soc.*, **107**, 2521 (1985).

73.   J.B. Jones and I.J. Jakovac, *Org.Synth.*, **63**, 10 (1984).

74.   I.J. Jakovac, H.B. Goodbrand, K.B. Lok and J.B. Jones, *J.Amer.Chem. Soc.*, **104**, 4659 (1982).

75.   D.M. Jerina, H. Ziffer and J.W. Daly, *J.Amer.Chem.Soc.*, **92**, 1056 (1970).

76.   C.-H. Wong, A. Pollak, S.D. McCurry, J.M. Sue, J.R. Knowles and G.M. Whitesides, *Methods Enzymol.*, **89**, 108 (1982).

77.   C.-H. Wong, S.L. Haynie and G.M. Whitesides, *J.Org.Chem.*, **47**, 5416 (1982).

78.   C. Auge, S. David, C. Mathieu and C. Gautheron, *Tetrahedron Lett.*, **25**, 1467 (1984).

79.   P.R. Rosevear, H.A. Nunez and R. Barker, *Biochem.*, **21**, 1421 (1982).

80.   M. Deluca and L.J. Cricka, *Arch.Biochem.Biophys.*, **226**, 285 (1983).

8 Stereoselective Free Radical Reactions

In the last few years, organic synthesis has been greatly enriched by advances made in free-radical chemistry largely due to the monumental works of Sir D.H.R. Barton, B. Giese and others as new methods have been developed to generate and stereoselectively trap short-lived radicals [1-10]. Regio- and chemo-selectivities of radical reactions can now be predicted to a greater extent on the basis of steric, polar and radical stabilization effects. However synthetic application of stereoselective free radical reactions has lagged far behind. This situation is being rapidly rectified.

Generally free radical reactions are considered by reaction type, i.e. addition, fragmentation, cyclization, rearrangement, etc. However, despite the diversity of free radical reactions, the number of practical methods available to conduct free radical reactions are few. In conducting free-radical chain reactions, the following are the main methods applied:

i)   The tin hydride method.
ii)   The mercury hydride method.
iii)   The fragmentation method.
iv)   The Barton (thiohydroxamate ester) method.
v)   The atom transfer method.

These methods are being increasingly supplemented by non-chain methods, such as redox processes and radical-radical coupling [1,11-20]. However before discussing the merits of these methods in stereoselective reactions, some basic concepts will be discussed.

8.1   Free Radical Chain Reactions

Most organic free radicals are reactive species and unlike their charged counterparts, they react with themselves by combination or disproportionation in a diffusion-controlled manner, i.e. the rate constants are about 10^9 L. mol^{-1} s^{-1}, and only a small temperature dependence is observed [21]. This requires that a low concentration of radicals should be maintained over the course of the reaction. This requirement is met by chain reactions. In order that a radical chain reaction be synthetically useful, a particular chain reaction must generate radicals selectively with sufficient lifespan to take part in the reaction. However, this lifetime must be strictly controlled by the nature of the chain transfer step, as unduly long-lived radicals may terminate the chain reaction. The chain-transfer step controls the generation of short- and long-lived radicals in solution and is an important part of synthetic planning, since its rate determines what reactions will (or will not) be allowed on intermediate radicals.

One of the most extensively applied reactions in synthesis is the trialkyltin hydride-mediated reduction of various organic functional groups, mostly halides [8,22]. Figure 8.1 shows the various steps involved in trialkyltin hydride-mediated radical reactions. In the first step, tri-n-butyltin radical abstracts a

halogen or another functional group affording an organic radical $A^{\bullet}$.

Step 1: $AX + {}^{n}Bu_3 Sn \xrightarrow{K_X} A^{\bullet} + {}^{n}Bu_3 SnX$

Step 2: $A^{\bullet} \xrightarrow{K_A} B^{\bullet}$

Step 3: $A^{\bullet} + {}^{n}Bu_3 SnH \xrightarrow{K_H} AH + {}^{n}Bu_3 Sn^{\bullet}$

Step 4: $B^{\bullet} + {}^{n}Bu_3 SnH \xrightarrow{K_{H'}} BH + {}^{n}Bu_3 Sn^{\bullet}$

Step 5: $B^{\bullet} \xrightarrow{K_D}$ non - radical products

Figure 8.1. Trialkyltin hydride-mediated radical reactions.

Radical $A^{\bullet}$ should have sufficient lifetime to react via a series of inter- or intra-molecular reactions (step 2) to form a new radical $B^{\bullet}$. The lifetimes of radicals $A^{\bullet}$ and $B^{\bullet}$ are determined by the rate of the chain-transfer steps 3 and 4. These rates are controlled by the rate constants for abstraction of hydrogen (K_H, $K_{H'}$) from the trialkyltin hydride and by the concentration of tin hydride. Step 5 signifies chain termination and formation of non-radical products, which is the last step in a radical reaction.

Since many of the relevant rate constants are known [23, 24], one can plan a synthesis based on radical reactions. One can also conduct a series of reactions between radical generation and chain transfer, thus allowing the construction of multiple carbon-carbon bonds.

From a practical point of view, nearly all free radical reactions can be grouped into two broad classes: atom (or group) abstraction and addition to multiple bonds. In the first class (designated as substitution, S_H2), a radical abstracts one atom or group from an organic molecule in a homolytic fashion to generate a new radical in which the odd electron is located at the site of the abstracted functionality (Fig. 8.2. A). Though the reaction is reversible, it is nearly never so on the time scale of a radical chain reaction. The rate of the reaction is largely controlled by its exothermicity while the direction of the reaction is determined by the relative bond strengths of the forming and breaking bonds. The most commonly abstracted groups are halogens, hydrogen, SR, SeR etc.

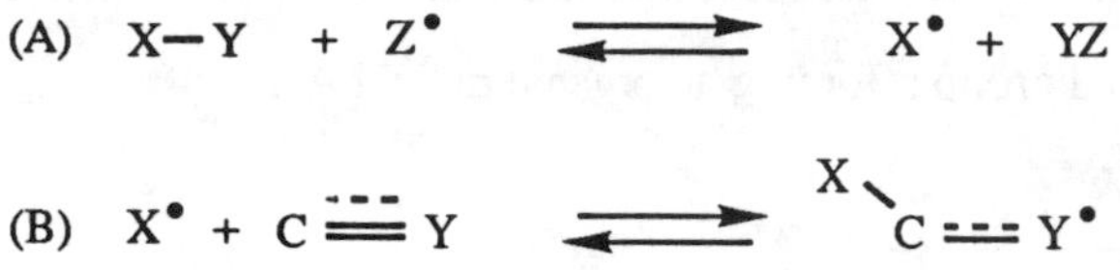

(A) $X{-}Y$ + $Z^{\bullet}$ $\rightleftharpoons$ $X^{\bullet}$ + YZ

(B) $X^{\bullet}$ + $C \overset{---}{=\!=} Y$ $\rightleftharpoons$ $\underset{C \,=\!=\!= Y^{\bullet}}{\overset{X}{\diagdown}}$

Figure   8.2. (A) Atom (group) abstraction.

(B) Addition to multiple bonds.

In the second class, a radical adds to an unsaturated group which may be a carbon-carbon, carbon-oxygen or carbon-nitrogen multiple bond (Fig. 8.2. B). The addition is, in principle, also reversible, the position of the equilibrium being controlled by the relative bond strengths as well as by the relative stabilities of the radical involved. Radical addition to a carbon-carbon multiple bond is particularly favorable since a $C{-}C$ σ bond is formed at the expense of a $C{-}C$ π-bond, which is energetically quite favorable.   The reactions are exothermic and pass through early transition states where the frontier molecular orbital theory is fully applicable [25]. Similarly, eliminations are quite useful in radical reactions with atoms or groups that form relatively weak bonds to carbon, and at the same time give rise to stable radicals. Such groups are R_3Sn, RS, RSe, and halogen.

Intramolecular radical addition reactions are particularly useful in synthesis for the construction of rings, while intermolecular radical addition reactions are becoming increasingly common in the stereoselective construction of the carbon frame work. Radical reactions can be compared to anionic counterparts when designing a particular synthesis. They are highly chemoselective, and functional groups containing $N{-}H$ or $O{-}H$ bonds require no protection. Steric crowding, particularly on the diradical center, is often tolerated [25].

8.1.1 The Tin Hydride Method

The tin hydride method has become the most commonly used method involving the use of free radicals to form carbon-carbon bonds [8]. Fig. 8.3 shows the tin hydride (1) mediated cyclization of 5-hexenyl radical (4) affording the five-membered ring compound (7) through the usual steps involving initiation and propagation. The chain carrier (2) is generated in an initiation step from (1). Azobisisobutyronitrile (AIBN) is commonly used as the initiator.

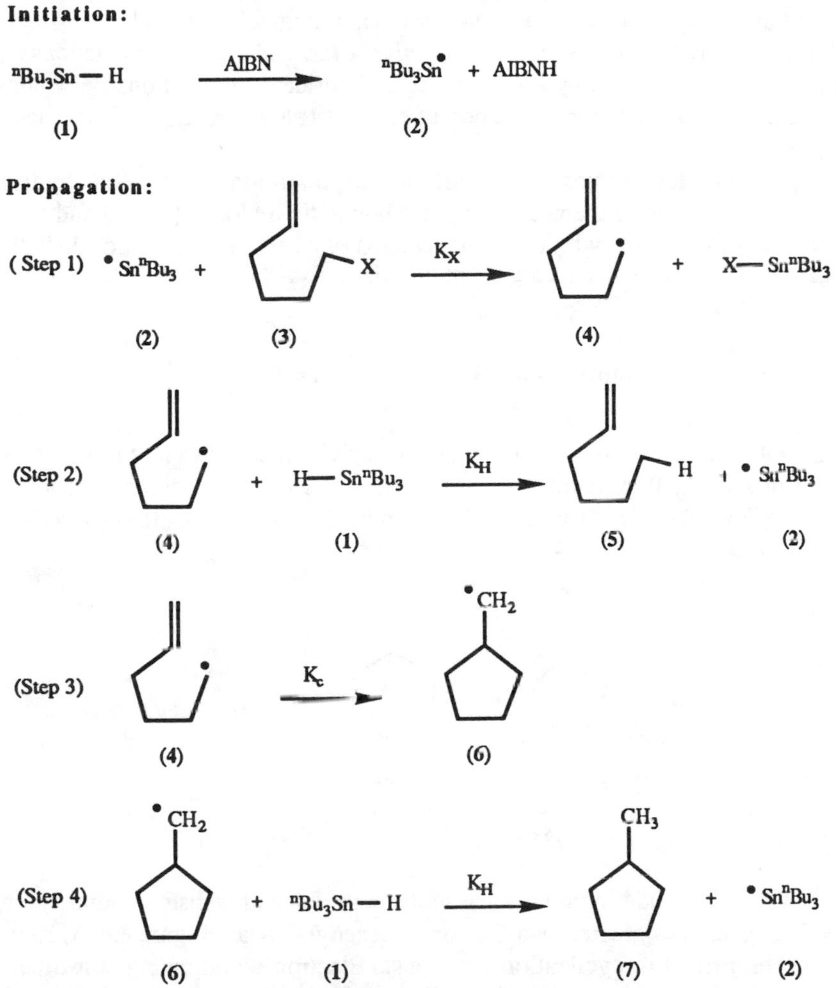
Initiation:
$^nBu_3Sn—H$
AIBN
$^nBu_3Sn^\bullet$ + AIBNH
(1)
(2)
Propagation:
(Step 1) $^\bullet Sn^nBu_3$ +
X
K_X
+ X— Sn^nBu_3
(2)
(3)
(4)
(Step 2)
+ H— Sn^nBu_3
K_H
H
+ $^\bullet Sn^nBu_3$
(4)
(1)
(5)
(2)
(Step 3)
K_c
$^\bullet CH_2$
(4)
(6)
(Step 4)
$^\bullet CH_2$
+ $^nBu_3Sn—H$
K_H
CH_3
+ $^\bullet Sn^nBu_3$
(6)
(1)
(7)
(2)

Figure 8.3. Cyclization of 5-hexenyl radical by the tin-hydride method.

The propagation sequence begins with atom or group abstraction by (2) from (3) to give the hexenyl radical (4) (step 1). In the following step (step 2), the radical (4) may abstract a hydrogen from (1) to afford the reduced product (5) and the radical (2). This a chain transfer step and is a second order reaction. Alternatively in a first order reaction (step 3), (4) may undergo cyclization in a 5-*exo*-fashion affording (6). In another chain transfer step, (6) may abstract hydrogen from tin hydride to give (7) and the chain carrying tin radical (2).

By careful adjustment of the tin hydride concentration, which controls the product distribution, a "window" of radical lifetime is obtained. The window is closed by raising the tin hydride concentration. Intermediate radicals have shorter lifetimes, as hydrogen abstraction is relatively fast and only rapid reactions can compete. Alternatively, by lowering the tin hydride concentration, the window is opened and longer lifetimes are permitted so that slower reactions can occur.

Various techniques are available to maintain a low tin hydride concentration, such as the use of polymer-bound tin hydrides [26-29] and the *in situ* generation of tin hydrides by the reaction of a catalytic amount of tin halide with a standard hydride reducing agent [$NaBH_4$ or $NaCNBH_3$) [30-32].

8.1.1.1   Intramolecular Radical Cyclizations

Intramolecular cyclization of substituted hexenyl radicals provides a useful tool for synthesising five-membered rings since 5-*exo*-cyclizations are generally preferred over 6-*endo* closures as illustrated by the formation of (6) and (8) from (4) in a ratio of 80:1 [33,34].

(4)   (6) 5-*exo*   (8) 6-*endo*

(6 : 8 = 50 : 1)

The tin hydride method is particularly useful with substrates undergoing cyclization at rates greater than the parent hexenyl radical. Figure 8.4(A) shows the general trend in cyclization reactions. Electron withdrawing substituents (EWG) that lower the LUMO of the alkene acceptor dramatically accelerate the cyclization of nucleophilic alkyl radicals. The presence of N or O at position 3 also greatly accelerates the cyclization. Substitution at positions 2-4 has a modest beneficial effect. Alkyl substituents on the radical bearing carbon 1 have little effect, provided that C-5 is not disubstituted. The presence of R group at position 5 as shown (in Fig. 8.4 (B)) reduces the rate of 5-*exo* cyclization significantly so that 6-*endo* cyclization may become competitive. Also, substitution at C-1 by a radical stabilizing group (Y=CO, C=C, heteroatom) slows down the cyclization. Radical stabilizing groups endocyclic to the forming ring are also rate-retarding [35,36].

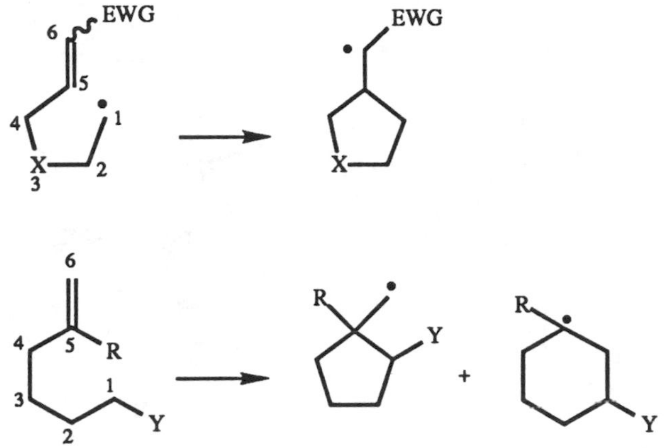

Figure 8.4. Effects of substituents on cyclization, EWG = electron withdrawing group.

The utility of intramolecular radical cyclizations for the formation of five-membered rings is evident from the ever-increasing number of publications in this area [37]. The reaction has been particularly useful in the synthesis of carbocycles from sugars with preservation of stereochemistry. For instance, the (Z)-ester (10) derived from D-ribose via (9) through a Wittig reaction and benzoylation is subjected to tin hydride-mediated cyclization to afford the cyclized products (11) and (12) in a ratio of 91:9 [38,39]. The corresponding (E)-ester affords a [1:1] mixture of products (11) and (12), indicating that the cyclization is substrate-dependent. Similarly, cyclization of the radical (14), derived from the D-glucopyranose derivative (13), afforded a single product (15) in high yield [39].

The results have been rationalized in terms of the transition state geometry in which the bulky groups are as far apart as possible. Thus in the last example the reaction is believed to proceed through a chair-like transition state (16) in which the pseudo-equatorial radical center attacks the butenyl group in a pseudo-equatorial position while the benzyloxy groups on the butenyl side chain occupy the favorable equatorial site.

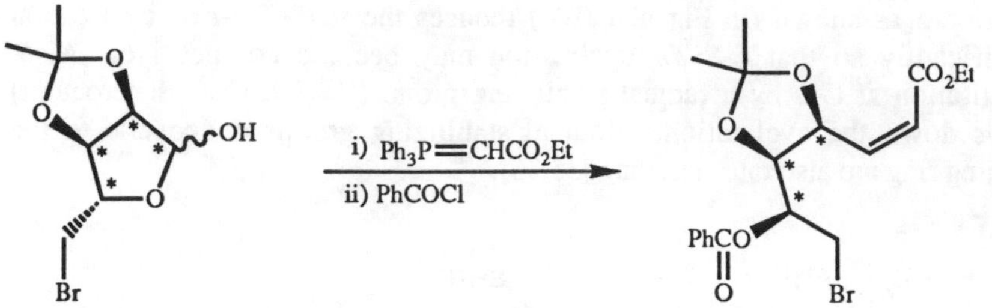

(9)

i) Ph$_3$P=CHCO$_2$Et
ii) PhCOCl

(10)

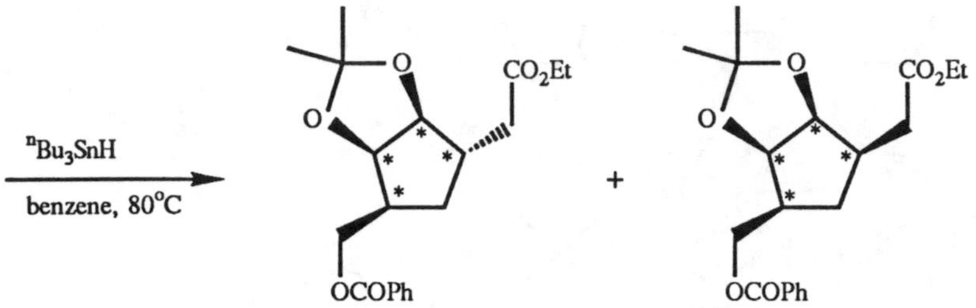

^{n}Bu$_3$SnH

benzene, 80°C

(11)

+

(12)

(11 : 12 = 91:9)

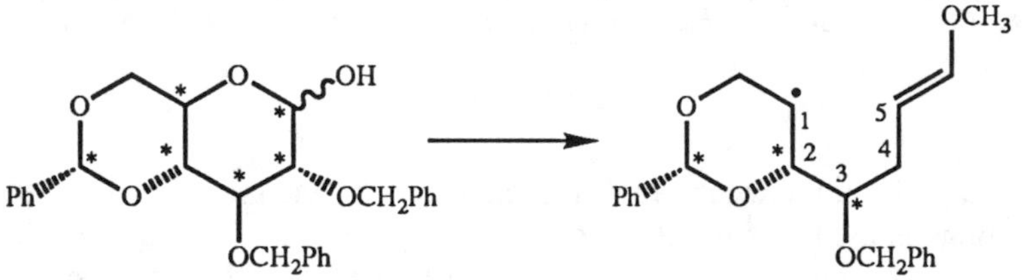

(13)

(14)

^{n}Bu$_3$SnH

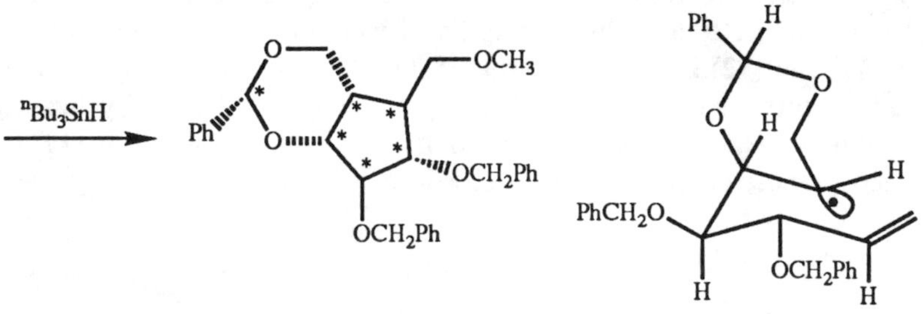

(15)

(16)

The transition state (16) also shows that the dioxane radical is in a flexible boat conformation in which the bulky phenyl and butenyl groups occupy the pseudo-equatorial positions, the lone-pair repulsion between the ring oxygen atom is minimized and the flag-pole repulsion which destabilizes the cyclohexane boat form is absent. Table 8.1 shows some examples of stereoselective tin hydride-mediated intramolecular radical cyclizations.

Table 8.1. **Stereoselective Tin Hydride Mediated Intramolecular Radical Cyclizations**

Substrate	Product [†]	Selectivity (% yield)	Reference
		(2α:2β = 4:1)	[40]
		(E : Z = 5:1)	[41]
		(3, 4 trans : cis = 85.15)	[42]
		95%	[43]
		85%	[44, 45]

(Table 8.1. contd.)

Substrate	Product [†]	Selectivity (% yield)	Reference

| | | 81% | [46] |

| | | 65 | [47] |

| | | 100% d.p. | [48] [+] |

| | | 73 | [49] |

| | | 65 | [50] |

[†] Only major isomer is shown.

[+] Cyclization followed by Jones oxidation.

8.1.1.2   Intermolecular Radical Additions

Although the stereochemistry of intramolecular radical reactions, particularly cyclizations and rearrangements, can be controlled to a large extent mainly due to an understanding of the steric and polar effects, and some understanding of radical stabilities, the factors controlling the stereochemistry in *intermolecular* radical additions, particularly for acyclic radicals and acyclic alkenes, are poorly understood [8]. This restricts the use of intermolecular radical reactions in comparison to other reaction types such as concerted reactions and those involving carbanion chemistry (Chapter 4) because of the exquisite control of stereochemistry possible with these other approaches [51-53].

Reactions of cyclic radicals with alkenes have been thoroughly studied by the tin hydride method [8]. These have shown that the various radicals produced within the chain have different selectivities but they do not lose their high reactivity.

Generally, the unpaired electron in carbon-centred free radicals occupies an orbital which has mainly p-character. Such radicals are π radicals, allowing attack from both sides. High stereoselectivity can only be achieved when (a) the radicals react with a preferred conformation, and (b) the radicals are attacked predominantly from one side. The effects of substituents R on the stereoselectivity of cyclic and vinyl radicals (17) and (18) respectively have been extensively studied. However acyclic radicals (19) have been studied only relatively recently [54].

(17)   (18)   (19)   (20)

Introduction of an electron withdrawing (EWG) substituent on the alkene such as (20) greatly accelerates additions, since alkyl radicals exhibit nucleophilic character. It has been shown that electron withdrawing substituents on C-1 in (20) exert mainly polar effects; substituents that lower the LUMO of the alkene dramatically accelerate the reaction of a nucleophilic alkyl radical [25]. Conversely addition of electrophilic radicals to alkenes is accelerated by electron-releasing groups, while substituents at C-2 in (20) exert both steric and polar effects. Alkyl and other non-activating groups at C-2 significantly retard the rate of addition of alkyl radicals while electron withdrawing groups at C-2 provide a modest acceleration.

The synthetic utility of tin hydride mediated intermolecular radical reactions is evident from the diastereoselective synthesis of *C*-glycopyranosides. For instance, α-D-glycopyranosyl bromide (21) reacts with acrylonitrile (22) affording (23) exclusively in 72% yield [55-57].

The formation of (23) is surprising in that the new C - C bond is axial. This has been explained by assuming that the glucosyl radical exists in the boat conformation (24) which follows from ESR measurements [58]. Thus (23) is formed by equatorial attack on (24). Similarly, the chloromalonate (25) adds to the enol ether (26) to give the product (27) stereoselectively in 56% yield [59].

Organotin compounds have been employed in radical chain reactions without involving hydrogen donors [61-63]. For instance the reaction of the pentose derivative (28) with allyltin (29) gives the pseudomonic acid derivative (32) in 72% yield [60].

(28)   (29)   (30)

(31)   (32)   (2)

The reaction is believed to proceed *via* the carbohydrate radical (30) which reacts with the allyltin (29) to give the adduct radical (31). This eliminates a trialkyltin radical by β-bond cleavage affording the product (32) while the trialkyltin radical regenerates the carbohydrate radical.

Similarly, the chloromalonate (33) adds to the enol ether (34) to give stereoselectively the product (35) in 56% yield [59].

(33)   (34)   (35)

The use of chiral auxiliaries attached to alkenes provide another way to effect stereoselective radical reactions [54, 64, 65]. Thus diastereomeric excess of 80:1 has been achieved recently with amide substituted alkenes [54]. For instance, maleic acid menthyl ester (36) reacts with 2-propanol (37) affording (S)-(38) in 62% d.e. [64].

(36)   (37)   (38) (S)

The radical (37) attacks the *re*-face of the substrate (36) affording the (*S*)-product as shown in Fig.8.5.

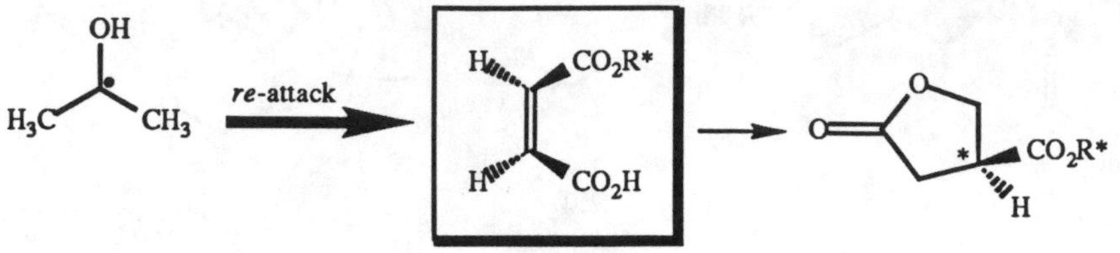

(37)   (36)   (38) (*S*)

Figure 8.5. Radical attack on *Re*-face of substrate, R*=menthyl.

Recently it has been demonstrated that control of stereochemistry at the α olefin center in radical addition reactions to acyclic alkenes is possible with C_2 symmetric chiral auxiliaries [54]. For instance , 4-oxopentenoic amide (39) reacts with cyclohexyl iodide (40) to give four products (41-44) in a ratio of 0.48 : 0.02 : 0.28 : 0.23 [54].

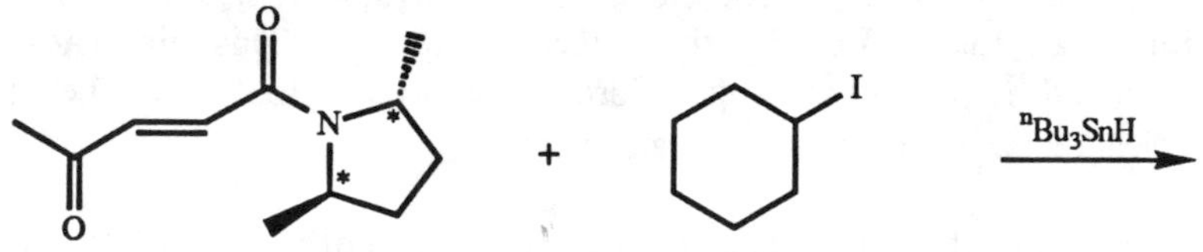

(39)   (40)

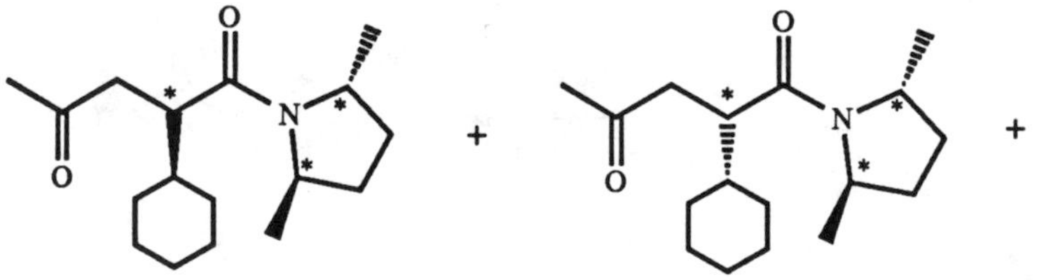

(41)   (42)

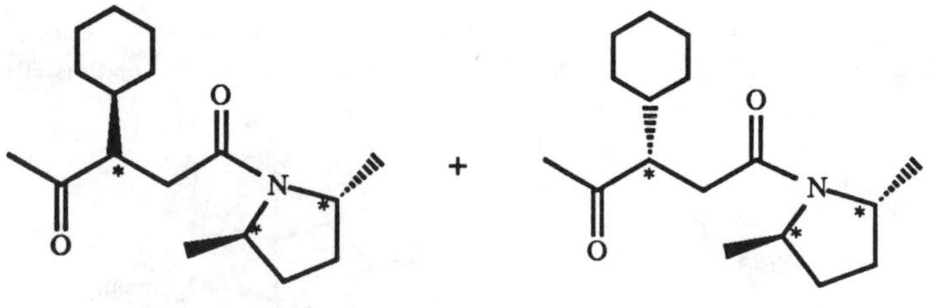

(43)   (44)

By employing mercury hydride method (which will be discussed in the next section) a diastereo-selectivity of 80:1 has been achieved in this reaction.

8.1.2   The Mercury Hydride Method

In addition to tin hydride-mediated radical reactions, organomercury compounds have proved particularly useful in radical chain reactions in recent years [54, 66-68]. Alkylmercury salts particularly halides and acetates, e.g. (45) when treated with boron or tin hydrides give organomercury hydrides (46) which *via* hydrogen abstraction afford the labile alkylmercury radical (47). This decomposes spontaneously to mercury and an alkyl radical which attacks the olefinic substrate (48) to give the adduct radical (49). The alkylmercury hydride plays the role of a hydrogen donor, trapping the adduct radicals (49) to give the product (50).

$$RHg\,X \xrightarrow{\text{NaBH}_4} RHgH$$

(45)   (46)

$$RHgH \longrightarrow RHg^{\bullet} \longrightarrow R^{\bullet} + Hg$$

(46)   (47)

(48)   (49)

(49)   RHgH   (50)

There are several advantages in the mercury method over the tin method. Mild reaction conditions (room temp, absence of light) are required in the mercury method. Furthermore the reaction requires a very short time (in the order of minutes) and the mercury can be separated without any problems. However only very reactive alkenes can be employed in the mercury method, since the alkylmercury hydride is a better hydrogen donor than trialkyltin hydride.

8.1.2.1   Intramolecular Cyclization Reactions

Organomercury compounds bearing β-hetero substituents undergo reductive coupling with electrophilic olefins upon treatment with sodium trimethoxyborohydride (the Giese reaction) [69-71]. The Giese reaction has been successfully employed in the intramolecular cyclization of 5-hexenyl radicals to give mainly five-membered rings [72,73]. For instance the diene (51) undergoes mercuration with mercuric acetate affording the presumed intermediate (53). This undergoes reductive cyclization with NaBH$_4$ to give the product (54) in 72% yield [72].

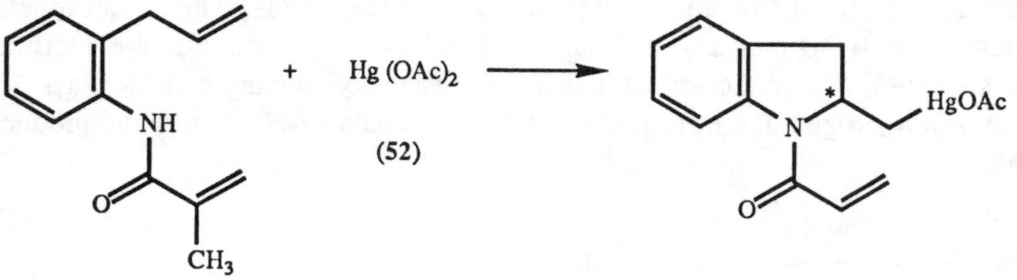

(52)

(51)   (53)

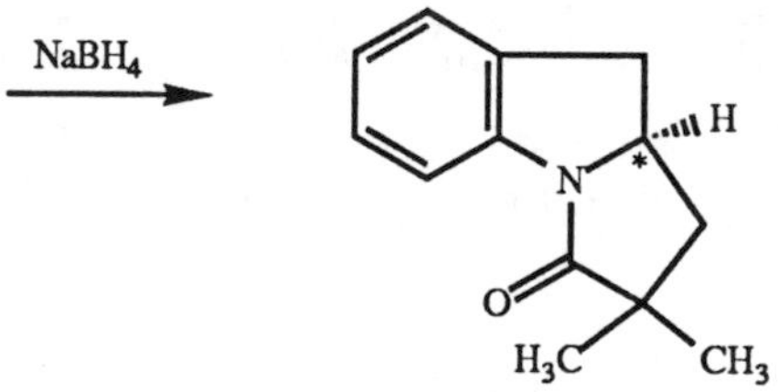

(54)

Similarly, the reaction of the substrate (55) with (52) followed by reductive cyclization afforded the product (56) in 60% yield and 63% d.e. [73].

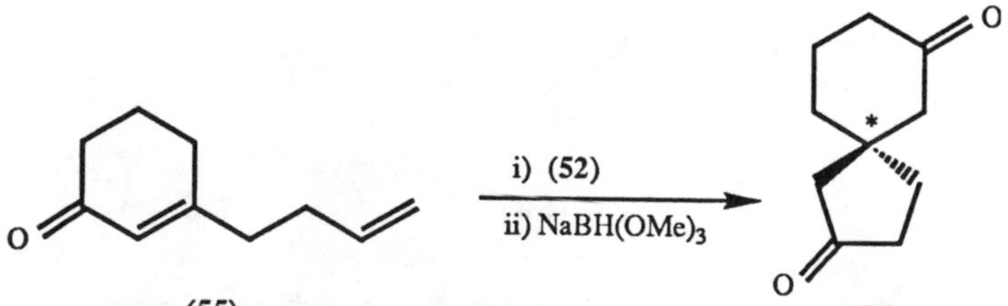

i) (52)
ii) NaBH(OMe)$_3$

(55)   (56)

The mercury method has also been successfully employed in prostaglandin synthesis [74]. For instance, treatment of the hydroperoxide (57) with mercury (II) chloroacetate (58) afforded the chloromercurial (59) which underwent facile radical cyclization to give predominantly the *cis*-product (60).

Similarly mercuric acetate (52) reacts with linalool (61) to give the monocyclic organomercury compound (62). The reaction proceeds by electrophilic attack of (52) on the electron-rich double bond in (61) followed by intramolecular attack by the hydroxyl group. Homolysis of the C—Hg bond gave the radical (63) which underwent reductive cyclization to give (64) in 52% yield [75].

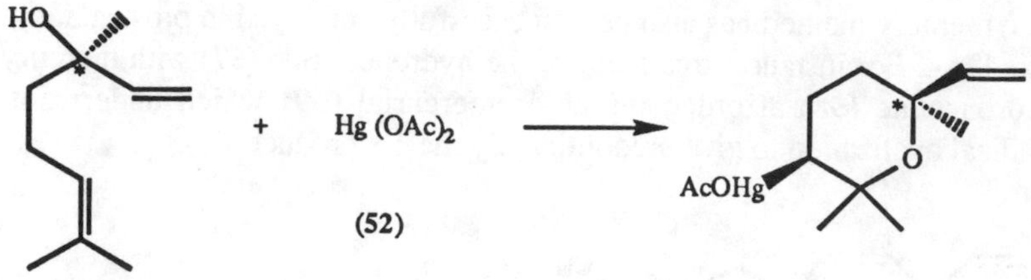

(61)   (52)   (62)

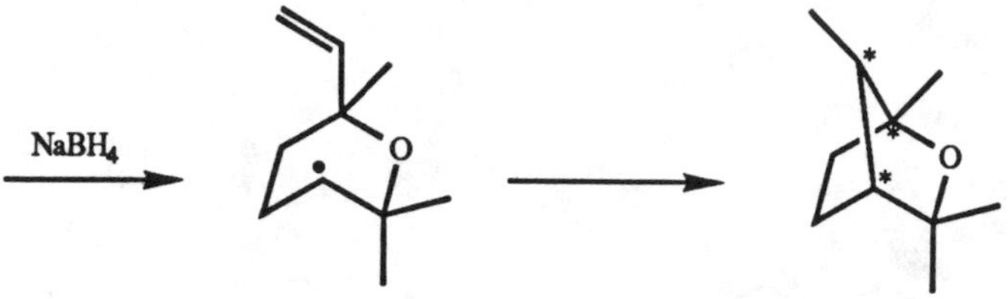

(63)   (64)

8.1.2.2   Intermolecular Radical Reactions

The versatility of the mercury method lies in the fact that organomercury compounds are easily accessible from various substrates such as alkenes, cyclopropanes and ketones.

8.1.2.2.1   Cyclic   Radicals

In the reactions mediated by the mercury method involving addition to β-substituted radicals , the *anti* : *syn* ratio depends on the substituents on both the radical and the radical trap [8, 76]. For instance, cyclopentene (65) reacts with Hg(OAc)$_2$ (52) in the presence of EtOH to give the mercurial adduct (66) which forms the cyclopentyl radical (67). Chloroacrylonitrile (66) adds preferably to the *anti* face of (67) to give the products (69) and (70) in an *anti* : *syn* ratio of 72:28 [77].

(65)   **(52)**   **(66)**   **(67)**

(67)   **(68)**   **(69)**   **(70)**

(69:70 = 72:28)

Table 8.2 shows the effects of substituents in both the radical and the radical trap on the selectivity of the reaction.

Table 8.2. Effects of Substituents on the Selectivity of Intermolecular Radical Reactions

R	X	Y	Z	A:B
OC$_2$H$_5$	H	Cl	CN	72:28
OC$_2$H$_5$	H	H	CN	77:23
OC$_2$H$_5$	H	H	CO$_2$CH$_3$	88:12
OC$_2$H$_5$	H	H	Ph	90:10
OC$_2$H$_5$	CO$_2$C$_2$H$_5$	H	CO$_2$C$_2$H$_5$	98:2
CH$_3$	H	H	CN	92:8
CH$_3$	CO$_2$C$_2$H$_5$	H	CO$_2$C$_2$H$_5$	98:2
NHCOCH$_3$	H	Cl	CN	>98:<2

It is evident from table 8.2 that substituents on the alkene have a profound influence on the stereoselectivity of the reaction, whereby both the reactivity and the shielding of the alkene play a role. Thus selectivity increases from 72:28 for β-ethoxy-cyclopentyl radical (67) to 90:10 when α-chloroacetonitrile is substituted by styrene, which reacts 200 times more slowly. When terminal alkenes react with the β–ethoxycyclopentyl (67), there is a direct relationship between decreasing reactivity and enhanced stereoselectivity. This is in accordance with the "reactivity-selectivity principle", which predicts that a slower reaction passes through a late-transition state in which steric hindrance plays a major role than in the faster reaction.

In the case of β-ethoxycyclopentyl radical (67), the stereoselectivity is also influenced by the solvent. Thus the stereoselectivity of addition of acrylonitrile to (67) at 20°C increases from 68:32 when the reaction is carried out in cyclohexane to 76:24 in THF, 77:23 in CH_2Cl_2 and 81:19 inCH_3CN [77]. Table 8.3 shows some examples of stereoselective radical reactions of cyclic substrates.

Table 8.3.   Stereoselective Radical Reactions of Cyclic Substrates

Substrate	Reagent	Product[t]	Anti : Syn	Reference
	$C_6H_{11}HgH$		94:6	[78]
	^tBuHgX		6:94 *syn : anti*	[79]
	i) B_2H_6 ii) $Hg(OAc)_2$ iii) $NaBH_4$,		53	[80]
	i) $Hg(OAc)_2$, CH_3OH ii) $R_4N^+BH_4^-$ iii)		90	[66]

(Table 8.3. contd.)

Substrate	Reagent	Product[†]	Anti : Syn	Reference
	i) Hg(OAc)₂, CH₃OH ii) NaBH₄, ⟍⟍CN		70	[66]
	i) Hg(NO₃)₂, CH₃CN ii) NaCl iii) NaBH₄, Cl⟍⟍CN iv) NaH		95:5	[76]
	i) Hg(OAc)₂, CH₃OH ii) NaBH₄, ⟍⟍CN		82	[81,82]
	i) Hg(OAc)₂, H₂O ii) NaBH₄, ⟍⟍CN		65	[81,82]
	i) N₂H₄ ii) HgO, Hg(OAc)₂ iii) NaBH₄, ⟍⟍CN		77	[83]

[†] Only the major isomer is shown.

8.1.2.2.2   Acyclic Substrates

There are only a few examples of stereoselective intermolecular radical reactions with acyclic alkenes [54]. However, so far, no cases of high stereoselectivity are known with terminal alkenes and it is therefore likely that the α-carbon must be substituted with a chiral auxiliary to affect high stereoselectivity [8, 54]. A decisive breakthrough was achieved with α, β-unsaturated amides [54].

For instance, the fumaric diamide (71) reacts with the *tert.* butyl radical by the mercury method to give the product (72) and (73) in a ratio of 98.8:1.2 [54].

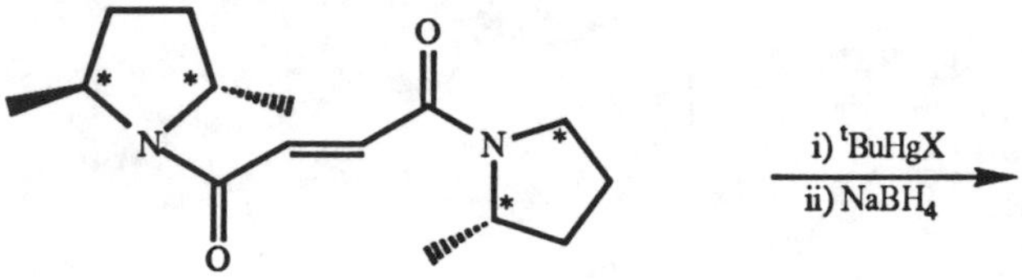

(71)

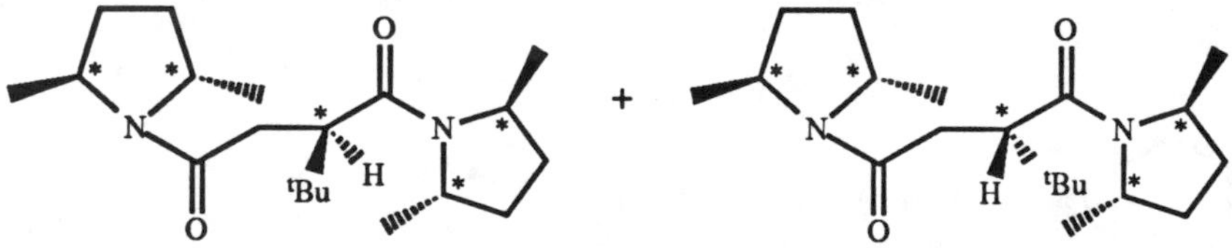

i) ^{t}BuHgX
ii) NaBH$_4$

(72)   +   (73)

(72:73 = 98.8 : 1.2

The high stereoselectivity observed with amides is due to the two stereogenic centres of the amine component. Rotation around the C - N bond does not change the conformation of the substrate because of the C_2 symmetry of the amines. The two methyl groups of the amine component shield the two sides of the alkene to different extent as depicted in (74).

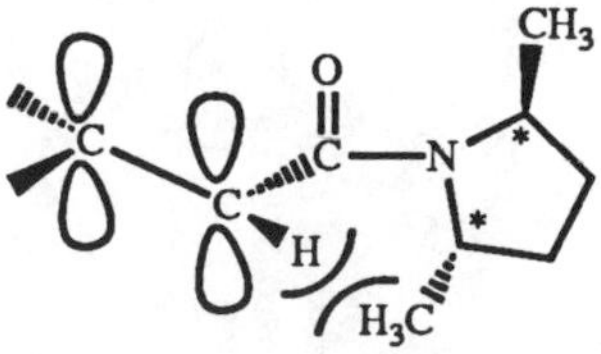

(74)

The α–methyl group is closer to the alkene than the β-one. Thus the attack on the olefinic carbon atom is directed to different degrees by the two methyl groups, and the radical addition is stereoselective.

8.1.3 The Fragmentation Method

The fragmentation method involves generation of chain transfer radical by a fragmentation reaction rather than by the process of hydrogenation as in the tin and mercury methods. As shown for the allyl (Fig. 8.6A) and vinyl (Fig.8.6B) substrates the C–X bond undergoes fragmentation affording the radical X• which is either itself the chain-transfer reagent or generates the chain-transfer reagent in a subsequent rapid reaction with a neutral molecule.

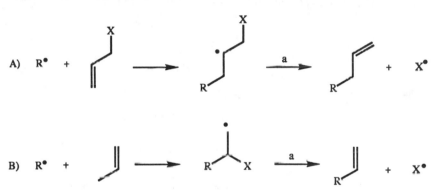

Figure 8.6. Radical reaction by fragmentation method, (A) Allyl substrates, B) Vinyl substrates; a = fragmentation.

An advantage of this method is that metal hydride is not required in the reaction for proton abstraction and hence the lifetimes of intermediate radicals are not limited by the rate of hydrogen atom abstraction. The net result of the reaction, shown in Fig.8.6, is an overall allylation or vinylation of a suitable radical precursor. In reactions involving vinylation, activating groups such as trialkyltin, cobaloxime or thiophenyl are employed in order to direct the radical addition to the X bearing carbon [84,85].

Free radical allylation with allylstannanes has become a powerful method for functionalization of organic substrates and although allylplumbanes can be used instead of allylstannanes, allylgermanes and allylsilanes are not often used, because of the slow cleavage of the C - Ge and C - Si bonds [86-88]. The key element in the success of the fragmentation method is due to (a) its ability to provide for intermediate radicals with longlife times which permit relatively slow reactions , and (b) the rapid transfer of the chain when the final radical is generated. For instance, allylation of the thioacyl derivative (75) with allyltri-n-butylstannane (76) in toluene at ambient temperature and with photochemical initiation afforded the product (77) in 93% yield [89].

Similarly, β-stannyl enone (78) reacts with the iodo acetal (79) to give the product (80) in 72% yield [90].

The fragmentation method has also been employed with success in the regioselective double vicinal carbon-carbon bond forming reactions of electron-deficient alkenes using allylic stannanes and organoiodo compounds [91].

For instance, the allylstannane (81) and methyl iodide react with the substrate (82) to give the product (83) in 87% yield. The allyl function from allylstanane was regioselectively introduced in the carbon α to the cyano group in 1,1-dicyano-2-phenyl ethane (82) while the methyl group from methyl iodide was introduced at the β-carbon.

8.1.4 The Barton (Thiohydroxamate Ester) Method

In contrast to the previous reactions involving trialkyltin or organomercury radicals as chain-transfer agents, a novel option is provided by the chemistry of thiohydroxamic acid esters, developed by Sir D.H.R. Barton and his co-workers [92-122]. Initially thioesters were employed as derivatives in the deoxygenation of secondary alcohols affording the corresponding alkanes in high yields [92].

For instance, O-cholestanyl thiobenzoate (87), prepared from 5α-cholestan-3β-ol (84) and (85) via the salt (86) reacted with tributylstannane affording the deoxygenation produced (88) in 70-75% yield [92].

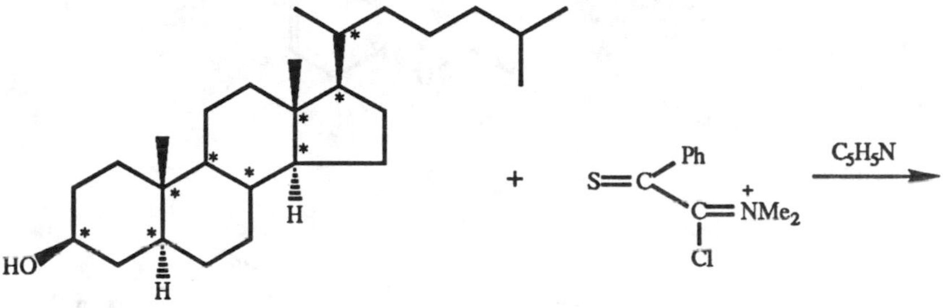

(84)   (85)

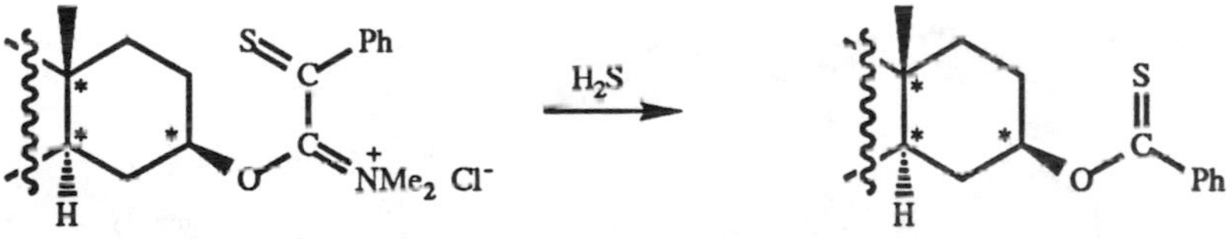

(86)   (87)

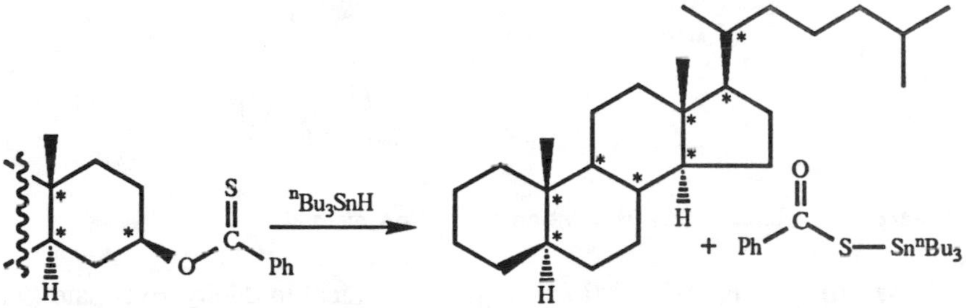

(87)   (88)   (89)

The driving force of the reaction is the thermodynamically favourable change in going from the thiocarbonyl (87) to the carbonyl compound (88). Analogous to the deoxygenation of secondary alcohols, a new method was devised for the radical decarboxylation of carboxylic acids involving O-esters of thiohydroxamic acids [95,97].

The Barton method employs N-hydroxypyridine-2-thione (90) for esterification of carboxylic acids or acid chlorides to give thiohydroxamic acid esters (91) which undergo efficient radical chain decarboxylation in the presence of tin hydride reagent to give nor-alkanes in high yields [95]. Fig.8.7 shows the steps involved in the radical decarboxylation by the Barton method.

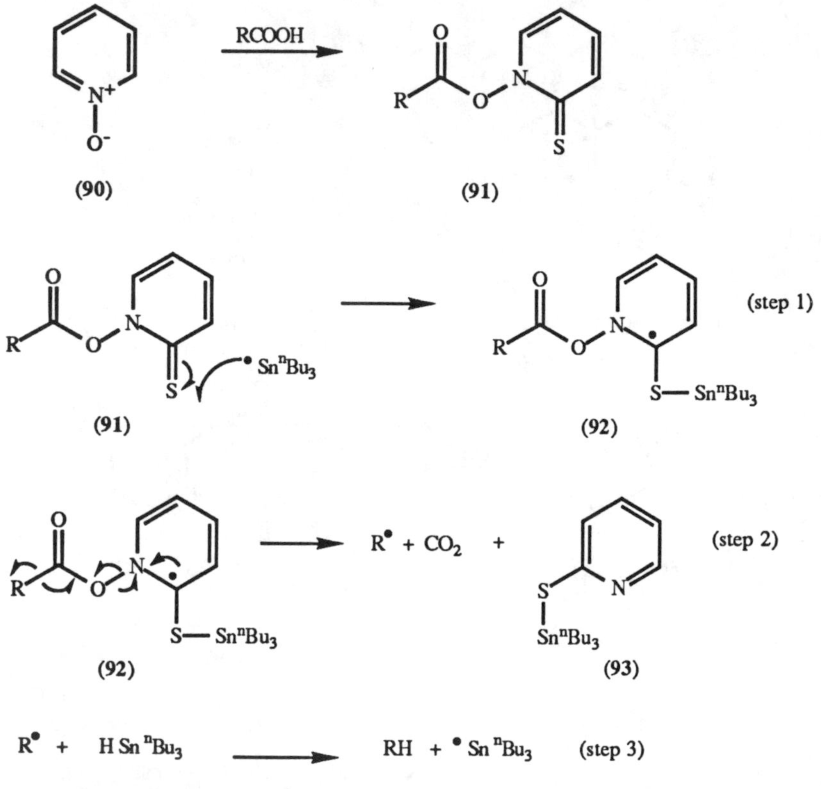

Figure 8.7. Radical decarboxylation by Barton method.

As shown in Fig.8.7, addition of stannyl radical to thiohydroxamate (91) produces (92) (step 1). This is followed by fragmentation of (92) with the loss of CO_2 to give the alkyl radical and (93) (step 2). The enthalpic driving force for this fragmentation is provided by the formation of CO_2 and the aromatization to the mercaptopyridine (93), an entropic driving force being provided by the formation of two molecules (CO_2 and 93) from one molecule (92). The real power of the Barton method lies in the fact that intermediate radicals (formed in step 2) can be intercepted by a variety of neutral molecules as shown in Fig.8.8.

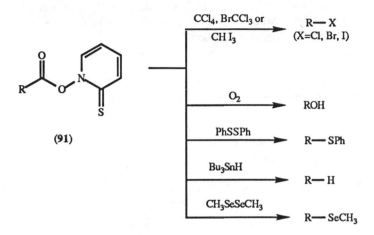

Figure 8.8. Some applications of the Barton method.

As shown in Fig.8.8, the Barton method is very general and a wide variety of transformations of intermediate radicals are possible. The method has been extended by Barton and his co-workers to various substrates, such as the radical induced fragmentation of thiocarbonates [92], 1,2-dixanthates [118], α-epoxy-xanthates [119], primary xanthates [120], tertiary thioformates [121], etc.

The Barton method has been extended to stereospecific carbon-carbon bond formation reactions [93,122]. Thus radical decarboxylative addition of protected derivatives of (+)-L-tartaric acid to activated alkenes leads to the overall substitution of the carboxy group with complete retention of configuration [122]. For instance, irradiation of the cyclic ester derivative (94) of the (R, R)-tartaric acid afforded the *trans* sulphide (95) in 78% yield [122].

(94)   hʋ
              —→
              -CO₂   (95)

8.1.5 The Atom Transfer Method

One of the fundamental reactions of organic free radicals is the addition of a reagent X - Y across a carbon-carbon double (or triple) bond as shown in Fig.8.9 [123]. This is also called the Kharasch addition.

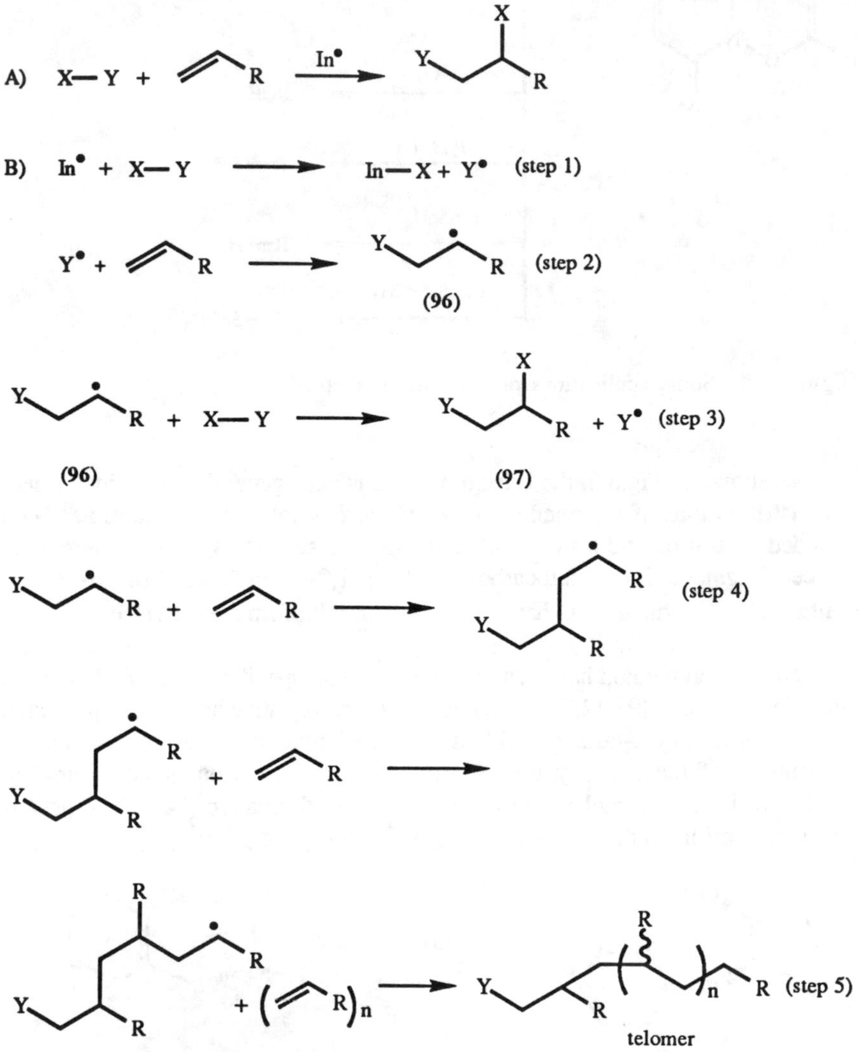

Figure 8.9. Addition by atom transfer (Kharasch addition), X = H or halogen; Y = C or heteroatom; In = initiator.

When Y is a carbon, X is usually H or halogen. The transformation exchanges one σ bond and one π bond for two σ bonds. A general mechanism of the reaction is shown in Fig.8.9 (B). Initiation is facilitated by the abstraction

of X to provide the initial radical Y• (step 1). Steps 2 and 3 are the propagation steps. Step 2 results in the formation of adduct (96). The crucial chain-transfer step (step 3) involves transfer of an atom or group X from the starting material

X - Y to give the adduct (97). The derived radical Y• is the chain-transfer agent. In effect, both chain transfer and site selective generation of the initial radical are combined into this single step. In methods that employ external reagents (such as tin hydride), both chain-transfer and site-selective generation of the initial

radical are accomplished in two separate steps. Since the radical Y• is generated directly by atom abstraction from the starting material X-Y, this class of reactions is called "atom transfer" or Kharasch additions. Because the chain transfer steps differ, similar reactions under atom transfer or tin hydride mediated procedures give totally different product ratios.

The basic requirements for atom transfer-mediated radical additions are (i) rigid exothermic addition (or cyclization or fragmentation) must convert a

relatively stable radical Y• to a less stable radical (96), and (ii) the donor species (X—Y() must be able to transfer X more rapidly than other competing reactions of (96), such as radical termination or telomerization (step 5).

8.1.5.1 Hydrogen Atom Transfer Addition and Cyclization

The peroxide-initiated addition of hydrogen bromide to alkenes affording (98) is one of the many examples of the well-known Kharasch addition reactions [124].

(98)

With the proper choice of reactants and reaction conditions, it is possible to employ hydrogen atom transfer in the formation of carbon-carbon bonds. Heteroatoms or groups which stabilize radicals serve to weaken and hence activate adjacent C - H bonds for participation in hydrogen atom transfer additions [1,125,126]. For instance, dimethylmalonate (99) reacts with vinyl acetate (100) in the presence of an initiator to give the product (101) in 65% yield [127].

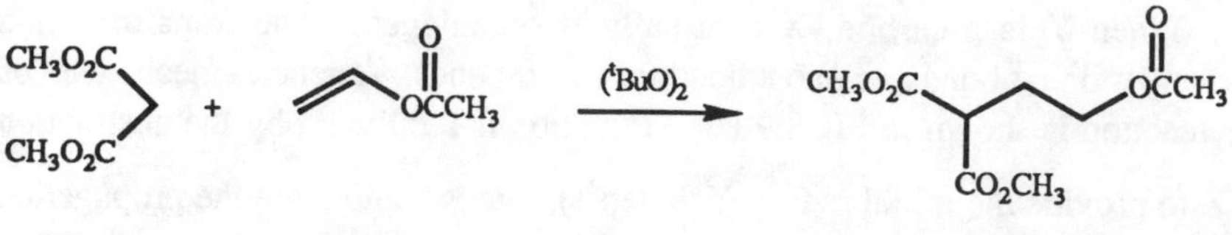

(99)   (100)   (101)

The reaction is started by the *tert.* butoxy radical which is formed by the homolytic cleavage of $(^tBuO)_2$ abstracts a hydrogen atom from dimethylmalonate in step 1, as shown in Fig.8.10. The malonyl radical (102) adds to vinyl acetate (100) generating radical (103) (step 2). In the subsequent chain-transfer step (step 3) hydrogen atom is transferred from malonate (99) to (103) affording the addition product (101).

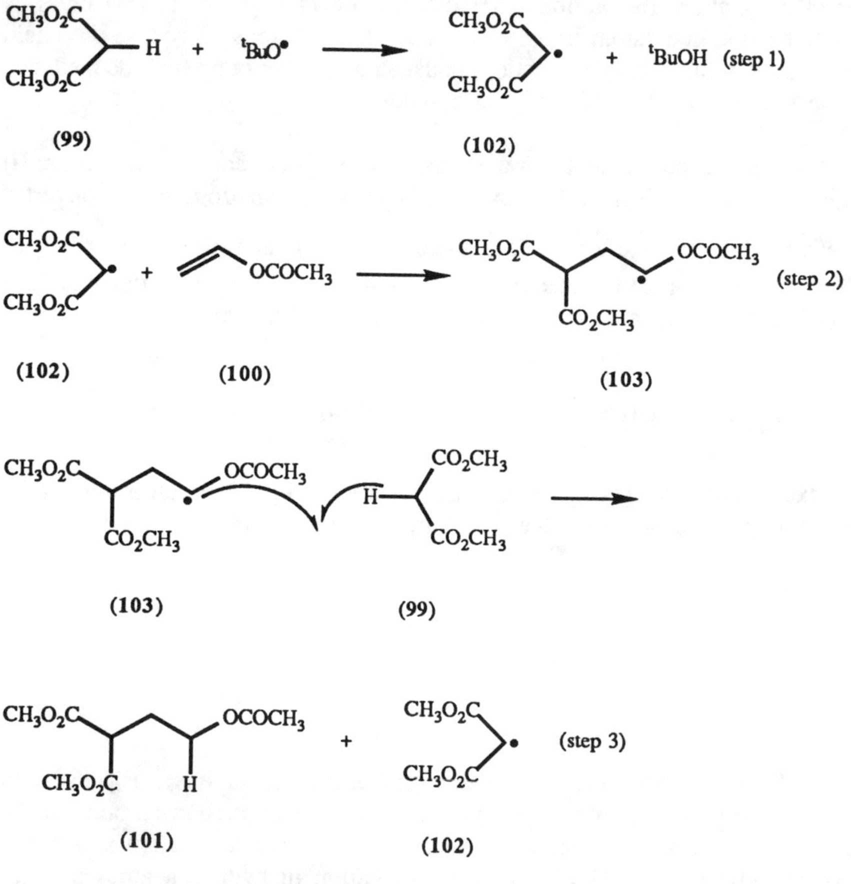

(99)   (102)

(102)   (100)   (103)

(103)   (99)

(101)   (102)

Figure 8.10. Mechanism of hydrogen atom transfer additions.

The excess of hydrogen atom donor (malonate (99) in this case) must be used to facilitate transfer of hydrogen atom to a carbon-centered radical (step 3). By conducting the reaction at higher temperature, polycondensation, or telomer formation is avoided.

The hydrogen atom transfer method has been employed in sequences containing addition and cyclization reactions [128,129]. For instance the per-ester (106) initiates the addition of acetaldehyde to the unsaturated ester (104) leading to the radical (107) which cyclizes to (108). This is followed by hydrogen atom transfer from another acetaldehyde molecule to give the product (109) in upto 57% yield [129].

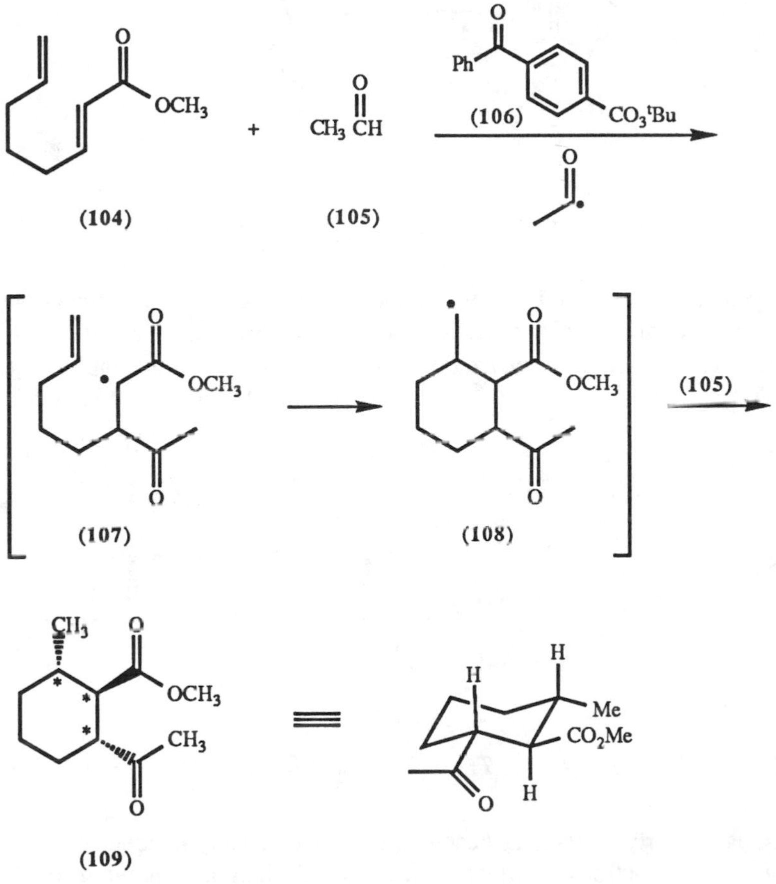

(104)   (105)

(107)   (108)   (105)

(109)

8.1.5.2   Halogen Atom Transfer

8.1.5.2.1   Halogen Atom Transfer Additions

The halogen atom transfer method is a unique and powerful method to generate and control intermediate radicals affording access to a wide variety of synthetic transformations [130]. The reactions may be catalyzed by metals, e.g. Pd, Al etc. [131,132]. For instance, reaction of phenyl tri-bromomethyl sulfone (110) with cyclopentene (111) afforded the *trans*-product (112) in 63-83% yield [133].

(110)   (111)   (112)

8.1.5.2.2   Halogen Atom Transfer Cyclizations

Radical cyclizations mediated by halogen atom transfer constitute an important route to synthetically useful reactions [134-139]. For instance, the α,α-dichloroester (113) undergoes cyclization in the presence of metal complex, e.g. [CpMo(CO)$_3$]$_2$ to give the products (114) and (115) in a ratio of (3.9:1) [137].

(113)   (114)   (115)

8.1.5.2.3   Halogen Atom Transfer Annulations

Halogens, particularly iodine, has been employed extensively in controlling the course of radical reactions [140]. For instance, irradiation of butynyl iodide (116) and the alkene (117) in benzene containing hexabutylditin afforded the products (118) and (119) in a ratio of 11:1 [140].

(116)   (117)   (118)   (119)

(118 : 119 = 11:1)

Similarly, iodomalonate (120) reacts with the allyl alcohol (121) in the presence of hexabutylditin to give the product (122) in 40% yield.

(120)   (121)   (122)

8.1.6 Heteroatom-Halogen Donors

The weak nature of the nitrogen-halogen and oxygen-halogen bonds can be exploited to serve as an atom transfer process in radical reactions [141-145]. For instance, the nitrogen-chlorine bond in chloroamine (123) can be cleaved homolytically by TiCl$_3$ followed by cyclization and chlorine atom transfer to give the product (127) in 60% yield [146].

The reaction sequence proceeds by the formation of a metal complex (124) which cyclizes to give the radical (125). This undergoes a second cyclization affording the radical intermediate (126) which abstracts a chlorine atom from (123) to give the product (127).

(123)   (124)

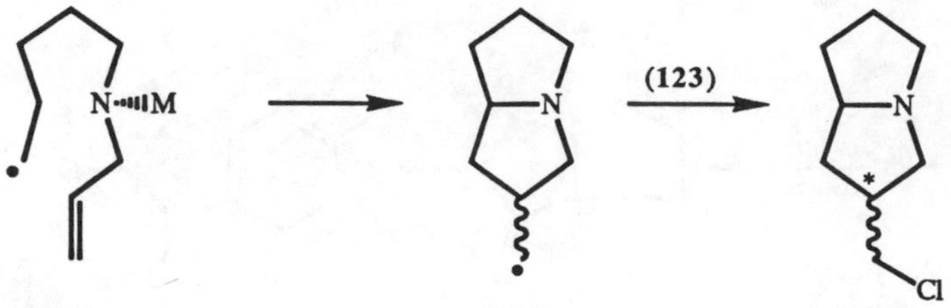
N···M
N
(123)
N
*
Cl

(125) (126) (127)

8.1.7 *Organocobalt Transfer Method*

The bond between carbon and cobalt is relatively weak and free radicals are generated by exposure of various cobaloximes and cobalamines to heat or light [147-152]. Hence several transformations are possible involving organocobalt group transfer [149, 150]. For instance, treatment of aryl iodide **(128)** with Co(salen) **(129)** afforded the complex **(130)**. Irradiation of (**130**) afforded the product (**131**). As shown in Fig. 8.11, the cobalt can be replaced by a number of functional groups, including hydroxyl, halogen, oxime, phenylthio and phenylseleno affording the products (**132-136**).

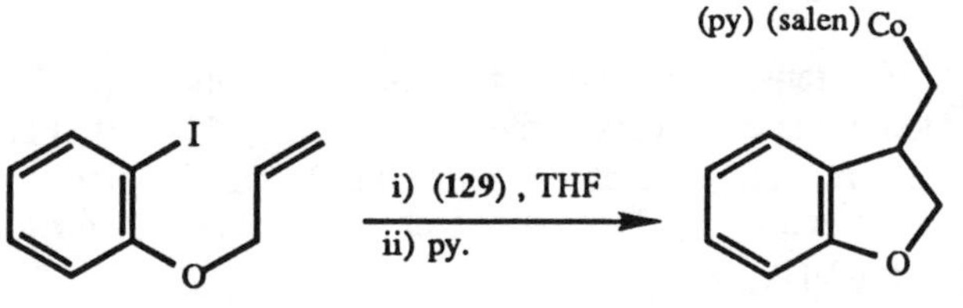
(py) (salen) Co
I
O
i) (129) , THF
ii) py.
O

(128) (130)

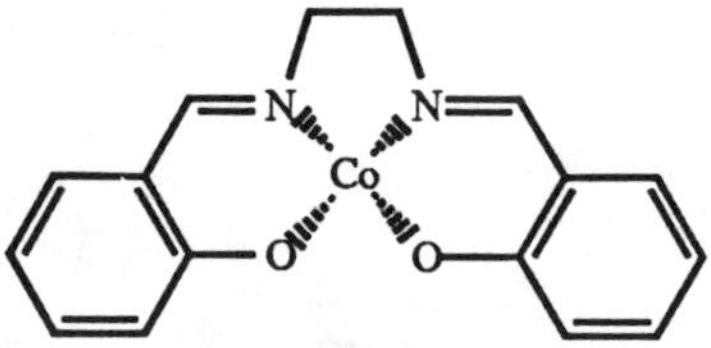
N
N
Co
O
O

(129)

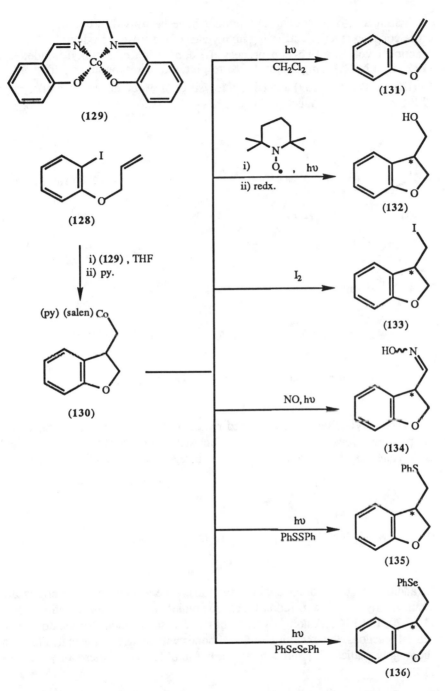

Figure 8.11. Organocobalt transfer method in synthesis, salen = (129), py = pyridine.

Similarly, organocobalt compounds have been used in the alkyl-alkenyl cross-coupling reactions affording high yields of the crossed products [152]. The reaction involves a two-step process: (i) conversion of an alkyl halide, e.g. (137) with NaCQ (dmg H)₂py (138) to give the corresponding pyridine alkyl Co(III) cobaloxime (139) and (ii) photolysis of (139) in the presence of styrene (140) to afford the product (141) in 95% yield [152].

(137)   (138)   (139)

(139)   (140)   (141) 95%

(dmg H = dimethyl glyoxime monanion, py = pyridine)

The alkyl-cobaloxime mediated reaction is unique among radical olefin coupling reactions because the olefin functionality is transformed into radical adduct after addition to the cobaloxime and is regenerated in the final product.

8.2   Non-Chain Radical Reactions

In addition to the radical chain   reactions, several non-chain reactions are synthetically useful, e.g. redox reactions, radical-radical chain reactions, etc., [1,10-20, 153,154]. Among a variety of metal-based methods for the non-chain radical reactions, the oxidation of C - H bonds with manganese acetate has been thoroughly studied [20] and is particularly useful for the synthesis of γ-lactones [155].

For instance, 1-octene (142) reacts with acetic acid in the presence of manganese acetate to give the lactone (145) in 74% yield [155]. The reaction proceeds via the complex radical (143) which is oxidized to the cation (144) by Mn(III). The cation subsequently cyclizes to give the product (145).

(142)   (143)

(144)   (M = metal)   (145)

Similarly, the aroylacetate (146) reacts with the styrene derivative (147) in the presence of manganese acetate to give, via the radical (148), the product (149) in 52% yield [11]. The method is useful for the facile synthesis of substituted tetralones. The presence of additional electron-withdrawing groups on the radical precursor greatly facilitate the oxidation and the possibility of tandem radical and cationic cyclizations provides powerful means for ring formation.

(146)   (147)

(148)   (149)

8.3   References

1.   B. Giese, "Radicals in Organic Synthesis: Formation of Carbon-Carbon Bonds", Pergamon Press, Oxford (1986).

2.   D.H.R. Barton and S.Z. Zard, *Pure Appl. Chem.*, **58**, 675 (1986).

3.   D.H.R. Barton and N. Ozbalik, *Phosphorus, Sulfur, Silicon*, **43**, 349 (1989).

4.   D.J. Hart, *Science* (Washington, D.C.), **223**, 883 (1984).

5.   D. Crich and L. Quintero, *Chem. Rev.*, **89**, 1413 (1989).

6.   A.L.J. Beckwith, *Tetrahedron*, **37** (18), 3073 (1981).

7.   W.P. Neumann, *Synthesis*, 665 (1987).

8.   B. Giese, *Angew. Chem.*, 969 (1989).

9.   M. Ramaiah, *Tetrahedron*, **43** (16), 3541 (1987).

10.   W.E. Fristad and J.R. Peterson, *J. Org.Chem.*, **50**, 10 (1985).

11.   F.Z. Yang, M.K. Trost and W.E. Fristad, *Tetrahedron Lett.*, **28**, 1493 (1987).

12. B.B. Snider, R. Mohan and S.A. Kates, *J. Org. Chem.*, **50**, 3659 (1985).

13. (a) W.E. Fristad and S.S. Hershberger, *J.Org.Chem.*, **50** 1026 (1985). (b) W.E. Fristad, J.R. Peterson and A.B. Ernst, *J.Org.Chem.*, **50**, 3143 (1985).

14. (a) E.J. Corey and M. Kang, *J.Amer.Chem.Soc.*, **106**, 5384 (1984). (b) E.J. Corey and A.K. Ghosh, *Tetrahedron Lett.*, **28**, 175 (1987). (c) E.J. Corey and A.K. Ghosh, *Chemistry Lett.*, 223 (1987).

15. E.J. Corey and A.W. Gross, *Tetrahedron Lett.*, **26**, 4291 (1985).

16. P. Breuilles and D. Uguen, *Tetrahedron Lett.*, **25**, 5759 (1984).

17. G. Midgley and C.B. Thomas, *J. Chem. Soc., Perkin. Trans II*, 1103 (1987).

18. C. Gardrat, *Synth. Commun.* 14, 1191 (1984).

19. N. Fujimoto, H. Nishino and K. Kurosawa, *Bull.Chem.Soc. Jpn.*, **59**, 3161 (1986).

20. W.E. Fristad, J.R. Peterson, A.B. Ernst and G.B. Urbi, *Tetrahedron*, **42**, 3429 (1986).

21. K.U. Ingold, "Free Radicals", (J.Kochi, ed.), Vol.1, p.37, Wiley, New York, (1973).

22. H.G. Kuivila, *Acc.Chem.Res.*, **1**, 299 (1968).

23. H. Fischer and H. Paul, *Acc.Chem.Res.*, **20**, 200 (1987).

24. D. Griller and K.U. Ingold, *Acc.Chem.Res.*, **13**, 193, 317 (1980).

25. B. Giese, *Angew. Chem.*, **22**, 753 (1983).

26. H. Schumann and B. Pachaly, *Angew. Chem.*, **20**, 1043 (1981).

27. Y. Ueno, K. Chino, M. Watanabe, O. Moriya, M. Okawara, *J.Amer.Chem.Soc.*, **104**, 5564 (1982).

28. Y. Ueno, O. Moriya, K. Chino, M. Watanabe and M. Okawara, *J. Chem. Soc., Perkin. Trans.I*, 1351 (1986).

29. N.M. Weinshenker, G.A. Gosby and J.Y. Wong, *J.Org.Chem.*, **40**, 1966 (1975).

30.   D.E. Bergbreiter and J.R. Blanton, *J.Org.Chem.*, **52**, 472 (1987).

31.   G. Stork and P.M. Sher, *J.Amer.Chem.Soc.*, **108**, 303 (1986).

32.   D.B. Gerth and B. Giese, *J.Org.Chem.*, **51**, 3726 (1986).

33.   J.M. Surzur, "Reactive Intermediates", (R.A. Abramovitch, ed.), Plenum Press, New York, (1982).

34.   A.L.J. Beckwith and K.U. Ingold, " Rearrangements in Ground and Excited States", (P.de Mayo, ed.), Academic Press, New York (1980).

35.   D.L.J. Clive and D.R. Cheshire, *J.Chem.Soc.Chem.Commun.*, 1520 (1987).

36.   A.L.J. Beckwith and S.A. Glover, *Aust.J.Chem.*, **40**, 157 (1987).

37.   For a comprehensive review on intermolecular radical cyclization, see: D.P. Curran, *Synthesis* 417, 489 (1988).

38.   C.S. Wilcox and L.M. Thomasco, *J.Org.Chem.*, **50**, 546 (1985).

39.   T.V. Rajan Babu, *J.Amer.Chem.Soc.*, **109**, 609 (1987).

40.   D.L.J. Clive and P.L. Beaulieu, *J.Chem.Soc.Chem.Commun.*,307 (1983).

41.   M.D. Bachi and E. Bosch, *Tetrahedron Lett.*, **27**, 641 (1986).

42.   N. Ono, H. Miyake, A.Kamimura, I. Hamamoto, R.Tamura and A.Kaji, *Tetrahedron*, **41**, 4013 (1985).

43.   M.Ladlow and G. Pattenden, *Tetrahedron Lett.*, **26**, 4413 (1985).

44.   R. Tsang and B. Fraser-Reid, *J.Amer.Chem.Soc.*, **108**, 8102 (1986).

45.   R. Tsang, J.K. Jr., Dickson, H. Pak., R. Walton and B. Fraser-Reid, *J.Amer.Chem.Soc.*, **109**, 3484 (1987).

46.   D.J. Hart and H.-C. Huang, *Tetrahedron Lett.*, **26**, 3749 (1985).

47.   K.C. Nicolaou, D.G. McGarry, P.K. Somers, C.A. Veale and G.T. Furst, *J.Amer.Chem.Soc.*, **109**, 2504 (1987).

48.   M. Ihara, N. Taniguchi, K. Fukumoto and T. Kametani, *J.Chem.Soc.Chem.Commun.* 1438 (1987).

49.   P. Dowd and S.-C. Choi, *J. Amer.Chem.Soc.*, **109**, 3493 (1987).

50.   G. Stork and M. Kahn, *J.Amer.Chem.Soc.*, **107**, 500 (1985).

51.   D.A. Evans, "Asymmetric Synthesis",(J.D. Morrison, ed.) Vol.3, p.2, Academic Press, Orlando (1984).

52.   C.H. Heathcock, "Asymmetric Synthesis", (J.D. Morrison, ed.), Vol.3, p.111, Academic Press, Orlando (1980).

53.   W. Oppolzer, *Angew.Chem.*, **23**, 876 (1984).

54.   N.A. Porter, D.M. Scott, I.J. Rosenstein, B. Giese, A Veit and H.G. Zeitz, *J.Amer.Chem.Soc.*, **113**, 1791 (1991).

55.   B. Giese and J. Dupuis, *Angew. Chem.*, **22**, 622 (1983).

56.   B. Giese , J. Dupuis and N. Nix, *Org.Synth.*, **65**, 236 (1987).

57.   R.M. Adlington, J.E. Baldwin, A. Basak and R.P. Kozyrod, *J.Chem.Soc. Chem.Commun.*, 944 (1983).

58.   J. Dupius, B. Giese, D. Ruegge, H.Fischer, H.-G. Korth and R. Sustmann, *Angew.Chem.*, **23**, 896 (1984).

59.   B.Giese, H. Orler and M. Leising, *Chem.Ber.*, **119**, 444 (1986).

60.   G.E. Keck, D.F. Kachensky and E.J. Enholm, *J.Org.Chem.*, **49**, 1462 (1984).

61.   G.E. Keck and J.B. Yates, *J.Amer.Chem.Soc.*, **104**, 5829 (1982).

62.   J.E. Baldwin, D.R. Kelly and C.B. Ziegler, *J.Chem.Soc.Chem.Commun.*, 133 (1984).

63.   R.R. Webb and S. Danishefsky, *Tetrahedron Lett.*, **24**, 1357 (1983).

64.   R. Vaßen, J. Runsink and H.-D. Scharf, *Chem.Ber.*, **119**, 3492 (1986).

65.   L. Horner and J. Klaus, *Liebigs Ann.Chem.*, 1232 (1979).

66.   B. Giese, *Angew.Chem.*, 553 (1985).

67.   B. Giese and J. Meister, *Chem. Ber.*, 2588 (1977).

68.   B. Giese and K. Groninger, *Tetrahedron Lett.*, **25**, 2743 (1984).

69.   B. Giese and K. Heuck, *Tetrahedron Lett.*, **21**, 1829 (1980).

70. B. Giese and K. Heuck, *Chem. Ber.*, **112**, 3759 (1979).

71. B. Giese and K. Heuck, *Chem.Ber.*, **114**, 1572 (1981).

72. S. Danishefsky and E. Taniyama, *Tetrahedron Lett.*, **24**, 15 (1983).

73. S. Danishefsky, S. Chackalamannil and B.J. Uang, *J.Org.Chem.*, **47**, 2231 (1982).

74. E.J. Corey, C. Shih, N.-Y. Shih and K. Shimoji, *Tetrahedron Lett.*, **25**, 5013 (1984).

75. Y. Matsuka, M. Kodama and S. Ito, *Tetrahedron Lett.*, 4081 (1979).

76. R. Henning and H. Urbach, *Tetrahedron Lett.*, **24**, 5343 (1983).

77. B. Giese, K. Heuck, L. Lenhardt and U. Luning, *Chem.Ber.*, **117**, 2132 (1984).

78. B. Giese and G. Kretzschmar, *Chem.Ber.*, **117**, 3175 (1984).

79. B. Giese and J. Meisner, *Tetrahedron Lett.*, 2783 (1977).

80. B. Giese and G. Kretzschmar, *Angew.Chem.*, **20**, 965 (1981).

81. B. Giese and W. Zwick, *Chem.Ber.*, **115**, 2526 (1982).

82. B. Giese and W. Zwick, *Chem.Ber.*, **116**, 1264 (1983).

83. B. Giese and U. Erfort, *Chem.Ber.*, **116**, 1240 (1983).

84. M.D. Johnson, *Acc.Chem.Res.*, **16**, 343 (1983).

85. A. Gaudemer, N. N. van Duong, N. Shah Karami, S. S. Achi, M.Frostin-Rio and D. Pujol, *Tetrahedron*, **41**, 4095 (1985).

86. G.E. Keck, E.J. Enholm, J.B. Yates and M.R. Wiley, *Tetrahedron*, **41**, 4079 (1985).

87. G.E. Keck and J.B. Yates, *J.Amer.Chem.Soc.*, **104**, 5829 (1982).

88. J.P. Light, M.Ridenour, L. Beard and J. Hershberger, *J.Organomet.Chem.*, **326**, 17 (1987).

89.   G.E. Keck, D.F.Kachensky and E.J. Enholm, *J.Amer.Chem.Soc.*, **50**, 4317 (1985).

90.   G.E. Keck and D.A. Burnett, *J.Org.Chem.*, **52**, 2959 (1987).

91.   K. Mizuno, M. Ikeda, S. Toda and Y. Otsuji, *J.Amer.Chem.Soc.*, **110**, 1288 (1988).

92.   D.H.R. Barton and S.W. McCombie, *J.Chem.Soc.Perkins Trans.I*, 1574 (1975).

93.   D.H.R. Barton, D.S. Gero, B.-S. Quiclet and M. Samadi, *Tetrahedron Lett.*, **30**(37), 4969 (1989).

94.   D.H.R. Barton and D. Crich, *Tetrahedron Lett.*, **26**, 757 (1985).

95.   D.H.R. Barton, D. Crich and W.B. Motherwell, *J.Chem.Soc.Chem.Commun.*, 939 (1983).

96.   D.H.R. Barton, D. Crich and W.B. Motherwell, *J.Chem.Soc.Chem.Commun.*, 242 (1984).

97.   D.H.R. Barton and G. Kretzschmar, *Tetrahedron Lett.*, **24**, 5889 (1983).

98.   D.H.R. Barton, D.Crich and G. Kretzschmar, *J.Chem.Soc.Perkin Trans.I*, 39 (1986).

99.   D.H.R. Barton and D. Crich, *J.Chem.Soc.Perkins Trans I*, 1603 (1986).

100.   D.H.R. Barton, D.Crich and P. Potier, *Tetrahedron Lett.*, **26**, 5943 (1985).

101.   D.H.R. Barton, D. Bridon, I. Fernandez-Picot and S.Z. Zard, *Tetrahedron*, **43**, 2733 (1987).

102.   D.H.R. Barton and D. Crich, *J.Chem.Soc.Chem.Commun.*, 774 (1984).

103.   D.H.R. Barton, D. Bridon and S.Z.Zard, *J.Chem.Soc.Chem.Commun.*, 1066 (1985).

104.   D.H.R. Barton and D. Crich, *J.Chem.Soc.Perkin Trans.I*, 1613 (1986).

105.   D.H.R. Barton, B.Garcia, H. Togo and S.Z. Zard, *Tetrahedron Lett.*, **27**, 1327 (1986).

106.   D.H.R. Barton, H. Togo and S.Z.Zard, *Tetrahedron*, **41**, 5507 (1985).

107. D.H.R. Barton, H.Togo and S.Z. Zard *Tetrahedron Lett.*, **26**, 6349 (1985).

108. D.H.R. Barton, D. Bridon and S.Z. Zard, *Heterocycles*, **25**, 449 (1987).

109. D.H.R. Barton and S.Z. Zard, *Tetrahedron*, **43**, 4297 (1987).

110. D.H.R. Barton, D. Bridon, Y. Herve, P. Potier, J. Thierry and S.Z. Zard, *Tetrahedron* **42**, 4983 (1986).

111. D.H.R. Barton, D.Crich and G.Kretzschmar, *Tetrahedron Lett.*, **25**, 1055 (1984).

112. D.H.R. Barton, B.Lacher and Z.S. Zard, *Tetrahedron Lett.*, **26**, 5939 (1985).

113. D.H.R. Barton, Y. Herve, P. Potier and J. Thierry, *J.Chem.Soc.Chem. Commun.*, 1298 (1984).

114. D.H.R. Barton, J.Guilhem, Y. Herve, P. Potier and J. Thierry, *Tetrahedron Lett.*, **28**, 1413 (1987).

115. D.H.R. Barton, D. Crich and W.B. Motherwell, *Tetrahedron Lett.*, **24**, 4979 (1983).

116. D.H.R. Barton, B. Lacher and S.Z.Zard, *Tetrahedron*, **43**, 4237 (1987).

117. D. Crich, *Aldrichimica Acta*, **20**, 35 (1987).

118. A.G.M. Barrett, D.H.R. Barton and R. Bielski, *J.Chem.Soc.Perkins Trans.I*, 2378 (1979).

119. D.H.R. Barton, R.S. Hay-Motherwell and W.B. Motherwell, *J.Chem.Soc. Perkins Trans I*, 2363 (1981).

120. D.H.R. Barton, W.B. Motherwell and A. Stange, *Synthesis*, 743 (1981).

121. D.H.R. Barton, W.Hartwig, R.S. Hay-Motherwell, W.B. Motherwell and A. Stange, *Tetrahedron Lett.*, 2019 (1982).

122. D.H.R. Barton, A. Gateau-Olesker, S.D. Gero, B. Lacher, C. Tachdjian and S.Z. Zard, *J.Chem.Soc.Chem.Commun.*, 1790 (1987).

123. E. Block, M. Aslam, V.Eswarakrishnan, K. Gebreyes, J. Hutchinson, R. Iyer, J.-A. Laffitte and A. Wall, *J.Amer.Chem.Soc.*, **108**, 4568 (1986).

124. W.F.Stacy and J.F. Harris, *Org.React.*, **13**, 150 (1963).

125. Y. Watanabe, Y. Tsiji and R. Takeuchi, *Bull.Chem.Soc.Jpn.*, **56**, 1428 (1983).

126. C. Walling and E.S. Huyser, *Org. React.*, **13**, 91 (1963).

127. D.H.R. Barton, D. Crich and W.B. Motherwell, *J.Chem.Soc.Chem.Commun.*, 119 (1984).

128. L.M. van der Linde and A.J.A. van der Weerdt, *Tetrahedron Lett.*, **25**, 1201 (1984).

129. P. Gottschalk and D.C. Neckers, *J.Org.Chem.*, **50**, 3498 (1985).

130. D. Bellus, *Pure Appl.Chem.*, **57**, 1827 (1985).

131. J. Tsuji, K. Sato and H. Nagashima, *Tetrahedron* **40** (21), 5003 (1985).

132. K. Maruoka, H. Sano, Y. Fukutani and H. Yamamoto, *Chemistry Lett.*, 1689 (1985).

133. D.L. Fields, Jr and H. Shechter, *J.Org.Chem.*, **51**, 3369 (1986).

134. H. Nagashima, K. Ara, H. Watamastu, and K. Itoh, *J.Chem.Soc.Chem. Commun.*, 518 (1985).

135. M. Mori, Y. Kubo and Y. Ban, *Tetrahedron Lett.*, **26**, 1519 (1985).

136. M. Mori, N. Kanda and Y. Ban, *J.Chem.Soc.Chem.Commun.*, 1375 (1986).

137. T.K.Hayes, A.J. Freyer, M. Parvez and S.M. Weinreb, *J. Org.Chem.*, **51**, 5501 (1986).

138. S. Takano, S. Nishizawa, M. Akiyama and K. Ogasawara, *Synthesis*, 949 (1984).

139. M. Mori, N. Kanda, I. Oda and Y. Ban, *Tetrahedron*, **41**, 5465 (1985).

140. D.P. Curran and M.H. Chen, *J.Amer.Chem.Soc.*, **109**, 6558 (1987).

141. P. Kovacic, M.K. Lowery and K.W. Field, *Chem.Rev.*, **70**, 639 (1970).

142. F. Minisci, *Synthesis*, 1 (1973).

143. R. Sutcliffe and K.U. Ingold, *J.Amer.Chem.Soc.*, **104**, 6071 (1982).

144. L. Stella, *Angew.Chem.*, **22**, 337 (1983).

145. A.L.J. Beckwith, R. Kazlauskas and M.R. Syner-Lyons, *J.Org.Chem.*, **48**, 4718 (1983).

146. J.M. Surzur and L. Stella, *Tetrahedron Lett.*, 2191 (1974).

147. G.N. Schrauzer, *Angew.Chem.*, **15**, 417 (1976).

148. E.G. Samsel and J.K. Kochi, *J.Amer.Chem.Soc.*, **108**, 4790 (1986).

149. V.F. Patel and G. Pattenden, *Tetrahedron Lett.*, **28**, 1451 (1987).

150. B.P. Branchaud, M.S. Meier, M.N. Malekzadeh, *J.Org.Chem.*, **52**, 212 (1987).

151. V.F. Patel and G. Pattenden, *J.Chem.Soc.Chem.Commun.*, 871 (1987).

152. B.P. Branchaud, M.S. Meier and Y. Choi, *Tetrahedron Lett.*, **29**, 167 (1988).

153. J.R. Peterson, R.S. Egler, D.B. Horsley and T.J. Winter, *Tetrahedron Lett.*, **28**, 6109 (1987).

154. A.B. Ernst and W.E. Fristad, *Tetrahedron Lett.*, **26**, 3761 (1985).

155. E.I. Heiba, R.M. Dessau and P.G. Rodewald, *J.Amer.Chem.Soc.*, **96**, 7977 (1974).

9 Miscellaneous Stereoselective Reactions

9.1 Asymmetric Cycloproponations

Asymmetric cyclopropanation, i.e. the stereoselective transfer of carbon ligands from chiral catalysts to alkenes resulting in asymmetric synthesis of cyclopropanes, constitutes one of the most useful methods for preparing biologically active compounds [1-3]. For instance pyrethroids, which are cyclopropane carboxylic acids, such as chrysanthemic acid (1) and its *cis*-dibromovinyl analogs, are among the most powerful insecticides known [4].

(1)   (2)

The key intermediate in the synthesis of pyrethroids is the (*1R*,3*R*)-hemicaronic aldehyde (6). This is prepared by the cyclopropanation of the oxazolidine (3) with the triphenylphosphorane derivative (4) affording the intermediate (5) exclusively in 60% yield. Removal of the chiral auxiliary affords (6) in 70% yield.

(3)   +   (4)

(5)   (6)

Chiral copper complexes have generally been employed in the asymmetric synthesis of cyclopropanes [5,6].

(9) $R_1 = CH_3$, $R_2 = $ 2-octyloxy -5-^{t}BuPh

For instance, 2-methyl-5,5,5-trichloro-2-pentene (7) reacts with 1-menthyl diazoacetate (8) in the presence of the chiral copper catalyst (9) to give the (R)-cis diastereomer (10) in 81.5% yield and 93% e.e [6].

(7)   (8)   (10) 93% e.e.

The dimethylcyclopropane carboxylic acid is an important intermediate in the synthesis of cilastatin (11), an enzyme inhibitor used in combination with the β-lactam antibiotic, imipenem [23].

(11)

Other catalytic systems have also been employed to effect asymmetric cyclopropanations [7-10]. For instance, styrene reacts with 2-diazodimedone (13) in the presence of (14) to afford the cyclopropane derivative (15) in 100% e.e [7].

(12)   (13)   (15)

The same reaction carried out in the presence of the immobilised chiral β-diketone **(16)** afforded the product **(15)** in 93% e.e. The results were reproducible even after three cycles of the catalyst.

(16)

Chiral iron-complexes have also been used in the asymmetric cyclopropanation reaction [11,12]. Thus the reagents **(17)** and **(18)**, which differ only in the configuration at iron, were prepared *in situ* and employed in the cyclopropanation of styrene, affording the diastereomeric products in high optical yields [11].

(17)   (18)

A mixture of **(17)** and **(18)** **(99:1)** afforded the product *trans*(1R,2R)-**(19)** in 88% e.e.

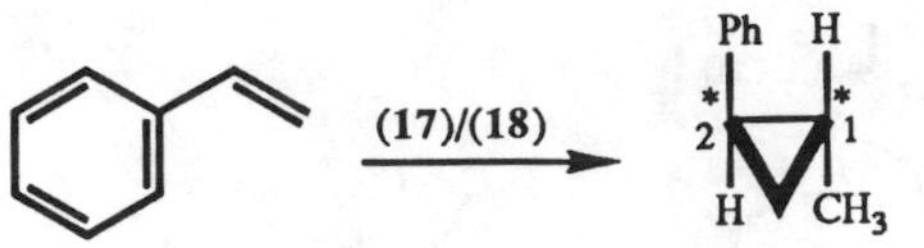

(12)

(19)
trans - (1R, 2R)
88% e.e.

In recent years chiral copper [13-18] and rhodium [19,20] catalysts have been used with success in the asymmetric cyclopropanation of α-keto-carbenoids. For instance the α-diazo-β-ketoester (20) reacts with the Aratani catalyst (21) to give the product (22) in 80% e.e. [18].

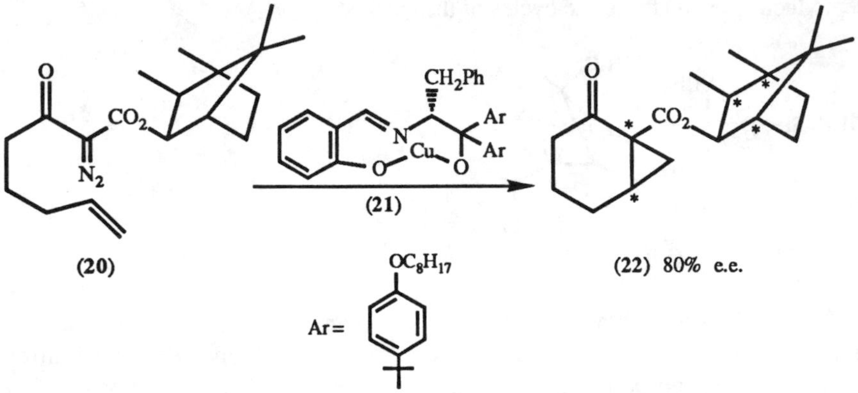

(20)

(21)

(22) 80% e.e.

$Ar=$

OC_8H_{17}

9.2   References

1.   A. Bernardi, C. Scolastico and R. Villa, *Tetrahedron Lett.*, **30**, 3733 (1989).

2.   T. Aratani, *Pure & Appl. Chem.*, **57**, 1839 (1985).

3.   T. Aratani, *J. Syn. Org.Chem., Jpn.*, **43**, 1134 (1985).

4.   (a) For a recent review on asymmetric cyclopropanations, see: A. Padwa and K.E. Krumpe, *Tetrahedron*, **48**(26), 5385 (1992).   (b) For a review on pyrethroid acids, see: D. Arlt, M. Jautelat and R. Lantzsch, *Angew.Chem.*, **20**, 703 (1981).

5.   H. Nozaki, H. Takaya, S. Moriuti and R. Noyori, *Tetrahedron*, **24**, 3655 (1968).

6.   T. Aratani, Y. Yoneyoshi and T. Nagase, *Tetrahedron Lett.*, **23** 685 (1982).

7.   S.A. Matline, W.J. Lough, L. Chan, D.M.H. Abram and Z. Zhou, *J.Chem.Soc.,Chem.Commun.*, 1038 (1984).

8. T. Aratani, Y. Yoneyoshi and T. Nagase, *Tetrahedron Lett.*, 1707 (1975).

9. A. Nakamura, A. Konishi, Y. Tatsuno and S. Otsuka, *J.Amer.Chem. Soc.*, **100**, 3443 (1978).

10. W.R. Moser, *J.Amer.Chem.Soc.*, **91**, 1135 (1969).

11. M. Brookhart, D. Timmers, J.R. Tucker and G.D. Williams, *J.Amer.Chem. Soc.*, **105**, 6721 (1983).

12. K.A.M. Kremer, P. Helquist and R.C. Kerber, *J.Amer.Chem.Soc.*, **103**, 1862 (1981).

13. T. Aratani, Y. Yoneyoshi and T. Nagase, *Tetrahedron Lett.*, 685 (1982).

14. H. Nozaki, S. Moriuti, H. Takaya and R. Noyori, *Tetrahedron Lett.*, 5239 (1966).

15. H. Fritschi, U. Leutenegger and A. Pfaltz, *Helv.Chim.Acta*, **71**, 1553 (1988).

16. R.E. Lowenthal, A. Abiko and S. Masamune, *Tetrahedron Lett.*, 6005 (1990).

17. D.A. Evans, K.A. Woerpel, M.M. Hinman, H. Nozaki and M.M. Faul, *J.Amer.Chem.Soc.*, **113**, 736 (1991).

18. W.G. Dauben, R.T. Hendricks, M.J. Luzzio and H.P. Ng, *Tetrahedron Lett.*, **31**, 6969 (1990).

19. M.P. Doyle, R.J. Pieter, S.F. Martin, R.E. Austin, C.J. Oalmann and P. Muller, *J.Amer.Chem.Soc.*, **113**, 1423 (1991).

20. M.P. Doyle, B.D. Brandes, A.P. Kazala, R.J. Pieters, M.B. Jarstfer, L.M. Watkins and C.T. Eagle, *Tetrahedron Lett.*, 6613 (1990).

Subject Index